METHODS IN MOLECULAR BIOLOGY

Series Editor
John M. Walker
School of Life and Medical Sciences
University of Hertfordshire
Hatfield, Hertfordshire, UK

For further volumes:
http://www.springer.com/series/7651

For over 35 years, biological scientists have come to rely on the research protocols and methodologies in the critically acclaimed *Methods in Molecular Biology* series. The series was the first to introduce the step-by-step protocols approach that has become the standard in all biomedical protocol publishing. Each protocol is provided in readily-reproducible step-by step fashion, opening with an introductory overview, a list of the materials and reagents needed to complete the experiment, and followed by a detailed procedure that is supported with a helpful notes section offering tips and tricks of the trade as well as troubleshooting advice. These hallmark features were introduced by series editor Dr. John Walker and constitute the key ingredient in each and every volume of the *Methods in Molecular Biology* series. Tested and trusted, comprehensive and reliable, all protocols from the series are indexed in PubMed.

Transgenesis

Methods and Protocols

Edited by

Thomas L. Saunders

Transgenic Animal Model Core, University of Michigan, Ann Arbor, MI, USA

Humana Press

Editor
Thomas L. Saunders
Transgenic Animal Model Core
University of Michigan
Ann Arbor, MI, USA

ISSN 1064-3745 ISSN 1940-6029 (electronic)
Methods in Molecular Biology
ISBN 978-1-0716-2989-5 ISBN 978-1-0716-2990-1 (eBook)
https://doi.org/10.1007/978-1-0716-2990-1

This Humana imprint is published by the registered company Springer Science+Business Media, LLC, part of Springer Nature.
The registered company address is: 1 New York Plaza, New York, NY 10004, U.S.A.

Preface

The field of transgenesis has been transformed and expanded by the application of CRISPR/Cas9 technology. The random integration of transgenes in the genome was the hallmark of animal transgenesis methods first described in the 1980s. At this time, the majority of new genetically engineered animal models rely upon gene targeting approaches instead of random integration. Random integration, with its attendant problems of local rearrangements and deletions, variegated transgene expression in tissues, and unexpected ectopic transgene expression is no longer the favored approach to transgenesis. The precise insertion of single copies of transgenes in specific locations of the genome that was previously achieved only with embryonic stem cell technology, somatic cell nuclear transfer, or the transplantation of primordial germ cells is now possible through the manipulation of zygotes with CRISPR. Transgene cassettes inserted with CRISPR into genes faithfully reproduce the expression patterns of targeted genes and avoid the problem of variegated or ectopic expression patterns. Of note, researchers still need to be concerned and control for unexpected complex rearrangements at or near integration sites when using CRISPR. Direct genome editing in zygotes with CRISPR tools has increased the number of valuable animal models used for basic and pre-clinical research, and extends to the creation of animals for agricultural applications.

Specific and precise genome edits can be produced with great facility with CRISPR. The efficiency of genome editing now brings other considerations to the forefront of the field. Questions to be addressed include initial experimental project design to minimize the loss of animal life when creating new genetic models. There is a need to identify carriers with targeted genes quickly and accurately. There is a new urgency for accurate genotyping because large numbers of potentially interesting founder animals have to be reduced to manageable numbers. Careful planning during project design can address these issues. Effective genotyping methods that extend beyond simple targeted PCR amplification and Sanger sequencing need to be implemented to control for correct gene editing. In the era of CRISPR genome editing, it can almost be taken for granted that cells and animals with gene-targeted modifications can be generated. Thus, the central problem has shifted from how does one create a new animal model to how to validate correct genome editing in animals. The application of NGS tools to identify gene edits and to verify transmission in subsequent generations requires exacting quality controls to ensure that edits are just what is intended without any unwanted genomic rearrangements or unexpected complexity. This volume includes information that focuses on genotyping and validation in addition to information on how to produce gene edited cells and animals for research. Future biomedical research advances will benefit from the use of precise gene targeting of transgenes in the genome as CRISPR technology supersedes earlier methods that relied on random transgene integration.

Ann Arbor, MI, USA *Thomas L. Saunders*

Contents

Contributors

TUIJA ALCANTAR • *Genentech, Inc., Department of Molecular Biology, South San Francisco, CA, USA*
ALASDAIR J. ALLAN • *The Mary Lyon Centre, MRC Harwell, Didcot, Oxon, UK*
NATASHA BACARRO • *Genentech, Inc., Department of Molecular Biology, South San Francisco, CA, USA*
DANIEL BIGGS • *Wellcome Centre for Human Genetics, University of Oxford, Oxford, UK*
MARIE-CHRISTINE BIRLING • *Université de Strasbourg, CNRS, INSERM, CELPHEDIA, PHENOMIN, Institut Clinique de la Souris, Illkirch, France*
ELIZABETH C. BRYDA • *Animal Modeling Core, University of Missouri, Columbia, MO, USA; Rat Resource and Research Center, University of Missouri, Columbia, MO, USA*
THORSTEN BUCH • *Institute of Laboratory Animal Science, University of Zurich, Zurich, Switzerland*
ROSIE K. BUNTON-STASYSHYN • *The Mary Lyon Centre, MRC Harwell, Didcot, Oxon, UK*
ROGER CAOTHIEN • *Genentech, Inc., Department of Molecular Biology, South San Francisco, CA, USA*
CHIANN-MUN CHEN • *Wellcome Centre for Human Genetics, University of Oxford, Oxford, UK*
Y. EUGENE CHEN • *Center for Advanced Models for Translational Sciences and Therapeutics, University of Michigan Medical Center, Ann Arbor, MI, USA*
GEMMA F. CODNER • *The Mary Lyon Centre, MRC Harwell, Didcot, Oxon, UK*
JON P. CONNELLY • *St. Jude Children's Research Hospital, Department of Cell & Molecular Biology, Memphis, TN, USA; St. Jude Children's Research Hospital, Center for Advanced Genome Engineering, Memphis, TN, USA*
XIAOXIA CUI • *Genome Engineering & Stem Cell Center, Department of Genetics, Washington University in St. Louis School of Medicine, St. Louis, MO, USA*
JEFFREY G. DANIEL • *Department of Pediatrics, University of Michigan, Ann Arbor, MI, USA*
BENJAMIN DAVIES • *Wellcome Centre for Human Genetics, University of Oxford, Oxford, UK; The Francis Crick Institute, London, UK*
DANIEL J. DAVIS • *Animal Modeling Core, University of Missouri, Columbia, MO, USA*
ALLISON C. FERGUSON • *Department of Pediatrics, University of Michigan, Ann Arbor, MI, USA*
ALEX FOWER • *The Mary Lyon Centre, MRC Harwell, Didcot, Oxon, UK*
SELIKA GARZA • *Rodent Genome Engineering Core, University of Texas Health San Antonio, San Antonio, TX, USA*
CHANNABASAVAIAH B. GURUMURTHY • *Mouse Genome Engineering Core Facility, University of Nebraska Medical Center, Omaha, NE, USA; Department of Pharmacology and Experimental Neuroscience, College of Medicine, University of Nebraska Medical Center, Omaha, NE, USA*
YANN HÉRAULT • *Université de Strasbourg, CNRS, INSERM, CELPHEDIA, PHENOMIN, Institut Clinique de la Souris, Illkirch, France*
ALEWO IDOKO-AKOH • *The Roslin Institute & Royal (Dick) School of Veterinary Studies, University of Edinburgh, Midlothian, UK*

Marques Jackson • *Genentech, Inc., Department of Molecular Biology, South San Francisco, CA, USA*
Boris Jerchow • *Novartis Institute for Biomedical Research (NIBR), Novartis Pharma AG, Basel, Switzerland*
Shigeru Kakuta • *Laboratory of Biomedical Science, Graduate School of Agricultural and Life Sciences, The University of Tokyo, Tokyo, Japan*
Bernhard Klinger • *Department of Animal Sciences, School of Life Sciences Weihenstephan, Technical University Munich, Freising, Germany*
Evgenea Kouranova • *Genome Engineering & Stem Cell Center, Department of Genetics, Washington University in St. Louis School of Medicine, St. Louis, MO, USA*
Ralf Kühn • *Max-Delbrück-Center for Molecular Medicine in the Helmholtz Association (MDC), Berlin, Germany*
Loic Lindner • *Université de Strasbourg, CNRS, INSERM, CELPHEDIA, PHENOMIN, Institut Clinique de la Souris, Illkirch, France*
Lauri G. Lintott • *The Centre for Phenogenomics, Toronto, ON, Canada; The Hospital for Sick Children, Toronto, ON, Canada*
Matthew Mackenzie • *The Mary Lyon Centre, MRC Harwell, Didcot, Oxon, UK*
Takashi Matsuwaki • *Department of Veterinary Physiology, Graduate School of Agricultural and Life Sciences, The University of Tokyo, Tokyo, Japan*
Michael J. McGrew • *The Roslin Institute & Royal (Dick) School of Veterinary Studies, University of Edinburgh, Midlothian, UK*
Hongsheng Men • *University of Missouri, Rat Resource and Research Center, Columbia, MO, USA*
Hiromi Miura • *Department of Molecular Life Science, Division of Basic Medical Science and Molecular Medicine, Tokai University School of Medicine, Kanagawa, Japan*
Lluis Montoliu • *National Centre for Biotechnology (CNB-CSIC) and Center for Biomedical Network Research on Rare Diseases (CIBERER-ISCIII), Madrid, Spain*
Ayaka Nakamura • *Support Center for Medical Research and Education, Tokai University, Kanagawa, Japan*
Brian Nakao • *Genentech, Inc., Department of Molecular Biology, South San Francisco, CA, USA*
Shilpa Narina • *St. Jude Children's Research Hospital, Department of Cell & Molecular Biology, Memphis, TN, USA; St. Jude Children's Research Hospital, Center for Advanced Genome Engineering, Memphis, TN, USA*
Lauryl M. J. Nutter • *The Centre for Phenogenomics, Toronto, ON, Canada; The Hospital for Sick Children, Toronto, ON, Canada*
Masato Ohtsuka • *Department of Molecular Life Science, Division of Basic Medical Science and Molecular Medicine, Tokai University School of Medicine, Kanagawa, Japan; The Institute of Medical Sciences, Tokai University, Kanagawa, Japan*
Raehum Paik • *Stanford University, Stanford, CA, USA*
Guillaume Pavlovic • *Université de Strasbourg, CNRS, INSERM, CELPHEDIA, PHENOMIN, Institut Clinique de la Souris, Illkirch, France*
Anna Pham • *Genentech, Inc., Department of Molecular Biology, South San Francisco, CA, USA*
David Preisinger • *Department of Animal Sciences, School of Life Sciences Weihenstephan, Technical University Munich, Freising, Germany*

SHONDRA M. PRUETT-MILLER • *St. Jude Children's Research Hospital, Department of Cell & Molecular Biology, Memphis, TN, USA; St. Jude Children's Research Hospital, Center for Advanced Genome Engineering, Memphis, TN, USA*

JUAN REYES JR • *Genentech, Inc., Department of Molecular Biology, South San Francisco, CA, USA*

BEATE RIEBLINGER • *Department of Animal Sciences, School of Life Sciences Weihenstephan, Technical University Munich, Freising, Germany*

MERONE ROOSE-GIRMA • *Genentech, Inc., Department of Molecular Biology, South San Francisco, CA, USA*

MASAHIRO SATO • *Department of Genome Medicine, National Center for Child Health and Development, Tokyo, Japan*

LAURENCE SCHAEFFER • *Université de Strasbourg, CNRS, INSERM, CELPHEDIA, PHENOMIN, Institut Clinique de la Souris, Illkirch, France*

ANGELIKA SCHNIEKE • *Department of Animal Sciences, School of Life Sciences Weihenstephan, Technical University Munich, Freising, Germany*

MARIE SEKIGUCHI • *Laboratory of Biomedical Science, Graduate School of Agricultural and Life Sciences, The University of Tokyo, Tokyo, Japan*

JORDAN A. SHAVIT • *Department of Pediatrics, University of Michigan, Ann Arbor, MI, USA; Department of Human Genetics, University of Michigan, Ann Arbor, MI, USA*

JUN SONG • *Center for Advanced Models for Translational Sciences and Therapeutics, University of Michigan Medical Center, Ann Arbor, MI, USA; Key Laboratory of Animal Cellular and Genetics Engineering of Heilongjiang Province, College of Life Science, Northeast Agricultural University, Harbin, China*

BARBARA J. STONE • *ParaTechs Corporation, Lexington, KY, USA*

LUCINDA TAM • *Genentech, Inc., Department of Molecular Biology, South San Francisco, CA, USA*

LYDIA TEBOUL • *The Mary Lyon Centre, MRC Harwell, Didcot, Oxon, UK*

MIA WALLACE • *Mouse Genetics Core, Department of Pediatrics, Washington University in St. Louis School of Medicine, St. Louis, MO, USA*

ZI TENG WANG • *Genome Engineering & Stem Cell Center, Department of Genetics, Washington University in St. Louis School of Medicine, St. Louis, MO, USA*

BENEDIKT WEFERS • *German Center for Neurodegenerative Diseases (DZNE), Munich, Germany; Helmholtz Zentrum München, German Research Center for Environmental Health, Institute of Developmental Genetics, Neuherberg, Germany*

J. MICHAEL WHITE • *Transgenic, Knockout and Microinjection Core, Department of Pathology & Immunology, Washington University in St. Louis School of Medicine, St. Louis, MO, USA*

THOMAS WINOGRODZKI • *Department of Animal Sciences, School of Life Sciences Weihenstephan, Technical University Munich, Freising, Germany*

WOLFGANG WURST • *German Center for Neurodegenerative Diseases (DZNE), Munich, Germany; Helmholtz Zentrum München, German Research Center for Environmental Health, Institute of Developmental Genetics, Neuherberg, Germany; Technische Universität München-Weihenstephan, Chair of Developmental Genetics, c/o Helmholtz Zentrum München, Neuherberg, Germany; Munich Cluster for Systems Neurology (SyNergy), Munich, Germany*

JIE XU • *Center for Advanced Models for Translational Sciences and Therapeutics, University of Michigan Medical Center, Ann Arbor, MI, USA*

DONGSHAN YANG • *Center for Advanced Models for Translational Sciences and Therapeutics, University of Michigan Medical Center, Ann Arbor, MI, USA*
CHARLES YU • *Genentech, Inc., Department of Molecular Biology, South San Francisco, CA, USA*
XINGE YU • *Department of Pediatrics, University of Michigan, Ann Arbor, MI, USA*
BRANKO ZEVNIK • *In vivo Research Facility, Faculty of Medicine and University Hospital Cologne, University of Cologne, Cologne, Germany; Cologne Excellence Cluster for Cellular Stress Responses in Aging-Associated Diseases (CECAD), Faculty of Medicine and University Hospital Cologne, University of Cologne, Cologne, Germany*
JIFENG ZHANG • *Center for Advanced Models for Translational Sciences and Therapeutics, University of Michigan Medical Center, Ann Arbor, MI, USA*

Chapter 1

Transgenesis and Genome Engineering: A Historical Review

Lluis Montoliu

Abstract

Our ability to modify DNA molecules and to introduce them into mammalian cells or embryos almost appears in parallel, starting from the 1970s of the last century. Genetic engineering techniques rapidly developed between 1970 and 1980. In contrast, robust procedures to microinject or introduce DNA constructs into individuals did not take off until 1980 and evolved during the following two decades. For some years, it was only possible to add transgenes, de novo, of different formats, including artificial chromosomes, in a variety of vertebrate species or to introduce specific mutations essentially in mice, thanks to the gene-targeting methods by homologous recombination approaches using mouse embryonic stem (ES) cells. Eventually, genome-editing tools brought the possibility to add or inactivate DNA sequences, at specific sites, at will, irrespective of the animal species involved. Together with a variety of additional techniques, this chapter will summarize the milestones in the transgenesis and genome engineering fields from the 1970s to date.

Key words Genetics, Genome editing, CRISPR, Transgenic, Knockout, YAC, BAC, ZFN, TALEN

1 Origin and Meaning of Fundamental Concepts in Transgenesis and Genome Engineering

In this introductory section, I will refer to the origin, meaning, and use of some common words in the field, such as transgene, transgenic, transgenesis, knockout, knock-in, knockdown, genetically modified organisms, genome engineering, genome editing, gene editing, and related terms.

1.1 Transgene, Transgenic, and Transgenesis

Transgenesis is a common term used nowadays to refer to any introduction and/or transmission of foreign genes in any organism. The prefix "trans" has a variety of meanings, including beyond, instead of, across of, from one side to another, from the other side, etc. But "trans" also qualifies the term adjacent to it, with the meaning of "pretending to be" and "behaving as." Therefore, transgene could also be interpreted as a DNA sequence beyond a

Thomas L. Saunders (ed.), *Transgenesis: Methods and Protocols*,
Methods in Molecular Biology, vol. 2631, https://doi.org/10.1007/978-1-0716-2990-1_1,

gene, behaving as a gene, entering into an organism, and crossing it.

Presently, the word "transgene" applies to any piece of DNA (from the same organism or, often, from another organism) that is introduced into an organism or into its genome, through a variety of techniques. Consequently, the adjective "transgenic" applies to any organism carrying a transgene. And, hence, the generic word "transgenesis" describes all methods and techniques used to introduce or deliver transgenes into organisms. In other words, transgenesis refers to the technologies applied to generate transgenic organisms.

The words "transgenic" and "transgene" did not appear in the scientific literature until 1982, through a review authored by Jon Gordon and Frank Ruddle [1], two articles published by Richard Palmiter and Ralph Brinster [2, 3], a comment by W Petri [4], and a report of mammalian cell transformation [5]. The first three references clearly were related with the production of the first transgenic mice. Hence, the word "transgenic," although years later popularized in the plant kingdom, began with its use in animals or animal cells.

Transgene and transgenic also derived from previous terms, coined in the 1970s, when the first techniques to introduce DNA pieces, or even entire or fragmented chromosomes, into cells became available. The first of such terms, namely, "transgenosis," appeared in 1973 [6], to describe the overall phenomenon of transfer, gene maintenance, transcription, translation, and function. This was illustrated by transferring some bacterial genes into plant cells [6, 7]. The term "transgenosis" did not become too popular and was only referred sparsely in some subsequent articles, in 1976 [8]. However, in the same year, another related term was first used in the literature: "transgenome" referring to the transfer of metaphase chromosomes as vehicles to introduce new genetic information into eukaryotic cells [9]. Transgenome was a word adopted by Frank Ruddle in 1977 to describe experiments involving the transfer of human chromosomes into rodent cells [10]. That publication was followed by two additional studies from the same laboratory, in 1978 [11, 12], one in 1979 [13] and yet another one in 1980 [14] further using the word transgenome to describe this type of experiments, eventually reviewed by Ruddle in 1981 [15]. These references likely set the path for the future pioneer use of the evolved words "transgene" and "transgenic" by Frank Ruddle himself in 1982 [1].

1.2 Knockout, Knock-in, and Knockdown

The common pugilistic term "knockout" or "knock-out" (usually simplified as KO) does not need much explanation. It describes the termination of a boxing match when a boxer is unable or is declared unable by the referee (as because of injuries) to continue the fight. In the field of genetically modified organisms, it explicitly refers to the inactivation of a given gene in the genome, through targeted or random mutagenesis techniques. It refers to the generation of null mutants. In the scientific literature it was first used in 1989, in a review by Bernstein and Breitman, where they described the first examples of genetic ablation in mice [16], after the seminal articles published by Mario Capecchi targeting genes for inactivation in mammalian cells [17], and in mouse embryonic stem (ES) cells [18], after Oliver Smithies described the first gene-targeting examples in mice [19]. The term "knockout" was further popularized through a Keystone meeting organized in 1991, entitled "Knock-in and knock-out. Transgenes, Development and Disease: A Keystone Symposium sponsored by Genentech and Immunex, Tamarron, CO, USA, January 12–18, 1991" [20] and beyond.

Also related with the "knockout" concept, there are two additional terms often used in the field: knock-in and knockdown. Knock-in refers to the introduction of a given mutation precisely in a gene of interest, thus providing the resulting gene with a new function. The first reported use of the "knock-in" term was also in 1991, at the same conference where the word "knockout" was eventually popularized within the mouse genetics community [20].

Knockdown refers to the partial inactivation of a gene, through a variety of methods, where the resulting animal might display a more subtle phenotype (hypomorph) not as strong as the one associated with a true knockout, which sometimes can be incompatible with life. It was originally described in combination with antisense oligonucleotide technology, mostly used in zebrafish where these modified antisense oligonucleotides were termed as morpholinos [21].

1.3 Genetically Modified Organisms and Related Terms

The term genetically modified organisms (abbreviated as GMO) was first used in 1992, 10 years after the first appearance of the word "transgenic," in an Australian law journal [22]. Somehow, the GMO concept was preferred in the regulation arena, because it encompassed any genetic modification in an organism, whether transgenic or knockout, and generated through targeted or random mutagenesis, using any type of vector or through chemical or physical mutagenic agents. It was first applied to microorganisms [23], but during the late 1990s, with the development of the first transgenic plants, the GMO debate was associated mainly with them [24]. An example of use of the GMO term can be found in the Directive 2001/18/EC of the European Parliament and of the Council of 12 March 2001 on the deliberate release into the environment of genetically modified organisms [25], where the following official (legal) definition of a GMO was included: "genetically modified organism (GMO)" means an organism, with the exception of human beings, in which the genetic material has been altered in a way that does not occur naturally by mating and/or natural recombination. Since then, GMO and transgenic terms have been often used as synonyms, which could be used indistinctly. Furthermore, the term GMO has been often used instead of transgenic when this latter word became irreversibly contaminated by a heavily polarized public debate and perceived as negative by the public opinion. However, it must be stated that this is not true. These two terms are not exchangeable. Indeed, the concept of GMO is much wider than that of transgenesis, whose use should be restricted to the introduction of new DNA pieces, whether from foreign or from the same organismal genome. GMOs also include gene inactivation and any other genetic alteration in the genome of an organism caused by agents, tools, or experimental methods. GMO can also be referred as GM animals (or GM plants). Some researchers have referred the use of "genetically altered" (GA) organisms or GA animals, meaning any type of genetic alteration. Curiously enough, the first use of such term in the scientific literature was also associated with law studies [26], although the references to GA animals continue nowadays [27].

Finally, another expression that is often used today to refer those animals generated using genome-editing tools is the term "genome-edited" (GE) organisms [28, 29], which also continues to be commonly used nowadays [30].

1.4 Genome Engineering

The term "genome engineering" was first used in 1993, in a review on site-specific recombinases and their use to modify at will the genome of cells or living organisms [31] Thereafter, it was adopted by Janet Rossant and Andras Nagy in 1995 to illustrate the powerful toolbox available to researchers to modify, at will, the mouse genome [32]. The origin of the genome engineering concept was much older, although not expressed explicitly like this. It stems from the time when the first recombinant DNA techniques were described in bacteria, using plasmids, in the 1970s, the origin of genetic engineering. Presently, taking also into account the genome-editing tools and genome-editing techniques [33], the concept of genome engineering refers to all known methods capable of modifying genomes at will.

1.5 Genome Editing or Gene Editing

The first reported use of the term "genome editing" was in 2003, associated with the use of site-specific recombinases, such as Cre or Flp [34]. A few years later, the term genome editing was applied to describe the use of zinc-finger nucleases (ZFNs) in specifically altering the genome of organisms [35]. Thereafter, genome-editing technology was applied also to subsequent programmable nucleases, such as transcription activator-like effector nucleases (TALEN), or, more recently, to the clustered regularly interspaced short palindromic repeats (CRISPR)-CRISPR-associated proteins (Cas) systems [36].

The term "gene editing" is also often used instead of "genome editing," and both are often replaced by "genome engineering," although, as indicated above, this latter term encompasses a wider meaning and extension. However, gene editing was first used to describe a different genetic process: the concept of RNA editing, originally described in some viruses [37]. Gene editing was later applied to describe a gene therapy approach involving the use of self-pairing, chimeric RNA/DNA oligonucleotides [38]. Eventually, the gene-editing concept was also adopted by the nascent genome-editing tools and was already applied to describe the uses and applications of ZFN [39].

2 Timeline of Transgenesis and Genome Engineering

What follows is a chronological personal account of relevant milestones, discoveries, and breakthroughs that helped shaping the fields of transgenesis and genome engineering.

2.1 1974

Most books accounting the history of transgenesis refer to classical 1980 (and beyond) microinjection experiments when describing the origins of the field. However, DNA manipulation had begun a decade earlier, with the origins of genetic engineering methods [40], and the generation of genetically modified animals as well. I think it is necessary to credit the pioneer studies and methods developed by the German embryologist and virologist Rudolf Jaenisch, whose initial and fundamental contributions to the field are often not recognized. In fact, in 1974, Rudolf Jaenisch, working in the laboratory of the mouse developmental biologist Beatrice Mintz, published that SV40 viral DNA could be detected in adult mice after injecting the genome of this virus in the blastocoel cavity of mouse blastocysts [41]. They found the viral DNA in several tissues of those mice, although they could not confirm whether this was a true DNA integration or an episomal event. This is the first report that must be listed as the origin of DNA manipulation efforts toward the generation of transgenic, hence genetically modified, animals.

2.2 1975

Rudolf Jaenisch reported in 1975 his pioneer work with another virus, the Moloney murine leukemia virus (M-MuLV). He used this virus to infect mouse preimplantation embryos at earlier stages (four- to eight-cell embryos), and, among the resulting mice born, he found some animals developing leukemia, hence unequivocally demonstrating the integration of this virus into the host mouse genome [42]. This was the second report illustrating how exogeneous DNA can be integrated in the mouse genome, hence resulting in the birth of transgenic mice.

2.3 1976

Another milestone reached by these pioneer Rudolf Jaenisch studies was achieved in 1976, when germline transmission of the M-MuLV virus DNA was eventually confirmed [43]. These experiments mark the foundation of animal transgenesis methods, using retroviruses, although the term transgenic itself would not be used until 1982 [1].

2.4 1980

The American researchers Jon Gordon and Frank Ruddle are often recognized as the pioneers of mouse transgenesis, even though their first studies in the field were published 6 years after the initial reports from Rudolf Jaenisch, who should be credited as the true pioneer of the field. However, it should be also recognized the fundamental contributions of Gordon and Ruddle. They contributed to the popularization of the initial mouse microinjection techniques, since their first publication, published in 1980, illustrating how nonviral DNA molecules from a common plasmid, pBR322, could be microinjected into the pronuclei of fertilized mouse oocytes and later be found in the resulting newborn mice [44].

It was also in 1980 when Mario Capecchi began using the microinjection technique to efficiently deliver DNA into mammalian cells in culture [45].

2.5 1981

Gordon and Ruddle reported in 1981 the effective germline integration of transgenes and, hence, the transmission of the microinjected DNA molecules to the progeny of founder transgenic mice [46].

In 1981, we first learned the pioneer studies of the mouse embryologist Ralph Brinster, working with the molecular biologist Richard Palmiter, which resulted in a fruitful and long-standing collaboration in the mouse transgenesis field. It was indeed in 1981 when Brinster and Palmiter published their first study, showing how a plasmid containing the thymidine kinase (TK) gene from herpes simplex virus (HSV), fused to the promoter and regulatory region of the mouse metallothionein-I gene, could be microinjected into the pronuclear of mouse embryos and resulted in the birth of several transgenic mice carrying the plasmid transgene [47]. In this initial publication, they also first reported the integration of multiple copies of transgene DNA organized as tandem arrays, a feature that was soon observed in all subsequent transgenic mouse experiments.

It was also in 1981 when the mouse ES cells, a group of pluripotent cells found in the inner cell mass of blastocysts contributing to all somatic and germline tissues of the developing embryo, were first isolated and described by Martin Evans and Matthew Kaufmann [48]. The isolation of mouse ES cells was independently confirmed, some months later, by Gail Martin, using a different and simpler method which, as a matter of fact, became the reference in the field [49].

2.6 1982

The year 1982 should be probably recognized as the year when the mouse transgenesis techniques became universally known and, likely, the year when many laboratories attempted reproducing the initial reports published by the pioneers in the field. In fact, in 1982, Brinster and Palmiter continued reporting features of the

HSV-TK transgenic mice, phenotyping the transgene proper gene regulation [50] and their transmission to offspring [2]. However, if 1982 must be highlighted in this historic account, it is because Brinster and Palmiter published the first most famous article in the mouse transgenic field [51]. They generated transgenic mice with a construct containing the promoter of the metallothionein-I (MT-I) gene fused to the rat growth hormone (rGH) gene. The resulting transgenic mice were dramatically (~2.5 times) bigger than their non-transgenic littermates, due to the activity of the microinjected transgene MT-I-rGH. The amazing picture of two littermates, one transgenic (bigger) and one non-transgenic (smaller), was chosen for the cover of the scientific journal *Nature*, where this experiment was published [51].

In 1982, Mario Capecchi also reported the first evidence of homologous recombination events upon directly microinjecting DNA into mammalian cells in culture [52].

2.7 1983

Gordon and Ruddle published in 1983 the first technical publication describing procedures for the generation of transgenic mice by pronuclear microinjection of one-cell stage embryos [53].

2.8 1984

Brinster and Palmiter published in 1984 their first comprehensive review accounting the generation of numerous transgenic mice [54]. Hammer, Palmiter, and Brinster should be underlined also for envisaging in 1984 the very first biomedical application of transgenesis, based on the partial correction of dwarfism in mutant mice with their MT-I-rGH transgene [55].

First attempts to microinject mouse embryonic teratocarcinoma (EC) cells into blastocysts by Allan Bradley were reported in 1984, with the birth of the first germline-transmitting mouse chimeras [56].

2.9 1985

Hammer, Palmiter, and Brinster, with the participation of other researchers, were the first in 1985 of applying the transgenic mouse techniques to other livestock species, thus making possible the birth of the first transgenic rabbits, sheep, and pigs by pronuclear microinjection, albeit with significantly lower efficiencies, as compared with those reported in mice [57]. It is also worth noting the 1985 review on transgenic mice published by Palmiter and Brinster [58] and their first comprehensive and systematic collection of parameters affecting the successful generation of transgenic mice, including DNA concentration, size, and form (supercoiled versus linear) of these DNA molecules, site of microinjection (pronucleus versus cytoplasm), the genetic background of the mouse strain, and also buffer composition [59].

The geneticist Oliver Smithies was investigating the mechanisms controlling homologous recombination, with the objective of being capable of replacing endogenous sequences by exogenous DNA constructs. With this aim, in 1985, he targeted an endogenous gene in mammalian cells with a plasmid construction previously digested with a suitable restriction enzyme cutting within the DNA sequence homologous to the murine chromosome. With this new approach, he and his collaborators achieved a significant efficiency, of about one recombinant clone per thousand transfected clones [60].

2.10 1986

In 1986, the teaching material of two practical courses on molecular embryology of the mouse, held at Cold Spring Harbor Laboratory in 1983 and 1984, were converted into the first edition of the book *Manipulating the Mouse Embryo: A Laboratory Manual* by Brigid Hogan, Frank Costantini, and Elizabeth Lacy [61]. Since 1986, thousands of researchers and generations of technicians have learned the basic methods of mouse transgenesis with this book, popularly known in the field as "The Bible." The subsequent editions that were published in the following years completed and updated the presently available technologies: second edition in 1994 by Brigid Hogan, Rosa Beddington (1956–2001), Frank Costantini, and Elizabeth Lacy [62]; third edition in 2003 by Andras Nagy, Marina Gertsenstein, Kristina Vintersten, and Richard Behringer [63]; and fourth and, to date, the last edition of this book series, published in 2014, by Richard Behringer, Marina Gertsenstein, Kristina Vintersten, and Andras Nagy [64].

In 1986, Palmiter and Brinster collected all their wisdom on the parameters affecting the successful generation of transgenic mice in a celebrated review that has remained valid and a reference in the field to our days [65].

Elizabeth Robertson and Allan Bradley, in Martin Evans' laboratory, reported in 1986 the first generation of germline-transmitting mouse chimeras using retroviral vectors inserted into mouse ES cells that had been injected into mouse blastocysts [66]. These pioneer experiments illustrated the uniqueness and potential of mouse ES cells as a tool for investigating developmental genetics of the mouse [67].

Mario Capecchi and collaborators, using direct microinjection methods in mammalian cells in culture, demonstrated in 1986 that targeted mutations could be introduced in predefined genes with high frequencies [17, 68], similar to those that Oliver Smithies had reported the year before [60].

2.11 1987

In 1987, Mario Capecchi integrated different concepts and methodological approaches and developed a technique for targeting specific genes in mouse ES cells through homologous recombination, using the positive-negative selection method he presented to the scientific community [18]. The first version of this method was designed to target selectable genes (e.g., HPRT), whose mutations could be selected positively in the presence of the antibiotic analogous G418 with an introduced neomycin-resistance (neoR) gene and, negatively, by growing the transformed cells in a culture medium containing the base analog 6-thioguanine, where only HPRT mutant clones would survive [18].

In 1987, a group of Japanese microbiologists first reported the presence of a curious set of repetitive DNA elements interspaced with unique sequences in the genome of a Gram-negative bacteria: *Escherichia coli* [69]. Similar DNA sequences were later found in Gram-positive mycobacteria [70] and eventually also in archaea [71]. These arrays of repetitive and unique DNA sequences would be known as CRISPR sequences and the effector genes adjacent to them as Cas (CRISPR-associated) proteins [72].

2.12 1988

Brinster and Palmiter reported the benefits of adding noncoding sequences, such as introns, in transgenic constructs, to increase the transgene expression efficiency [73]. It was a first sign of a revolution involving the use of large pieces of noncoding genomic DNA to consolidate transgene expression, which would be uncovered 5 years later with the use of artificial chromosome type of transgenes [74].

An alternative gene inactivation strategy was documented in 1988, upon combining a gene encoding a toxic product under the control of promoter DNA elements targeting a given cell type, as a classical transgenic application, overexpressing transgenes in a subset of cells or tissues of the host. Those were gain-of-function experiments and included cell-specific ablation studies, such as fusing tissue-specific promoter sequences with the diphtheria toxin gene [75].

Mario Capecchi published in 1988 an improved version of his positive-negative method, which could be applied also for non-selectable genes. The enrichment for homologous recombination events in mouse ES cells was driven by a neoR cassette surrounded by homologous sequences with the endogenous locus, where the random integration was negatively selected by the HSV-tk gene in the construct, outside the region of homology, whose integration rendered the transformed clone sensitive to nucleoside analogues such as FIAU or ganciclovir, resulting in the interference with DNA replication [76]. This was a robust universal method for targeting any murine gene in mouse ES cells by homologous recombination that became the reference method for many years. As it was included in this seminal publication, this was "a general strategy for targeting mutations to non-selectable genes" [76].

2.13 1989

Ralph Brinster demonstrated, in 1989, with a massive microinjection effort, that it is possible to target a genomic locus with exogenously provided homologous DNA transgenes. Brinster microinjected more than 10,000 mouse embryos and generated more than 500 transgenic mice with a genomic construct before encountering a single mouse founder compatible with an integration by homologous recombination [77].

It was also in 1989 when the first mutant mice generated from gene-targeted ES cells with the Hprt gene inactivated were obtained by Oliver Smithies [19]. These were the first knockout mice to be reported in the scientific literature. Mario Capecchi published in the same year a first review on producing targeted alterations in the mouse genome by homologous recombination techniques [78].

2.14 1990

This was the first year when different mouse mutations obtained by the new gene-targeting method by homologous recombination in mouse ES cells began accumulating, led by the first mouse mutation, in the Int-1 gene derived in Mario Capecchi's laboratory [79] and the same mouse mutation obtained in Allan Bradley's laboratory [80]. These pioneer papers were followed by many other similar experiments. Exactly 10 years after the first transgenic mouse microinjection experiments, a new technology boosted the functional genomics approaches in the mouse. After 1990, literally, thousands of mutant mice (KO) were generated, phenotyped, and archived. The majority of all these KO mice produced over the years have been cryopreserved by at least one of the several worldwide distributed repositories, and, therefore, they are all searchable through the International Mouse Strain Resource (IMSR) (http://www.findmice.org/) [81].

2.15 1991

In 1991, the second most famous transgenic mouse (after the giant transgenic mice reported by Palmiter and Brisnter about 10 years earlier [51]) was reported by the laboratory of Robin Lovell-Badge. Transgenic mice carrying a transgene with the Sry gene resulted in the development of males, irrespective of their chromosomal karyotype, thereby demonstrating the master role of this gene to induce testis differentiation and subsequent male development [82]. The picture of a transgenic mouse genotypically female and phenotypically male made it to the front cover of the *Nature* issue where these impressive results were published.

Another famous transgenic animal, this time not a mouse but a sheep, was also reported in 1991 from researchers at the Roslin Institute. Tracy, a transgenic sheep, was generated using a transgene overexpressing a protein found in human plasma (alpha-1-antitrypsine) in the sheep mammary gland [83]. This sheep and its progeny produced this human recombinant protein in the milk at very high concentrations.

2.16 1992

Transgenic mice often showed variable and ectopic transgene expression, also associated with undesired chromosomal position effects, linked with the genome site where the transgene had been integrated. The lack of important and normally unknown noncoding genome DNA sequences was the most common explanation for the usual suboptimal performance of transgenes. The use of large genomic fragments (~100 to ~1000 kb) appeared as a putative solution for inserting an entire locus into the host genome, encompassing all known and unknown DNA regulatory elements required for the faithful expression pattern of the transgene, similar to that of the endogenous locus. In the early 1990s, yeast artificial chromosomes (YACs) were the only vectors capable of accommodating such large DNA fragments [84].

In 1992, Günther Schütz's laboratory was the first to report the generation of transgenic mice with a relatively modest YAC (35 kb), carrying a tyrosinase (Tyr) transgene, a minigene, with fragments derived from the mouse genome and from cDNA [85]. A few months earlier that year, Rudolf Jaenisch's laboratory had succeeded introducing a larger (150 kb) YAC carrying the Col1a1 locus into mouse fibroblasts in culture [86].

2.17 1993

In 1993, three groups, working independently, succeeded in reporting the pioneer generation of transgenic mice carrying large YACs. Two of these laboratories were led by Günther Schütz [87] and Rudolf Jaenisch [88], and the third one was a company, Cell Genesys, Inc. [89]. The three publications appeared in March 1993 and reported three different methods. The Schütz laboratory used standard pronuclear microinjection of isolated YAC DNA (~250 kb) purified in the presence of polyamines [87]. The Jaenisch laboratory introduced a YAC DNA (~150 kb) into mouse ES cells via lipofection and derived germline-transmitting chimeras from them [88]. And Genesys, Inc., decided to fuse yeast spheroplasts carrying a YAC (~670 kb), and the rest of yeast chromosomes, with mouse ES cells from which transgenic mice were also derived [89].

Later in 1993, several other groups also succeeded in generating YAC transgenic mice [90–93], mostly using the microinjection protocol, which eventually became the preferred method. Also in 1993, the first review on the subject was published with a title summarizing the achievements made: "YAC transgenes: bigger is probably better" [94]. In fact, a popular review published in 2001 by Montoliu and Giraldo, reporting the many YAC transgenic mice generated to that date, indicated that the transfer of these YACs correlated with robust and optimal expression levels of the transgenes, irrespective of their location in the host genome, with position-independent and copy number-dependent transgene expression demonstrated in most cases [74].

It was also in 1993 when the first transgenic mouse carrying a P1-derived artificial chromosome (PAC) was reported [95]. PACs, in contrast to YACs, were circular and seemingly more stable and easier to handle [74].

2.18 1994

The gene-targeting approach in mouse ES cells by homologous recombination was, in 1994, already the method of choice for producing hundreds of specific mutations in selected genes at will. However, this strategy had some unexpected problems. A number of these KO mouse projects were not productive, because the resulting homozygous mutants were embryonic lethal or died shortly after birth, rendering impossible any further analysis [96]. Therefore, an alternative method was required to overcome the embryonic lethality observed in some (about 30%) of knockout mouse attempts.

Such desired method was developed by Klaus Rajewsky in 1994 [97]. In this pioneer publication, they reported the use of a prokaryotic site-specific recombination system, the Cre-loxP system, composed of a Cre recombinase and a target loxP sequence, a short (34 bp) directional DNA fragment recognized by Cre recombinase and used to remove any DNA sequence found in between two consecutive loxP sequences in cis and oriented in the same direction. The approach was smart and innovative. Two loxP sites had to be cloned in a DNA construct surrounding the target locus to be inactivated, usually inside consecutive introns, delimiting an exon ready to be excised. Thereafter, using ES cells and homologous recombination, such construct was integrated in the mouse genome, thereby generating the so-called "floxed" ("flanked by loxP sites") allele. This resulting allele was ready to be inactivated as soon as it was exposed to Cre recombinase. This nuclease was usually brought in through a standard transgene, in a different mouse, driving the expression of the Cre recombinase in a cell type or tissue of choice. Thereafter, upon mating the floxed mouse with the transgenic Cre mouse, the selected gene would only be inactivated where loxP sites and Cre recombinase coincided, thereby generating a tissue- or cell-specific knockout. Rajewsky and his collaborators demonstrated this method with a T-cell-specific knockout of a DNA polymerase gene beta, whose ubiquitous deletion was embryonic lethal [97]. This method would be known as "conditional mutagenesis" or "conditional mutants" [98].

Conditional mutagenesis boosted mouse knockout projects and multiplied the number of experiments and phenotypes one could derive from a single floxed mouse allele, upon exposing them to a variety of transgenic mice expressing Cre recombinase gene in different tissues or cell types [98]. The function of mouse genes required for proper embryo development could be investigated in the brain [99] or the immune system [100], thanks to this conditional mutagenesis approach.

The ease by which these conditional mutant mice carrying floxed alleles could be generated eventually triggered the foundation of several international initiatives aiming to generate and phenotype mouse knockout lines representative for all mouse

(~20,000) genes. Different consortia were launched around the world, including the International Knockout Mouse Consortium (IKMC) [101] and the European Conditional Mouse Mutagenesis Program (EUCOMM) [102], but eventually all projects merged into the International Mouse Phenotyping Consortium (IMPC) [103], currently active through its web portal (https://www.mousephenotype.org/), where up to 7824 mouse knockout lines have been already generated and phenotyped (October 2021 release) and made available to anyone interested.

2.19 1995

The conditional mutagenesis approach triggered additional innovations. The site-specific recombinases could be transformed into inducible recombinases upon fusing their coding regions with that of hormone-binding domains from nuclear receptors. This approach enabled the activation of Cre recombinase activity at will, on a timely manner, through the administration of hormone analogues, instrumentally used to release the recombinase, otherwise bound to chaperones in the cytoplasm, and allow its translocation to the nucleus, where the expected recombinase activity could take place. The first inducible Cre recombinase was produced by Pierre Chambon's laboratory in 1995 [104]. Other than the Cre/loxP system, additional site-specific recombinases (e.g., Flp, Dre, PhiC31) have been described and used in genetically modified animals for controlling the activation/inactivation of targeted mutations in selected genes [105].

In 1995, Ryuzo Yanagimachi, from his laboratory in Hawaii, established a new reproductive biology technique in mice: intracytoplasmic sperm injection (ICSI) [106]. It is widely assumed that most artificial reproduction methods were first tested in laboratory animals and, thereafter, adapted for human use. This is not the case for ICSI, whose pioneer use in human fertility clinics was first reported in 1992 [107]. Three years later, this useful, but most technically challenging, procedure was made available in mice. This success was achieved, thanks to a complementary technique, piezo injection, which softens the introduction of the ICSI needle into the cytoplasm of mouse unfertilized oocytes, which are more fragile than the human unfertilized oocytes.

It was also in 1995 when researchers reported the benefits for homologous recombination of using the yeast meganuclease, Sce-I. This extremely rare cutting enzyme was used to trigger single double-strand breaks (DSB) in the yeast genome, where homologous recombination events were observed at very high frequency [108]. This was the first generation of genome-editing tools, although with null flexibility. Recombinant meganucleases will not be made available until several years later [109, 110], once the second generation of genome-editing tools, namely, the Zn-finger nucleases (ZFN), had been already described [111].

2.20 1996

The year 1996 should be considered as the beginning of a new era of animal biotechnology. In particular, it represents the beginning of new developments in livestock biotechnology. In March 1996, the team led by Ian Wilmut at the Roslin Institute in Edinburg (Scotland, UK) first published his success in sheep cloned by nuclear transfer, using nuclei from a cultured sheep cell line featuring totipotent for nuclear transfer (TNT) capacity [112]. The resulting two sheep born, Moran and Megan, perhaps they were not as famous as the next cloned sheep that would be born at Roslin in 1996, but they deserve to be credited as the first mammalian examples that were cloned from an established cultured cell line. This outstanding achievement was based on the pioneer efforts of John Gurdon in the late 1950s and early 1960s on the cloning of frogs by nuclear transfer using nuclei from somatic cells [113, 114] and the first cloned sheep obtained by Willadsen in 1986 using embryonic blastomeres from an eight-cell sheep embryo fused with enucleated unfertilized eggs [115].

However, 1996 should be also remembered in this account of milestones in transgenesis and genome engineering, as the year when Dolly, the most famous cloned sheep, was born. In fact, Dolly was born on July 5, 1996, at the Roslin Institute (https://dolly.roslin.ed.ac.uk/facts/the-life-of-dolly/), although the world did not hear about it until February 1997 [116].

2.21 1997

Dolly the sheep, the first mammal cloned from an adult cell, was published in *Nature* in February 1997 [117]. The publication detailing the birth of Dolly, derived from nuclei of a sheep cell line established from mammary gland tissue fused to enucleated sheep oocytes, was the product of an extraordinarily rare event. Almost 300 sheep embryos were reconstituted by nuclear transfer before one of them reached to term and resulted in the birth of Dolly. That paper immediately generated enormous expectations [118] as well as fears and ethical controversies of all kinds [119].

The cloning procedure by somatic cell nuclear transfer (SCNT), originally described in sheep for the generation of Dolly, was reproduced in several species in the following years, each time with some species-specific methodological adaptations. The series of cloned animals included cows [120, 121], mice [122], goats [123], pigs [124], rats [125], rabbits [126], horses [127], dogs [128], and other species.

Other types of artificial chromosomes, beyond YACs, were explored. YACs were often unstable and chimeric, and its manipulation and modification required specific yeast culture skills, not universally present in all molecular biology laboratories. Besides PACs, already tested in 1993, bacterial artificial chromosomes (BACs) were first used for transgenesis in 1997. BACs (and PACs) were circular DNA vectors and offered a smaller cloning capacity (<300 kb) as compared to YACs. However, BACs were easier to handle with the standard bacterial skills available in all molecular biology laboratories [74]. The first transgenic animal generated with a BAC was published in 1997 [129]. BAC transgenesis very soon became the method of choice for producing transgenic animals carrying large genome fragments. This was mainly due to the ease by which these vectors could be manipulated and the development of efficient homologous recombination methods in bacteria in 1999 [130] by the laboratory of A. Francis Stewart, later known as BAC recombineering, for homologous recombination in *E. coli* [131].

A comprehensive overview of BAC transgenic mice was published by Thomas Saunders in 2009 [132]. In this review, they analyzed 707 transgenic founder mice generated with 86 BAC transgenes. They concluded that most (80%), but not all, BAC constructs resulted in transgenes with the expected expression pattern. In addition, a careful analysis was required to discriminate between intact BAC transgenes and fragmented BAC constructs integrated, prone to unexpected expression results (similarly to what had been observed before with YACs) [74].

In 1997, another novel technology was presented for mammalian transgenesis: the use of transposons. The first transposition system to succeed in vertebrates for transgenesis was Sleeping Beauty (SB), whose original DNA sequence, full of mutations accumulated through the evolution, was rescued in vitro, thereby resurrecting this DNA mobile element from the salmon genome, and proven to operate in mammalian cells [133]. The SB transposition system rapidly was shown to efficiently trigger genetic modifications in several mammalian species [134]. In mice, the SB system was proven for insertional mutagenesis and confirmed to be transmitted through the germline [135]. The development of hyperactive SB transposition systems enhanced their functionality for insertional mutagenesis [136], and its application was demonstrated in mice, rats, and rabbits [137]. However, some problems were observed upon using transposons for transgenesis. Chromosomal rearrangements were not rare at the transposed genomic locations, which could alter the expected expression patterns of transgenes. This is one of the known limitations of transposons in transgenesis [138].

The relevance of SCNT methods, presented early in 1997, was promptly recognized. The cell nuclei used to reconstruct embryos could also be derived from genetically modified cells in culture. Hence, the resulting animal would be a clone as well as a transgenic animal. The first transgenic and cloned animals of this kind were also sheep, also produced at the Roslin Institute in 1997 [139]. This seminal publication for the livestock community illustrated a more robust and efficient way of generating transgenic farm animals, as compared with the poor efficiencies observed by DNA pronuclear microinjection [57]. A similar success was reported in cows in 1998, with transgenic and cloned calves born from genetically modified fetal fibroblasts [140], and pigs [141].

2.22 1998

In 1998, 17 years after isolating the mouse ES cells, a similar pluripotent cell type was isolated from human blastocysts. These first human embryonic stem cells (HESC) were described by J.A. Thomson [142].

The combination of these HESC with cloning methods, described the year before, opened the Pandora's box [143]. In humans, reproductive cloning was banned in most countries, whereas therapeutic cloning was conceived as a potential treatment for uncurable diseases [144]. Reproductive cloning for human beings (routinely applied in other animal species) triggered many ethical dilemmas. On the other hand, therapeutic cloning generated many expectations for its potential application in regenerative medicine. However, therapeutic cloning was never achieved in humans but rather only described in a seminal paper from Rudolf Jaenisch's laboratory in 2002, demonstrating the complexity of this process in mice [145].

In 1998, the laboratory of Ryuzo Yanagimachi succeeded adapting the ICSI techniques to rats [146]. Their success in mice [106] and rats [146] popularized the use of piezo injectors for ICSI purposes.

Also in 1998, we discovered the RNA interference method, initially developed in nematodes, to alter the expression of some genes by interfering with its translation. The rationale behind this approach was the production of antisense RNA sequences, which would form double-strand RNA molecules upon pairing with their complementary messenger RNAs (mRNAs), eventually triggering an endogenous mechanism of clearance that would result in the knockdown or silencing of the selected gene. Andrew Fire and Craig Mello described this technique in the worm *Caenorhabditis elegans* in a seminar article in 1998 [147]. Their pioneer efforts were recognized in 2006 with a Nobel Prize in Physiology or Medicine which they were awarded jointly.

The RNA interference concept offered a method for silencing or inactivating the expression of genes without interrupting the genome. The resulting animals were known as knockdown, in contrast to knockout animals (where the selected gene had been inactivated) because often the expression of the interfered gene was significantly reduced, but it did not disappear. Different RNA molecules had been used for this purpose, including antisense RNA and distinct subtypes of RNA interference (RNAi, siRNA, shRNA, miRNA, etc.) molecules in transgenic animals. In zebrafish, the use of modified oligonucleotides (morpholinos) has been widely used to produce gene-specific knockdowns [148]. However, the use of proper controls and experimental conditions for RNA interference approaches in zebrafish must be considered for optimal results [149].

2.23 1999

In 1999, Teruhiko Wakayama and collaborators reported an alternative, and efficient, method for producing transgenic mice achieved through a combination of ICSI procedures using partially damaged sperm heads exposed to DNA molecules [150]. ICSI transgenesis was validated in other species, including nonhuman primates [151]. The ICSI transgenesis method was also extended for delivering large transgenes, such as BACs [152] or YACs [153] in mice, although with a lower efficiency for the integration of intact DNA constructs [153]. ICSI was also instrumental for cloning mice [122].

The combination of ICSI with DNA delivery is a type of sperm-mediated gene transfer (SMGT) method, although the origins of SMGT were highly controversial. The Italian researcher Claudio Spadafora announced in 1989 the use of sperm as a vector for introducing DNA into mouse eggs to generate transgenic mice [154]. That experiment was not reproducible in many laboratories, likely related to some undefined genetic factors whose role was essential for the success of this technique [155, 156]. The SMGT method was also used to generate transgenic pigs by the same Italian group [157]. However, contradictory results were obtained by other laboratories [158], thus suggesting that the SMGT method was not a robust technique.

2.24 2000

The SCNT was not only instrumental for producing transgenic livestock at higher efficiencies. The genetic manipulations could also include the inactivation of genes. Thanks to SCNT methods, it was possible to produce knockout animals in different species, beyond mice, overcoming the lack of bona fide ES cells, unknown in farm animals, through an alternative route.

Gene targeting in livestock was achieved in year 2000 using SCNT techniques. Once again, the Roslin Institute team described the first knockout sheep, inactivating the COL1A1 locus in sheep fibroblasts and thereafter using their nuclei to reconstitute sheep embryos by cloning [159]. A similar approach was applied in 2002 for generating the first knockout pigs carrying a targeted mutation in the gene encoding alpha-1,3-galactosyltransferase, to be used for xenotransplantation purposes [160]. Hence, SCNT methods have been, since the beginning of the twenty-first century until the revolution boosted by the arrival of genome-editing tools, the reference methods for regularly producing transgenic and knockout animals in mammalian species other than mice [161].

2.25 2001

ZFNs were first envisaged in 2001, as chimeric proteins, containing DNA binding domains, associated with several zing finger modules, and the DNA endonuclease activity from a bacterial enzyme,

Fok I, producing DSB at the genome locations dictated by the zinc-finger domains [162]. Similar to yeast meganucleases, the recognition of DNA sequences in ZFNs was again in charge of a protein, whose amino acid sequence targeted the desired DNA sequence. Therefore, a new ZFN had to be generated each time for each gene to be targeted. Furthermore, the technological platform for creating ZFNs was proprietary, and any potential user had to purchase these engineered ZFNs or establish a collaboration with the company owning the license for using this technology.

2.26 2002

The introduction of retroviruses for animal transgenesis had been already demonstrated by Rudolf Jaenisch in the 1970s. However, the use of viral vectors for shuttling genes of interest into animal genomes was not implemented until more than 30 years later. In 2002, the laboratory of David Baltimore reported the successful use of lentiviral vectors carrying reporter genes to generate transgenic mice and rats at much higher efficiencies than those obtained with DNA pronuclear microinjection techniques [163]. These lentiviral vectors were immediately applied to produce transgenic livestock and resulted in similar high efficiencies [164]. Lentivirus offered many advantages, including its use in a wider variety of species and mouse strains, where DNA pronuclear injection method was very inefficient or even not possible, and the high number of founder animals this technique could generate. In contrast, lentiviral vectors had a limited cloning capacity (~7 kb) and were often associated with multiple integration sites, thus making the establishment and characterization of stable transgenic animal lines a challenging project. Nowadays, the use of lentivirus for delivering genes to cells or for transient experiments in animals is probably the method of choice, ensuring a higher number of transgenic cells or newborns. Furthermore, the microinjection of lentivirus is easier, because viral particles can be directly microinjected under the zona pellucida. The known limitations of lentiviral transgenesis have probably prevented a wider success for this method. Nonetheless, lentiviral transgenesis belongs to the toolbox for genetically modifying animals, and, in fact, this method can rescue a transgenic experiment where other techniques have failed [165].

2.27 2005

In 2005, Francis Mojica published a pioneer paper suggesting that CRISPR-Cas systems are involved in bacterial immunity against foreign genetic elements, including bacteriophages and plasmids [166]. This article launched the research interest on this prokaryotic defense strategy, eventually leading to the establishment of CRISPR-Cas systems as genome-editing tools [167].

2.28 2006

The ethical dilemmas associated with therapeutic cloning in human beings were mainly related with the destruction of embryos (blastocysts), a step required to obtain patient-specific HESC from which to differentiate the expected somatic cells for repairing the damage tissue. These dilemmas were resolved in 2006 by Shinya Yamanaka, with a seminal publication where he reported the identification of the four transcription factors that were required to transform and reprogram any somatic cell into a pluripotent cell [168]. These findings launched the establishment of inducible pluripotent stem cells (iPS) which could be used in combination with transgenic and knockout techniques in mice [169]. Since then, several attempts have tried to establish iPS cells in livestock with mixed success [170].

2.29 2007

Martin Evans, Mario Capecchi, and Oliver Smithies (1925–2017) were awarded jointly the Nobel Prize in Physiology or Medicine in 2007, for their discoveries of principles for introducing specific gene modifications in mice by the use of ES cells [171].

2.30 2008

In 2008, the group of Austin Smith established defined cell culture media conditions to obtain pluripotent mouse ES cells, reproducibly and in a robust manner [172]. The use of three inhibitors (3i) in the culture cell media triggered an innate program for self-replication that did not require extrinsic instructions. Later the same year, the Austin Smith laboratory applied these findings on rat blastocysts and succeeded in the isolation of bona fide ES cells, thereby extending the possibility to apply gene-targeting techniques in rats [173].

2.31 2009

In 2009, Aron Geurts and collaborators first demonstrated the utility of ZFNs for specifically inactivating genes in animals, generating the first knockout rats with very high efficiencies, at a two-digit level, never seen before with ES cells and homologous recombination approaches in mice [174]. This experiment triggered the use of ZFNs for targeting genes in subsequent years and extended their use to a great variety of species including rats [175], mice [176], cattle [177], and pigs [178].

The use of ZFNs was particularly remarkable in livestock. Numerous animals with additional properties, such as increased resilience to diseases, were generated. Examples of these applications were ZFN genome-edited cattle with increased resistance to mastitis [179] or ZFN genome-edited pigs carrying a warthog-derived genetic variant meant to confer resistance to African swine fever [180].

2.32 2011

ZFNs, although useful and powerful, were essentially available from commercial sources only. This limitation triggered the development of alternative genome-editing tools that could be engineered in every single molecular biology laboratory. The TALENs were the next genome-editing tools developed, combining prokaryotic transcription factors of bacteria that infect plant cells fused with the same Fok I restriction enzyme domain used before with ZFNs. In 2011, the gene knockout reported with TALEN was described in rtas by Ignacio Anegon [181]. This initial success and the advantage that TALENs could be prepared in all laboratories quickly prompted many laboratories worldwide to embrace TALENs, leaving behind ZFNs. These new genome-editing tools, already the third generation, after yeast meganucleases and ZFNs, were used for generating genome-edited mice [182], and their benefits were extended to livestock [183], including cattle [184], pigs [185], and goats [186].

In 2011, Shirley Pease and Thomas Saunders edited another reference book on advanced transgenic technologies with the collaboration of many other colleagues, united through the International Society for Transgenic Technologies (ISTT) [187].

2.33 2012

In 2012, Shinya Yamanaka and John Gurdon were jointly awarded the Nobel Prize in Physiology or Medicine, for the discovery that mature cells can be reprogrammed to become pluripotent [188]. Surprisingly, none of the researchers of the Roslin Institute, responsible for the generation of Dolly the sheep, the first mammal to be cloned from adult cells, were included in this award.

In 2012, Emmanuelle Charpentier and Jennifer Doudna published a disruptive article where they described the compounds of a CRISPR-Cas9 defense system from *Streptococcus pyogenes* and proposed they could be used as universal genome-editing tools [189]. This publication was a game changer. ZFNs and TALENs already permitted targeted modification of genes through the generation of DSB and their subsequent repair by the endogenous cellular machinery. But ZFNs and TALENs, with their methodologic differences, were still too cumbersome and difficult to prepare. In contrast, CRISPR-Cas tools were easier to develop, cheaper, and highly versatile [190].

CRISPR-Cas genome-editing tools were fully programmable. In their simplest constitution, they were formed by a DNA endonuclease (usually Cas9) and a small RNA molecule (synthetic guide RNA, sgRNA) that would drive the Cas9 to generate a DSB precisely at the gene whose DNA sequence was complementary to the sgRNA [189, 190].

2.34 2013

It was not until early 2013 when the suggestive proposal made by Charpentier and Doudna became a reality. CRISPR-Cas9 reagents were first demonstrated to operate as efficient genome-editing tools on mammalian cells by two laboratories, working independently, led by Feng Zhang [191] and George Church [192].

In 2013, the first genome-edited animals were produced. Zebrafish was the first animal species to be genetically modified by CRISPR-Cas9 tools in the laboratory of Keith Joung [193]. Thereafter, Rudolf Jaenisch boosted again the field upon publishing the first CRISPR-Cas9 genome-edited mice [194]. CRISPR-Cas9 tools were extraordinarily powerful for genome engineering in mice, allowing the generation of multiple simultaneous gene inactivation, as well as the precise introduction of mutations at desired loci. The benefits were expanded later in 2013 to the production of conditional mutant alleles with remarkable efficiency [195].

It was also in 2013 when the first Cas9 variants were reported. The first variant, Cas9 nickase, was unable to produce DSB but still able to cut one single strand for apparently increased specificity [196]. The combination of two Cas9 nickases, targeting complementary DNA strands, was suggested to produce outputs that would simulate DSB. Thereafter, researchers reported the dead-Cas9 (dCas9) variant, with both nuclease domains inactivated, unable to cut but still able to target a genomic locus [197]. The association of dCas9 modules with a variety of effectors (activators, repressors) boosted the application of genome-editing tools to alter epigenetic chromatin marks [198, 199].

2.35 2014

In 2014, CRISPR-Cas9 methods impacted on livestock. This novel technology boosted the production of genome-edited animals for biomedical and biotechnological applications in nearly all species, with a much higher efficiency. Alone, by delivering CRISPR-Cas9 reagents inside one-cell embryos, or in combination with SCNT methods, by first using CRISPR-Cas9 on cultured cells and thereafter using their nuclei to derive the corresponding genome-edited animals, the CRISPR tools have been instrumental for the generation and analysis of many new genetically modified animals in multiple species.

In 2014, the first example of CRISPR-Cas9 use in mammalian species beyond mice was the publication of genome-edited pigs [200], followed by sheep [201], cattle [202], rabbits and goats [203], and dogs [204], among numerous other genome-edited animal species.

2.36 2015

In 2015, the laboratories of Joseph Miano [205] and Lluis Montoliu [206] expanded the range of CRISPR-Cas9 applications to the functional analysis of noncoding DNA regulatory elements in the mouse genome. The combined use of two CRISPR-Cas9 complexes targeting two close DNA sequences in cis allowed the efficient deletion of any DNA sequence existing in between. With CRISPR-Cas9 strategies, it was easier to target DNA regulatory elements and excise them to investigate their relevance. Furthermore, CRISPR-Cas technology allowed testing the association of some alterations in noncoding genome DNA regulatory elements with diseases [207].

In 2015 as well, the laboratory of George Church explored the limits of CRISPR-Cas9 tools for inactivating multiple genes simultaneously. They reported the successful inactivation of all inserted porcine endogenous retroviruses, PERVs, in the pig genome (>60 copies) by a CRISPR-Cas9 strategy, first in porcine cells [208] and later on, further using the genome-edited cells through SCNT, in genome-edited pigs [209]. A similar multiple mutation approach was undertaken by Angelika Schnieke (the researcher whose PhD project led to creating Dolly the sheep [117]), thereby generating genome-edited pigs suitable for xenotransplantation purposes [210].

In 2015, a new method for delivering CRISPR-Cas9 reagents into mouse embryos was presented, involving the use of direct electroporation procedures and, hence, avoiding pronuclear microinjection [211]. This technical innovation was further explored and reproduced in many laboratories and animal species, including livestock [212], and has become the reference and default method nowadays [213].

2.37 2016

The year 2016 began with three reports published in *Science* from three laboratories, working independently, showing how CRISPR-Cas9 strategies could be applied to remove the pathogenic exon carrying a mutation in a mouse model of Duchenne muscle dystrophy, thereby documenting pioneer preclinical examples of what could be a novel wave of innovative gene therapy approaches based on genome-editing tools [214–216]. Thereafter, numerous similar approaches were reported with analogous preclinical developments for a variety of congenital diseases.

In 2016, David Liu also presented the first version of his base editors, a combination of dCas9 or Cas9 nickase with a deaminase domain, capable of transforming a cytidine into a thymidine (C◇T), avoiding DSB and, hence, with less on-target effects (mosaicism) [217]. This first base editor was known as CBE. Base editing was another disruptive step in the short but intense evolution of CRISPR-derived genome-editing variants.

2.38 2017

In 2017, David Liu presented the second version of his base editors (ABE), this time permitting the adenine to guanine (A◇G) transformation, also avoiding DSB and allowing precise modification of selected nucleotides without on-target effects [218].

2.39 2018

In 2018, David Liu presented updated and improved versions of CBE and ABE for more efficient and more specific base editing [219].

Also in 2018, we learned one of the first successful uses of lipid nanoparticles (LNPs) to deliver CRISPR-Cas9 reagents (RNA molecules) encapsulated in vivo, intravenously, through a single administration, in mice and rats [220].

2.40 2019

In 2019, David Liu surprised again the scientific community with a new CRISPR chimeric tool, the prime editor, combining a Cas9 nickase with a longer RNA guide (pegRNA) directing in its 3′ end the synthesis of the complementary DNA strand, thanks to the presence of a reverse transcriptase domain [221].

2.41 2020

Emmanuelle Charpentier and Jennifer Doudna were awarded jointly the 2020 Nobel Prize in Physiology or Medicine for the development of a method for genome editing [222].

A further evolution of ABE base editors is released, with higher therapeutic potential [223].

2.42 2021

In 2021, Joseph Miano showed how primer editors outperformed classical CRISPR-Cas9 approaches in mice, generating comparable editing efficiencies but in the absence of on-target effects (mosaicism) [224].

It has been in 2021 when the benefits of base editing have become apparent in a variety of successful therapeutic approaches in preclinical animal models of Hutchinson-Gilford progeria [225], sickle cell anemia [226], and reduced cholesterol plasma levels [227], among other examples.

One of the latest reviews on the different CRISPR-Cas variants has been published in 2021, describing the benefits and limitations of classical CRISPR-Cas9 approaches, Cas9 nickases and dCas9, Cas9 associated with transposases, base editors, and prime editors [228].

3 Concluding Remarks

The transgenesis and genome engineering fields have evolved significantly over the last 50 years, since the pioneer experiments reported in the 1970s, using retroviruses infecting mouse embryos, to the latest CRISPR variant tools presented in 2021. This progression will continue advancing. I am sure novel and more innovative technologies will be made available to the scientific community. Probably, the prokaryotes will be again the source of yet unknown tools that would further improve our toolbox for manipulating genomes at will in laboratory animals and, eventually, in human beings, for therapeutic applications. If I would have to summarize the word illustrating the early years of transgenesis, I would choose "randomness." Researchers had no control of where the transgenes were landing and the associated consequences of integrating in this or that genomic location. Likewise, the most appropriate word for illustrating the last years of transgenesis and genome engineering would be "precision." This is what CRISPR tools have brought us, the possibility to decide where precisely in the genome we will be making our desired DNA modification.

In summary, the initial random procedures have been replaced by precise tools that trigger the desired alteration only at the chosen locus. Still, some technical limitations and uncertainties must be overcome, and, therefore, protocols need to be further evolved and improved. I have no idea what the next 50 years of DNA modification will uncover, in terms of new techniques, but what I am certain is that the evolution of transgenesis and genome engineering methods will continue generating surprises and interesting results.

Acknowledgments

This work was supported by the Spanish Ministry of Science and Innovation (MICINN) through the Grant [RTI2018-101223-B-I00] to L.M.

References

1. Gordon JW, Ruddle FH (1982) Germ line transmission in transgenic mice. Prog Clin Biol Res 85 Pt B:111–124
2. Palmiter RD, Chen HY, Brinster RL (1982) Differential regulation of metallothionein-thymidine kinase fusion genes in transgenic mice and their offspring. Cell 29:701–710
3. Palmiter RD, Brinster RL et al (1982) Dramatic growth of mice that develop from eggs microinjected with metallothionein-growth hormone fusion genes. Nature 300:611–615
4. Petri W (1982) Transgenic organisms and development. Nature 299:399–400
5. Ingles CJ, Shales M (1982) DNA-mediated transfer of an RNA polymerase II gene: reversion of the temperature-sensitive hamster cell cycle mutant TsAF8 by mammalian DNA. Mol Cell Biol 2:666–673
6. Doy CH, Gresshoff PM, Rolfe BG (1973) Biological and molecular evidence for the transgenosis of genes from bacteria to plant cells. Proc Natl Acad Sci U S A 70:723–726
7. Doy CH, Gresshoff PM, Rolfe BG (1973) Time-course of phenotypic expression of Escherichia coli gene Z following transgenosis in haploid Lycopersicon esculentum cells. Nat New Biol 244:90–91
8. Giles KL, Whitehead H (1976) Uptake and continued metabolic activity of azotobacter within fungal protoplasts. Science 193: 1125–1126
9. McBride OW, Athwal RS (1976) Genetic analysis by chromosome-mediated gene transfer. In Vitro 12:777–786
10. Fournier RE, Ruddle FH (1977) Stable association of the human transgenome and host murine chromosomes demonstrated with trispecific microcell hybrids. Proc Natl Acad Sci U S A 74:3937–3941
11. Miller CL, Ruddle FH (1978) Co-transfer of human X-linked markers into murine somatic cells via isolated metaphase chromosomes. Proc Natl Acad Sci U S A 75:3346–3350
12. McBride OW, Burch JW, Ruddle FH (1978) Cotransfer of thymidine kinase and galactokinase genes by chromosome-mediated gene transfer. Proc Natl Acad Sci U S A 75:914–918
13. Fournier RE, Juricek DK, Ruddle FH (1979) Somatic cell genetic analysis of transgenome integration. Somatic Cell Genet 5:1061–1077
14. Klobutcher LA, Miller CL, Ruddle FH (1980) Chromosome-mediated gene transfer results in two classes of unstable transformants. Proc Natl Acad Sci U S A 77:3610–3614
15. Scangos G, Ruddle FH (1981) Mechanisms and applications of DNA-mediated gene transfer in mammalian cells – a review. Gene 14:1–10
16. Bernstein A, Breitman M (1989) Genetic ablation in transgenic mice. Mol Biol Med 6: 523–530
17. Thomas KR, Capecchi MR (1986) Introduction of homologous DNA sequences into mammalian cells induces mutations in the cognate gene. Nature 324:34–38
18. Thomas KR, Capecchi MR (1987) Site-directed mutagenesis by gene targeting in mouse embryo-derived stem cells. Cell 51: 503–512
19. Koller BH, Hagemann LJ, Doetschman T et al (1989) Germ-line transmission of a planned alteration made in a hypoxanthine phosphoribosyltransferase gene by homologous recombination in embryonic stem cells. Proc Natl Acad Sci U S A 86:8927–8931
20. Roemer K, Johnson PA, Friedmann T (1991) Knock-in and knock-out. Transgenes, development and disease: a keystone symposium sponsored by Genentech and Immunex, Tamarron, CO, USA, January 12–18, 1991. New Biol 3:331–335
21. Ekker SC (2000) Morphants: a new systematic vertebrate functional genomics approach. Yeast 17:302–306
22. Neilson R (1992) Proposals for the future regulation of biotechnology in Australia. Melb Univ Law Rev 18:692–698
23. Mahaffee WF, Kloepper JW (1997) Bacterial communities of the rhizosphere and endorhiza associated with field-grown cucumber plants inoculated with a plant growth-promoting rhizobacterium or its genetically

modified derivative. Can J Microbiol 43:344–353
24. Somerville C (2000) The genetically modified organism conflict. Plant Physiol 123:1201–1202
25. Jensen KK, Gamborg C, Madsen KH et al (2003) Making the EU "risk window" transparent: the normative foundations of the environmental risk assessment of GMOs. Environ Biosaf Res 2:161–171
26. Fishman RE (1998) Patenting human beings: do sub-human creatures deserve constitutional protection? Am J Law Med 15:461–482
27. Birling MC, Fray MD, Kasparek P et al (2021) Importing genetically altered animals: ensuring quality. Mamm Genome. https://doi.org/10.1007/s00335-021-09908-x
28. Lee HJ, Lee HC, Han JY (2015) Germline modification and engineering in avian species. Mol Cells 38:743–749
29. Chenouard V, Brusselle L, Heslan JM et al (2016) A rapid and cost-effective method for genotyping genome-edited animals: a heteroduplex mobility assay using microfluidic capillary electrophoresis. J Genet Genomics 43:341–348
30. Bunton-Stasyshyn RK, Codner GF, Teboul L (2021) Screening and validation of genome-edited animals. Lab Anim. https://doi.org/10.1177/00236772211016922
31. Kilby NJ, Snaith MR, Murray JA (1993) Site-specific recombinases: tools for genome engineering. Trends Genet 9:413–421
32. Rossant J, Nagy A (1995) Genome engineering: the new mouse genetics. Nat Med 1:592–594
33. Grand Moursel L, Visser M, Servant G et al (2021) CRISPRing future medicines. Expert Opin Drug Discov 16:463–473
34. Stark WM, Akopian A (2003) Designer recombinases: tools to cut and paste genomic DNA sequences. Discov Med 3:34–35
35. Miller JC, Holmes MC, Wang J et al (2007) An improved zinc-finger nuclease architecture for highly specific genome editing. Nat Biotechnol 25:778–785
36. Modell AE, Lim D, Nguyen TM et al (2022) CRISPR-based therapeutics: current challenges and future applications. Trends Pharmacol Sci 43:151–161
37. Jacques JP, Hausmann S, Kolakofsky D (1994) Paramyxovirus mRNA editing leads to G deletions as well as insertions. EMBO J 13:5496–5503
38. Bertoni C, Morris GE, Rando TA (2005) Strand bias in oligonucleotide-mediated dystrophin gene editing. Hum Mol Genet 14:221–233
39. Mani M, Kandavelou K, Dy FJ et al (2005) Design, engineering, and characterization of zinc finger nucleases. Biochem Biophys Res Commun 335:447–457
40. Sinsheimer RL (1977) Recombinant DNA. Annu Rev Biochem 46:415–438
41. Jaenisch R, Mintz B (1974) Simian virus 40 DNA sequences in DNA of healthy adult mice derived from preimplantation blastocysts injected with viral DNA. Proc Natl Acad Sci U S A 71:1250–1254
42. Jaenisch R, Fan H, Croker B (1975) Infection of preimplantation mouse embryos and of newborn mice with leukemia virus: tissue distribution of viral DNA and RNA and leukemogenesis in the adult animal. Proc Natl Acad Sci U S A 72:4008–4012
43. Jaenisch R (1976) Germ line integration and Mendelian transmission of the exogenous Moloney leukemia virus. Proc Natl Acad Sci U S A 73:1260–1264
44. Gordon JW, Scangos GA, Plotkin DJ et al (1980) Genetic transformation of mouse embryos by microinjection of purified DNA. Proc Natl Acad Sci U S A 77:7380–7384
45. Capecchi MR (1980) High efficiency transformation by direct microinjection of DNA into cultured mammalian cells. Cell 22:479–488
46. Gordon JW, Ruddle FH (1981) Integration and stable germ line transmission of genes injected into mouse pronuclei. Science 214:1244–1246
47. Brinster RL, Chen HY, Trumbauer M, Senear AW, Warren R, Palmiter RD et al (1981) Somatic expression of herpes thymidine kinase in mice following injection of a fusion gene into eggs. Cell 27:223–231
48. Evans MJ, Kaufman MH (1981) Establishment in culture of pluripotential cells from mouse embryos. Nature 292:154–156
49. Martin GR (1981) Isolation of a pluripotent cell line from early mouse embryos cultured in medium conditioned by teratocarcinoma stem cells. Proc Natl Acad Sci U S A 78:7634–7638
50. Brinster RL, Chen HY, Warren R et al (1982) Regulation of metallothionein--thymidine kinase fusion plasmids injected into mouse eggs. Nature 296:39–42
51. Palmiter RD, Brinster RL, Hammer RE et al (1982) Dramatic growth of mice that develop from eggs microinjected with

metallothionein-growth hormone fusion genes. Nature 300:611–615
52. Folger KR, Wong EA, Wahl G et al (1982) Patterns of integration of DNA microinjected into cultured mammalian cells: evidence for homologous recombination between injected plasmid DNA molecules. Mol Cell Biol 2: 1372–1387
53. Gordon JW, Ruddle (1983) Gene transfer into mouse embryos: production of transgenic mice by pronuclear injection. Methods Enzymol 101:411–433
54. Brinster RL, Palmiter RD (1984) Introduction of genes into the germ line of animals. Harvey Lect 80:1–38
55. Hammer RE, Palmiter RD, Brinster RL (1984) Partial correction of murine hereditary growth disorder by germ-line incorporation of a new gene. Nature 311:65–67
56. Bradley A, Evans M, Kaufman MH et al (1984) Formation of germ-line chimaeras from embryo-derived teratocarcinoma cell lines. Nature 309:255–256
57. Hammer RE, Pursel VG, Rexroad CE Jr et al (1985) Production of transgenic rabbits, sheep and pigs by microinjection. Nature 315:680–683
58. Palmiter RD, Brinster RL (1985) Transgenic mice. Cell 41:343–345
59. Brinster RL, Chen HY, Trumbauer ME et al (1985) Factors affecting the efficiency of introducing foreign DNA into mice by microinjecting eggs. Proc Natl Acad Sci U S A 82: 4438–4442
60. Smithies O, Gregg RG, Boggs SS et al (1985) Insertion of DNA sequences into the human chromosomal beta-globin locus by homologous recombination. Nature 317:230–234
61. Hogan B, Costantini F, Lacy E (1986) Manipulating the mouse embryo: a laboratory manual.. Cold Spring Harbor Laboratory Press, 1st edn. Cold Spring Harbor, New York
62. Hogan B, Beddington R, Costantini F, Lacy E (1994) Manipulating the mouse embryo: a laboratory manual. Cold Spring Harbor Laboratory Press, 2nd edn. Cold Spring Harbor Laboratory Press, New York
63. Nagy A, Gerstenstein M, Vintersten K, Behringer R (2003) Manipulating the mouse embryo: a laboratory manual. Cold Spring Harbor Laboratory Press, 3rd edn. Cold Spring Harbor, New York
64. Behringer R, Gerstenstein M, Vintersten K, Nagy A (2014) Manipulating the mouse embryo: a laboratory manual. Cold Spring Harbor Laboratory Press, 4th edn. Cold Spring Harbor, New York
65. Palmiter RD, Brinster RL (1986) Germ-line transformation of mice. Annu Rev Genet 20: 465–499
66. Robertson E, Bradley A, Kuehn M, Evans M (1986) Germ-line transmission of genes introduced into cultured pluripotential cells by retroviral vector. Nature 323:445–448
67. Bradley A, Robertson E (1986) Embryo-derived stem cells: a tool for elucidating the developmental genetics of the mouse. Curr Top Dev Biol 20:357–371
68. Thomas KR, Folger KR, Capecchi MR (1986) High frequency targeting of genes to specific sites in the mammalian genome. Cell 144: 419–428
69. Ishino Y, Shinagawa H, Makino K et al (1987) Nucleotide sequence of the iap gene, responsible for alkaline phosphatase isozyme conversion in Escherichia coli, and identification of the gene product. J Bacteriol 169:5429–5433
70. Hermans PW, van Soolingen D, Bik EM et al (1991) Insertion element IS987 from Mycobacterium bovis BCG is located in a hot-spot integration region for insertion elements in Mycobacterium tuberculosis complex strains. Infect Immun 59:2695–2705
71. Mojica FJ, Juez G, Rodríguez-Valera F (1993) Transcription at different salinities of Haloferax mediterranei sequences adjacent to partially modified PstI sites. Mol Microbiol 9: 613–621
72. Jansen R, Embden JD, Gaastra W et al (2002) Identification of genes that are associated with DNA repeats in prokaryotes. Mol Microbiol 43:1565–1575
73. Brinster RL, Allen JM, Behringer RR, Gelinas RE, Palmiter RD (1988) Introns increase transcriptional efficiency in transgenic mice. Proc Natl Acad Sci U S A 85:836–840
74. Giraldo P, Montoliu L (2001) Size matters: use of YACs, BACs and PACs in transgenic animals. Transgenic Res 10:83–103
75. Behringer RR, Mathews LS, Palmiter RD, Brinster RL (1988) Dwarf mice produced by genetic ablation of growth hormone-expressing cells. Genes Dev 2(4):453–461
76. Mansour SL, Thomas KR, Capecchi MR (1988) Disruption of the proto-oncogene int-2 in mouse embryo-derived stem cells: a general strategy for targeting mutations to non-selectable genes. Nature 336:348–352
77. Brinster RL, Braun RE, Lo D, Avarbock MR (1986) Targeted correction of a major histocompatibility class II E alpha gene by DNA microinjected into mouse eggs. Proc Natl Acad Sci U S A 86:7087–7091

78. Capecchi MR (1989) Altering the genome by homologous recombination. Science 244: 1288–1292
79. Thomas KR, Capecchi MR (1990) Targeted disruption of the murine int-1 proto-oncogene resulting in severe abnormalities in midbrain and cerebellar development. Nature 346:847–850
80. McMahon AP, Bradley A (1990) The Wnt-1 (int-1) proto-oncogene is required for development of a large region of the mouse brain. Cell 62:1073–1085
81. Eppig JT, Motenko H, Richardson JE (2015) The International Mouse Strain Resource (IMSR): cataloging worldwide mouse and ES cell line resources. Mamm Genome 26: 448–455
82. Koopman P, Gubbay J, Vivian N et al (1991) Male development of chromosomally female mice transgenic for Sry. Nature 351:117–121
83. Wright G, Carver A, Cottom D (1991) High level expression of active human alpha-1-antitrypsin in the milk of transgenic sheep. Biotechnology (N Y) 9:830–834
84. Rubock MJ, Larin Z, Cook M (1990) A yeast artificial chromosome containing the mouse homeobox cluster Hox-2. Proc Natl Acad Sci U S A 87:4751–4755
85. Schedl A, Beermann F, Thies E et al (1992) Transgenic mice generated by pronuclear injection of a yeast artificial chromosome. Nucleic Acids Res 20:3073–3077
86. Strauss WM, Jaenisch R (1992) Molecular complementation of a collagen mutation in mammalian cells using yeast artificial chromosomes. EMBO J 11:417–422
87. Schedl A, Montoliu L, Kelsey G et al (1993) A yeast artificial chromosome covering the tyrosinase gene confers copy number-dependent expression in transgenic mice. Nature 362:258–261
88. Strauss WM, Dausman J, Beard C (1993) Germ line transmission of a yeast artificial chromosome spanning the murine alpha 1 (I) collagen locus. Science 259:1904–1907
89. Jakobovits A, Moore AL, Green LL (1993) Germ-line transmission and expression of a human-derived yeast artificial chromosome. Nature 362:255–258
90. Choi TK, Hollenbach PW, Pearson BE et al (1993) Transgenic mice containing a human heavy chain immunoglobulin gene fragment cloned in a yeast artificial chromosome. Nat Genet 4:117–1123
91. Lamb BT, Sisodia SS, Lawler AM et al (1993) Introduction and expression of the 400 kilobase amyloid precursor protein gene in transgenic mice [corrected]. Nat Genet 5: 22–30
92. Davies NP, Rosewell IR, Richardson JC et al (1993) Creation of mice expressing human antibody light chains by introduction of a yeast artificial chromosome containing the core region of the human immunoglobulin kappa locus. Biotechnology (N Y) 11:911–914
93. Gaensler KM, Kitamura M, Kan YW (1993) Germ-line transmission and developmental regulation of a 150-kb yeast artificial chromosome containing the human beta-globin locus in transgenic mice. Proc Natl Acad Sci U S A 90:11381–11385
94. Forget BG (1939) YAC transgenes: bigger is probably better. Proc Natl Acad Sci U S A 90: 7909–7911
95. Linton MF, Farese RV Jr, Chiesa G (1993) Transgenic mice expressing high plasma concentrations of human apolipoprotein B100 and lipoprotein(a). J Clin Invest 92:3029–3037
96. Zhuang Y, Soriano P, Weintraub H (1994) The helix-loop-helix gene E2A is required for B cell formation. Cell 79:875–884
97. Gu H, Marth JD, Orban PC (1994) Deletion of a DNA polymerase beta gene segment in T cells using cell type-specific gene targeting. Science 265:103–106
98. Plück A (1996) Conditional mutagenesis in mice: the Cre/loxP recombination system. Int J Exp Pathol 77:269–278
99. Tronche F, Kellendonk C, Kretz O (1999) Disruption of the glucocorticoid receptor gene in the nervous system results in reduced anxiety. Nat Genet 23:99–103
100. Sharma S, Zhu J (2014) Immunologic applications of conditional gene modification technology in the mouse. Curr Protoc Immunol 2(105):10.34.1–10.34.13
101. Bradley A, Anastassiadis K, Ayadi A et al (2012) The mammalian gene function resource: the International Knockout Mouse Consortium. Mamm Genome 23:580–586
102. Friedel RH, Seisenberger C, Kaloff C et al (2007) EUCOMM–the European conditional mouse mutagenesis program. Brief Funct Genomic Proteomic 6:180–185
103. Brown SD, Moore MW (2012) The International Mouse Phenotyping Consortium: past and future perspectives on mouse phenotyping. Mamm Genome 23:632–640
104. Metzger D, Clifford J, Chiba H et al (1995) Conditional site-specific recombination in mammalian cells using a ligand-dependent

chimeric Cre recombinase. Proc Natl Acad Sci U S A 92:6991–6995
105. Birling MC, Gofflot F, Warot X (2009) Site-specific recombinases for manipulation of the mouse genome. Methods Mol Biol 561:245–263
106. Kimura Y, Yanagimachi R (1995) Intracytoplasmic sperm injection in the mouse. Biol Reprod 52:709–720
107. Rosenwaks Z, Pereira N (2017) The pioneering of intracytoplasmic sperm injection: historical perspectives. Reproduction 154:F71–F77
108. Choulika A, Perrin A, Dujon B et al (1995) Induction of homologous recombination in mammalian chromosomes by using the I-SceI system of Saccharomyces cerevisiae. Mol Cell Biol 15:1968–1973
109. Epinat JC, Arnould S, Chames P et al (2003) A novel engineered meganuclease induces homologous recombination in yeast and mammalian cells. Nucleic Acids Res 31: 2952–2962
110. Smith J, Grizot S, Arnould S (2006) A combinatorial approach to create artificial homing endonucleases cleaving chosen sequences. Nucleic Acids Res 34:e149
111. Fernández A, Josa S, Montoliu L (2017) A history of genome editing in mammals. Mamm Genome 28:237–246
112. Campbell KH, McWhir J, Ritchie WA et al (1996) Sheep cloned by nuclear transfer from a cultured cell line. Nature 38:64–66
113. Gurdon JB, Elsdale TR, Fischberg M (1958) Sexually mature individuals of Xenopus laevis from the transplantation of single somatic nuclei. Nature 182:64–65
114. Gurdon JB (1962) Adult frogs derived from the nuclei of single somatic cells. Dev Biol 4: 256–273
115. Willadsen SM (1986) Nuclear transplantation in sheep embryos. Nature 320:63–65
116. Callaway E (2016) Dolly at 20: the inside story on the world's most famous sheep. Nature 534:604–608
117. Wilmut I, Schnieke AE, McWhir J (1997) Viable offspring derived from fetal and adult mammalian cells. Nature 385:810–813
118. Hodgson J (1997) Dolly opens a farm full of possibilities. Nat Biotechnol 15:306
119. Harris J (1997) "Goodbye Dolly?" The ethics of human cloning. J Med Ethics 23:353–360
120. Cibelli JB, Stice SL, Golueke PJ (1998) Cloned transgenic calves produced from non-quiescent fetal fibroblasts. Science 280:1256–1258
121. Kato Y, Tani T, Sotomaru Y (1998) Eight calves cloned from somatic cells of a single adult. Science 282:2095–2098
122. Wakayama T, Perry AC, Zuccotti M et al (1998) Full-term development of mice from enucleated oocytes injected with cumulus cell nuclei. Nature 294:369–374
123. Baguisi A, Behboodi E, Melican DT (1999) Production of goats by somatic cell nuclear transfer. Nat Biotechnol 17:456–461
124. Polejaeva IA, Chen SH, Vaught TD et al (2000) Cloned pigs produced by nuclear transfer from adult somatic cells. Nature 407:86–90
125. Zhou Q, Renard JP, Le Friec G (2003) Generation of fertile cloned rats by regulating oocyte activation. Science 302:1179
126. Challah-Jacques M, Chesne P, Renard JP (2003) Production of cloned rabbits by somatic nuclear transfer. Cloning Stem Cells 5:295–299
127. Galli C, Lagutina I, Crotti G et al (2003) Pregnancy: a cloned horse born to its dam twin. Nature 424:635
128. Lee BC, Kim MK, Jang G et al (2005) Dogs cloned from adult somatic cells. Nature 437: 641
129. Yang XW, Model P, Heintz N (1997) Homologous recombination based modification in Escherichia coli and germline transmission in transgenic mice of a bacterial artificial chromosome. Nat Biotechnol 15:859–865
130. Muyrers JP, Zhang Y, Testa G et al (1999) Rapid modification of bacterial artificial chromosomes by ET-recombination. Nucleic Acids Res 27:1555–1557
131. Testa G, Zhang Y, Vintersten K et al (2003) Engineering the mouse genome with bacterial artificial chromosomes to create multipurpose alleles. Nat Biotechnol 21:443–447
132. Van Keuren ML, Gavrilina GB, Filipiak WE et al (2009) Generating transgenic mice from bacterial artificial chromosomes: transgenesis efficiency, integration and expression outcomes. Transgenic Res 18:769–785
133. Ivics Z, Hackett PB, Plasterk RH et al (1997) Molecular reconstruction of sleeping beauty, a Tc1-like transposon from fish, and its transposition in human cells. Cell 91:501–510
134. Izsvák Z, Ivics Z, Plasterk RH (2000) Sleeping beauty, a wide host-range transposon vector for genetic transformation in vertebrates. J Mol Biol 302:93–102
135. Dupuy AJ, Clark K, Carlson CM (2002) Mammalian germ-line transgenesis by transposition. Proc Natl Acad Sci U S A 99:4495–4499

136. Zayed H, Izsvák Z, Walisko O et al (2004) Development of hyperactive sleeping beauty transposon vectors by mutational analysis. Mol Ther 9:292–304
137. Katter K, Geurts AM, Hoffmann O et al (2013) Transposon-mediated transgenesis, transgenic rescue, and tissue-specific gene expression in rodents and rabbits. FASEB J 27:930–941
138. Geurts AM, Collier LS, Geurts JL et al (2006) Gene mutations and genomic rearrangements in the mouse as a result of transposon mobilization from chromosomal concatemers. PLoS Genet 29:e156
139. Schnieke AE, Kind AJ, Ritchie WA et al (1997) Human factor IX transgenic sheep produced by transfer of nuclei from transfected fetal fibroblasts. Science 278:2130–2133
140. Cibelli JB, Stice SL, Golueke PJ et al (1998) Transgenic bovine chimeric offspring produced from somatic cell-derived stem-like cells. Nat Biotechnol 16:642–646
141. Polejaeva IA (2001) Cloning pigs: advances and applications. Reprod Suppl 58:293–300
142. Thomson JA, Itskovitz-Eldor J, Shapiro SS et al (1998) Embryonic stem cell lines derived from human blastocysts. Science 282:1145–1147
143. Burley J (1999) The ethics of therapeutic and reproductive human cloning. Semin Cell Dev Biol 10:287–294
144. Colman A, Kind A (2000) Therapeutic cloning: concepts and practicalities. Trends Biotechnol 18:192–196
145. Rideout WM 3rd, Hochedlinger K, Kyba M et al (2002) Correction of a genetic defect by nuclear transplantation and combined cell and gene therapy. Cell 109:17–27
146. Dozortsev D, Wakaiama T, Ermilov A et al (1998) Intracytoplasmic sperm injection in the rat. Zygote 6:143–147
147. Fire A, Xu S, Montgomery MK, Kostas SA et al (1998) Potent and specific genetic interference by double-stranded RNA in Caenorhabditis elegans. Nature 39:806–811
148. Nasevicius A, Ekker SC (2000) Effective targeted gene 'knockdown' in zebrafish. Nat Genet 26:216–220
149. Stainier DYR, Raz E, Lawson ND et al (2017) Guidelines for morpholino use in zebrafish. PLoS Genet 13:e1007000
150. Perry AC, Wakayama T, Kishikawa H et al (1999) Mammalian transgenesis by intracytoplasmic sperm injection. Science 284:1180–1183
151. Chan AW, Luetjens CM, Dominko T et al (2000) TransgenICSI reviewed: foreign DNA transmission by intracytoplasmic sperm injection in rhesus monkey. Mol Reprod Dev 56:325–328
152. Perry AC, Rothman A, de las Heras JI et al (2001) Efficient metaphase II transgenesis with different transgene archetypes. Nat Biotechnol 19:1071–1073
153. Moreira PN, Giraldo P, Cozar P et al (2004) Efficient generation of transgenic mice with intact yeast artificial chromosomes by intracytoplasmic sperm injection. Biol Reprod 71: 1943–1947
154. Lavitrano M, Camaioni A, Fazio VM et al (1989) Sperm cells as vectors for introducing foreign DNA into eggs: genetic transformation of mice. Cell 57:717–723
155. Maione B, Lavitrano M, Spadafora C et al (1998) Sperm-mediated gene transfer in mice. Mol Reprod Dev 50:406–409
156. Smith K, Spadafora C (2005) Sperm-mediated gene transfer: applications and implications. BioEssays 27:551–562
157. Lavitrano M, Busnelli M, Cerrito MG et al (2006) Sperm-mediated gene transfer. Reprod Fertil Dev 18:19–23
158. García-Vázquez FA, Ruiz S, Grullón LA et al (2011) Factors affecting porcine sperm mediated gene transfer. Res Vet Sci 91:446–453
159. McCreath KJ, Howcroft J, Campbell KH et al (2000) Production of gene-targeted sheep by nuclear transfer from cultured somatic cells. Nature 405:1066–1069
160. Lai L, Kolber-Simonds D, Park KW et al (2002) Production of alpha-1,3-galactosyltransferase knockout pigs by nuclear transfer cloning. Science 295:1089–1092
161. Polejaeva IA, Campbell KH (2000) New advances in somatic cell nuclear transfer: application in transgenesis. Theriogenology 53:117–261
162. Bibikova M, Carroll D, Segal DJ et al (2001) Stimulation of homologous recombination through targeted cleavage by chimeric nucleases. Mol Cell Biol 21:289–297
163. Lois C, Hong EJ, Pease S et al (2002) Germline transmission and tissue-specific expression of transgenes delivered by lentiviral vectors. Science 295:868–872
164. Fässler R (2004) Lentiviral transgene vectors. EMBO Rep 5:28–29
165. Park F (2007) Lentiviral vectors: are they the future of animal transgenesis? Physiol Genomics 31:159–173

166. Mojica FJ, Díez-Villaseñor C, García-Martínez J et al (2005) Intervening sequences of regularly spaced prokaryotic repeats derive from foreign genetic elements. J Mol Evol 60:174–182
167. Mojica FJM, Montoliu L (2016) On the origin of CRISPR-Cas technology: from prokaryotes to mammals. Trends Microbiol 24:811–820
168. Takahashi K, Yamanaka S (2006) Induction of pluripotent stem cells from mouse embryonic and adult fibroblast cultures by defined factors. Cell 126:663–676
169. Hanna J, Markoulaki S, Schorderet P et al (2008) Direct reprogramming of terminally differentiated mature B lymphocytes to pluripotency. Cell 133:250–264
170. Su Y, Zhu J, Salman S et al (2020) Induced pluripotent stem cells from farm animals. J Anim Sci 98:skaa343
171. Vogel G (2007) Nobel prizes. A knockout award in medicine. Science 318:178–179
172. Ying QL, Wray J, Nichols J et al (2008) The ground state of embryonic stem cell self-renewal. Nature 453:519–523
173. Buehr M, Meek S, Blair K (2008) Capture of authentic embryonic stem cells from rat blastocysts. Cell 135:1287–1298
174. Geurts AM, Cost GJ, Freyvert Y (2009) Knockout rats via embryo microinjection of zinc-finger nucleases. Science 325:433
175. Rémy S, Tesson L, Ménoret S et al (2010) Zinc-finger nucleases: a powerful tool for genetic engineering of animals. Transgenic Res 19:363–371
176. Meyer M, de Angelis MH, Wurst W et al (2007) Gene targeting by homologous recombination in mouse zygotes mediated by zinc-finger nucleases. Proc Natl Acad Sci U S A 107:15022–15026
177. Yu S, Luo J, Song Z (2011) Highly efficient modification of beta-lactoglobulin (BLG) gene via zinc-finger nucleases in cattle. Cell Res 21:1638–1640
178. Hauschild J, Petersen B, Santiago Y (2011) Efficient generation of a biallelic knockout in pigs using zinc-finger nucleases. Proc Natl Acad Sci U S A 108:12013–12017
179. Liu X, Wang Y, Tian Y et al (2014) Generation of mastitis resistance in cows by targeting human lysozyme gene to β-casein locus using zinc-finger nucleases. Proc Biol Sci 281: 20133368
180. Lillico SG, Proudfoot C, King TJ et al (2016) Mammalian interspecies substitution of immune modulatory alleles by genome editing. Sci Rep 6:21645
181. Tesson L, Usal C, Ménoret S, Leung E, Niles BJ, Remy S, Santiago Y, Vincent AI, Meng X, Zhang L, Gregory PD, Anegon I, Cost GJ et al (2011) Knockout rats generated by embryo microinjection of TALENs. Nat Biotechnol 29:695–696
182. Panda SK, Wefers B, Ortiz O et al (2013) Highly efficient targeted mutagenesis in mice using TALENs. Genetics 195:703–713
183. Carlson DF, Tan W, Lillico SG et al (2012) Efficient TALEN-mediated gene knockout in livestock. Proc Natl Acad Sci U S A 109: 17382–17387
184. Wu H, Wang Y, Zhang Y et al (2015) TALE nickase-mediated SP110 knockin endows cattle with increased resistance to tuberculosis. Proc Natl Acad Sci U S A 112:E1530–E1539
185. Xin J, Yang H, Fan N et al (2013) Highly efficient generation of GGTA1 biallelic knockout inbred mini-pigs with TALENs. PLoS One 8:e84250
186. Cui C, Song Y, Liu J et al (2015) Gene targeting by TALEN-induced homologous recombination in goats directs production of β-lactoglobulin-free, high-human lactoferrin milk. Sci Rep 5:10482
187. Pease S, Saunders TL (eds) (2011) Advanced protocols for animal transgenesis. An ISTT Manual, Springer, Berlin
188. Colman A (2013) Profile of John Gurdon and Shinya Yamanaka, 2012 Nobel laureates in medicine or physiology. Proc Natl Acad Sci U S A 110:5740–5741
189. Jinek M, Chylinski K, Fonfara I, Hauer M, Doudna JA, Charpentier E (2012) A programmable dual-RNA-guided DNA endonuclease in adaptive bacterial immunity. Science 337(6096):816–821
190. Seruggia D, Montoliu L (2014) The new CRISPR-Cas system: RNA-guided genome engineering to efficiently produce any desired genetic alteration in animals. Transgenic Res 23:707–716
191. Cong L, Ran FA, Cox D et al (2013) Multiplex genome engineering using CRISPR/Cas systems. Science 339:819–823
192. Mali P, Yang L, Esvelt KM et al (2013) RNA-guided human genome engineering via Cas9. Science 339:823–826
193. Hwang WY, Fu Y, Reyon D et al (2013) Efficient genome editing in zebrafish using a CRISPR-Cas system. Nat Biotechnol 31: 227–229
194. Wang H, Yang H, Shivalila CS et al (2013) One-step generation of mice carrying mutations in multiple genes by CRISPR/Cas-mediated genome engineering. Cell 153: 910–918

195. Yang H, Wang H, Shivalila CS et al (2013) One-step generation of mice carrying reporter and conditional alleles by CRISPR/Cas-mediated genome engineering. Cell 154: 1370–1379
196. Ran FA, Hsu PD, Lin CY et al (2013) Double nicking by RNA-guided CRISPR Cas9 for enhanced genome editing specificity. Cell 154:1380–1389
197. Qi LS, Larson MH, Gilbert LA et al (2013) Repurposing CRISPR as an RNA-guided platform for sequence-specific control of gene expression. Cell 152:1173–1183
198. Gilbert LA, Larson MH, Morsut L et al (2013) CRISPR-mediated modular RNA-guided regulation of transcription in eukaryotes. Cell 154:442–451
199. Doudna JA, Charpentier E (2014) Genome editing. The new frontier of genome engineering with CRISPR-Cas9. Science 346: 1258096
200. Hai T, Teng F, Guo R et al (2014) One-step generation of knockout pigs by zygote injection of CRISPR/Cas system. Cell Res 24: 372–375
201. Crispo M, Mulet AP, Tesson L et al (2015) Efficient generation of Myostatin knock-out sheep using CRISPR/Cas9 technology and microinjection into zygotes. PLoS One 210: e0136690
202. Wang Z (2015) Genome engineering in cattle: recent technological advancements. Chromosom Res 23:17–29
203. Guo R, Wan Y, Xu D (2016) Generation and evaluation of Myostatin knock-out rabbits and goats using CRISPR/Cas9 system. Sci Rep 6:29855
204. Amoasii L, Hildyard JCW, Li H et al (2018) Gene editing restores dystrophin expression in a canine model of Duchenne muscular dystrophy. Science 362:86–91
205. Han Y, Slivano OJ, Christie CK et al (2015) CRISPR-Cas9 genome editing of a single regulatory element nearly abolishes target gene expression in mice--brief report. Arterioscler Thromb Vasc Biol 35:312–315
206. Seruggia D, Fernández A, Cantero M et al (2015) Functional validation of mouse tyrosinase non-coding regulatory DNA elements by CRISPR-Cas9-mediated mutagenesis. Nucleic Acids Res 43:4855–4867
207. Lupiáñez DG, Kraft K, Heinrich V et al (2015) Disruptions of topological chromatin domains cause pathogenic rewiring of gene-enhancer interactions. Cell 161:1012–1025
208. Yang L, Güell M, Niu D et al (2015) Genome-wide inactivation of porcine endogenous retroviruses (PERVs). Science 350:1101–1104
209. Niu D, Wei HJ, Lin L et al (2017) Inactivation of porcine endogenous retrovirus in pigs using CRISPR-Cas9. Science 357:1303–1307
210. Fischer K, Kraner-Scheiber S, Petersen B et al (2016) Efficient production of multi-modified pigs for xenotransplantation by 'combineering', gene stacking and gene editing. Sci Rep 6:29081
211. Qin W, Dion SL, Kutny PM et al (2015) Efficient CRISPR/Cas9-mediated genome editing in mice by zygote electroporation of nuclease. Genetics 200:423–430
212. Lin JC, Van Eenennaam AL (2021) Electroporation-mediated genome editing of livestock zygotes. Front Genet 12:648482
213. Fernández A, Morín M, Muñoz-Santos D et al (2020) Simple protocol for generating and genotyping genome-edited mice with CRISPR-Cas9 reagents. Curr Protoc Mouse Biol 10:e69
214. Long C, Amoasii L, Mireault AA et al (2016) Postnatal genome editing partially restores dystrophin expression in a mouse model of muscular dystrophy. Science 351:400–403
215. Nelson CE, Hakim CH, Ousterout DG et al (2016) In vivo genome editing improves muscle function in a mouse model of Duchenne muscular dystrophy. Science 351:403–407
216. Tabebordbar M, Zhu K, Cheng JKW et al (2016) In vivo gene editing in dystrophic mouse muscle and muscle stem cells. Science 351:407–411
217. Komor AC, Kim YB, Packer MS et al (2016) Programmable editing of a target base in genomic DNA without double-stranded DNA cleavage. Nature 533:420–424
218. Gaudelli NM, Komor AC, Rees HA et al (2017) Programmable base editing of A•T to G•C in genomic DNA without DNA cleavage. Nature 551:464–471
219. Koblan LW, Doman JL, Wilson C et al (2018) Improving cytidine and adenine base editors by expression optimization and ancestral reconstruction. Nat Biotechnol 36:843–846
220. Finn JD, Smith AR, Patel MC et al (2018) A single administration of CRISPR/Cas9 lipid nanoparticles achieves robust and persistent in vivo genome editing. Cell Rep 22:2227–2235
221. Anzalone AV, Randolph PB, Davis JR et al (2019) Search-and-replace genome editing without double-strand breaks or donor DNA. Nature 576:149–157

222. Ledford H, Callaway E (2020) Pioneers of revolutionary CRISPR gene editing win chemistry Nobel. Nature 586:346–347
223. Gaudelli NM, Lam DK, Rees HA et al (2020) Directed evolution of adenine base editors with increased activity and therapeutic application. Nat Biotechnol 38:892–900
224. Gao P, Lyu Q, Ghanam AR et al (2021) Prime editing in mice reveals the essentiality of a single base in driving tissue-specific gene expression. Genome Biol 22:83
225. Koblan LW, Erdos MR, Wilson C et al (2021) In vivo base editing rescues Hutchinson-Gilford progeria syndrome in mice. Nature 589:608–614
226. Newby GA, Yen JS, Woodard KJ et al (2021) Base editing of haematopoietic stem cells rescues sickle cell disease in mice. Nature 595: 295–302
227. Musunuru K, Chadwick AC, Mizoguchi T et al (2021) In vivo CRISPR base editing of PCSK9 durably lowers cholesterol in primates. Nature 593:429–434
228. Newby GA, Liu DR (2021) In vivo somatic cell base editing and prime editing. Mol Ther 29:3107–3124

Chapter 2

Practical Application of the 3Rs in Rodent Transgenesis

Thorsten Buch, Boris Jerchow, and Branko Zevnik

Abstract

The principles of the 3Rs (replace, reduce, refine), as originally published by Russell and Burch, are internationally acclaimed guidelines for meeting ethical and welfare standards in animal experimentation. Genome manipulation is a standard technique in biomedical research and beyond. The goal of this chapter is to give practical advice on the implementation of the 3Rs in laboratories generating genetically modified rodents. We cover 3R aspects from the planning phase through operations of the transgenic unit to the final genome-manipulated animals. The focus of our chapter is on an easy-to-use, concise protocol that is close to a checklist. While we focus on mice, the proposed methodological concepts can be easily adapted for the manipulation of other sentient animals.

Key words Genome manipulation, Transgenesis, Genome editing, 3Rs (Replace, Refine, Reduce), Animal welfare, Harm-benefit analysis

1 Introduction

Genetic perturbation of protein networks is an important tool for understanding cellular physiology in steady state and disease. The mouse has been the first mammal to be routinely accessible for genetic manipulation, allowing the investigation of the roles of its genes, pathways, and genomic elements. Because of its abundant use, most genetic tools were initially developed for the mouse and later adapted for use in other species. Thus, the generation of genetically modified mice has become a routine technique that is widely used in basic and preclinical research. Because of it being a forerunner, animal welfare aspects surrounding genetic manipulation techniques have been most thoroughly assessed in the mouse. In this methodological guideline for improvement of animal welfare, notably based on the 3R concept [1], we will hence restrict ourselves largely to the mouse. Many state-of-the-art techniques that we will present for the amelioration of the experimental impact on animal well-being can be easily adapted to other species as well. A detailed discussion is given at the end. We will not introduce any

Thomas L. Saunders (ed.), *Transgenesis: Methods and Protocols*,
Methods in Molecular Biology, vol. 2631, https://doi.org/10.1007/978-1-0716-2990-1_2,

of the methods for which we make recommendations on how to optimize compliance with the 3Rs. Rather we want to refer the reader to the other chapters of this book and other excellent literature [2, 3]. Our aim is to provide easy-to-use checklists for the practitioner. These checklists are rather extensive and based on our own interpretation of available information. Especially in the context of animal welfare assessment, room for interpretation exists depending on regional moral values and national legislation. Also, certain techniques are preferred over others by specific researchers, animal welfare officers, and government offices. We recommend that these decisions be based on available literature and hence should be considered evidence-based. Nevertheless, we found the literature in some aspects lacking evidence to truly allow an informed decision. More research on animal welfare aspects appears to be necessary.

The here-presented protocol has been structured according to the workflow of a typical facility for the genetic manipulation of rodents. We do not indicate which of the 3Rs a particular advice refers to, as this is generally clear, e.g., an increase in efficiency of a process usually results in the use of fewer animals.

2 Materials

The use of high-quality material and reagents, from known and tested sources, is pivotal for optimizing experimental success and therefore minimizes the number of animals used in the procedure until successful line establishment. Optimal general laboratory supplies will be found in the specialized methods section of this book. A non-exhaustive list of resources, including websites, is provided with a special focus on provision of 3R adherence. When writing a research protocol, the inclusion of such resources should be included but should be adapted to the individual protocol:

- Information on genomes, genes, and phenotypes: Mouse Genome Informatics (MGI), http://www.informatics.jax.org/; https://www.mousephenotype.org/.
- Information on existing mouse models: International Mouse Strain Resource (IMSR), http://www.findmice.org.
- Genome browsers: Ensembl (https://www.ensembl.org/index.html), UCSC (https://genome.ucsc.edu).
- Critical incident reporting: CIRS-LAS, https://www.cirs-las.de/home.
- >8 *g* of nesting material suitable for mice to build a full dome nest [4].

- Ordering mice, ES cells, and/or targeting vectors [5]: The International Mouse Phenotyping Consortium (IMPC), https://www.mousephenotype.org/help/mouse-production/ordering-products/.
- goGermline sterile host embryos [6], https://www.ozgene.com/gogermline-knockout-and-knock-in-mice/.
- Kit Eazygote, zygote-stage frozen embryos for manipulation by injection or electroporation, https://janvier-labs.com/en/elevage/kit-frozen-embryos/.
- CARD HyperOva (Superovulation Reagent for mouse [7]), https://www.cosmobiousa.com/products/card-hyperova.
- Genetically sterile males (replacing vasectomies): CD1;B6D2-Tg(Prm1-EGFP)#Ltku/H; order: https://www.infrafrontier.eu/search?keyword=EM:12662; or Gapdhs^{tm1Dao} [8].
- Nonsurgical embryo transfer devices:
  - NSET: https://paratechs.com.
  - TCET: http://www.elimspringsbiotech.com/.

3 Methods

3.1 General

1. Ensure the facility generates an adequate number of lines per year to take advantage of the efficiency of scale. Consider centralizing efforts at your institution, between institutions and outsourcing. As a general rule, if stud males and sterile males are used on average less than once per week, centralization measures such as fusion with another facility should be evaluated.
2. Follow relevant literature and expert information to identify and develop efficient and effective state-of-the-art protocols and procedures. Attend conferences in the field and become a member of relevant mailing lists (e.g., ISTT mailing list).
3. All personnel must be properly trained and undergo continuous education.
4. Maintain detailed documentation for outcome evaluation:
   (a) Animal numbers must be recorded across all steps of a process.
   (b) Embryo number and quality must be recorded across all steps of a process.
   (c) If performance is low, measures must be taken to identify and solve apparent issues. Refer to the respective sections below for detailed minimal success standards.
   (d) Critical incidents and errors should be reported, preferably publicly, using reporting systems such as the German CIRS-LAS system (https://www.cirs-las.de/home).

5. Maintain optimal animal husbandry:
   (a) Husbandry conditions must adhere to internationally accepted standards (temperature, humidity, light cycle, cage size and occupancy, etc.), taking into account local applicable legislation.
   (b) High-energy food should be provided after the first trimester of pregnancy and during lactation (*see* **Note 1**).
   (c) At least 8 *g* of adequate, high-quality nesting material must be provided [9].
   (d) Hygiene status must be maintained in accordance with international standards such as those recommended by FELASA [10].
   (e) Animals should be acclimatized before use. However, acclimatization may not be possible in some cases (e.g., with prepubertal superovulation).
   (f) Non-aversive handling such as cup and tunnel handling should be performed.
   (g) Positive conditioning of animals should be considered to reduce stress as much as possible.
   (h) Disturbance of animals must be minimized and especially avoided for recipient females.
   (i) Animals must not be single-housed over longer periods without a specific reason. Consider companion animals when single-housing cannot be avoided.
   (j) Valid legislation and respective guidelines on humane euthanasia must be strictly followed.
6. Quality controlled materials must be employed for the manipulation procedure, including:
   (a) Media
   (b) Embryo culture equipment (incubators)
   (c) High-quality and highly purified macromolecules (DNA, RNA, proteins)
7. Use properly maintained, state-of-the-art inverted microscopes, including cooled injection tables and ancillary equipment for micromanipulation.
8. For endonuclease-based modifications, consider electroporation over microinjection for delivery of single-stranded DNA molecules (ssODN) smaller than 500 nucleotides, in order to reduce the number of embryos needed for manipulation and thus the number of donors and recipients. More complex genome alterations using larger DNAs, however, are mostly inefficient and increase the overall number of donors and recipients needed.

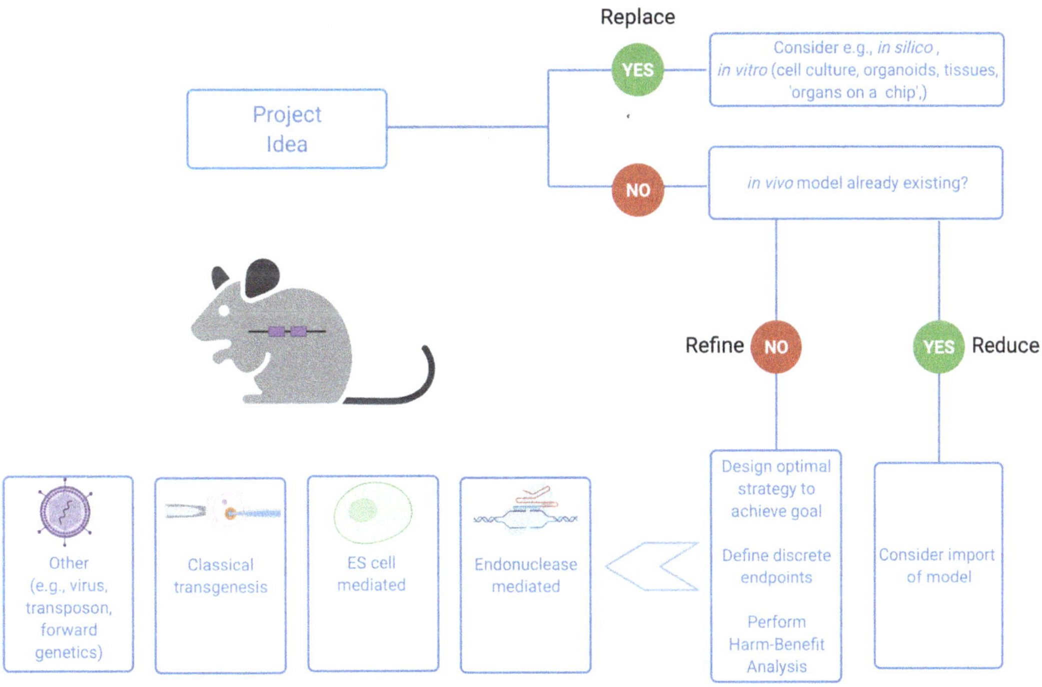

Fig. 1 3Rs decision process on a transgenic model

9. Consider in situ electroporation directly into embryos within the oviduct if applicable and properly trained. This sophisticated method eliminates the need to breed and kill embryo donors.

3.2 Choosing the Best Experimental Method

The decision on a specific transgenic model to be created by the transgenesis unit is the result of a scientific evaluation of the technical possibilities and the experimental objective. The decision process must involve all 3Rs. For a graphical illustration, *see* Fig. 1.

3.2.1 Non-sentient or Nonanimal Model

Assess if the use of nonanimal-based models or non-sentient animal models is possible before a project for model generation by transgenic methods is initiated. Evaluate alternative technologies alone or in combination:

(a) In silico modeling via bioinformatics and computing
(b) Simple cell culture models based on ex vivo material
(c) Simple cell culture models based on cell lines
(d) Embryonic or induced pluripotent stem cell-based assays
(e) Organoid and other 3D organ models
(f) Organ slices
(g) Organs-on-a-chip systems
(h) *Drosophila*, *Caenorhabditis*, and other invertebrate models

3.2.2 Existing Models

Assess reduction options if replacement is not an option. Evaluate the following:

(a) Screen database and literature for existing models.

(b) Check suitability of existing models for your purpose according to the following parameters:

(i) Does the type of mutation suit the experimental purpose?

(ii) Are the background strain genetics known and appropriate [11]?

(c) Evaluate importing versus de novo establishment according to the following parameters:

(i) Does the hygiene status allow direct import?

(ii) What harm or stress does a respective transport of live animals inflict (*see* **Note 2**)?

(iii) Are cryopreserved sperm or embryos available?

(iv) How many animals will be involved in harmful procedures (e.g., surgeries)?

(v) How many animals will be involved in total?

3.2.3 Type of Mutation

(a) Create inducible mutants by using recombination systems such as Cre-loxP or controllable expression systems such as CreER or Tet-On/Off when there is an indication that a harmful phenotype is likely with a conventional ubiquitous mutation. Limit harmful phenotypes to the tissue of interest or a defined time period whenever possible.

(b) Use an in vitro intermediate step in ES cells if there is a risk that inefficient manipulation of embryos could lead to high animal numbers:

(i) Evaluate the success probability of the in vivo (embryo) methods.

(ii) Assess the molecular design of the targeted gene alteration with respect to success probability (*see* **Note 3**).

(iii) Follow accepted standards, such as the recommendations for the development of knockout alleles in protein-coding genes [12].

3.2.4 Harm-Benefit Analysis (HBA)/Weighing of Interests

(a) Perform a HBA (also referred to as "weighing of interests") before engaging in a project to generate a genetically modified animal model:

(i) Generally, HBA needs to be performed with regard to national laws, local ethical standards, and animal procedure guidelines. The two parts of the report of the AALAS-FELASA Working Group on Harm-Benefit Analysis may help with this endeavor [13, 14] (*see* **Note 4**).

Table 1
Factors in a HBA for a genome modification project

Pros	**Increase gain**
Gain of knowledge	Provide model to the scientific community
Elucidation of physiological processes (involved in disease)	Provide detailed information in publication (ARRIVE guidelines)
Modeling disease, humanized animal models	Optimize translatability
Drug development	Publish open access
Preclinical drug testing	Provide data according to FAIR principles
(Preclinical) compound safety, pharmacokinetics	
Cons	**Decrease harm**
Pain: injections, surgery	Minimize number of injections
Death: animal use per se, animal numbers to be considered	Use best practice methods, e.g., adequate anesthesia and analgesia
Stressful holding and handling conditions	Use humane method for euthanasia
The inherent worth of the animal (sometimes called dignity)	Limit animal numbers as much as possible; avoid necessity for backcrossing
Harm: temporal or permanent negative effect on physical or mental condition, such as obesity, amputation, infertility, disturbance of behavior, etc., caused by the intended genetic modification	Provide adequate nesting and enrichment; avoid single-housing whenever possible; consider non-aversive handling methods; ensure adequate hygiene
	Consider use of inducible modifications to minimize number of affected animals; balance harm against increase in animal numbers
	Define line-specific mitigation strategies as soon as a harmful phenotype occurs

(ii) Assess the anticipated severity of the model in the planning phase of the project (Table 1) (*see* **Note 5**):

1. Include veterinarians, animal welfare experts, and scientists who are experts in the research field in the evaluation.
2. Consult publications on similar models as well as databases on GM animals and human conditions for information on potential phenotypes.
3. Consider genetic background effects.
4. Consider additional genome manipulations that may increase severity.

(iii) Choose the method with the least potential for unintended mutations, depending on the requirements of the scientific question.

(b) Make a Go/No-Go decision before the start of the project. Document how you weighed the potential benefit of the model against the potential harm inflicted.

(c) Choose optimal protocols and methods to generate the GM model in question for further reduction of the 3Rs footprint.

(d) In case of outsourcing, assess welfare standards of external partners before commencing the project.

3.3 Transgenic Techniques

3.3.1 Classical Transgenesis

In classical transgenesis, after microinjection into zygotes, a transgene inserts into the host genome in an uncontrollable fashion [15]. Evaluate whether the given research question cannot be better answered by precise GM employing ES cell technology or genome editing via CRISPR/Cas (*see* **Note 6**):

1. Consider a transposon system for integration if size requirements can be met [16].
2. Generate new transgenic lines on the background of interest, if possible. This obviates the later need for backcrossing.
3. Meet the minimal standards for efficiency [17]:
   (a) >70% of injected zygotes should survive microinjection.
   (b) >80% of injected zygotes should develop to the two-cell stage.
   (c) > 8% of transferred zygotes should result in live-born pups.
   (d) Of founder animals, 8–16% should be transgenic, in case of average size transgenes (*see* **Note 7**).
4. Examine several transgenic founder lines resulting from a given experiment for expression (*see* **Note 8**).
5. Choose for further analysis founding lines that show the desired expression profile without unwanted side effects causing unnecessary harm.
6. Identify the precise transgene location and integration within the genome by, e.g., targeted locus amplification (TLA), Samplix Xdrop®, or sequencing.
7. Backcross the founder animals to establish lines with one stable integration each (*see* **Note 9**).
8. Cryopreserve independent lines. This allows further control experiments [18].
9. Consider alternatives to classical DNA microinjection for increasing efficiency. Among others, consider the following:
   (a) Transduction with lentiviral vectors.
   (b) Using transposons can make integration of BACs at least five times more efficient. Note, however, that size limitations apply.

(c) The use of recombinases such as Flp and Cre or integrases such as Φ31 together with sequence-matched constructs and one-cell embryos carrying the respective sites for recombination allows very efficient targeted integration. However, this approach requires additional lines in the colony [19–21].

3.3.2 ES Cell Mutagenesis

1. Perform extensive characterization of the mutation already in ES cells by, e.g., Southern blotting, PCR, TLA, Samplix Xdrop®, and sequencing technologies. Characterization should also be performed when ES cells are obtained from repositories [22].
2. Ensure pluripotency of ES cells by adhering to stringent quality control measures:
   (a) Maintain a master cell bank with germline-tested wild-type ES cells.
   (b) Tightly adhere to optimized cell culture conditions.
   (c) Use only karyotypically normal ES cell batches.
3. Use ES cells derived from the background strain to be used for planned research.
4. Analyze gene-targeted ES clones for aneuploidies before generating chimeras by chromosome count or PCR-based methods. Use only those ES clones for injection with more than two thirds displaying a correct chromosome count of 40 X/Y.
5. Choose appropriate combinations of ES cell line and host embryos for optimal germline transmission rate (*see* **Note 10**):
   (a) ES cells derived from inbred strains (e.g., C57BL/6) can be combined with host embryos also derived from inbred strains (e.g., B6(Cg)-$Tyr^{c\text{-}2J}$ or BALB/c) but not with those derived from outbred strains.
   (b) ES cell lines from F1 hybrids (e.g., B6D2F1 or B6129F1) can be successfully combined with outbred embryos (e.g., CD1).
   (c) Consider using host embryos lacking the ability to develop sperm (*see* **Note 11**).
6. Inject up to 50 embryos/ES clone.
7. Meet the minimal standards for efficiency:
   (a) >90% of manipulated embryos should survive.
   (b) Use approximately three recipients per ES clone.
   (c) >50% of recipients should establish a pregnancy.
   (d) Birth rates should reach 25–50% of embryos transferred.
   (e) Roughly 50% of pups born should be chimeric with an ES cell coat color contribution of 70% or more.

8. Breed max. three chimeras per ES clone for germline transmission (*see* **Note 12**):
   (a) Germline transmission should be obtained from about 40–100% of chimeras mated and 50–70% of ES cell clones injected.
   (b) Terminate parallel experiments with chimeras of different ES cell clones harboring the same mutations immediately upon obtaining germline transmission (sperm from these males can be frozen as a backup if it should be needed later).

3.3.3 Endonucleases

1. For efficiency reasons, consider the use of the CRISPR/Cas system over Zn-fingers or TALEN systems.
2. Use only gRNAs with high cutting efficiency and low off-target probability. Evaluate gRNAs by:
   (a) Use of dedicated algorithms such as those reviewed in [23]
   (b) Testing cutting efficiency in vitro
   (c) Multiplexing of gRNAs against the target
   (d) Manipulation of a small number (20–30) of zygotes and analysis of mutations, e.g., by sequencing, either directly or after in vitro development into blastocysts [24] (*see* **Note 13**)
3. Reduce the chances of undesired mosaic founder animals [25]:
   (a) Use Cas9 ribonucleoprotein instead of Cas9 mRNA.
   (b) Lower the concentration of Cas9.
4. Consider increasing the efficiency of the CRISPR/Cas9 homology-directed repair system (*see* **Note 14**) by methods such as:
   (a) Chemical stabilization of donor DNA and gRNAs [26, 27]
   (b) Silent mutations in template DNA [28]
   (c) Optimization of the distance between cutting site and mutation target [28]
   (d) Design of asymmetric donor single-stranded oligodeoxynucleotides relative to the PAM site [28]
   (e) Cas9 variants [29]
5. Meet the minimal standards for efficiency:
   (a) >70% of embryos should survive manipulation with endonucleases.
   (b) >80% of manipulated embryos should reach the two-cell stage after overnight culture.

6. Do not rely solely on analysis of founder animals for verification of the expected mutation; definitive genotype confirmation should be performed in subsequent generations (*see* **Note 15**):
   (a) Confirm the presence of the correct mutation through a thorough quality control screening in G1 animals.
   (b) Exclude most likely aberrant and off-target mutations in G1 animals (*see* **Note 16**).
7. Consider adding additional mutations by genome editing on preexisting animal models instead of intercrossing separately generated mutations.

3.4 Donor Females

1. Choose donor strain such that mutations are introduced into the background, on which subsequent research will be carried out. For the reduction of animal numbers, backcrossing should be avoided.
2. Consider superovulation. It will increase embryo yield per donor and therefore reduce the number of animals involved.
   (a) Consult literature before titrating hormones to optimize superovulation yields for individual strains.
   (b) Consider hyperovulation [7, 30] followed by IVF to generate embryos for manipulation (this can be especially useful when using donors which are in short supply).
   (c) Meet the minimal standards for efficiency:
      (i) >80% superovulated donor females should be plug positive after mating (>50% without superovulation making use of the Whitten effect) [31].
      (ii) A superovulated female should produce at least 30 zygotes, with more than 70% of them intact.
      (iii) The number of injectable morulae or blastocysts may vary between five (for strains like BALB/c) and ten (for strains like C57BL/6) (*see* **Note 17**).
   (d) Consider in vitro fertilization instead of natural mating. However, male mice have to be killed to collect sperm. If possible, use archived sperm to reduce the number of donor males to be killed.
   (e) Try to cryopreserve excess embryos for future use or to use them for tests and optimizations.
   (f) Consider purchasing cryopreserved embryos when your own colony, including stud males, is underused.
   (g) Do not superovulate for in utero manipulation.

3.5 Surgical Procedures

1. Use aseptic conditions.
2. Provide preemptive analgesia. We recommend subcutaneous administration of 0.1 mg/kg buprenorphine 30 minutes before the start of the procedure.
3. Provide anesthesia with isoflurane/O_2 or ketamine/xylazine (*see* **Note 18**).
4. After initiation of the anesthesia, 50 μL 0.5% bupivacaine in NaCl can be injected subcutaneously at the cutting site as infiltration anesthesia.
5. Support body temperature maintenance by the use of a heat mat (or similar).
6. Apply eye ointment to protect eyes from dehydration.
7. Shave and disinfect the surgical area.
8. For bilateral embryo transfer, perform only a single dorsal skin cut.
9. After embryo transfer, stitch the peritoneum and close the skin with a clamp or tissue adhesive. Clamps should be removed 7 days after the surgery.
10. Apply analgesia immediately after surgery (while still under narcosis), the same evening, and the next morning. This may be a subcutaneous injection of 10 mg/kg bodyweight carprofen. Analgesia using slow-release buprenorphine may be considered, if available.
11. Monitor animal health and absence of pain for at least the first 3 days after surgery by, e.g.:
    (a) The Mouse Grimace Scale [32, 33]
    (b) Posture, e.g., signs of abdominal pressings (belly pressing), twitching, and writhing
12. If the animals continue to show signs of pain, analgesia must continue (*see* **Note 19**). Predefine humane endpoints for termination of the experiment.

3.6 Embryo Transfer

3.6.1 Sterile Males

1. Favor the use of genetically sterile males over surgical sterilization.
2. In case of surgical sterilization (vasectomy or epididymectomy):
    (a) Use males at an early age (typically 6 weeks) so that the animals can be used for as long as possible.
    (b) Access the vas deferens or the epididymis via the scrotum instead of the abdomen.
    (c) Avoid test matings. Instead, start using sterile males as early as possible for mating to embryo-transfer dams and record the outcomes. Exclude fertile males, in the rare event of failed surgery.

(d) In the case of using genetically sterile males:

(i) Consider the use of dominant genetic male sterility (***for source of animals***, *see* Subheading 2) or ensure alternative use of surplus animals when recessive male sterility is used [8].

(ii) Consider using extra females (produced from the breeding to generate sterile males) as surrogate dams for training purposes or sentinel animals for your health monitoring program.

(iii) Alternatively, consider using natural hybrid sterility by crossing *Mus musculus domesticus* females with *Mus musculus musculus* males. All male offspring are sterile. However, you will need to maintain two parental strains to breed interspecies hybrids (***for source of animals***, *see* Subheading 2).

3. Record copulation success per male.
4. Replace males according to declining mating performance and not at fixed intervals.

3.6.2 Surrogate Dams and Embryo Implantation

1. Choose dams from a mouse strain with a good record of reproduction and rearing of newborns such as CD1(ICR), Swiss Webster, NMRI, or F1 hybrid strains (e.g., B6CBAF1 or B6D2F1).
2. Ensure that sexual maturity is reached (6–8 weeks old, depending on strain).
3. Ensure that animals are in the optimal weight and age range.
4. Synchronize females by exposing them to male pheromones (Whitten effect):
   (a) Place dams on bedding of a male before their planned use.
   (b) Mate females 48 hours after exposure.
5. Mate sufficient dams to match the expected number of embryos to be transferred.
6. Identify dams in proestrus or estrus for mating with sterile males [34].
7. Non-plugged females should be used repeatedly in their next estrous cycles.
8. Consider transcervical instead of surgical embryo transfer to avoid surgery (***for source of materials***, *see* Subheading 2).
9. For transcervical implantation:
   (a) Implant embryos transcervically from the morula stage onward (*see* **Note 20**).
   (b) This is done by implantation in a non-anesthetized female using a speculum and a pipette.

10. Avoid individual housing:
    (a) Place surrogate dams in groups.
    (b) After confirming the gravidity of the dams, reduce to two animals per cage to avoid crowding after birth (*see* **Note 21**).
11. The reuse of females upon weaning of their litter for a second embryo transfer should be considered [35].

3.7 Identification of Harmful Phenotypes

1. Assess whether the new mutation leads to a harmful phenotype by analyzing animals according to Table 2.

Table 2
Template for welfare assessment for harmful phenotypes as a result of genome modification

Mortality	Consider necropsy to investigate the cause of death
Reproduction data	Litter size Infertile pairs Care by dam/cannibalism Death between birth and weaning Frequency of gravidity
Preweaning animals	Size Coloration Size differences Food intake (milk spot)
General condition	Weight/body condition score Food/water intake Skin/fur condition Senses (sight, hearing, balance) Body orifices Externally visible deformities Respiration/breathing Abnormal posture
Behavior and motor functions	Apathy Jumpiness Stereotypic behavior/barbering Aggression and bite wounds Self-mutilation Nest construction and nest condition Reaction to handling
Clinical symptoms	Tremors Seizures/convulsions Lameness Writhing Discharge (ocular, nasal) Prolapse (rectal, penis, vaginal) Tumors Malformations
Project-specific indicators	Depending on the model generated

2. The number of animals assessed should be sufficient to provide statistically significant results, but animals should not be bred for the sole purpose of a welfare assessment.
3. Assessments should be carried out after birth, at weaning and at a mature age over their entire lifetime.
4. Define specific strategies to reduce the number of affected individuals or to mitigate the harmful phenotype, e.g., by intensified care and husbandry.
5. Welfare assessments should be repeated on a regular basis until harmful phenotypes are clearly identified or ruled out.
6. Assessments should be documented and the information, including mitigation strategies or termination criteria, reported to other potential users, together with a general description of the GM line.

3.8 Beyond Rodents

Obviously, genome modification is not limited to rodents. Wherever researchers work with sentient animals, the measures outlined here must be considered to optimize animal welfare during a transgenic project. The sentient animals to which these stringent welfare requirements apply generally include vertebrates, cephalopods, and certain crustaceans (e.g., decapods). When working with non-rodents, we recommend taking our list of recommendations and adapting them to species-specific needs.

4 Notes

1. High caloric water gel is an alternative (www.clearh2o.com).
2. Take strain and line characteristics, such as harmful phenotypes, into consideration.
3. In the case of complex genome alterations, especially homologous recombination of large genome segments, the success rates of direct generation in embryos are sometimes low and can lead to the use of excessive numbers of animals.
4. While most people agree on what interventions and consequences need to be considered, there is inherent disagreement about the individual weight of each factor. While we must leave the weighing to the reader, Zintzsch et al. provide extensive guidance [36, 37].
5. The actual severity of the phenotype of the GM model must be assessed and documented and should be published after its generation.
6. Increase the probability of obtaining the desired expression pattern and strength by the use of optimized expression constructs for large vectors (BACs, YACs) [38].

7. To increase efficiency, the use of a transposon system can be considered if construct size does not exceed the maximum payload of those systems [16].
8. Integrated large constructs may be fragmented. In addition, transgenes may integrate into functional regions of the genome with deleterious effects [39, 40].
9. The transgene may have been integrated at multiple genomic locations that segregate upon breeding, resulting in variable phenotypes.
10. As a rule, ES cells must be at least as vital as the host embryo, to allow sufficient participation in embryonic development.
11. A mouse line with improved germline transmission has been developed under the name "goGermline" [6]. Male animals developing from goGermline embryos lack germ cells due to testicular atrophy and are therefore infertile. Therefore, chimeras generated with such host embryos are fertile only if ES cells take part in development and generate functional male sperm. This ensures that when the chimeras are mated, only offspring resulting from the injected ES cells are produced. Moreover, it is possible to distinguish between fertile and sterile males by palpating their testes before mating. However, the males must be anesthetized prior to palpation; otherwise, they will not tolerate the procedure, and there is a risk that the testes will be retracted into the abdominal cavity.
12. Combinations of ES cell lines and host embryos resulting in the expression of different coat colors (e.g., "black" C57BL/6 ES cells with "white" BALB/c embryos) are often used to identify chimeras with a substantial contribution of ES cells by coat color. The strain for subsequent mating should be identical to the ES cell strain to obtain a pure genetic background.
13. The improved predictability of successful genome editing comes at the cost of using more zygotes for pre-screening. However, in comparison, the increased number of donor animals needed to obtain the test embryos will likely be considerably less than the number of additional animals that would be needed to repeat an unsuccessful editing procedure. In addition, surplus colony animals that might otherwise be killed may primarily be used for such purposes, as can be surplus embryos from other experiments, provided that they are of the same strain background used to design the sgRNA.
14. Injection of single-stranded DNA (ss) or supercoiled plasmid vectors into embryos at the G2 cell cycle stage can significantly increase the efficiency of homology-mediated repair in CRISPR/Cas9 experiments [41–43]. Thus, it is possible to use CRISPR/Cas9 directly in embryos for precise integrations. Constructs can have a size of up to about 11 kilobases (kb).

15. In case of large phenotype screens, it is advisable to screen for different mutations in founders.
16. Aberrant mutations include off-target mutations, on-target insertions or deletions at the mutation site, and duplications or integration of templates away from the target site. Techniques used may include, e.g., Southern blotting, PCR, TLA, Samplix Xdrop®, or sequencing technologies.
17. Significantly lower numbers have to be expected for non-superovulated females.
18. Isoflurane anesthesia through a face mask requires a relatively rigid fixation of the narcotized mouse, which can complicate the microsurgical embryo transfer using a binocular microscope.
19. The analgesic carprofen may be given per os in the 3 days following surgery by voluntary intake, a particularly refined method [44, 45]. To facilitate this protocol, the animals must be accustomed to the oral intake. This can be done, for instance, by offering them sweetened condensed milk mixed with water (3:10) from a 200 μL micropipette over the 2 days before surgery. On the following days, the animals voluntarily drink a drug dissolved in condensed milk.
20. Multiday culture of manipulated embryos (at least up to the morula stage) before transcervical transfer may negatively affect viability and birth rate.
21. Consider that litter size after embryo transfer is usually smaller than after mating. Reduce chances of neglect and associated death of newborns by the joint rearing of the offspring of two surrogate dams.

Acknowledgments

We thank Ronald Naumann, Jan Parker-Thornburg, Thomas Rülicke, and Anne Zintzsch for their valuable input during the writing of the manuscript. Figure 1 was created with Biorender.com.

References

1. Russell WMS, Burch RL (1959) The principles of humane experimental technique. Methuen, London
2. Behringer R, Gertsenstein M, Nagy KV, Nagy A (2014) Manipulating the mouse embryo: a laboratory manual. Cold Spring Harbor Laboratory Press
3. Pease S, Saunders TL (2011) Advanced protocols for animal transgenesis: an ISTT manual. Springer, Berlin Heidelberg
4. Hess SE, Rohr S, Dufour BD et al (2008) Home improvement: C57BL/6J mice given more naturalistic nesting materials build better nests. J Am Assoc Lab Anim Sci 47:25–31
5. Birling M-C, Yoshiki A, Adams DJ et al (2021) A resource of targeted mutant mouse lines for 5,061 genes. Nat Genet 53:416–419. https://doi.org/10.1038/s41588-021-00825-y
6. Koentgen F, Lin J, Katidou M et al (2016) Exclusive transmission of embryonic stem

cell-derived genome through the mouse germline. Genesis 54:326–333. https://doi.org/10.1002/dvg.22938
7. Takeo T, Nakagata N (2015) Superovulation using the combined administration of inhibin antiserum and equine chorionic gonadotropin increases the number of ovulated oocytes in C57BL/6 female mice. PLoS One 10: e0128330. https://doi.org/10.1371/journal.pone.0128330
8. Garrels W, Wedekind D, Wittur I et al (2018) Direct comparison of vasectomized males and genetically sterile Gapdhs knockout males for the induction of pseudopregnancy in mice. Lab Anim 52:365–372. https://doi.org/10.1177/0023677217748282
9. Gaskill BN, Gordon CJ, Pajor EA et al (2013) Impact of nesting material on mouse body temperature and physiology. Physiol Behav 110–111:87–95. https://doi.org/10.1016/j.physbeh.2012.12.018
10. FELASA working group on revision of guidelines for health monitoring of rodents and rabbits, Mähler Convenor M, Berard M et al (2014) FELASA recommendations for the health monitoring of mouse, rat, hamster, Guinea pig and rabbit colonies in breeding and experimental units. Lab Anim 48:178–192. https://doi.org/10.1177/0023677213516312
11. Dobrowolski P, Fischer M, Naumann R (2018) Novel insights into the genetic background of genetically modified mice. Transgenic Res 27: 265–275. https://doi.org/10.1007/s11248-018-0073-2
12. Wefers B, Wurst W, Kühn R (2011) Design and generation of gene-targeting vectors. Curr Protoc Mouse Biol 1:199–211. https://doi.org/10.1002/9780470942390.mo100179
13. Laber K, Newcomer CE, Decelle T et al (2016) Recommendations for addressing harm-benefit analysis and implementation in ethical evaluation – report from the AALAS-FELASA working group on harm-benefit analysis – part 2. Lab Anim 50:21–42. https://doi.org/10.1177/0023677216642397
14. Brønstad A, Newcomer CE, Decelle T et al (2016) Current concepts of harm-benefit analysis of animal experiments – report from the AALAS-FELASA working group on harm-benefit analysis – part 1. Lab Anim 50:1–20. https://doi.org/10.1177/0023677216642398
15. Gordon JW, Scangos GA, Plotkin DJ et al (1980) Genetic transformation of mouse embryos by microinjection of purified DNA. Proc Natl Acad Sci U S A 77:7380–7384. https://doi.org/10.1073/pnas.77.12.7380
16. Rostovskaya M, Naumann R, Fu J et al (2013) Transposon mediated BAC transgenesis via pronuclear injection of mouse zygotes. Genes N Y N 2000 51:135–141. https://doi.org/10.1002/dvg.22362
17. Fielder TJ (2011) Transgenic production benchmarks. In: Pease S (ed) Advanced protocols for animal transgenesis: an ISTT manual. Springer-Verlag, Berlin, pp 81–97
18. Hart-Johnson S, Mankelow K (2021) Archiving genetically altered animals: a review of cryopreservation and recovery methods for genome edited animals. Lab Anim:00236772211007306. https://doi.org/10.1177/00236772211007306
19. Shmerling D, Danzer C-P, Mao X et al (2005) Strong and ubiquitous expression of transgenes targeted into the beta-actin locus by Cre/lox cassette replacement. Genes N Y N 2000 42: 229–235. https://doi.org/10.1002/gene.20135
20. Tasic B, Hippenmeyer S, Wang C et al (2011) Site-specific integrase-mediated transgenesis in mice via pronuclear injection. Proc Natl Acad Sci U S A 108:7902–7907. https://doi.org/10.1073/pnas.1019507108
21. Ohtsuka M, Miura H, Mochida K et al (2015) One-step generation of multiple transgenic mouse lines using an improved Pronuclear Injection-based Targeted Transgenesis (i-PITT). BMC Genomics 16:274. https://doi.org/10.1186/s12864-015-1432-5
22. Bradley A, Anastassiadis K, Ayadi A et al (2012) The mammalian gene function resource: the International Knockout Mouse Consortium. Mamm Genome 23:580–586. https://doi.org/10.1007/s00335-012-9422-2
23. Cui Y, Xu J, Cheng M et al (2018) Review of CRISPR/Cas9 sgRNA design tools. Interdiscip Sci 10:455–465. https://doi.org/10.1007/s12539-018-0298-z
24. McBeath E, Parker-Thornburg J, Fujii Y et al (2020) Rapid evaluation of CRISPR guides and donors for engineering mice. Genes Basel 11. https://doi.org/10.3390/genes11060628
25. Mehravar M, Shirazi A, Nazari M, Banan M (2019) Mosaicism in CRISPR/Cas9-mediated genome editing. Dev Biol 445:156–162. https://doi.org/10.1016/j.ydbio.2018.10.008
26. Lee K, Mackley VA, Rao A et al (2017) Synthetically modified guide RNA and donor DNA are a versatile platform for CRISPR-Cas9 engineering. eLife 6:e25312. https://doi.org/10.7554/eLife.25312
27. Renaud J-B, Boix C, Charpentier M et al (2016) Improved genome editing efficiency

and flexibility using modified oligonucleotides with TALEN and CRISPR-Cas9 nucleases. Cell Rep 14:2263–2272. https://doi.org/10.1016/j.celrep.2016.02.018

28. Paquet D, Kwart D, Chen A et al (2016) Efficient introduction of specific homozygous and heterozygous mutations using CRISPR/Cas9. Nature 533:125–129. https://doi.org/10.1038/nature17664
29. Kim N, Kim HK, Lee S et al (2020) Prediction of the sequence-specific cleavage activity of Cas9 variants. Nat Biotechnol 38:1328–1336. https://doi.org/10.1038/s41587-020-0537-9
30. Hasegawa A, Mochida K, Inoue H et al (2016) High-yield superovulation in adult mice by anti-inhibin serum treatment combined with estrous cycle synchronization. Biol Reprod 94:21. https://doi.org/10.1095/biolreprod.115.134023
31. Whitten WK (1958) Modification of the oestrous cycle of the mouse by external stimuli associated with the male; changes in the oestrous cycle determined by vaginal smears. J Endocrinol 17:307–313
32. Leach MC, Klaus K, Miller AL et al (2012) The assessment of post-vasectomy pain in mice using behaviour and the Mouse Grimace Scale. PLoS One 7:e35656. https://doi.org/10.1371/journal.pone.0035656
33. Miller AL, Kitson GL, Skalkoyannis B et al (2016) Using the mouse grimace scale and behaviour to assess pain in CBA mice following vasectomy. Appl Anim Behav Sci 181:160–165. https://doi.org/10.1016/j.applanim.2016.05.020
34. Byers SL, Wiles MV, Dunn SL, Taft RA (2012) Mouse estrous cycle identification tool and images. PLoS One 7:e35538. https://doi.org/10.1371/journal.pone.0035538
35. Kolbe T, Palme R, Touma C, Rülicke T (2012) Repeated use of surrogate mothers for embryo transfer in the mouse. Biol Reprod:86. https://doi.org/10.1095/biolreprod.111.092445
36. Zintzsch A, Noe E, Reissmann M et al (2017) Guidelines on severity assessment and classification of genetically altered mouse and rat lines. Lab Anim 51:573–582. https://doi.org/10.1177/0023677217718863
37. Zintzsch A, Noe E, Grimm H (2020) Navigating uncertainties: how to assess welfare and harm in genetically altered animals responsibly—a practical guideline. Animals 10:857. https://doi.org/10.3390/ani10050857
38. Moltó V-G, Montoliu L, Pease S, Saunders T (2011) Designing transgenes for optimal expression. In: Advanced protocols for animal transgenesis: an ISTT manual. Springer, Berlin, pp 81–97
39. Goodwin LO, Splinter E, Davis TL et al (2019) Large-scale discovery of mouse transgenic integration sites reveals frequent structural variation and insertional mutagenesis. Genome Res 29:494–505. https://doi.org/10.1101/gr.233866.117
40. Chiang C, Jacobsen JC, Ernst C et al (2012) Complex reorganization and predominant non-homologous repair following chromosomal breakage in karyotypically balanced germline rearrangements and transgenic integration. Nat Genet 44:390–397, S1. https://doi.org/10.1038/ng.2202
41. Miura H, Quadros RM, Gurumurthy CB, Ohtsuka M (2018) Easi-CRISPR for creating knock-in and conditional knockout mouse models using long ssDNA donors. Nat Protoc 13:195–215. https://doi.org/10.1038/nprot.2017.153
42. Chu VT, Weber T, Graf R et al (2016) Efficient generation of Rosa26 knock-in mice using CRISPR/Cas9 in C57BL/6 zygotes. BMC Biotechnol 16:4. https://doi.org/10.1186/s12896-016-0234-4
43. Gu B, Posfai E, Gertsenstein M, Rossant J (2020) Efficient generation of large-fragment knock-in mouse models using 2-cell (2C)-homologous recombination (HR)-CRISPR. Curr Protoc Mouse Biol 10:e67. https://doi.org/10.1002/cpmo.67
44. Ingrao JC, Johnson R, Tor E et al (2013) Aqueous stability and oral pharmacokinetics of meloxicam and carprofen in male C57BL/6 mice. J Am Assoc Lab Anim Sci 52:553–559
45. Scarborough J, Mueller F, Arban R et al (2020) Preclinical validation of the micropipette-guided drug administration (MDA) method in the maternal immune activation model of neurodevelopmental disorders. Brain Behav Immun 88:461–470. https://doi.org/10.1016/j.bbi.2020.04.015

Chapter 3

Genetic and Molecular Quality Control of Genetically Engineered Mice

Lauri G. Lintott and Lauryl M. J. Nutter

Abstract

Genetically engineered mice are used as avatars to understand mammalian gene function and develop therapies for human disease. During genetic modification, unintended changes can occur, and these changes may result in misassigned gene-phenotype relationships leading to incorrect or incomplete experimental interpretations. The types of unintended changes that may occur depend on the allele type being made and the genetic engineering approach used. Here we broadly categorize allele types as deletions, insertions, base changes, and transgenes derived from engineered embryonic stem (ES) cells or edited mouse embryos. However, the methods we describe can be adapted to other allele types and engineering strategies. We describe the sources and consequ ences of common unintended changes and best practices for detecting both intended and unintended changes by screening and genetic and molecular quality control (QC) of chimeras, founders, and their progeny. Employing these practices, along with careful allele design and good colony management, will increase the chance that investigations using genetically engineered mice will produce high-quality reproducible results, to enable a robust understanding of gene function, human disease etiology, and therapeutic development.

Key words Mouse, Genome editing, Genetic engineering, Quality control, ES cells, Cas9, Genotyping, Germline transmission

1 Introduction

Genetically engineered mice are used as avatars to explore development and disease of nearly every mammalian organ system. These mice are generated by adding, removing, and/or changing DNA sequences in the genome, and the established mouse line is then used to annotate the pathophysiological consequences of the genetic change(s). The alleles generated can be categorized into one of four general types, deletions, insertions, base changes, and transgenes. In mice, genetic engineering can be done in embryonic stem (ES) cells [1, 2] or in embryos, usually zygotes [3–6] or two-cell embryos [7]. Establishing a mouse line requires that the engineered allele is transmitted to progeny, and this is accomplished

Thomas L. Saunders (ed.), *Transgenesis: Methods and Protocols*,
Methods in Molecular Biology, vol. 2631, https://doi.org/10.1007/978-1-0716-2990-1_3,

by generating and breeding ES cell chimeras [2, 8] or breeding founder mice generated from embryo manipulation. Derived progeny is then subjected to quality control (QC) to confirm the desired genetic change(s) has (have) been transmitted (see below).

Appropriate QC mitigates the risk of confounding results, which could lead to incorrect interpretations about pathophysiological observations in newly developed mouse lines. QC can be broadly categorized as genetic QC and molecular QC. The former includes assays to ensure the desired allele was accurately engineered and reduces the risk of co-transmission of unwanted or unintended genetic changes, while the latter includes additional assays to confirm that the introduced genetic change(s) have the expected effect(s) on gene expression. Each allele type is associated with unintended outcomes (Table 1) dependent on the strategy used to engineer the allele. In turn, for each unintended outcome, there are assays that can identify their occurrence (Table 2). Additionally, for unintended outcomes that are difficult to identify or remedy, good experimental design can mitigate the risk that gene-phenotype relationships are misassigned (*see* Subheading 3.12).

Here, we describe approaches for screening founders generated using locus-specific endonucleases, chimeras generated by ES cell targeting, and founders generated by random transgenesis, as well as strategies for screening and genetic QC of their progeny. These approaches focus on the commonly produced allele types at The Centre for Phenogenomics Model Production Core which has generated >1300 new genetically engineered mouse lines over the past 10 years. We present these approaches in the order we find is the most efficient progression to eliminate the most common incorrect alleles or mice early in the process. While not the focus of this manuscript, allele design does influence, and sometimes dictate, the ease with which mice can be screened and QC'd. We also discuss approaches for molecular QC and colony maintenance and cohort production.

2 Allele Design Considerations

The scientific questions being asked will often dictate the allele type needed in the mouse line being developed, but sometimes a choice of allele types is available. There is no perfect allele choice, perfect production method, nor even perfect allele design. Each comes with their caveats with respect to both intended outcome and potential confounds (Table 1). For example, there are several approaches to generating a putative null allele, such as inserting stop cassettes (e.g., selection markers or reporter trap cassettes [13]) into a gene, deleting one or more exons with or without the insertion of reporter or selection cassettes, introducing a premature termination codon (PTC) with a targeted change, or

Table 1
Common unintended outcomes from genetic engineering

Intended change	Genetic engineering method	Possible unintended or confounding outcomes
Random transgenes	DNA fragment introduced into zygotes by microinjection or viral transduction	Sequence changes at insertion site [9] Intragenic transgene insertion (e.g., [10]) Partial transgene insertion Transgene mutation Non-physiologic transgene expression Position effect modulation of expression Transgene silencing [11]
	DNA fragment introduced into ES cells by electroporation, transfection, or viral transduction, with or without antibiotic or other selection	Sequence changes at insertion site [9] Intragenic insertion (e.g., [10]) Partial transgene insertion Transgene mutation Non-physiologic transgene expression Position effect modulation of expression Transgene silencing [11] Passenger mutations [12]
Targeted insertion[a]	Targeting vector introduced into ES cells by electroporation or transfection[b] and antibiotic selection	Partial targeting vector insertion [13–15] Random targeting vector insertion [14, 15] Targeting vector mutation Aberrant splicing Off-target expression changes [16] Passenger mutations [12]
	Repair template and locus-specific endonuclease[c] introduced into zygote or two-cell embryos by microinjection	Partial repair template insertion [17] Repair template mutation Random target vector insertion Aberrant splicing Off-target expression changes [16] Off-target mutations

(continued)

Table 1
(continued)

Intended change	Genetic engineering method	Possible unintended or confounding outcomes
Targeted deletion	Targeting vector introduced into ES cells by electroporation or transfection[b] and antibiotic selection	Partial targeting vector insertion [13–15] Random targeting vector insertion [14, 15] Targeting vector mutation Aberrant splicing Off-target expression changes [16] Passenger mutations [12]
	Locus-specific endonuclease[c] introduced into zygotes by microinjection or electroporation	On-target DNA rearrangements [18, 19] Random reintegration of deleted DNA [4] Aberrant splicing [20] Off-target expression changes [16] Off-target mutations
Targeted base changes[d]	Oligonucleotide repair template and locus-specific endonuclease[c] introduced into zygotes by microinjection or electroporation	Incomplete changes [17] or unwanted mutations at target site Aberrant splicing [20] Random repair template insertion Off-target expression changes [16] Off-target mutations
	Targeting vector introduced into ES cells by electroporation or transfection[b] and antibiotic selection	Partial targeting vector insertion [14, 15] Random targeting vector insertion [14, 15] Targeting vector mutation Aberrant splicing Off-target expression changes [16] Passenger mutations [12]

[a]For knock-ins less than 100 base pairs (e.g., epitope tags, loxP sites), it is now routine to use single-strand oligonucleotides with 40- to 70-base homology arms flanking the sequence to be introduced in conjunction with a locus-specific endonuclease (e.g., *S. pyogenes* Cas9 [3]) in mouse zygotes. Larger changes rely on plasmid-based, virus-based, or synthetic (single- or double-stranded DNA) targeting vectors or repair templates and are done in either ES cells [1] or mouse embryos [5, 7]

[b]ES cell targeting may include the use of a locus-specific endonuclease to improve targeting efficiency [1]. Off-target mutagenesis [21], or unexpected on-target changes [18, 19], may occur when this is done. Using ribonucleoprotein to limit the duration of endonuclease activity rather than plasmid-based expression systems may mitigate these risks

[c]Presently, *S. pyogenes* Cas9, an RNA-guided endonuclease derived from the bacterial CRISPR/Cas system [22, 23], is the most used locus-specific endonuclease for mouse genome engineering. TAL-effector nucleases and zinc-finger nucleases have also been used [24–26]

[d]For changes less than ~50 base pairs, it is now routine to use single-strand oligonucleotides and locus-specific endonucleases in mouse zygotes [3]

Table 2
QC assays to evaluate potential confounds due to unintended outcomes of genetic engineering

Unintended outcome	Confound(s)	QC assay(s)
Aberrant splicing	Target gene aberrant splicing excludes one or more engineered exons to maintain an ORF escaping NMD and producing a protein	Molecular QC—evaluate target gene transcript structure by reverse transcription (RT)-PCR[a] [27] or long-read RNA sequencing[a]
Non-physiologic transgene expression	Phenotypic changes due to dominant-negative (titration) or ectopic expression not reflective of the transgene's physiological role	Molecular QC—determine transgene expression levels by digital PCR[a] [28] or RNA sequencing[a]; determine expression patterns by reporter gene expression[a] [29], in situ hybridization[a] or RNAscope[a] [30], Western blot[a] [31], and/or immunostaining[a] (e.g., [32])
Off-target expression changes	Disruption of enhancers or promoters at or near introduced sequence changes alters linked or unlinked gene expression	Molecular QC—evaluate transcription of neighboring genes by digital PCR[a] or global transcription by RNA sequencing[a]
On-target DNA rearrangements	CN or structure of target locus may change affecting gene function	Genetic QC—sequence target allele and target locus CN GLT test breeding will eliminate events that occur on the homolog in founder mice Molecular QC—evaluate target gene transcript expression, structure, and/or sequence
Partial or mutated target vector or repair template insertion	Changes allele product function. In knock-in alleles, this could abrogate or change the function of the knock-in DNA; in knockout alleles, the target gene could maintain some or all function(s)	Genetic QC knock-in—fully sequence the insert by long-range PCR and sequencing, WGS[a], or TLA[a] [33, 34] Genetic QC knockout—validate LOA by digital PCR[a] [28] or Southern blot[a] [35] Molecular QC—validate gain or loss of desired transcript(s) or protein (s) by digital PCR[a] [28], RT-PCR and sequencing[a], or RNA sequencing[a], reporter gene expression if present[a] [29], in situ hybridization[a] or RNAscope[a] [30], Western blot[a] [31], immunostaining[a] (e.g., [32]), and/or functional assay
Partial or mutated transgene insertion	Transgene not expressed or mutant transcript is expressed, thereby abrogating or altering transcript or protein stability and/or function	Genetic QC—fully sequence the transgene by long-range PCR, WGS[a], or TLA[a] [33, 34]

(continued)

Table 2 (continued)

Unintended outcome	Confound(s)	QC assay(s)
Passenger mutations Off-target mutations	Misassociation of phenotypes with intended change rather than passenger or off-target mutation	Colony management—establish lines by backcrossing to standard genetic strains obtained from reputable commercial suppliers. Derive experimental and control cohorts from crosses producing all relevant genotypes within the same breeding colony
Random target vector or repair template insertion	Insertional mutagenesis at other loci may affect expression and/or function of endogenous genes or extraneous expression of vector or repair template sequences	Genetic QC—determine vector backbone absence by PCR and TCN by digital PCR[a] [28] or Southern blot[a] [35]
Reintegration of deleted DNA	Insertional mutagenesis at other loci may affect expression and/or function of nontarget genes or result in extraneous expression of sequences targeted for deletion	Genetic QC—validate LOA by digital PCR[a] [28] or Southern blot[a] [35]
Sequence changes at transgene insertion site or intragenic insertion	Loss or rearrangement of DNA sequence or insertional mutagenesis may affect expression and/or function of endogenous genes	Genetic QC—identify the transgene insertion site using WGS[a] or TLA[a] [33, 34]
Transgene silencing	Phenotype associations change or disappear	Molecular QC—assess transgene expression frequently across generations, e.g., by digital PCR[a] [28]

[a]These protocols are not described in this chapter and can typically be accessed through either academic core or commercial service providers. Alternatively, references to published protocols are provided

introducing a frameshifting insertion and/or deletion (INDEL). Inserting stop cassettes, without deleting downstream gene sequences, can result in splicing around the cassette and expression of the wild-type gene product. INDEL or deletion alleles that introduce PTCs may undergo aberrant splicing to restore the open reading frame (ORF) [36–38] or escape nonsense-mediated decay (NMD) [39] resulting in proteins that may retain some function (hypomorphic or hypermorphic), act as dominant-negative (antimorphic) proteins, or gain new function (s) (neomorphic). Similarly, human genes can be expressed in the mouse by inserting transgenes randomly, by inserting the ORF for expression at a safe harbor locus (e.g., *Gt(ROSA)26Sor*), or by full-gene humanization. Transgene expression may not recapitulate the endogenous physiologic expression and may be silenced over time [11]; safe harbor knock-ins may not be expressed in all the appropriate cells/tissues and may not have the same RNA stability as the endogenous gene; and fully humanized genes may not be correctly expressed either because transcriptional and posttranscriptional regulatory sequences are not conserved or because protein-protein interactions are disrupted due to nonconservation. Additionally, all alleles necessarily disrupt both strands of the target DNA, potentially disrupting functional units on the nontarget strand (e.g., regulatory elements, antisense transcription units), which may confound phenotype interpretation [40]. An understanding of common determinants of NMD [41], transgene silencing [11], alternative splicing, transcriptional regulation, protein-protein interactions, and translation initiation all assists in allele design and predicting the outcome of genetic engineering. However, for many genes, these determinants are not well understood. Indeed, the purpose of generating the mouse line is often to define one or more aspects of gene regulation and/or function. As a result, allele design must rely on some assumptions about gene and allele structure, and the line between molecular QC and hypothesis testing may be blurred.

2.1 Deletion Alleles

Knockout alleles are the most common type of deletion allele and are intended to abrogate gene function. In protein-coding genes, knockout alleles should be designed to remove a critical region. A critical region is one or more exons common to all annotated full-length protein-coding transcripts within the first half of the ORF that, when removed, will introduce a frameshift and PTC [42]. To prevent unintended splicing alterations that could restore or create protein function, preserve the splice acceptors and donors of exons flanking the deletion by leaving at least 100 base pairs of the intron sequences adjacent to the remaining exons intact. To optimize the likelihood that residual transcripts will be targeted for NMD, the introduced PTC should be at least 50 bases 5′ of the last splice junction(s), be at least 150 bases from the start codon, and not occur in a large (>400 bases) exon [39]. Some gene structures, e.g., single-exon genes or genes with all exons in the same frame, require alternative approaches, such as deletion of the entire ORF or deletion within one exon (intra-exon) or between two exons (inter-exon) to introduce a frameshift.

While there remains much to be learned about protein-coding gene function, even less is understood about the noncoding genome. Deletion alleles are powerful tools with which to interrogate the function of the noncoding genome.

The use of RNA-guided nucleases (e.g., Cas9 [4, 43, 44], Cas12a [45, 46]) to generate deletion alleles is the most efficient and straightforward method in use. Advantages over ES cell-based production methods include shortened production timelines, ease of reagent production with both guide RNAs and endonucleases commercially available, no need for selection cassettes, ease of screening and genetic QC, and the need, usually, for fewer animals to establish the line. While there is a concern about off-target mutagenesis mediated by RNA-guided nucleases, it appears to be rare in vivo [47–50], and the use of ES cells does not reduce the risk of unwanted mutations being transmitted to subsequent generations since most ES cells acquire sequence changes and copy number (CN) variation during culture [12]. In the absence of a repair template, the double-strand breaks introduced by endonucleases may be repaired by either nonhomologous end joining (NHEJ) or microhomology-mediated end joining (MMEJ) to generate a deletion allele [43, 51, 52]. Double-strand breaks introduced during late S and G2 phases of the cell cycle may be repaired by homology-directed repair or homologous recombination using the replicated sister chromatid [53]. This type of repair would be undetectable using assays for deletion alleles. Analysis of large-scale production programs showed a median founder rate of 6.2% (founders per embryos transferred) after electroporation of Cas9 endonuclease into mouse zygotes [54]. Thus, homology-directed repair or homologous recombination does not prevent the recovery of deletion alleles.

Deletion alleles in genes or of noncoding genomic segments that are homozygous (HOM) lethal or haplo-insufficient can be difficult to produce directly in embryos as efficient editing can prevent any founders from being born. Alternative approaches for these genes include intentional production of mosaic animals (e.g., by injecting reagents into a single blastomere of a two-cell embryo) or generating conditional alleles (see below). The Mouse Genome Informatics (MGI [55]) database provides an excellent resource to identify lethal genes prior to production efforts, and if data is not available for the mouse gene, consider information about human cell essentiality [54, 56, 57] and mutation frequency in population-based sequence datasets (e.g., gnomAD [58]).

Finally, before engineering a new null allele, it is prudent to determine if a null allele line is already available. Putative null alleles, along with baseline phenotyping data, for more than 9000 genes are publicly available through the efforts of the International Mouse Phenotyping Consortium [59]. Coupled with the efforts of scientists around the world, null alleles for most mouse genes have been generated [60]. Using existing mouse lines is ethically sound, and existing phenotype data provides a foundation for subsequent discoveries. However, knowledge of how lines were QC'd is essential, and additional QC may be needed to confirm abrogation of gene expression and/or function before undertaking studies on imported lines [61]. The protocols described below and elsewhere [14, 28] can assist with evaluating the QC that may be required for imported lines.

2.2 Insertion Alleles

Insertion alleles include both targeted and random integration of exogenous DNA. Knock-in alleles introduce exogenous DNA into a specific locus in the genome, while transgenic alleles (see below) result from random integration of exogenous DNA into the genome. There are many types of knock-in alleles including reporter alleles, epitope tags, targeted transgenics, and conditional knockouts. Gene humanization replaces the mouse gene with its human ortholog (a knockout/knock-in).

Reporter alleles are used to delineate expression of the target gene by inserting a reporter cassette, such as a fluorescent protein, within the transcription unit. The cassette may insert the reporter ORF within the target gene ORF or may place a splice acceptor-reporter within an intron. Null reporter alleles have the reporter cassette and a poly(A) sequence inserted at or near the endogenous start codon. They are designed to terminate target gene transcription, both abrogating gene function and reporting on gene expression patterns. Another strategy uses internal ribosome entry sites (IRES [62]) or 2A peptides [63–65] to direct co-expression of the target and reporter genes. Insertion of reporter ORFs can affect transcript stability and/or may disrupt regulatory or other functional sequences. Thus, validation of the allele as a faithful reporter of gene expression is needed (*see* Subheading 3.13).

Epitope tags are short polypeptide sequences that enable the detection of the tagged protein with an anti-epitope antibody and are useful when antibodies specific to the protein being studied are not available. Tagging proteins can disrupt structure or function or may interfere with protein-protein interactions; this can be mitigated by including a flexible linker [66, 67] between the epitope tag and the protein of interest.

Targeted transgenics are used to insert an expression cassette into a so-called "safe harbor" locus. The most used safe harbor locus is *Gt(ROSA)26Sor* (MGI: 104735 [68–70]). *Igs2* (MGI: 5461148, also known as H11 [71]) and *Igs7* (MGI: 5490612, also known as TIGRE [72]) are less commonly used but are useful for co-expression of targeted transgenic alleles. Safe harbor loci enable expression of inserted genes in virtually every cell type and tissue and are used to direct either ubiquitous or tissue-/temporal-specific gene expression dependent on the exogenous promoter used in the knock-in cassette.

Conditional knockout alleles are knock-ins of recombinase target sites within a protein-coding gene. These alleles are used to investigate tissue- or developmental stage-specific gene function. In these alleles, a critical region is flanked by site-specific recombinase target sites, most commonly loxP sites that are the target of the phage recombinase Cre [73]. However, FRT sites, the target of the yeast Flp recombinase [74, 75], and rox sites, the target of Dre recombinase [76, 77], are also used. Combinations of recombinase target sites and/or conditional alleles can be used to precisely define gene expression within cell populations or tissue regions [78].

When loxP sites are in the same orientation, Cre-mediated recombination deletes the loxP-flanked (floxed) DNA, and gene function is disrupted; when loxP sites are in opposite orientation, DNA inversion rather than deletion will occur. LoxP sites should be positioned to mitigate the risk of disrupting expression or splicing of the pre-Cre allele. This is done by avoiding annotated regulatory regions, often in the first intron of a gene. In addition, regions conserved across species, especially distantly related species such as human and mouse, should be avoided as they are more likely to contain functionally conserved sequences. Finally, loxP sites are usually located at least 150–200 base pairs from splice donors or acceptors to leave sequences necessary for splicing intact.

LoxP sites can be added using site-specific endonucleases and two short oligonucleotide templates, one for each loxP site, or using a single long template that includes both loxP sites and the intervening sequence. LoxP sites can also be added by gene targeting in ES cells with the addition of a selection cassette (usually flanked by FRT sites for subsequent removal) to the targeting vector. The homology arms on a single long template for endonuclease-mediated insertion or in an ES cell targeting vector are usually longer than those used with oligonucleotide templates and, relative to the two-template approach, may require longer amplicons to confirm the insertion location and more sequencing to ensure no unintended mutations were introduced during DNA repair. Genotyping and QC methods of the resulting mice are dictated by the allele generation approach. All approaches can result in the insertion of single loxP sites, or when using endonuclease-based approaches loxP sites in *trans* [14, 79–81]. When this occurs, a second round of endonuclease/template treatment, or selection of alternate ES cell clones, may recover *cis*-located loxP sites and a bona fide conditional allele. The reported success of these approaches varies, with different model production (transgenic) facilities having preferred approaches.

Other types of conditional alleles not discussed in detail here include conditional by inversion (COIN) alleles [82], artificial introns [83], or conditional point mutations [84, 85]. The genetic and molecular QC required for these alleles can be adapted from those described for conditional knockout and knock-in alleles.

LoxP conditional alleles are only useful when a suitable Cre-expressing line is available. While not discussed here, QC of Cre-expressing lines is critical [9, 86] as are breeding strategies that generate control animals that have only the conditional allele or the Cre allele. When using inducible Cre alleles, such as Cre-ER and its derivatives with tamoxifen-dependent nuclear localization [87] or tetracycline-inducible or tetracycline-repressible promoters [88], animals of all genotypes, with and without treatment, should be experimentally assessed. Results from these control groups can lead to serendipitous discoveries [89].

2.3 Base Change Alleles

Base changes, often referred to as point mutations, are used to investigate the functional consequences of nucleotide variants that are not deletions or insertions of DNA sequences. Most often base change alleles alter one or more codons within a protein-coding sequence but can also be used to modify splice donor or acceptor sequences to alter splicing (and perhaps introduce PTCs) or to modify noncoding sequences that may impact gene expression, DNA conformation, and/or nuclear organization. In most model production facilities, it is now routine to use RNA-guided nucleases (e.g., [3]) to generate alleles with up to 60 bases changed. In this approach, a short oligonucleotide repair template ($\leq$ 200 bases) and an appropriate RNA-guided nuclease are introduced into embryos. The double-strand breaks introduced by the nuclease may be repaired by either homology-directed repair using the supplied template or by NHEJ or MMEJ [43, 51, 52]. Additional sequence changes to prevent the nuclease recutting the correctly repaired DNA (desired allele) are often included in the repair template. However, since both coding and silent mutations may have unintended consequences on transcription, splicing, and/or translation (e.g., [90–92]), molecular QC is essential to verify that any base changes have the predicted consequences on gene expression and/or RNA processing.

2.4 Transgenic Alleles

Transgenic alleles result from random integration of exogenous DNA into the mouse genome. These alleles are typically generated by pronuclear microinjection of a linear DNA template into zygotes, although electroporation into ES cells and subsequent chimera generation is also used. Bacterial sequences in the plasmid backbone can cause transgene silencing [93, 94] and should be removed from the transgene prior to injection, although some modified plasmid sequences enable sustained expression [95]. Transgenic alleles are often comprised of concatemeric insertions of DNA. Repeat-induced silencing [11] as well as excessive transgene expression due to multi-copy insertions can both confound analysis. CN can be reduced if a recombinase target site was included at one end of the transgene, and this may reactivate transgene expression [11] or reduce expression to approximate physiological levels. The recombinase target site to include in the transgene construct depends on the anticipated downstream use of the mouse line. For example, if generating a Cre-expressing transgenic line or if planning to cross the transgenic line with Cre-dependent conditional alleles and associated Cre-expressing lines, use FRT or F3 sites targeted by Flp recombinase [96] or rox sites targeted by Dre recombinase [76]. Transgene expression is also influenced by genomic context [97], a phenomenon known as position effect variegation. Thus, to conclusively identify genotype-phenotype relationships, multiple independent transgene founder lines must be analyzed.

An alternative approach to generating transgenic lines is to use transposable elements, such as PiggyBac or Sleeping Beauty [98]. Transgenic lines are derived by pronuclear microinjection of the transgene flanked by appropriate transposon terminal repeats along with the transposase mRNA [99]. The use of transposable elements generally avoids concatemeric insertions but does not preclude multiple independent insertions in a single founder mouse or potential position effects on transgene expression. Similarly, lentiviral vectors are also used to generate transgenic mice, albeit with multiple proviral insertion sites [100]. Coincubation of zona-free embryos with virus obviates the need for pronuclear injection. Targeted knock-ins (described above) can be used to avoid multimeric insertion, multiple independent insertions, and position effect variegation. However, random transgenic lines are often faster to produce and can be used to efficiently screen a variety of constructs.

3 Methods

Below, we provide stepwise procedures for founder (or chimera, if coat-color markers are not available) and N1 screening and genetic QC of common allele types. We define founders as mice born with a mutation that is a consequence of the attempted genetic manipulation, even if the mutation was not the intended outcome. N1 mice are the offspring of a backcross between founders or chimeras and strain-matched wild-type (WT) mice. We use "screening" to refer to methods to identify founders or N1 mice with the desired mutation and "genetic QC" to refer to methods designed to ensure that the desired mutation was accurately created. Where possible, screening is done using straightforward and unambiguous end-point PCR. Genetic QC involves allele sequencing and often CN assessment. However, depending on the mutation, sequencing may be necessary for screening as well as genetic QC. Since the exact screening and genetic QC protocols needed will depend on the desired allele type and the strategy used to generate the allele, we present stepwise protocols in the order that is the most efficient for producing fully QC'd mouse lines for each allele type.

3.1 Generation of Chimeras and Founders

To generate chimeras, genetically engineered ES cells are injected into or aggregated with host embryos to generate chimeric embryos that are transferred into pseudopregnant recipients for parturition and birth [8]. Chimeras are usually generated with ES cells encoding a different coat color than the host embryo, e.g., agouti- or black-coat ES cells with black, albino, or agouti embryos, as appropriate. Coat color contribution of the ES cell to the resultant chimera is used as a surrogate for ES cell germline contribution. Typically, >50% coat color chimeras are used for germline transmission (GLT) test breeding. In the absence of coat color markers, genetic screens (e.g., quantitative PCR to detect the desired allele [14]) are used to identify chimeras for breeding. When ES cells are used for genetic engineering, genetic QC must be done on the ES cell clones to select the appropriate clone from which to generate chimeras as well as on the derived mice [14, 15].

Founders born from embryos treated with locus-specific endonucleases or injected with DNA to produce transgenic alleles are screened using PCR, as described below, to identify those with the desired change(s) for GLT test breeding. Founders are often mosaic with the genotyping done on somatic tissue biopsies as a surrogate for germline surveillance. When selecting founders for GLT test breeding, full genetic QC is usually unnecessary since the alleles present in somatic tissue may not represent the alleles transmitted to the next generation. When multiple suitable founders are identified, quantitative PCR and/or additional genetic QC such as CN assessment can identify founders with the highest frequency of the desired allele; however, somatic tissue quantities may not always represent what is present in the germline.

3.2 GLT Test Breeding

Transmission of the engineered allele to subsequent generations is necessary to establish a new mouse line. GLT test breeding is established by crossing the chimera or founder to commercially purchased WT mice of the desired genetic background. The principles for selecting which chimeras and founders to breed are described in each of the relevant sections below. Since most ES cells, and therefore most fertile chimeras, are male (XY), GLT test breeding of chimeras usually involves a backcross or outcross to commercial female mice. Transgenic or endonuclease-derived founders may be either male or female and are backcrossed to commercial WT mice of the opposite sex.

When coat color chimeras are bred, GLT of the ES cell genome may be ascertained by the progeny's coat color, if an appropriate cross is used (e.g., 129 ES cells in B6 embryos crossed to B6 with agouti-coat color pups or B6 ES cells in albino embryos crossed to albino B6 mice (*Tyr*^{c}) with black-coat color pups, indicating transmission of the ES cell genome). While coat color is a simple way to identify GLT of the ES cell genome, this is only appropriate when the strain used for breeding matches the desired genetic background of the strain being produced. If the desired strain background is not used, the mice born during GLT test breeding have a mixed background are not useful, and this is not consistent with the principles of the 3R's [101].

To minimize the number of animals bred and produced during GLT test breeding, single chimeras or founders may be bred sequentially, only breeding additional chimeras or founders as needed. While this is effective with endonuclease-engineered founders, it does not work as well with chimeras. Partial transmission is often observed from chimeras generated by blastocyst injection, with GLT frequently not observed in the first litters. As a result, chimeras may age beyond their useful breeding span while awaiting complete (i.e., multi-litter) GLT test breeding results from other chimeras. Additionally, for random transgenic lines and for lines derived from ES cells, analysis of multiple lines originating from different transgenic founders or ES cell clones is considered best practice to account for potential position effects or passenger mutations, respectively.

In contrast, analysis of independent endonuclease-derived lines is not currently an established standard. This may in part be due to direct production in mouse embryos rather than cultured cells. Cultured ES cells typically accumulate significant passenger mutations over time [12], whereas embryos are under stricter genetic selection and should be produced from mouse colonies with high-quality genetic monitoring. In addition, endonucleases are generally used for targeted rather than random mutagenesis so position effects are not a consideration. The rationale for analyzing independent endonuclease-derived lines is to control for endonuclease-mediated off-target events, but these events are rare in mouse embryos [47–50]. To analyze independent endonuclease-derived lines, each line must be generated using endonucleases with different targeting specificities (e.g., different guide RNAs for RNA-guided nucleases) as only then would the spectrum of potential endonuclease-mediated off-target mutagenesis events be independent. Risks of confounding results associated with unlinked passenger or off-target mutations can be mitigated by good colony management practices (*see* Subheading 3.12).

GLT of the desired allele may fail because ES cells or mutant cells in founders failed to contribute to the germline by random chance or because the engineered allele reduces either fertility or viability. The International Mouse Phenotyping Consortium found that ~30% of knockout alleles are lethal or subviable [102] and ~5% of knockout alleles affect the fertility of one or both sexes. If the mutation reduces viability, no or very few pups with the desired allele may survive. Fertility effects include mutations that are incompatible with functional germ cell formation or that have negative effects on breeding behavior. Additionally, phenotypes may be sex specific [103]. If backcrosses fail to generate carrier progeny, additional crosses using chimeras with a lower contribution of ES cells may rescue behavioral phenotypes. Similarly, using endonuclease-derived founders that appear to be heterozygous (HET) rather than homozygous (HOM) or of the opposite sex may enable GLT when HOM founders do not. Before concluding GLT has failed, each chimera is bred to generate at least 42 pups, and each transgenic or endonuclease-derived founder is bred to generate at least 28 pups without the desired mutation.

For each allele type, specific strategies to identify progeny with the desired mutation are described in their respective sections below. Appropriate molecular QC should also be performed before projects are undertaken to ensure the results of the genetic modification are the same as those intended for a given allele.

3.3 Genomic DNA Isolation for Screening and QC

Founders and pups born from GLT test breeding are identified (e.g., with ear notches or ear tags) and tissue biopsied (e.g., ear or tail biopsies) according to protocols approved by local animal care guidelines and regulations. DNA is then prepared using protocols compatible with downstream analysis. For PCR amplicons less than 2 kb in length, crude DNA preparations (e.g., [104, 105] or commercial kits such as Sigma REDExtract-N-Amp or KAPA Mouse Genotyping Kit) are often sufficient. For PCR amplicons greater than 2 kb in length and quantitative PCR applications, purification of DNA by organic extraction [105] or commercial kits (e.g., Macherey-Nagel NucleoSpin Tissue) is usually necessary to remove PCR inhibitors that prevent efficient amplification of long amplicons and/or accurate quantitation.

3.4 Assay Design

Most assays used for screening and QC of mice are endpoint PCR assays. Southern blot is still used for ES cell QC [35] in some facilities, but obtaining sufficient high-quality genomic DNA for Southern blot from small tissue biopsies from live mice is often not possible. All PCR primers are designed with the same annealing temperature to allow PCR protocols to be standardized. There are many primer-design tools available; we routinely use Primer3Plus [106, 107] (https://www.primer3plus.com/index.html) with a Max Poly-X of 3 and GC clamp of 1 and the primer specifications noted in Table 3.

When possible, design PCR assays to generate amplicons less than 2 kb long as these can usually be amplified from crude DNA preparations. Many genes are part of gene families, have conserved functional domains, and/or have pseudogenes; check that chosen

Table 3
Recommended changes to Primer3 default settings

	Min	Opt	Max
Primer size	22	25	28
Primer Tm	58	60	64
Primer GC%	45	50	54

primer pairs do not produce unintended amplicons using a tool such as GEAR-Genomics Silica (https://www.gear-genomics.com) or NCBI primer-BLAST [108]. We use Silica to confirm that each primer pair is predicted to produce only a single amplicon in the mouse genome and to avoid primers with off-target annealing temperatures within 5 °C of the on-target annealing temperature.

CN, loss-of-allele (LOA), and allelic-discrimination assays for either digital PCR or quantitative real-time PCR should be designed using appropriate software or by the commercial source from which the assays are sourced (e.g., Integrated DNA Technologies, Thermo Fisher Scientific). Suppliers use different probe chemistries and lengths so designs from different suppliers may not be interchangeable. Quantitative real-time PCR assays require a standard curve or CN control and provide relative quantitation [109], whereas digital PCR can provide absolute quantitation [110]. The choice of methodology will depend on the availability of equipment and controls as well as the application. Assays for exogenous DNA templates may not have existing genomic controls or a cell or mouse line with known CNs available, which are needed for relative quantification. Thus, digital PCR is better suited for transgene CN determination and identification of exogenous insertions of unique (non-genomic) DNA into the genome.

CN assays are used to detect extra unwanted repair template insertions or to quantify the number of transgenes integrated (template copy number, TCN). TCN assay primers and probes should target within the repair template or transgene, at least 10 bp away from the ends of the template. LOA assays are used to confirm that genomic sequences targeted for deletion are not reinserted elsewhere in the genome [4] and/or that a locus has been disrupted by insertion of the knock-in allele [14]. An alternative to CN and LOA assays for detecting unwanted insertions is whole genome sequencing (WGS). However, short-read sequencing could lead to misalignment of reads to the endogenous locus for deletion segments.

3.5 Amplicon Sequencing

Amplicons are sequenced as part of screening when diagnostic PCRs cannot be designed (e.g., for short stretches of base changes or insertions) and during genetic QC to identify or confirm the definitive allele sequence. If the distance between the PCR primers and the site of the desired change is <500 bp, then the PCR primers can be used for sequencing provided there are no polynucleotide or di- or trinucleotide repeat stretches (>8) between the primer and the site of the sequence change. If such stretches exist, design a nested primer for sequencing that is at least 100 bp from the site of the sequence change and does not have the repeat (s) between it and the sequence change. If the desired change is >500 bp from the PCR primers, then nested primers that are at least 150 bp from the desired change can be used for sequencing.

Prepare amplicons for sequencing according to your preferred sequencing facility's recommendations. If DNA fragment analysis indicates a single amplicon and the sequencing primer is one that was used for amplification, it is usually sufficient to simply dilute the amplicon ten- to 20-fold in water before sequencing. For amplicons that require sequencing with nested primers, or if sequence chromatograms are messy, digest the PCR with Exo-CIP (NEB E1050) to remove the PCR primers before diluting the amplicon for sequencing. If chromatograms are messy after Exo-CIP digestion, it is likely the PCR produced two or more distinct amplicons that are similar in size. Sequence traces can be deconvoluted using online tools such as ICE [111] or DECODR [112]. If deconvolution fails, then subclone the amplicons using your preferred PCR cloning kit and sequence several clones. When DNA fragment analysis reveals amplicons of different sizes, gel purify the desired amplicon(s) with commercial gel purification kits (e.g., Macherey-Nagel 740609, NEB T1020) or other protocol (e.g., [113, 114]) prior to sequencing.

3.6 Endonuclease-Mediated Deletion Alleles

Deletion alleles are now widely produced using RNA-guided endonucleases such as Cas9 and Cpf1/Cas12a in embryos rather than ES cell targeting. To generate a deletion allele, two guide RNAs with appropriate specificity [49] whose protospacer sequences flank the region to be deleted are complexed with their cognate endonuclease (Fig. 1a) and introduced into zygotes [4] by microinjection or electroporation. The QC protocols presented below are for deletion alleles generated using this strategy, but they are equally applicable to deletion alleles generated with zinc-finger or TAL-effector nucleases targeting the same relative locations with respect to a desired deletion. Strategies for screening and QC of mice generated from ES cells are provided in Subheading 3.10:

1. Design assays to screen founders and screen and QC N1 mice. Two primer pairs are sufficient to identify mice with the desired deletion allele (Fig. 1a).
   (a) The wild-type (wt) primers flank one of the two endonuclease cut sites such that the annealing site for one primer is in the segment to be deleted (i.e., absent from the deletion allele). This primer pair will produce an amplicon from only the wild-type allele.
   (b) The deletion (del) primer annealing sites flank the segment to be deleted and are at least 150 bp upstream and downstream of the predicted deletion junction. This ensures that when the deletion amplicon is sequenced, at least 100 bp of sequence on either side of the deletion junction can be observed. Additionally, positioning the del primers this distance from the cut sites will mitigate

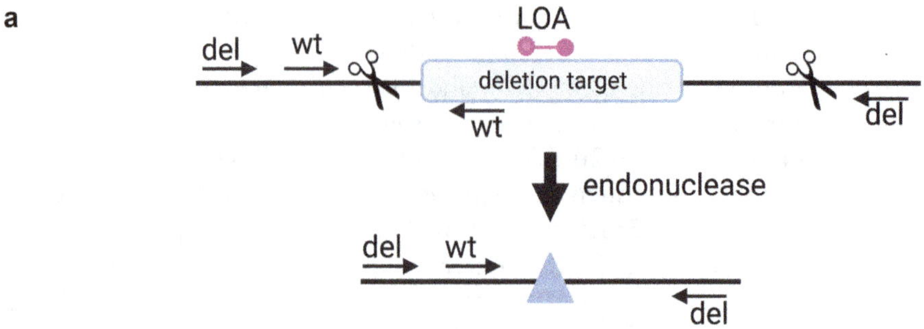

b

Assay	Primers		Amplicons by allele type		Sequence
	Fwd	Rev	wild-type	deletion	
wt	wt-F	wt-R	wt	none	no
del	del-F	del-R	wt	deletion	in N1

Fig. 1 Endonuclease-mediated deletion allele genotyping schematic. (**a**) Primer locations before and after the deletion (blue bar) is introduced. The wt primers flank one of the two endonuclease cut sites and only produce an amplicon from the wild-type allele. The del primers flank both cut sites and will produce a smaller amplicon from the deletion allele than from the wild-type allele. The scissors mark the endonuclease cut sites. The triangle marks the deletion junction following DNA repair, usually by NHEJ or MMEJ. Arrows pointing to the right are forward primers and pointing to the left are reverse primers. The LOA assay (probe shown in pink) recognizes the deleted sequence. (**b**) Assays for endonuclease-mediated deletion alleles. In HET or mosaic animals, both the wild-type and deletion allele amplicons are produced. For X-linked genes, mosaic male founders will have both alleles, but HEM males will only have one allele. The sequence column specifies which amplicons should be sequence for genetic QC. wt, wild-type; del, deletion; NHEJ, nonhomologous end joining; MMEJ, microhomology-mediated end joining; LOA, loss of allele; HET, heterozygous; HEM hemizygous; QC, quality control

losing the primer binding sites during DNA repair, which can delete sequence beyond the cut site. Unexpected on-target mutagenesis [19, 115, 116], including the deletion of 100s or 1000s of bases beyond the cut site, can occur, and in these cases the del assay may not produce an amplicon because one of the del primer annealing sites is deleted.

(c) The LOA assay is designed to detect the deleted genomic segment (Subheading 3.4). The assay should be designed as close to the middle of the segment as possible to mitigate the risk that exonuclease activity within the cell will remove primer or probe binding sites prior to reinsertion. For excised fragments larger than 50 kb, multiple LOA assays may provide additional assurance.

2. Identify founders using the del and wt assays (Fig. 1b).

(a) With the del assay, founders harboring the deletion allele will produce the smaller deletion amplicon and, if mosaic or HET and PCR conditions permit, may also produce the larger wt amplicon. The deletion amplicon may vary in size if additional sequence upstream or downstream of the cut sites was deleted or novel sequences inserted during DNA repair.

(b) With the wt assay, the wt amplicon is produced if the wild-type allele is present. The wt amplicon size may vary if the DNA double-strand break at either endonuclease target site was imprecisely repaired without introducing a deletion. Founders are usually mosaic so the wt assay acts as a positive control for DNA quality; however, a HOM founder will not produce a wt amplicon.

3. Select founder mice for GLT test breeding. Backcross founders with the correct size deletion to commercially purchased WT mice of the desired genetic background. For most deletion alleles, a single founder will generate the allele of interest [54]. Founders not selected for breeding can be maintained in case needed for subsequent breeding. *See* Subheading 3.2 for more information.

4. Identify N1 mice with the deletion allele using the wt and del assays. When GLT is successful, N1 mice for autosomal genes or female N1 mice for X-linked genes are HET, and the del assay will produce the deletion amplicon and the wt amplicon if PCR conditions permit. For male N1 mice with X-linked alleles, the del assay will only produce the deletion amplicon. For N1 mice without the deletion allele, the wt assay will produce the wt amplicon, and, if PCR conditions allow, the del assay will produce the wt amplicon.

5. Genetic QC of N1 mice with the deletion allele.

(a) Sequence the deletion amplicon to identify the deletion junction and allele sequence. Sequence at least 100 bp upstream and downstream of the deletion junction. If DNA flanking the endonuclease cut sites was deleted, it may be necessary to redesign the del primers to ensure sufficient sequence flanking the deletion junction can be observed.

(b) To ensure that the excised DNA fragment was not inserted elsewhere in the genome, check the CN of the excised fragment using either quantitative real-time or digital PCR [28, 117] using WT female mice as two-copy controls. The CN assay should target near the middle of the excised DNA fragment (Fig. 1a) to mitigate the risk of loss of the primer or probe binding sites resulting from DNA chew back that can happen. Mice pass QC if the deleted region has a CN of one for autosomal segments in HET mice or X-linked segments in HET females and a CN of 0 in HEM males for X-linked regions.

6. N1 mice pass QC if:
   (a) The LOA assay indicates that the deleted sequence was not reinserted elsewhere in the genome.
   (b) The allele sequence indicates that the expected region was deleted and no exogenous DNA was inserted (note that the deletion junction may have additional bases added or deleted during DNA repair).
7. Genotype QC'd colonies. If needed, the del primers can be redesigned to target the deletion boundaries more closely than recommended for screening. While genotyping with only the del assay is possible, we recommend using both the wt and the del assay. In the del PCR, the smaller deletion amplicon can outcompete the larger wt amplicon in HET mice which leads to misidentification of HET as HOM. The wt assay will help avoid such errors because, in an independent PCR, it will be produced from HET mice but not from HOM mice. Similarly, while it might be tempting to genotype using only the wt assay, the lack of an amplicon could indicate a HOM mouse, or it could be the result of poor-quality DNA resulting in failed PCR amplification.

3.7 Endonuclease-Mediated Insertion Alleles

The insertion size dictates the allele engineering approach used for newly generated alleles. Smaller insertions (60–100 bases long) are efficiently generated using an endonuclease and a commercially synthesized single-stranded DNA template up to 200 bases long, including 40- to 70-base homology arms [3]. Larger insertions can be introduced using repair templates with longer homology arms. Long single-stranded DNA templates [5], biotinylated double-stranded DNA templates [118], adeno-associated virus [119, 120], or circular plasmids [6] are all used as templates for larger insertions along with appropriate RNA-guided endonucleases in zygotes or two-cell embryos. Insertions can also be engineered using ES cell targeting [1]; *see* Subheading 3.10:

1. Design assays to screen founders and screen and QC N1 mice (Fig. 2).
   (a) The wild-type (wt) assay primers flank the site of insertion, with one or both primers located beyond the ends of the template homology arms, if possible.
   (b) Design an insert (in-wt) assay if the insert is long enough (≥20 bases) to enable the design of a primer (in) that does not anneal to the wild-type allele and pairs with a wt primer.
   (c) If neither of the wt primers anneal beyond the end of the repair template homology arms, design long-range (lr) primers. The lr primers flank the region used in the repair template and are designed for use in conjunction with the in primers (lr-in assay).

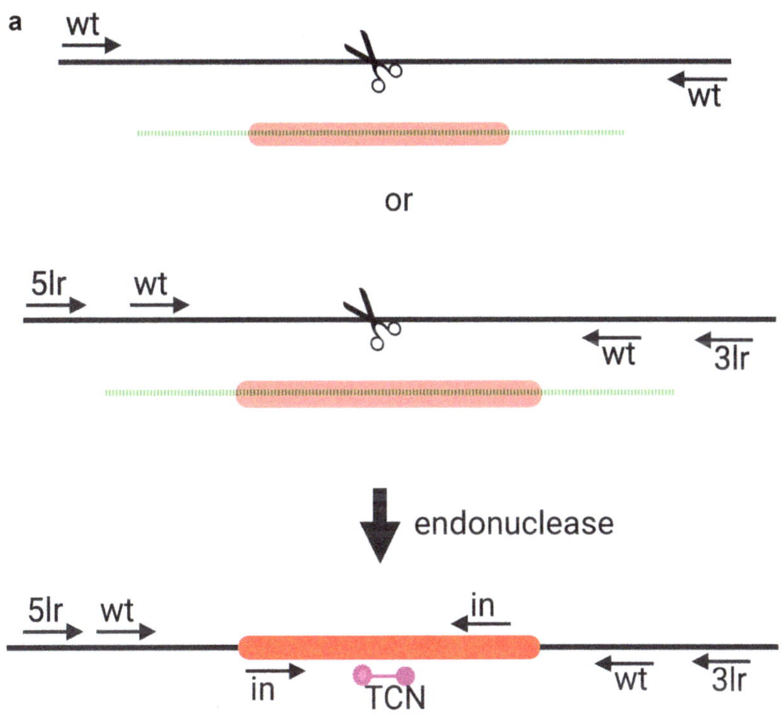

b

	Primers		Amplicons		
Assay	**Fwd**	**Rev**	**wild-type**	**insertion**	**Sequence**
wt	wt-F	wt-R	wt	wt + insertion	yes*
wt-in**	wt-F	in-R	none	wt-in	yes***
in-wt**	in-F	wt-R	none	in-wt	yes***
5lr****	wt-F or 5lr	in-R	none	5 lr	yes***
3lr****	in-F	wt-R or 3lr	none	3 lr	yes***

*if the insertion is small, subclone and sequence the amplicon

** generally one of these is sufficient for genotyping purposes

*** only necessary if the insertion amplicon cannot be sequenced

****5lr and 3lr only needed if wt primers do not flank the template sequence

Fig. 2 Endonuclease-mediated insertion allele genotyping schematic. (**a**) Primer locations before and after endonuclease-mediated insertions. If practical, the wt primer pair flanks the sequence included in the repair template (green dotted line) that includes an insert (red bar) flanked by homology arms; if the repair template is longer than 2 kb, then lr primers that flank the repair template are needed. The in primer(s) anneal to the insert sequence and are used with wt or lr primers to assess if the insert is in the correct location. The TCN is a digital PCR assay (probe shown in pink) that uniquely recognizes the insert sequence. Arrows pointing to the right are forward primers and pointing to the left are reverse primers. (**b**) Assays for endonuclease-generated insertions. The wt assay produces an amplicon from the wt allele and not from the insertion allele unless PCR conditions permit its amplification. In the latter case, the insertion allele produces a wt + insert amplicon. If the repair template is >2 kb, the 5 lr and/or a 3 lr assay will confirm that the inserted sequence is present at the correct locus. The sequence column indicates what amplicons should be sequenced for genetic QC. wt, wild type; del, deletion; in, insert; TCN, template copy number; HET, heterozygous; lr, long range; QC, quality control

   (d) Design a TCN assay that uniquely recognizes the inserted DNA for CN determination using digital PCR (*see* Subheading 3.4).

2. Identify founders using endpoint PCR.

   (a) For insertions without an in-wt assay, insertions may be detected by a size change of the wt amplicon. Since INDEL alleles can also change amplicon size, sequence the wt amplicon to definitively identify founders. Mosaic founders may produce multiple chromatogram traces that can be difficult to deconvolute by eye. ICE [111] or DECODR [112] can help identify founders that carry the correct insertion sequence. Alternatively, subclone the wt amplicon using a commercial kit (e.g., Thermo Fisher Scientific K457501) and sequence several clones.

   (b) For insertions with an in-wt assay, use the wt and in-wt assays to identify founders (Fig. 2). The wt assay will produce a smaller amplicon from the wild-type allele and a larger amplicon if the insertion is present and PCR conditions permit. The in-wt assay(s) detects the insertion directly but, if the wt primer is within the homology arm, will not confirm location. In this case, perform lr assays.

   (c) If the wt primer in the wt-in assay is within the template, use the corresponding 5lr-in and/or 3lr-in assays to verify that the target vector has inserted into the correct genomic locus.

   (d) An optional TCN screen may be done on founders with the insertion to identify tandem insertions and/or off-target insertions of the repair template. A CN >2 indicates that extra copies of the template have inserted into the founder genome. Ideally, founders with TCN of 2 or less for autosomal alleles or X-linked alleles in females or 1 or less for sex-linked alleles in males should be used for GLT test breeding. However, since founders are often mosaic, segregation of off-target insertions may occur during GLT test breeding.

3. Select up to five founder mice for GLT test breeding (Subheading 3.2). Select founders of both sexes that appear HET or HEM for the desired insertion allele (i.e., produced correctly sized amplicons for all assays) for backcrossing to commercially purchased WT mice of the desired genetic background.

4. Screen N1 mice using the same assays used to identify the knock-in allele in founders. When GLT is successful, N1 mice will produce the in-wt amplicon. For autosomal genes or X-linked genes in female N1 mice, the wt assay will produce both the wt and wt + insertion amplicons, PCR conditions allowing (Fig. 2). Male N1 mice for sex-linked genes will only

produce the wt + insertion amplicon. N1 mice without the insertion allele will produce the wt-sized amplicon and lack the wt + insertion amplicon.

5. Perform N1 genetic QC on mice that have the knock-in allele.
   (a) Sequence the complete insertion, the homology arms and at least 100 bp flanking the ends of the homology arms.
   (b) If the wt primers flank the insertion beyond the ends of the homology arms, gel purify and sequence the wt amplicon that contains the insertion to obtain complete sequence (*see* **Note 1**).
   (c) If the wt primers do not flank the insertion beyond the ends of the homology arms, or this amplicon is not efficiently amplified by PCR, sequence overlapping in-wt amplicons to obtain the complete sequence (*see* **Note 1**).
   (d) Use the TCN assay to ensure that the repair template was not inserted elsewhere in the genome (*see* **Note 2** for alternatives and **Note 3** for interpretation).
   (e) If a plasmid template was used, check that the plasmid backbone was not integrated in the genome using an endpoint PCR assay that targets the backbone.
6. A mouse line fully passes QC if:
   (a) The sequence matches the expected insertion.
   (b) There are no alterations to the flanking sequence.
   (c) The CN of the inserted sequence is 1 in HET and HEM.
   (d) There is no evidence that the plasmid backbone (if used) was integrated.

   If the inserted sequence is different than expected, the mouse line may still pass QC if the change is not predicted to alter function, e.g., when a change in a protein-coding region does not alter the amino acid sequence. In these cases, it is prudent to perform molecular QC to ensure the difference does not have unexpected consequences on transcription, splicing, and/or translation (e.g., [90–92]).

   Mice with extra inserted copies of the template sequence or plasmid backbone may be rescued if those insertions are unlinked to the targeted knock-in locus. To test if the extra copies are unlinked, backcross N1 mice with commercially purchased WT mice of the desired genetic background and identify N2 mice with single copy insertions using the TCN assay. Alternatively, screen and QC additional N1 mice to identify a lineage with single copy insertions.
7. Genotype knock-in mouse lines using the wt and in-wt PCR assays. Mice with only the smaller wt amplicon are wild type, mice with both the smaller wt amplicon and the in-wt amplicon

are HET, and mice with only the larger wt amplicon (when PCR conditions allow its amplification) and the in-wt amplicon are HOM or HEM for males with sex-linked knock-in alleles.

3.8 Endonuclease-Mediated Conditional Alleles

Conditional knockout alleles, a special case of insertion alleles, are usually engineered by inserting two loxP sites flanking a critical region (*see* Subheading 2). Screening and N1 QC strategies depend on whether two short templates, each encoding a single loxP site, or a single long template that includes both loxP sites and the loxP-flanked (floxed) DNA is used:

1. Design assays for founder and N1 screening (Fig. 3).
   (a) Design assays to detect loxP site insertion at the target locus. Assay design is simplified if sequences for universal primers are included adjacent to the loxP sites (Fig. 3b). If universal primer sites are not included with the loxP sites, design gene-specific primers that overlap the loxP sites such that the 3′ end of the primer is contained within the loxP site and amplification is dependent on the presence of the loxP site.
   (b) The proximal loxP (ploxP), inserted 5′ to the floxed DNA, is detected by the ploxP assay that has one primer (ploxP-R) specific to the ploxP site and a gene-specific primer (pwt-F) at least 150 bp 5′ of the insertion site, outside of the homology arm if the resulting amplicon is <2 kb.
   (c) The distal loxP (dloxP), inserted 3′ relative to the floxed DNA, is detected by the dloxP assay that has one primer (dloxP-F) specific to the dloxP site and a gene-specific primer (dwt-R) at least 150 bp 3′ of the insertion site, outside of the homology arm if the resulting amplicon is <2 kb.
   (d) The lox-lox assay amplifies the DNA between loxP sites when they are located on the same homolog (in *cis*) and uses primers specific to the ploxP (ploxP-F) and dloxP (dloxP-F).
   (e) If using two oligonucleotide templates to insert the loxP sites, design two TCN assays to check that each loxP site was only inserted once. If the probe targets the loxP site, the same probe will detect both loxPs and the loxPs can be independently verified by designing primers that anneal to the unique flanking regions that were included in the template(s) used to insert the loxP sites. If a single long template was used to insert the loxP sites, a single TCN targeting the repair template should be sufficient. TCN primers should be at least 10 bp internal to the ends of the template sequence.

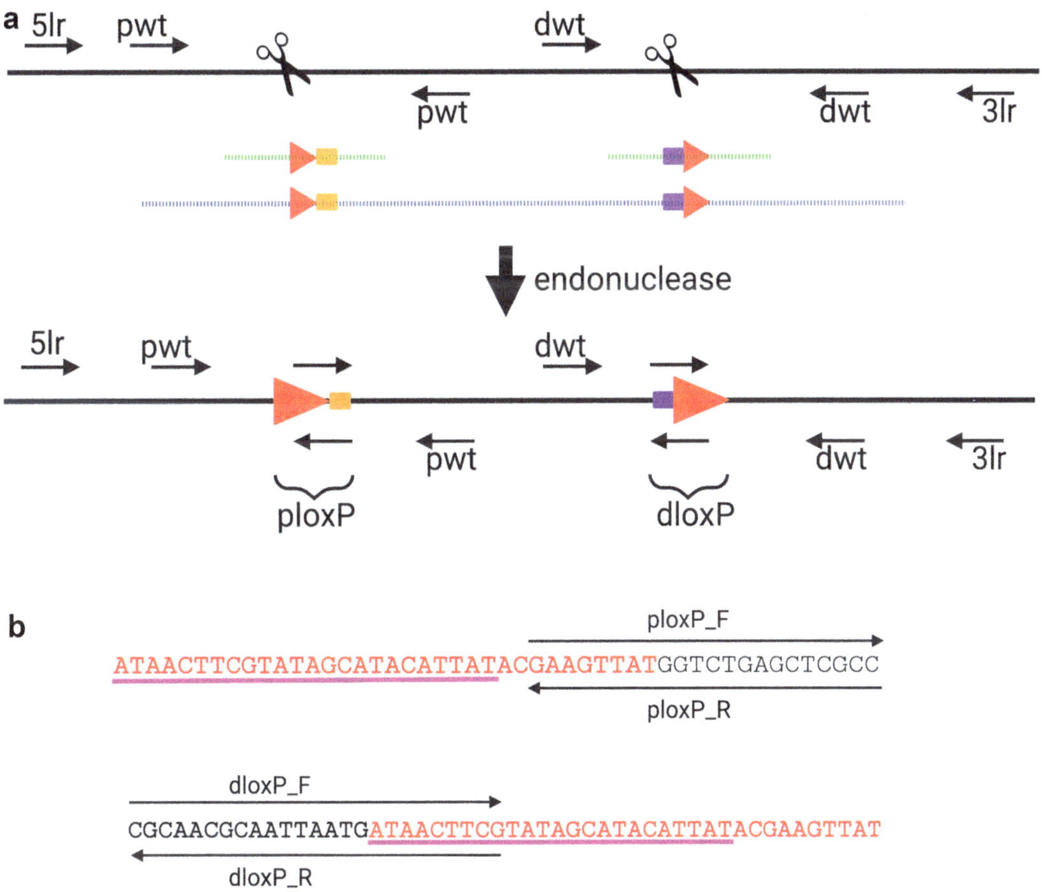

c

Assay	Primers: Fwd	Primers: Rev	Amplicons by allele type: wild-type	ploxP	dloxP	deletion	cis	sequence
pwt	pwt-F	pwt-R	wt	wt+	nr	none	nr	no
dwt	dwt-F	dwt-R	wt	nr	wt+	none	nr	no
ploxP	pwt-F	ploxP-R	none	ploxP	nr	none	nr	no
dloxP	dloxP-F	dwt-R	none	nr	dloxP	none	nr	no
del	pwt-F	dwt-R	wt	wt+	wt+	deletion	nr	no
lox-lox	ploxP-F	dloxP-R	none	nr	nr	none	lox-lox	no
5lr	pwt-F or 5lr	dloxp-R	none	nr	5lr	none	nr	ploxP & exon
3lr	ploxP-F	dwt-R or 3lr	none	3lr	nr	none	nr	dloxP & exon

Fig. 3 Endonuclease-mediated conditional allele genotyping schematic. (**a**) Primer locations before and after loxP insertions. The pwt and dwt primer pairs flank the proximal and distal loxP sites (red triangles), respectively. Short sequences for genotyping purposes are included adjacent to and/or overlapping the loxP sites (orange box represents sequence specific to the ploxP and the purple box represents sequence specific to the dloxP). The loxP-specific primers are used in conjunction with the wt primers to identify ploxP- and dloxP-containing alleles. If two short templates (green dotted lines) are used to insert the loxP sites, then the wt primers can flank the template sequence and successful amplification using a wt primer, and a loxP primer indicates that the loxP is inserted in the correct location. If a single long template (blue dotted line) is used and the wt primers are within the template homology arms, then the lr primers are needed to determine if the loxP sites are inserted in the correct location. Arrows pointing to the right are forward primers and pointing to the

(f) Design gene-specific assays with primers flanking each of the loxP insertion sites (Fig. 3a). These assays act as positive control assays for amplification of the target DNA and can facilitate interpretation of zygosity (see below). To limit the number of primers needed, reuse the gene-specific primers used for detecting the loxP sites.

The pwt (pwt-F to pwt-R) and dwt (dwt-F to dwt-R) assays flank the ploxP and dloxP inserts, respectively. If short (<201 bp) templates were used to insert the loxPs, design these primers to anneal with sequence that flanks that used in the templates. If a single long template was used, such that the pwt and dwt amplicons would be longer than 2 kb, the pwt and dwt primers may anneal within the template sequence to make amplification more reliable.

(g) If primers pwt-F and dwt-R are outside the template homology arms, then these primers are used in the 5 lr and 3 lr assays. However, if either is within their respective template, then design 5 lr and/or 3 lr primers that flank the template sequence. The 5 lr and 3 lr primers are used in conjunction with the dloxP-R and ploxP-F primers, respectively, to ensure the loxP sites were inserted in the desired location.

2. Identify founders using endpoint PCR. Founders are screened in two steps to reduce the number of assays performed by eliminating mice without the desired allele as soon as possible.

2.1. Identify founders with both loxP sites using the first five PCR assays in Fig. 3c.

(a) The ploxP and dloxP assays identify mice that have one or both loxP sites inserted.

Fig. 3 (continued) left are reverse primers. (**b**) Proximal and distal loxP sites and primers used at TCP. LoxP sites are in red, and added sequences to generate annealing sites for specific PCR primers in the repair templates are in black. Sequence targeted by the CN probe (TaqMan® sourced from Thermo Fisher Scientific) is underlined in pink. ploxP- and dloxP-specific primers are indicated by arrows above and below the sequence. (**c**) Assays for conditional alleles. For each assay, the wt amplicon is the size produced from the wild-type allele, and the wt+ size is the wild-type size plus the size of the loxP insert. The ploxP and dloxP assays produce an amplicon if their respective loxP sites were inserted. The del assay amplicon is produced if the DNA between the endonuclease cut sites is deleted. The lox-lox assay produces an amplicon if both loxP sites are inserted in *cis* (on the same homologue). nr indicates the assay is not relevant to that allele type. Mosaic animals, depending on the assay, can produce three or more (if INDELS were created) amplicons, while HET mice produce, at most, two amplicons per assay. For X-linked genes, mosaic male founders may produce three or more alleles, but HEM N1 males will have one allele. pwt, proximal wild type; dwt, distal wild type; ploxP, proximal loxP; dloxP, distal loxP; lr, long range; del, deletion; HET, heterozygous; INDEL, insertion and/or deletion; HEM, hemizygous

(b) The pwt and dwt assays are positive controls to ensure that the target DNA is amplifiable and that the resultant amplicons are the correct size with and/or without loxP sites inserted. If founders are mosaic or HET for loxP insertions, the pwt and dwt assays will produce two amplicons that differ by the size of the inserted loxP site (34 bp) and any universal primer sequence. Depending on amplicon size and fragment analysis methodology, this doublet may or may not be readily identifiable. Unexpected pwt and dwt amplicon sizes may indicate that INDELs were introduced at the cut sites.

(c) Finally, the del assay identifies mice with a deletion between the two endonuclease cut sites. The presence of the deletion allele in a founder may preclude the amplification of the pwt or dwt amplicons that represent the wild-type allele. The presence of only the pwt and dwt amplicons with loxP insertions should not be misinterpreted as HOM insertion if the del assay is positive.

(d) Mice that have both loxP sites are candidates for additional screening. Ideally, choose mice that produce the correct pwt and dwt amplicon sizes; however, as founders are mosaic, mice may still have correct loxP insertions even if the wt assays produce incorrectly sized amplicons.

2.2. Identify founders that have *cis*-located loxP sites using the lox-lox assay (Fig. 3c). If the pwt_F and dwt_R primers are outside the repair template homology arms, then founders that are lox-lox, ploxP, and dloxP positive likely have the loxP sites in *cis* and in the correct genomic location. If the pwt_F and dwt_R primers anneal within the template sequence, use the 5 lr and 3 lr assays to ensure the loxP sites are inserted at the correct location. At this point, founders with *cis*-located ploxP and dloxP pair inserted in the correct genomic location may be backcrossed to produce N1s. However, you may wish to sequence the loxP sites to ensure they were fully inserted [81, 121]. *See* Fig. 3c for which amplicons to sequence.

3. Select up to five founder mice with correctly located *cis*-inserted loxP sites for GLT test breeding (Subheading 3.2). Select HET founders for backcrossing to commercially purchased WT mice of the desired genetic background.

4. Screen N1 mice for GLT. If the lox-lox amplicon is shorter than 2 kb, screen N1 mice using the lox-lox assay. If the lox-lox assay is greater than 2 kb in length, screen with either the ploxP

or the dloxP assay. In both cases, include either the pwt or dwt assay as a positive control for DNA quality.

5. Perform genetic QC on loxP-positive N1 mice.
   (a) If only one loxP site was checked in the screening step, run the lox-lox assay to ensure both loxP sites are present.
   (b) Use the amplicons from the 5 lr and 3 lr assays to sequence the loxP sites, the region included in the repair template (s) and at least 100 bp of sequence flanking the ends of the homology arms for all templates used.
   (c) Sequence exons and ~100 bp of intron sequence flanking each exon in the 5 lr and 3 lr amplicons to ensure no unexpected on-target sequence changes (*see* **Note 1**). When long templates are used, full sequencing of the template segment, or restriction mapping, can provide additional assurance that no unexpected changes occurred during DNA repair.
   (d) Check that the repair template(s) was (were) not inserted elsewhere in the genome using a TCN assay with digital PCR (*see* **Note 2** for alternatives and **Note 3** for interpretation). For longer templates, multiple TCN assays may be used.
6. Mice pass QC if:
   (a) All sequences are as expected.
   (b) The loxP sites co-segregated from founder to N1.
   (c) The CN for each loxP template is equal to one.

   If the sequences are different than expected, the mouse line may still pass QC if the change(s) is not expected to alter function, such as when a change in a protein-coding region does not alter the amino acid sequence. However, it is prudent to perform molecular QC to ensure any differences do not have unexpected consequences on transcription, splicing, and/or translation (e.g., [90–92]).

   Mice with extra copies of the template sequence may be rescued if the insertions are unlinked to the floxed locus. To test if the extra copies are unlinked, backcross N1 mice with commercially purchased WT mice of the desired genetic background and identify N2 mice without extra copies of the repair template(s) using the TCN assay(s). Alternatively, screen and QC additional N1 mice to identify mice without extra copies of the repair template(s).
7. Genotype conditional mouse lines using one of the loxP assays and its corresponding wt assay (Fig. 3c). For alleles where loxP sites are not tightly linked (e.g., separated by >1 cM), use both loxP assays to confirm co-segregation. Use a wt assay to identify

HOM versus HET animals; in HET animals, the wt assay will produce both amplicons, while, in HOMs, only the larger wt amplicon will be produced. To reliably differentiate the wt doublet, wt assays may be redesigned after QC to generate shorter amplicons (i.e., with annealing sites within the template) to more readily distinguish size differences of ≥34 base pairs. Alternatively, TCN assays targeting the loxP sites could be used to assess zygosity with two copies indicating HOM and one copy indicating HET.

3.9 Endonuclease-Mediated Sequence Change Alleles

Sequence change alleles have one or a few co-located bases in the DNA changed without adding or deleting any DNA and are efficiently introduced using RNA-guided endonucleases and a short DNA template (≤200 bases). Screening founders and N1 mice for sequence change alleles using PCR and sequencing is advantageous since it provides the definitive allele sequence at the screening stage. Other PCR-based methods for screening such alleles include high-resolution melt (HRM) analysis, T7 endonuclease assays, and restriction digest polymorphisms. Each of these approaches requires sequencing to ensure that changes in melting temperature, endonuclease sensitivity, and/or restriction digest fragment size are associated only with the intended changes. The method below describes using PCR and sequencing to identify sequence change alleles generated with endonucleases and short DNA templates.

1. Design assays for screening founders and N1 mice for N1 QC.
   (a) The wt assay amplifies at least 300 bp of sequence flanking the site of the sequence change (Fig. 4a). The wt primer pair flanks the region included in the repair template and, for QC purposes, each primer should anneal at least 150 bp away from the sequence change.
   (b) Design a TCN assay. The primers and probe anneal with the template sequence, and the primers should be at least 10 bp internal to the ends of the template. The probe may recognize only the variant allele or may recognize only the wild-type allele.
2. Identify founders by sequencing the wt amplicon. Founders are often mosaic so inspect the sequence chromatogram to identify the appropriate double peaks at the expected mutation positions (Fig. 4b). Endonuclease editing will often also generate INDEL alleles so several different traces may be observed from a single mosaic founder, and deconvoluting these traces by eye is difficult. Tools such as ICE [111] and DECODR [112] can be helpful for trace deconvolution (Fig. 4c).
3. Choose up to five founder mice for GLT test breeding (Subheading 3.2). Select founders of both sexes that appear HET for the sequence change (i.e., have the wild-type and the

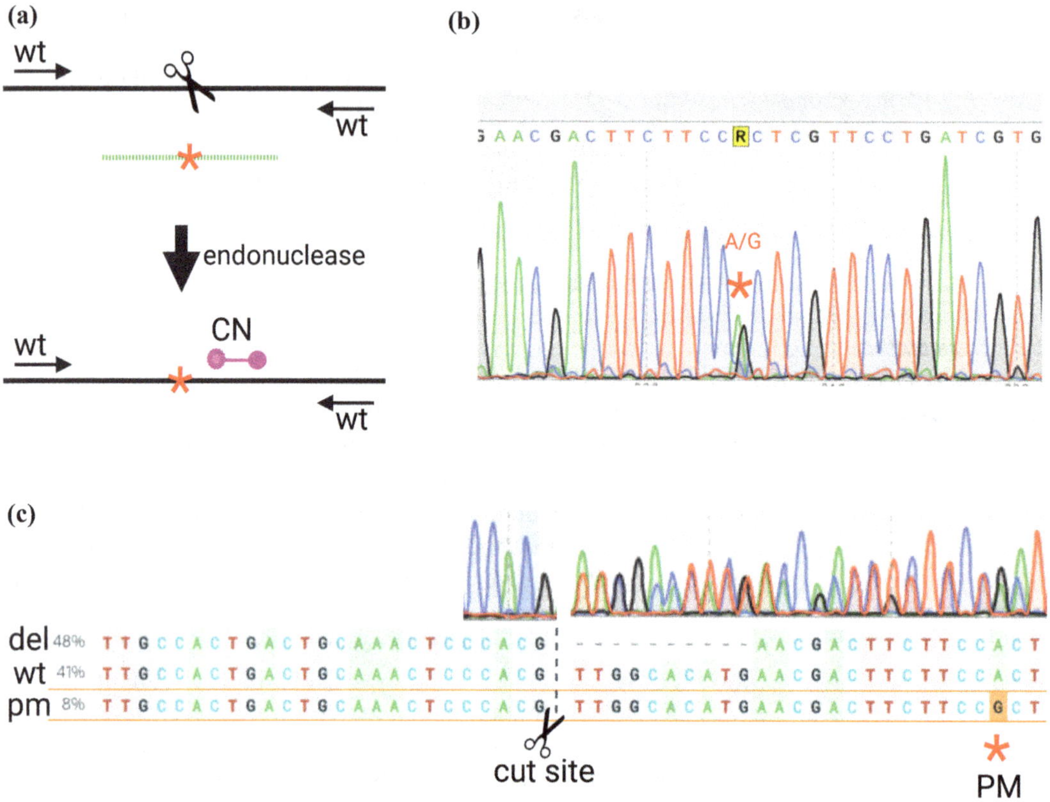

Fig. 4 Endonuclease-mediated sequence change allele genotyping schematic. (**a**) The wt PCR primers flank the template sequence ensuring the region amplified is in the desired genomic location. Black line is the target allele, red asterisk marks the sequence change location, and green dotted line is the repair template. The CN probe (in pink) anneals to a region included in the template but not to the sequence changes, i.e., it recognizes both the wild-type and sequence change alleles. Arrows pointing to the right are forward primers and pointing to the left are reverse primers. (**b**) A founder chromatogram showing the two peaks at the target site that correspond to the wild-type and changed bases. (**c**) ICE analysis output indicating that the founder with mixed traces has at least three different alleles including wild-type, a 10-bp INDEL and the desired sequence change (PM). wt, wild type; CN, copy number; del, deletion; INDEL, insertion and/or deletion; PM, point mutation

desired sequence change in the sequencing traces or, if using a deconvolution tool, those with the desired sequence at a frequency of 25–75%). If possible, avoid breeding compound HET mice (i.e., have INDELS along with the desired sequence change) as sequence change alleles are frequently designed to generate putative disease-associated alleles. INDELS may produce PTCs that can give rise to lethal null alleles in disease-associated alleles [122]. Thus, compound HET mice may not survive to produce litters. Similarly, depending on the expected allele-associated phenotype, mice HOM for the sequence change may be difficult to breed. However, HOM founders may be selected for GLT test breeding if the sequence change does not have negative effects on fertility, fecundity, viability, or breeding behavior.

4. Identify N1s harboring the base changes by sequencing the wt amplicons. Since N1s are HET, it is relatively easy to identify chromatogram double peaks indicative of heterozygous bases (Fig. 2b). Sequence beyond the ends of the template homology arms or at least 100 bp of sequence upstream and downstream of the target base changes, whichever is longer.
5. Genetic QC of N1 mice. Use the TCN assay to ensure that the template was not inserted elsewhere in the genome (*see* **Note 2** for alternatives and **Note 3** for interpretation).
6. Mice pass QC if:
   (a) The allele sequence matches wild type except for the introduced base changes.
   (b) No extra copies of the repair template are integrated into the genome.

   Mice with extra copies of the template sequence may be rescued if those insertions are unlinked to the targeted locus. To test if the extra copies are unlinked, backcross N1 mice with commercially purchased WT mice of the desired genetic background and identify N2 mice with the correct TCN. Alternatively, screen and QC additional N1 mice to identify N1 mice with the correct TCN.
7. Genotype fully QC'd mouse line. To simplify genotyping fully QC'd lines, establish a real-time or digital PCR allelic-discrimination assay with two probes, one specific for each allele and each with a different fluorochrome (*see* **Note 4**).

3.10 ES Cell-Derived Alleles

While most mouse alleles are now generated using endonuclease-mediated genome editing, ES cells are still a source of many alleles. For example, knockout and conditional allele mice may be generated using the 1000s of publicly available targeted ES cells [42, 59]. Additionally, complex alleles such as large knock-in alleles and conditional alleles may be generated using ES cell targeting. ES cell targeting vectors usually include a selection cassette, such as for G418 (neomycin) resistance, making nearly every ES cell-derived allele a variant of an insertion allele. Even if using an endonuclease to improve the efficiency of targeting [1], selection is still necessary to identify those ES cell clones that have incorporated the targeting cassette.

Genetic QC of mice generated from ES cells requires more steps relative to those generated by endonuclease editing, as the longer homology arms (usually a total of 10 kb in length) used for targeting make screening mice with primers that flank the arms impractical—thus, screening and target site validation are independent steps. Targeted ES cells are subjected to full QC (e.g., [14, 15,

35]) prior to generating chimeras. However, it is critical to fully QC HET N1 mice born from GLT test breeding of chimeras as there are many examples of multiple genotypes being transmitted from chimeras with fully QC'd ES [14, 15].

In their simplest form, deletion alleles engineered by ES cell targeting have a selection marker replacing the deleted genomic DNA (Fig. 1b). The strategy described here is for mice generated from ES cells with deletion alleles and applies to the deletion alleles (e.g., tm1) produced by the International Knockout Mouse Consortium [42, 123]. Alternate protocols are used for the International Knockout Mouse Consortium knockout-first alleles (e.g., tm1a) [42, 123] whose QC is described elsewhere [14].

Insertion or knock-in alleles are generated in ES cells using target vectors similar to those used to generate deletion alleles but with additional sequences for insertion along with the selection cassette. Some allele designs include a reporter gene, which is spliced to the upstream exons of the target gene so that the reporter is expressed under the control of the target gene's promoter. Others are designed to express variant alleles (e.g., point mutations), humanized sequences (e.g., replacing single exons up to entire genes and/or regulatory sequences with their human orthologous sequence), recombinase target sites (e.g., for conditional alleles), recombinases, or other functional products. Knock-in alleles may be targeted to specific genes or to safe harbor loci.

ES cell targeting allows the generation of very complex alleles limited only by one's imagination and diligence in generating the appropriate targeting vectors and ES cell clones. The strategies described here are for simple alleles with a single site of insertion and must be extended as appropriate for more complex alleles using the overarching principles of validating target location and integrity along with targeting vector integrity and CN. For example, many of the same QC assays described for endonuclease-mediated conditional alleles can be applied to ES cell-derived conditional alleles with allowances for the longer homology arms used for ES cell targeting vectors and the presence of a selection cassette:

1. Design assays to screen and QC N1 mice (Fig. 5).
   (a) The wt primers:

      For deletion alleles, one wt primer anneals within the intended deletion segment, and the other primer anneals outside of the deletion segment but within the vector homology arms. This assay will only amplify the wild-type allele.

      For insertion alleles, the wt assay primer pair flanks the insertion and produces a larger amplicon from the knock-in allele than from the wild-type allele.

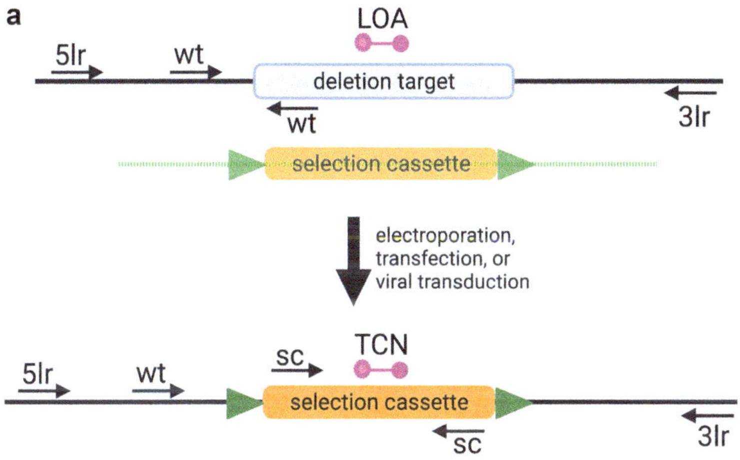

b

Assay	Primers		Amplicons by allele type		Sequence
	Fwd	Rev	wild-type	deletion	
wt	wt-F	wt-R	wt	none	no
wt-sc	wt-F	sc-R	none	wt-sc	no
5lr	5lr	sc-R	none	5lr-sc	in N1
3lr	sc-F	3lr	none	3lr-sc	in N1

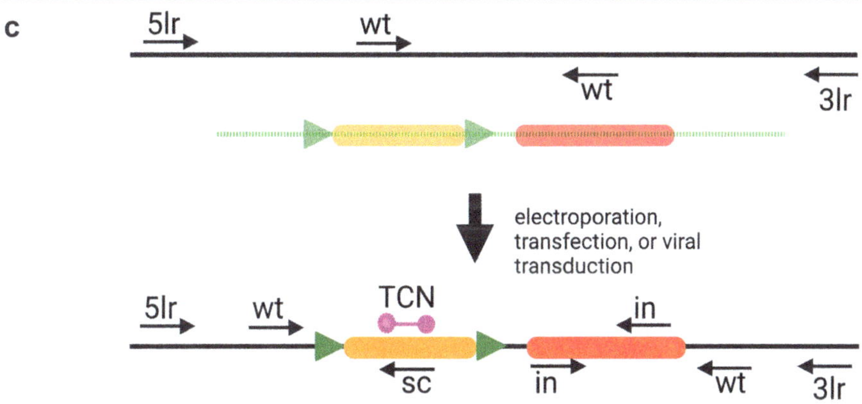

d

Assay	Primers		Amplicons		Sequence
	Fwd	Rev	wild-type	insertion	
wt	wt-F	wt-R	wt	wt + insertion*	no
wt-sc	wt-F	sc-R	none	wt-sc	insertion
in-wt	in-F	wt-R	none	in-wt	insertion
5lr	5lr	in-R	none	5 lr	no
3lr	in-F	3lr	none	3 lr	no

* often too large to efficiently amplify

Fig. 5 ES cell-generated allele genotyping schematic. (**a**) Deletion allele. The wt primers flank one side of the intended deletion junction, such that the wt assay only produces a product from the wild-type alleleas one wt primer anneals to the deleted (light blue bar) region. The lr primers flank the region included in the repair

(b) The selection-cassette (sc) primers anneal to the inserted selection cassette.

(c) The insertion (in) primer(s) anneal to the inserted sequence(s), distinct from the selection cassette, in knock-in alleles.

(d) The long-range (lr) primers flank the vector homology arms. These primers are used in conjunction with either the sc or in primers to check that the selection cassette or the insertion sequence was inserted in the correct genomic location.

(e) Quantitative assays:

For deletion alleles, the LOA assay targets within the deletion segment.

For insertion alleles, the LOA spans the insertion target site such that the assay will only detect the wild-type allele.

(f) For all allele types, design at least one TCN. For some vectors, multiple TCN assays along the length of the vector may be needed, particularly if they have multiple functional sequence segments (e.g., multiple selection cassettes), are very long, or have regions of homology with the target site within the desired insertion cassette.

Fig. 5 (continued) template that includes a selection cassette (yellow bar) flanked by homology arms (broken green line). The homology arms commonly total 10,000 bp. The sc primers anneal the selection cassette. The lr primers are used with sc primers to check that the selection cassette is inserted in the correct locus. The green triangles are recombinase sites for removal of the selection cassette. Arrows pointing to the right are forward primers and pointing to the left are reverse primers. (**b**) Assays for deletion mutations generated by ES cell targeting. For each assay, the wt amplicon is the size produced from the wild-type allele; the wt-sc, 5lr-sc, and 3lr-sc will only be produced if the selection cassette replaces the deleted wild-type sequence. In HET or chimeric animals, both the wild-type and deletion allele amplicons are produced. For X-linked genes, male chimeras may have both wild-type and deletion alleles, but HEM males will only have one allele. The sequence column specifies which amplicons should be sequenced as part of genetic QC. (**c**) Targeted insertions. The wt primers flank the insertion site within the homology arms. The in primers recognize the inserted DNA sequence (red bar) and the sc primers the selection cassette (yellow bar) needed for ES cell targeting. The 5 lr and 3 lr primers anneal to the target locus beyond the ends of the template homology arms. The TCN is a digital PCR assay that uniquely recognizes the insert sequence. The green triangles are recombinase sites for removal of the selection cassette. Arrows pointing to the right are forward primers and pointing to the left are reverse primers. (**d**) Assays for insertion mutations generated by ES cell targeting. The wt assay produces an amplicon from the wt allele and not from the insertion allele unless PCR conditions permit its amplification. In the latter case, the insertion allele produces a wt + insertion amplicon. The wt-sc and in-wt will be produced from all ES cells with the repair template incorporated into the ES cell genome, whether at the targeted locus or randomly. The 5 lr and 3 lr amplicons will only be amplified if the insertion is integrated at the targeted locus. wt, wild type; lr, long range; sc, selection cassette; HET, heterozygous; HEM, hemizygous; in, insertion; QC, quality control; TCN, template copy number

2. Identify high-contribution (>50%) chimeras for breeding. Most ES cells are male (XY); thus, fertile chimeras with high ES cell contribution are usually male. Some high-contribution chimeras are female, but these are usually derived from ES cells that have lost their Y chromosome during culture (X0) or were derived from female embryos that escaped sex conversion. The methods described below use somatic tissue observations or assays to estimate ES cell germline contribution. Sperm genotyping and subsequent IVF are only recommended if chimeras do not produce offspring when mated with WT females [4].

   2.1. If available, identify high-contribution chimeras using coat color markers. This can be done when the coat color specified by the ES cells (agouti for 129-derived lines or black for B6-derived lines, except for the JM8A lines which are agouti [124]) is different than that specified by the host embryos (black for B6 embryos or albino (white) for outbred CD-1, ICR, or Swiss Webster or B6-*Tyr*^{c} embryos). Male chimeras with >50% of their coat determined by the ES cell genome are used for breeding.

   2.2. If coat color markers are not available, determine which chimeras have the highest ES cell contribution using the LOA assay. Correct targeting on one homolog generates a HET ES cell line. If a chimera is 100% ES cell-derived, the LOA CN will be 1 for autosomal genes or 0 for X-linked genes in male chimeras. Select chimeras with CNs closest to that of a HET or HEM mouse, as these will have the highest contribution of ES cells. Use WT female mice as two-copy controls for LOA assays.

3. Select chimeras for GLT test breeding (Subheading 3.2). Choose three to five high-contribution chimeras for backcrossing to commercially purchased WT mice of the desired genetic background.

4. N1 screening and genetic QC.

   4.1 Use LOA assays to confirm either the deletion of the intended genomic segment or a correctly targeted insertion (*see* **Note 3** for interpretation).

   4.2 Use the 5 lr and 3 lr assays to verify that the selection and/or insertion sequence has inserted into the correct genomic locus (*see* **Note 1** for alternatives). Correctly targeted vectors, without any unexpected on-target sequence changes, will generate amplicons that are the correct size (*see* **step 4.4**).

   4.3 Use TCN assays to ensure the vector has integrated only once in the genome. Vectors may insert off target and also in tandem copies at the correct target site, though this is relatively rare in ES cell-based targeting (*see* **Note 2** for alternatives and **Note 3** for interpretation).

4.4 Sequence and/or restriction map the long-range amplicons to validate that the targeted allele has only the desired changes (*see* **Note 1** for alternatives). For amplicons >2 kb, sequencing exons and splice acceptor and donor regions (~150 bp of intronic sequence on either side of the exon) should be sufficient. Sequence any functional regions within insertions, such as splice acceptors or reporters. Restriction mapping confirms no large-scale rearrangements within the homology arms. All sequences and restriction fragments should match those expected.

4.5 Check for plasmid backbone integrations using PCR with primers specific to the plasmid backbone (e.g., targeting the origin of replication or bacterial antibiotic selection marker) to ensure the plasmid backbone did not insert into the genome. No amplicon should be detected in mice that pass QC. A positive control can be made by spiking the targeting vector into wild-type genomic DNA.

5. Mice fully pass QC if:
   (a) The LOA and TCN assays produce the correct CN.
   (b) The sequences match to what is expected, allowing for strain background polymorphisms.
   (c) Restriction mapping matches what is expected.
   (d) No plasmid backbone is detected.

   Mice that have extra copies of the targeting vector or integration of the plasmid backbone may be rescued if those insertions are unlinked to the targeted locus. To test if the extra copies are unlinked, backcross N1 mice with commercially purchased WT mice of the desired genetic background and identify N2 mice with the correct TCN. Alternatively, screen and QC additional N1 mice to identify N1 mice with the correct TCN.

   While the targeted allele sequence will ideally match the strain background's wild-type sequence except for the intended changes, some sequence variants may be acceptable if they are known polymorphisms or if the variant is not predicted to generate a functional change. Identify sequence variants in background strains by sequencing a WT littermate or by checking for annotated polymorphisms in Ensembl [125] or at the European Variant Archive (https://www.ebi.ac.uk/eva/). If a sequence change does not change the amino acid sequence and is not predicted to alter splicing or regulatory sequences, it may not alter function; however, such changes should be subjected to molecular QC to ensure there are no effects on expression.

6. Prior to selection cassette removal, fully QC'd mouse lines can be genotyped using the wt assay and the wt-sc assay.
7. If applicable, selection cassettes can be removed by crossing the newly generated mouse line with germline Cre- or Flp-deleter lines (e.g., [74]) depending on the recombinase target sites flanking the selection cassette. Alternative approaches include using transduction of cell-permeable Cre [126] or microinjection of Flp mRNA into preimplantation embryos. Endpoint PCR for the selection cassette confirms its removal. After selection cassette removal, design a new assay that flanks the deletion (del) in knockout mice to detect the deletion allele or a wt-in assay to detect the targeted allele in knock-in mice. Mice are then genotyped using the wt and del or wt-in assay, as appropriate. HET mice will produce both the wt and smaller del or wt-in amplicons. HOM mice will produce only the smaller del or wt-in amplicon, and WT mice will produce the wt amplicon and the larger del amplicon. The smaller del and wt-in amplicons will not be present.

3.11 Transgenes

Transgenes, usually comprised of a promoter and ORF, are randomly inserted into the mouse genome by ES cell electroporation or zygote microinjection. DNA fragment insertion often alters endogenous sequence at the integration site [9] beyond what is expected from a simple insertion. While the use of transposons or lentiviral vectors may avoid structural changes at the integration site, insertion can still disrupt genome functional elements. Thus, best practices include identifying and sequencing the insertion site to identify mutagenesis associated with transgene insertion. Position effect variegation occurs regardless of the transgene insertion method, and therefore several founder lines should be assessed both to identify the most suitable lines with regard to transgene expression and to confirm that phenotypes are consistent across lines:

1. Design assays for founder and N1 screening and for N1 QC (Fig. 6).
   (a) Design a transgene (tg)-specific PCR assay that targets the transgene and not the mouse genome. When a heterologous promoter is used to drive mouse gene expression, one primer anneals to the promoter region and one to the coding region. Alternatively, amplify across an intron-exon boundary such that the endogenous mouse gene will produce a larger amplicon than the transgene, or if the transgene is derived from other species, design primers to species-specific regions.
   (b) Design a TCN assay specific to the transgene.

a

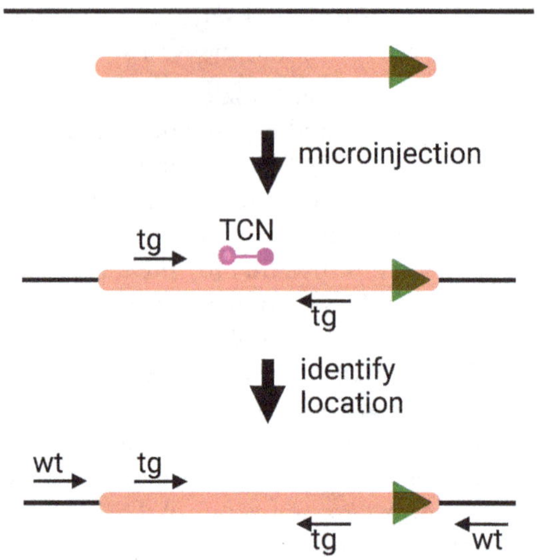

b

Assay	Primers Fwd	Primers Rev	Amplicons wild-type	Amplicons transgene	Sequence
tg	tg-F	tg-R	none	tg	in N1*
wt	wt-F	wt-R	wt	wt + tg**	no
wt-tg***	wt-F	tg-R	none	wt-tg	in N1*
tg-wt***	tg-F	wt-R	none	tg-wt	in N1*

* if the transgene was not sequenced during localization

** if PCR conditions allow

*** generally one of these is sufficient for genotyping purposes

Fig. 6 Genotyping schematic for transgenes. (**a**) Transgene allele. Transgenes (red bar) are randomly inserted into the genome, and founder animals are screened using tg primers that uniquely bind to the transgene sequence. A recombinase target site (green triangle) located near the end of the transgene enables recombinase-mediated CN reduction of transgene concatemers, if needed. TCN (probe shown in pink) is a digital PCR assay that uniquely recognizes the transgene sequence. The transgene is localized within the genome as part of N1 QC, and wt primers are designed that flank the transgene insertion site. (**b**) Assays for transgene alleles. The wt amplicon is only produced from the wild-type allele unless PCR conditions permit amplification across the inserted transgene. After QC is complete, wt-tg or tg-wt assays enable allele-specific genotyping of the transgene allele. tg, transgene; TCN, template copy number; wt, wild type; QC, quality control

2. Identify founders for GLT breeding using the tg assay. At this point, TCN assays may be used to estimate transgene CN; however, this assay is of limited use as it is possible that not all transgene copies will co-segregate in subsequent generations.
3. Select founder mice for GLT test breeding. Choose up to five mice for backcrossing to commercially purchased WT mice of the desired genetic background.

4. Screen N1 mice using the tg assay.
5. QC N1 mice.
   (a) Use the TCN assay to determine transgene CN by digital PCR.
   (b) Use targeted locus amplification (TLA) [34] to locate the insertion site of the transgene. Examine surrounding genomic sequence to determine the extent of genomic DNA changes and whether these changes and/or the site of insertion is predicted to disrupt functional regions. If transposons or lentiviruses were used to generate the transgenic mouse, PCR strategies with transposon-specific primers and random anchored primers (e.g., [127]) may be used to identify the insertion site(s).
6. Establish the transgenic lines and evaluate transgene expression to select the lines appropriate for study (*see* Subheading 3.13).
7. Genotyping established lines. For each transgenic line, design an allele-specific assay with one primer annealing to the transgene and the other the flanking genomic sequence (wt-tg or tg-wt, Fig. 5). This assay differentiates between different transgenic alleles, whereas internal transgene primers do not. Differentiating between different transgenes is essential as multiple founder lines must be evaluated to control for position effects.

3.12 Colony Maintenance and Cohort Production

Good experimental design requires close attention to colony maintenance and experimental cohort production strategies. Following best practices reduces the risk of off-target mutations, passenger mutations, or new naturally occurring mutations leading to inaccurate interpretations of genotype-phenotype relationships. Passenger mutations, such as CN variation, occur in virtually every ES cell line and clone [12] and will change with passage number (time in culture). The risk of off-target mutagenesis in endonuclease-mediated engineering depends on the target site specificity of the endonuclease used (e.g., [49]). Several algorithms (e.g., [128, 129]) are available to identify potential off-target sites for various endonucleases, but since these generally use the reference genome (from a single C57BL/6J mouse) for these predictions, they may not be comprehensive. The most comprehensive way to identify passenger or off-target mutations is WGS. However, WGS is not without caveats as a single individual does not represent the whole population (e.g., all N1 mice that pass QC), nor does it account for the introduction and accumulation of new mutations due to natural genetic processes as the line is bred and maintained. Further, since off-target or passenger mutations are most likely to occur in noncoding regions, their impact can be difficult to interpret. When establishing a line, additional backcrosses to commercially purchased mice will reduce the frequency of passenger or off-target mutations originating from the genetic engineering process. Each backcross will reduce the frequency of unlinked passenger or off-target mutations originating from the ES cell genome or founders by half, on average.

To establish a new genetically engineered line after genetic QC, use N1 mice with the exact same engineered sequence. Usually, at least one additional backcross is needed to expand the line before intercrossing to generate HOM mice for experimentation. Breeding is a significant source of genetic variation, with estimates of up to 25–30 new sequence variants (e.g., single nucleotide variants, indels) each generation [130]. Thus, closed colonies can quickly become distinct from the parental strain from which they were derived. Colony management practices that include backcrosses to commercially purchased mice to replenish the breeding stock or cryopreservation of foundation stock [131] will assist in maintaining the desired genetic background. The maintenance of the desired genetic background with backcrossing to commercially purchased mice is dependent on the genetic monitoring and QC performed by the supplier of the parental WT strain. Most commercial suppliers have publicly available information about their genetic monitoring program.

The variation introduced with breeding is random, and such random variation may affect the phenotypes being studied in as little as one generation; consequently, both experimental and control cohorts should be produced from the same breeding colony (i.e., sets of breeding pairs generating all genotypes required). This production method ensures that experimental and control animals share their genetic background, to the extent possible, and the only systematic difference(s) are those alleles being genotyped.

Other sources of variation include passenger mutations introduced during ES cell culture [12] and off-target mutations introduced by endonucleases [132, 133], although the latter appear to be rare during mouse genome editing [47–50]. Historically, to reduce the risk of passenger mutations confounding phenotype interpretations, genetically engineered lines were established from two independent ES cell clones. While this can control for independent passenger mutations acquired during targeting, these ES cell clones still share the passenger mutations acquired in the parental cell line used for targeting. For endonuclease-derived lines, independent lines could be obtained using endonucleases with different target sequences so that their sequence-dependent off-target mutagenesis profiles would be different. However, independent lines maintained as closed colonies will "drift" apart due to natural mutations, and genetic modifiers may accumulate in one line but not the other regardless of the production method or the initial passenger or off-target mutation profiles. Risks of confounding results associated with unlinked passenger or off-target mutations can be mitigated by obtaining both experimental and control animals from the same breeding colony (i.e., by producing both HOM and WT animals from HET x HET intercrosses or both HET or hemizygous (HEM) and WT animals from HET x WT and WT x HET backcrosses).

While endonuclease-mediated genome editing is usually done in the genetic background required for subsequent studies, there are limited strain backgrounds among ES cells used for genetic engineering. If the original engineered allele is in a background different from that needed for study, the strain background can be changed by producing a congenic line through backcrossing to WT mice of the desired background. Speed congenics [134–136] reduces the number of backcrosses and mice needed to generate congenic lines, consistent with the reduction and refinement principles of the 3R's. Commercial services to assist with speed congenics include miniMUGA [137] at Transnetyx and speed congenic services at the Jackson Laboratory and Charles River.

Once sufficient mice for a new line are available, cryopreservation is strongly recommended to prevent line loss due to disaster or genetic drift. Sperm cryopreservation requires only two or three males when the line has been produced in a congenic or isogenic background that is commercially available, since the oocyte donors for cryorecovery can be commercially sourced. If a mixed background or noncommercial background is used, then embryo cryopreservation from mating between males and females within the colony will safeguard the line.

3.13 Molecular QC

Even with careful allele design followed by genetic QC, the outcome of genetic manipulations must be functionally validated. Examples abound of unexpected outcomes (summarized in Table 1 and reviewed in [28]). Differentiating between unintended (i.e., phenotypes resulting from genomic changes rather than a consequence of the engineered gene's function) and intended (i.e., phenotypes resulting from functional changes to the engineered gene) outcomes is challenging but may open new areas of investigation. For example, observed phenotypes could be due to changes to regulatory sequences of a nontarget gene, rather than a direct consequence of alterations to the targeted gene. For protein-coding genes, protein expression is the gold standard for assessing gene expression; however, specific antibodies may not be available, and antibody-independent assays may not be available or accessible. Additionally, changes to noncoding transcripts and their potential functional roles should be considered. Long-read and short-read transcriptomic profiling can yield information about expression of the mutant allele and also may reveal unexpected changes in the expression of other genes. Alternatively, use quantitative PCR to assess transcriptional changes in one or a few specific genes, or, in some applications, endpoint reverse transcription PCR (RT-PCR) may be sufficient.

The minimum molecular QC required for knockout alleles is to confirm ablation of gene expression. For insertions, confirm that the insertion is expressed, and for sequence changes check that the altered sequence is expressed and associated transcripts are processed as expected. Confirm wild-type expression of conditional alleles prior to Cre excision and ablation of expression after Cre excision. Finally, check that transgenes are expressed. For protein-coding genes, use Western blots or immunostaining when appropriate antibodies are available. When proteins cannot be directly assayed, evaluate mRNA expression using endpoint PCR (not quantitative), real-time PCR, digital PCR, or long-read RNAseq (best due to assessment of splicing isoforms).

If antibodies are not available for a knockout allele (or post-Cre conditional alleles), and if NMD is expected [39, 138, 139], check that no mRNA is detectable in knockout HOMs. If the knockout transcript escapes NMD, a transcript missing the deleted region will be detected. HET mice will have the wild-type mRNA but should not have a message equivalent in size to the deletion transcript. Sequence any mutant transcripts to confirm that functional protein cannot be generated from those transcripts (i.e., there is a frameshift and a PTC).

Expression of genes close to the targeted locus can be affected by changes at the target site. This could be due to removal or disruption of critical promoter or enhancer sequences or due to regulatory feedback loops in which the targeted gene or transgene participates [16]. Expression of neighboring genes can be evaluated using real-time PCR, digital PCR, or RNAseq (for unbiased transcriptome assessment). Because the effects on neighboring gene expression can be unintended, such as deletion of critical regulatory sequences, or a functional consequence, such as reduced or increased expression of regulatory proteins, the line between molecular QC and phenotypic analysis may be blurred. Subsequent studies to differentiate between these possibilities may be required.

3.14 Conclusion

Genetically engineered mouse lines are powerful tools for assessing genotype-phenotype relationships and can serve as avatars to understand mammalian health and disease. Critical to their value is the affirmation that the genetic engineering produced the intended sequence changes, that colony management and cohort production strategies produce animals that faithfully represent the desired change(s) within the same genetic background, and that the genetic changes produced the intended molecular outcomes. While the line between molecular QC and phenotype may blur, the data are critical to interpretation of experimental outcomes and understanding mammalian biology. Investment in thorough QC up front will pay dividends throughout the study period.

4 Notes

1. Targeted long-read sequencing, such as Cas9-assisted nanopore sequencing [140], can be used rather than PCR amplicon sequencing.
2. WGS is an alternative approach to determine off-target insertion of the repair template; however, the additional costs and computational analyses needed may not be warranted if appropriate colony management and cohort production practices are used.
3. In N1 mice, the expected CN for TCN assays depends on what sequence the assay targets. If the assay targets template sequence that is common to both the wild-type and mutant alleles, a CN of 2 (1.8–2.2) is expected for autosomal alleles and sex-linked alleles in females, and a CN of 1 (0.8–1.2) is expected for sex-linked genes in hemizygous males. If the probe anneals to sequence unique to the template (e.g., anneals to an exogenous sequence or a base change allele), the CN for autosomal and sex-linked alleles should be 1 (0.8–1.2).
4. Alternate approaches to genotyping with allelic-discrimination assays include high-resolution melt (HRM) analysis, T7 endonuclease assays, and restriction fragment polymorphisms. The first two assays are not sequence specific, so there is a chance that natural mutations may not be detected within the colony or that natural mutations may result in misidentification of mice. Restriction fragment polymorphisms may be correlated directly with the desired sequence change, but more often these are intentionally introduced as "silent" changes in coding sequences adjacent to the targeted change. Again, this could result in natural mutations being undetected at the targeted change.

Acknowledgments

The authors would like to thank their colleagues at The Centre for Phenogenomics for their assistance with implementing the assays described in this manuscript. The authors received salary support funding from Genome Canada and Ontario Genomics OGI-137 and the Canada Foundation for Innovation Major Science Initiatives (MSI) Fund 35534. Figures were created with BioRender.com.

References

1. Gertsenstein M, Mianne J, Teboul L et al (2020) Targeted mutations in the mouse via embryonic stem cells. Methods Mol Biol 2066:59–82
2. Gertsenstein M, Nutter LM, Reid T et al (2010) Efficient generation of germ line transmitting chimeras from C57BL/6N ES cells by aggregation with outbred host embryos. PLoS One 5:e11260
3. Gertsenstein M, Nutter LMJ (2018) Engineering point mutant and epitope-tagged alleles in mice using Cas9 RNA-guided nuclease. Curr Protoc Mouse Biol 8:28–53
4. Gertsenstein M, Nutter LMJ (2021) Production of knockout mouse lines with Cas9. Methods 191:32–43
5. Quadros RM, Miura H, Harms DW et al (2017) Easi-CRISPR: a robust method for one-step generation of mice carrying conditional and insertion alleles using long ssDNA donors and CRISPR ribonucleoproteins. Genome Biol 18:92
6. Abe T, Inoue KI, Furuta Y et al (2017) Pronuclear microinjection during S-phase increases the efficiency of CRISPR-Cas9-assisted knockin of large DNA donors in mouse zygotes. Cell Rep 31:107653
7. Gu B, Gertsenstein M, Posfai E (2020) Generation of large fragment knock-in mouse models by microinjecting into 2-cell stage embryos. Methods Mol Biol 2066:89–100
8. Behringer RR, Gertsenstein M, Nagy K et al (2014) Manipulating the mouse embryo: a laboratory manual, 4th edn. Cold Spring Harbor Laboratory Press, Cold Spring Harbor, New York
9. Goodwin LO, Splinter E, Davis TL et al (2019) Large-scale discovery of mouse transgenic integration sites reveals frequent structural variation and insertional mutagenesis. Genome Res 29:494–505
10. Odeh H, Hagiwara N, Skynner M et al (2004) Characterization of two transgene insertional mutations at pirouette, a mouse deafness locus. Audiol Neurootol 9:303–914
11. Garrick D, Fiering S, Martin DI et al (1998) Repeat-induced gene silencing in mammals. Nat Genet 18:56–59
12. Liang Q, Conte N, Skarnes WC et al (2008) Extensive genomic copy number variation in embryonic stem cells. Proc Natl Acad Sci U S A 105:17453–17456
13. Skarnes WC, Rosen B, West AP et al (2011) A conditional knockout resource for the genome-wide study of mouse gene function. Nature 474:337–342
14. Ryder E, Gleeson D, Sethi D et al (2013) Molecular characterization of mutant mouse strains generated from the EUCOMM/KOMP-CSD ES cell resource. Mamm Genome 24:286–294
15. Birling MC, Yoshiki A, Adams DJ et al (2021) A resource of targeted mutant mouse lines for 5,061 genes. Nat Genet 53:416–449
16. West DB, Engelhard EK, Adkisson M et al (2016) Transcriptome analysis of targeted mouse mutations reveals the topography of local changes in gene expression. PLoS Genet 12:e1005691
17. Paix A, Folkmann A, Goldman DH et al (2017) Precision genome editing using synthesis-dependent repair of Cas9-induced DNA breaks. Proc Natl Acad Sci U S A 114: E10745–E10E54
18. Boroviak K, Fu B, Yang F et al (2017) Revealing hidden complexities of genomic rearrangements generated with Cas9. Sci Rep 7: 12867
19. Blayney J, Foster EM, Jagielowicz M et al (2020) Unexpectedly high levels of inverted re-insertions using paired sgRNAs for genomic deletions. Methods Protoc 3:53
20. Tuladhar R, Yeu Y, Tyler Piazza J et al (2019) CRISPR-Cas9-based mutagenesis frequently provokes on-target mRNA misregulation. Nat Commun 10:4056
21. Kim S, Kim D, Cho SW et al (2014) Highly efficient RNA-guided genome editing in human cells via delivery of purified Cas9 ribonucleoproteins. Genome Res 24:1012–1019
22. Doudna JA, Charpentier E (2014) Genome editing. The new frontier of genome engineering with CRISPR-Cas9. Science 346(6213):1258096
23. Jiang F, Doudna JA (2017) CRISPR-Cas9 structures and mechanisms. Annu Rev Biophys 46:505–529
24. Gaj T, Gersbach CA, Barbas CF 3rd (2013) Zfn, talen, and CRISPR/Cas-based methods for genome engineering. Trends Biotechnol 31:397–405
25. Chen YG, Ciecko AE, Khaja S et al (2020) UBASH3A deficiency accelerates type 1 diabetes development and enhances salivary gland inflammation in nod mice. Sci Rep 10:12019

26. Biagosch CA, Vidali S, Faerberboeck M et al (2021) A comprehensive phenotypic characterization of a whole-body Wdr45 knock-out mouse. Mamm Genome 35:332–349
27. Beverly SM (2001) Enzymatic amplification of RNA by PCR (RT-PCR). Curr Protoc Mol Biol Chapter 15:Unit 15.5
28. Lindner L, Cayrou P, Rosahl TW et al (2021) Droplet digital PCR or quantitative PCR for in-depth genomic and functional validation of genetically altered rodents. Methods 191: 107–119
29. Adams NC, Gale NW (2006) High resolution gene expression analysis in mice using genetically inserted reporter genes. In: Pease S, Lois C (eds) Mammalian and avian transgenesis — new approaches. Principles and practice. Springer Berlin Heidelberg, Berlin, Heidelberg, pp 131–172
30. Mead TJ, Apte SS (2020) Expression analysis by RNAScope™ in situ hybridization. In: Apte SS (ed) Adamts proteases: methods and protocols. Springer, New York, pp 173–178
31. Ni D, Xu P, Gallagher S (2016) Immunoblotting and immunodetection. Curr Protoc Mol Biol 114:8.10.1–8.10.36
32. Goldstein M, Watkins S (2008) Immunohistochemistry. Curr Protoc Mol Biol Chapter 14:Unit 14.6
33. de Vree PJ, de Wit E, Yilmaz M et al (2014) Targeted sequencing by proximity ligation for comprehensive variant detection and local haplotyping. Nat Biotechnol 32:1019–1025
34. Cain-Hom C, Splinter E, van Min M et al (2017) Efficient mapping of transgene integration sites and local structural changes in cre transgenic mice using targeted locus amplification. Nucleic Acids Res 45:e62
35. Codner GF, Erbs V, Loeffler J et al (2021) Universal Southern blot protocol with cold or radioactive probes for the validation of alleles obtained by homologous recombination. Methods 191:59–67
36. Kapahnke M, Banning A, Tikkanen R (2016) Random splicing of several exons caused by a single base change in the target exon of CRISPR/Cas9 mediated gene knockout. Cell 5:45
37. Lalonde S, Stone OA, Lessard S et al (2017) Frameshift indels introduced by genome editing can lead to in-frame exon skipping. PLoS One 12:e0178700
38. Smits AH, Ziebell F, Joberty G et al (2019) Biological plasticity rescues target activity in CRISPR knock outs. Nat Methods 16:1087–1093
39. Lindeboom RGH, Vermeulen M, Lehner B et al (2019) The impact of nonsense-mediated mRNA decay on genetic disease, gene editing and cancer immunotherapy. Nat Genet 51: 1645–1151
40. Ghosh A, Chakrabarti R, Shukla PC (2021) Inadvertent nucleotide sequence alterations during mutagenesis: highlighting the vulnerabilities in mouse transgenic technology. J Genet Eng Biotechnol 19:30
41. Supek F, Lehner B, Lindeboom RGH (2021) To nmd or not to nmd: nonsense-mediated mRNA decay in cancer and other genetic diseases. Trends Genet 37:657–668
42. Bradley A, Anastassiadis K, Ayadi A et al (2012) The mammalian gene function resource: the International Knockout Mouse Consortium. Mamm Genome 23:580–586
43. Wang H, Yang H, Shivalila CS et al (2013) One-step generation of mice carrying mutations in multiple genes by CRISPR/Cas-mediated genome engineering. Cell 153: 910–918
44. Modzelewski AJ, Chen S, Willis BJ et al (2018) Efficient mouse genome engineering by CRISPR-EZ technology. Nat Protoc 13: 1253–1274
45. Kim Y, Cheong SA, Lee JG et al (2016) Generation of knockout mice by Cpf1-mediated gene targeting. Nat Biotechnol 34:808–810
46. Hur JK, Kim K, Been KW et al (2016) Targeted mutagenesis in mice by electroporation of Cpf1 ribonucleoproteins. Nat Biotechnol 34:807–808
47. Iyer V, Shen B, Zhang W et al (2015) Off-target mutations are rare in Cas9-modified mice. Nat Methods 12:479
48. Willi M, Smith HE, Wang C et al (2018) Mutation frequency is not increased in CRISPR-Cas9-edited mice. Nat Methods 15:756–758
49. Anderson KR, Haeussler M, Watanabe C et al (2018) CRISPR off-target analysis in genetically engineered rats and mice. Nat Methods 15:512–514
50. Peterson KA, Khalouei S, Woodd JA et al (2021) Whole genome analysis for 163 guide RNAs in Cas9 edited mice reveals minimal off-target activity. bioRxiv 2021.08.11.455876
51. Yang H, Wang H, Shivalila CS et al (2013) One-step generation of mice carrying reporter and conditional alleles by CRISPR/Cas-mediated genome engineering. Cell 154: 1370–1379
52. Allen F, Crepaldi L, Alsinet C et al (2019) Predicting the mutations generated by repair

of Cas9-induced double-strand breaks. Nat Biotechnol 37:64–72

53. Hustedt N, Durocher D (2016) The control of DNA repair by the cell cycle. Nat Cell Biol 19:1–9
54. Elrick H, Peterson KA, Wood JA et al (2021) The production of 4,182 mouse lines identifies experimental and biological variables impacting Cas9-mediated mutant mouse line production. bioRxiv. 2021: 2021.10.06.463037
55. Eppig JT, Richardson JE, Kadin JA et al (2015) Mouse Genome Informatics (MGI): reflecting on 25 years. Mamm Genome 26: 272–284
56. Tsherniak A, Vazquez F, Montgomery PG et al (2017) Defining a cancer dependency map. Cell 170:564–576
57. Meyers RM, Bryan JG, McFarland JM et al (2017) Computational correction of copy number effect improves specificity of CRISPR-Cas9 essentiality screens in cancer cells. Nat Genet 49:1779–1784
58. Karczewski KJ, Francioli LC, Tiao G et al (2020) The mutational constraint spectrum quantified from variation in 141,456 humans. Nature 581:434–443
59. Koscielny G, Yaikhom G, Iyer V et al (2014) The International Mouse Phenotyping Consortium web portal, a unified point of access for knockout mice and related phenotyping data. Nucleic Acids Res 42(Database issue): D802–D809
60. Peterson KA, Murray SA (2021) Progress towards completing the mutant mouse null resource. Mamm Genome 33:123–134
61. Birling MC, Fray MD, Kasparek P et al (2021) Importing genetically altered animals: ensuring quality. Mamm Genome 33:100–107
62. Bochkov YA, Palmenberg AC (2006) Translational efficiency of EMCV IRES in bicistronic vectors is dependent upon IRES sequence and gene location. BioTechniques 41:283–284, 6, 8 passim
63. Kim JH, Lee SR, Li LH et al (2011) High cleavage efficiency of a 2a peptide derived from porcine teschovirus-1 in human cell lines, zebrafish and mice. PLoS One 6:e18556
64. Tang W, Ehrlich I, Wolff SB et al (2009) Faithful expression of multiple proteins via 2a-peptide self-processing: a versatile and reliable method for manipulating brain circuits. J Neurosci 29:8621–8629
65. Trichas G, Begbie J, Srinivas S (2008) Use of the viral 2a peptide for bicistronic expression in transgenic mice. BMC Biol 6:40
66. Chichili VPR, Kumar V, Sivaraman J (2013) Linkers in the structural biology of protein–protein interactions. Protein Sci 11:153–156
67. Klein JS, Jiang S, Galimidi RP et al (2014) Design and characterization of structured protein linkers with differing flexibilities. Protein Eng Des Sel 27:325–330
68. Soriano P (1999) Generalized lacZ expression with the ROSA26 Cre reporter strain. Nat Genet 21:70–71
69. Soriano P, Friedrich G, Lawinger P (1991) Promoter interactions in retrovirus vectors introduced into fibroblasts and embryonic stem cells. J Virol 65:2314–2319
70. Zambrowicz BP, Imamoto A, Fiering S et al (1997) Disruption of overlapping transcripts in the ROSA beta geo 26 gene trap strain leads to widespread expression of beta-galactosidase in mouse embryos and hematopoietic cells. Proc Natl Acad Sci U S A 94: 3789–3794
71. Tasic B, Hippenmeyer S, Wang C et al (2011) Site-specific integrase-mediated transgenesis in mice via pronuclear injection. Proc Natl Acad Sci U S A 108:7902–7907
72. Zeng H, Horie K, Madisen L et al (2008) An inducible and reversible mouse genetic rescue system. PLoS Genet 4:e1000069
73. Nagy A (2000) Cre recombinase: the universal reagent for genome tailoring. Genesis 26: 99–109
74. Birling MC, Dierich A, Jacquot S et al (2012) Highly-efficient, fluorescent, locus directed Cre and Flpo deleter mice on a pure C57BL/6N genetic background. Genesis 50:482–489
75. Wu Y, Wang C, Sun H et al (2009) High-efficient Flpo deleter mice in C57BL/6J background. PLoS One 4:e8054
76. Anastassiadis K, Fu J, Patsch C et al (2009) Dre recombinase, like Cre, is a highly efficient site-specific recombinase in E. coli, mammalian cells and mice. Dis Model Mech 2:508–515
77. Han X, Zhang Z, He L et al (2021) A suite of new Dre recombinase drivers markedly expands the ability to perform intersectional genetic targeting. Cell Stem Cell 28:1160–76 e7
78. Woods JP, Srinivasan RS, Olson L (2021) Sweet Dre-ams about intersectional genetics. Cell Stem Cell 28:989–990
79. Gurumurthy CB, O'Brien AR, Quadros R et al (2019) Reproducibility of CRISPR-Cas9 methods for generation of conditional mouse alleles: a multi-center evaluation. Genome Biol 20:171

80. Shang R, Zhang H, Bi P (2021) Generation of mouse conditional knockout alleles in one step using the i-GONAD method. Genome Res 31:121–130
81. Lanza DG, Gaspero A, Lorenzo I et al (2018) Comparative analysis of single-stranded DNA donors to generate conditional null mouse alleles. BMC Biol 16:69
82. Mandalos N, Saridaki M, Harper JL et al (2012) Application of a novel strategy of engineering conditional alleles to a single exon gene, Sox2. PLoS One 7:e45768
83. Guzzardo PM, Rashkova C, Dos Santos RL et al (2017) A small cassette enables conditional gene inactivation by CRISPR/Cas9. Sci Rep 7:16770
84. Gilet JG, Ivanova EL, Trofimova D et al (2020) Conditional switching of Kif2a mutation provides new insights into cortical malformation pathogeny. Hum Mol Genet 29:766–784
85. Capulli M, Costantini R, Sonntag S et al (2019) Testing the Cre-mediated genetic switch for the generation of conditional knock-in mice. PLoS One 14:e0213660
86. Heffner CS, Herbert Pratt C, Babiuk RP et al (2012) Supporting conditional mouse mutagenesis with a comprehensive Cre characterization resource. Nat Commun 3:1218
87. Feil R, Brocard J, Mascrez B et al (1996) Ligand-activated site-specific recombination in mice. Proc Natl Acad Sci U S A 93:10887–10890
88. Backman CM, Zhang Y, Malik N et al (2008) Generalized tetracycline induced Cre recombinase expression through the ROSA26 locus of recombinant mice. J Neurosci Methods 176:16–23
89. Maani N, Sabha N, Rezai K et al (2018) Tamoxifen therapy in a murine model of myotubular myopathy. Nat Commun 9:4849
90. Nielsen KB, Sorensen S, Cartegni L et al (2007) Seemingly neutral polymorphic variants may confer immunity to splicing-inactivating mutations: a synonymous SNP in exon 5 of MCAD protects from deleterious mutations in a flanking exonic splicing enhancer. Am J Hum Genet 80:416–342
91. Biro JC (2008) Correlation between nucleotide composition and folding energy of coding sequences with special attention to wobble bases. Theor Biol Med Model 5:14
92. Doktor TK, Schroder LD, Andersen HS et al (2014) Absence of an intron splicing silencer in porcine Smn1 intron 7 confers immunity to the exon skipping mutation in human Smn2. PLoS One 9:e98841
93. Suzuki M, Kasai K, Saeki Y (2006) Plasmid DNA sequences present in conventional herpes simplex virus amplicon vectors cause rapid transgene silencing by forming inactive chromatin. J Virol 80:3293–3300
94. Chen ZY, He CY, Meuse L et al (2004) Silencing of episomal transgene expression by plasmid bacterial DNA elements in vivo. Gene Ther 11:856–864
95. Lu J, Zhang F, Fire AZ et al (2017) Sequence-modified antibiotic resistance genes provide sustained plasmid-mediated transgene expression in mammals. Mol Ther 25:1187–1198
96. Schaft J, Ashery-Padan R, van der Hoeven F et al (2001) Efficient Flp recombination in mouse ES cells and oocytes. Genesis 31:6–10
97. Kvon EZ, Zhu Y, Kelman G et al (2020) Comprehensive in vivo interrogation reveals phenotypic impact of human enhancer variants. Cell 180:1262–1271
98. Chew SK, Rad R, Futreal PA et al (2011) Genetic screens using the piggyBac transposon. Methods 53:366–721
99. Carlson DF, Geurts AM, Garbe JR et al (2010) Efficient mammalian germline transgenesis by cis-enhanced Sleeping Beauty transposition. Transgenic Res 20:29–45
100. Ikawa M, Tanaka N, Kao WW et al (2003) Generation of transgenic mice using lentiviral vectors: a novel preclinical assessment of lentiviral vectors for gene therapy. Mol Ther 8:666–673
101. Russell WMS, Burch R (1959) The principles of humane experimental technique. Methuen, London
102. Dickinson ME, Flenniken AM, Ji X et al (2016) High-throughput discovery of novel developmental phenotypes. Nature 537:508–554
103. Karp NA, Mason J, Beaudet AL et al (2017) Prevalence of sexual dimorphism in mammalian phenotypic traits. Nat Commun 8:15475
104. Truett GE, Heeger P, Mynatt RL et al (2000) Preparation of PCR-quality mouse genomic DNA with hot sodium hydroxide and tris (hotshot). BioTechniques 29(52):54
105. Green MR, Sambrook J (2012) Molecular cloning: a laboratory manual. Cold Spring Harbor Laboratory Press, Cold Spring Harbor, New York
106. Untergasser A, Cutcutache I, Koressaar T et al (2012) Primer3--new capabilities and interfaces. Nucleic Acids Res 40:e115
107. Untergasser A, Nijveen H, Rao X et al (2007) Primer3plus, an enhanced web interface to primer3. Nucleic Acids Res 35(Web Server issue):W71–W74

108. Ye J, Coulouris G, Zaretskaya I et al (2012) Primer-blast: a tool to design target-specific primers for polymerase chain reaction. BMC Bioinformatics 13:134
109. D'Haene B, Vandesompele J, Hellemans J (2010) Accurate and objective copy number profiling using real-time quantitative PCR. Methods 50:262–270
110. Kanagal-Shamanna R (2016) Digital PCR: principles and applications. Methods Mol Biol 1392:43–50
111. Conant D, Hsiau T, Rossi N et al (2012) Inference of CRISPR edits from Sanger trace data. CRISPR J 5:123–130
112. Bloh K, Kanchana R, Bialk P et al (2021) Deconvolution of complex DNA repair (DECODR): establishing a novel deconvolution algorithm for comprehensive analysis of CRISPR-edited Sanger sequencing data. CRISPR J 4:120–131
113. Vaux DL (1992) Rapid recovery of DNA from agarose gels. Trends Genet 8:81
114. Boyle JS, Lew AM (1995) An inexpensive alternative to glassmilk for DNA purification. Trends Genet 11:8
115. Bortesi L, Zhu C, Zischewski J et al (2016) Patterns of CRISPR/Cas9 activity in plants, animals and microbes. Plant Biotechnol J 14: 2203–2216
116. Kosicki M, Tomberg K, Bradley A (2018) Repair of double-strand breaks induced by CRISPR-Cas9 leads to large deletions and complex rearrangements. Nat Biotechnol 36: 765–771
117. Mianne J, Codner GF, Caulder A et al (2017) Analysing the outcome of CRISPR-aided genome editing in embryos: screening, genotyping and quality control. Methods 15:121–122
118. Gu B, Posfai E, Gertsenstein M et al (2020) Efficient generation of large-fragment knock-in mouse models using 2-cell (2c)-homologous recombination (HR)-CRISPR. Curr Protoc Mouse Biol 10:e67
119. Mizuno N, Mizutani E, Sato H et al (2018) Intra-embryo gene cassette knockin by CRISPR/Cas9-mediated genome editing with adeno-associated viral vector. iScience 9:286–297
120. Yoon Y, Wang D, Tai PWL et al (2018) Streamlined ex vivo and in vivo genome editing in mouse embryos using recombinant adeno-associated viruses. Nat Commun 9: 412
121. Codner GF, Mianne J, Caulder A et al (2018) Application of long single-stranded DNA donors in genome editing: generation and validation of mouse mutants. BMC Biol 16: 70
122. Cacheiro P, Munoz-Fuentes V, Murray SA et al (2020) Human and mouse essentiality screens as a resource for disease gene discovery. Nat Commun 11:655
123. Austin CP, Battey JF, Bradley A et al (2004) The knockout mouse project. Nat Genet 36: 921–924
124. Pettitt SJ, Liang Q, Rairdan XY et al (2009) Agouti C57BL/6N embryonic stem cells for mouse genetic resources. Nat Methods 6: 493–495
125. Lilue J, Doran AG, Fiddes IT et al (2018) Sixteen diverse laboratory mouse reference genomes define strain-specific haplotypes and novel functional loci. Nat Genet 50: 1574–1583
126. Kim K, Kim H, Lee D (2009) Site-specific modification of genome with cell-permeable Cre fusion protein in preimplantation mouse embryo. Biochem Biophys Res Commun 388:122–126
127. Ruf S, Symmons O, Uslu VV et al (2011) Large-scale analysis of the regulatory architecture of the mouse genome with a transposon-associated sensor. Nat Genet 43:379–386
128. Bae S, Park J, Kim JS (2014) Cas-OFFinder: a fast and versatile algorithm that searches for potential off-target sites of Cas9 RNA-guided endonucleases. Bioinformatics 30:1473–1475
129. Listgarten J, Weinstein M, Kleinstiver BP et al (2018) Prediction of off-target activities for the end-to-end design of CRISPR guide RNAs. Nat Biomed Eng 2:38–47
130. Uchimura A, Higuchi M, Minakuchi Y et al (2015) Germline mutation rates and the long-term phenotypic effects of mutation accumulation in wild-type laboratory mice and mutator mice. Genome Res 25:1125–1134
131. Taft RA, Davisson M, Wiles MV (2006) Know thy mouse. Trends Genet 22:649–653
132. Hsu PD, Scott DA, Weinstein JA et al (2013) DNA targeting specificity of RNA-guided Cas9 nucleases. Nat Biotechnol 31:827–832
133. Zhang XH, Tee LY, Wang XG et al (2015) Off-target effects in CRISPR/Cas9-mediated genome engineering. Mol Ther Nucleic Acids 4:e264
134. Andrews KR, Hunter SS, Torrevillas BK et al (2021) A new mouse SNP genotyping assay for speed congenics: combining flexibility, affordability, and power. BMC Genomics 22: 378

135. Visscher PM (1999) Speed congenics: accelerated genome recovery using genetic markers. Genet Res 74:81–85
136. Wong GT (2002) Speed congenics: applications for transgenic and knock-out mouse strains. Neuropeptides 36:230–236
137. Sigmon JS, Blanchard MW, Baric RS et al (2020) Content and performance of the mini-MUGA genotyping array: a new tool to improve rigor and reproducibility in mouse research. Genetics 216:905–930
138. Popp MW, Maquat LE (2014) The dharma of nonsense-mediated mRNA decay in mammalian cells. Mol Cells 37:1–8
139. Hoek TA, Khuperkar D, Lindeboom RGH et al (2019) Single-molecule imaging uncovers rules governing nonsense-mediated mRNA decay. Mol Cell 75:324–339
140. Gilpatrick T, Lee I, Graham JE et al (2020) Targeted nanopore sequencing with Cas9-guided adapter ligation. Nat Biotechnol 38: 433–438

Chapter 4

Genotyping Genome-Edited Founders and Subsequent Generation

Matthew Mackenzie, Alex Fower, Alasdair J. Allan, Gemma F. Codner, Rosie K. Bunton-Stasyshyn, and Lydia Teboul

Abstract

Targeted nucleases allow the production of many types of genetic mutations directly in the early embryo. However, the outcome of their activity is a repair event of unpredictable nature, and the founder animals that are produced are generally of a mosaic nature. Here, we present the molecular assays and genotyping strategies that will support the screening of the first generation for potential founders and the validation of positive animals in the subsequent generation, according to the type of mutation generated.

Key words Genotyping, Allele QC, CRISPR-Cas9, Genome editing, Embryo/zygote engineering, Genetic engineering, Mouse models, Transgenesis

1 Introduction

Genome editing applies targeted nucleases, such as CRISPR-Cas9, to increase the efficiency of targeted genome modifications compared to classical homologous recombination methods. As a result, it is now possible to carry out gene editing in vivo, and techniques to modify genes directly in the early embryo have become widely used. However, this shift in methodology now means that the burden of genotyping and allele validation is placed in live animals, as opposed to in cultured embryonic stem cells (ESCs). Furthermore, directly engineering in vivo has removed the need for exogenous selection markers, which facilitate genotyping of engineered ESCs and the mice derived from them.

The repertoire of achievable alleles has also expanded and may now range from small and subtle modifications such as single-nucleotide variants (SNVs) and small insertions and deletions (indels) to larger, multi-kilobase-scale deletions, insertions, or replacements, thus requiring new genotyping methods. It is also important to consider that the outcome of any gene-editing

Thomas L. Saunders (ed.), *Transgenesis: Methods and Protocols*,
Methods in Molecular Biology, vol. 2631, https://doi.org/10.1007/978-1-0716-2990-1_4,

experiment will probably include many unintended variants, and the genotyping methods must be capable of not only identifying the desired allele but also detecting unintended mutations which should be excluded [1]. As we shall describe in this chapter, the strategies required to genotype and validate alleles vary according to the category of allele. The use of nucleases also introduces the opportunity for off-target activity: nuclease activity at loci with a similar sequence to that of the intended target, which can result in indels and other modifications. These potential off-target effect loci are, at least somewhat, predictable and should be screened for unwanted modification in all cases, regardless of the category of intended allele.

One of the most profound differences in the screening and validation of gene-edited alleles, compared to alleles generated by classical engineering in ESCs, is the pattern of cellular and genetic makeup of the first two generations of mice: G0 mice born from gene editing are commonly mosaic [2–4], meaning that they are composed of two or more cell populations/lineages each derived from independent repair events (and/or the original unedited cells). These different populations may be present in differing proportions across the tissues (Fig. 1). Mosaicism is crucially different from the chimeric G0 mice created via traditional genome engineering in ESCs; every chimeric pup in a litter is composed of a combination of the same two cell types, those of the ESC and the host blastocyst, each having a known and predefined genotype. This difference is further highlighted in the G1 progeny when a G0 founder has been backcrossed to a wild-type (WT) mouse. Common to both mosaic gene-edited and chimeric gene-engineered G0 founders, the presence of the desired allele at G0 does not guarantee germ line transmission. In ESC-based genome engineering, assessing G1 offspring for germ line transmission of the targeted allele involves identifying only two potential, predefined, genotypes. By contrast, a mosaic gene-edited G0 founder may transmit multiple allelic variants, resulting in pups from the same litter having different mutant genotypes (Fig. 1c), including, in some instances, alleles that were not detected in the biopsy used for the G0 genotyping.

These differences and complexities demand novel approaches for genotyping and require careful record keeping of pedigrees to track the segregation of alleles across the first two generations and separate the distinct lines founded at G1. The aim of G0 analysis is to perform a screen that identifies potential founders carrying the desired allele [4]. It is only once individual alleles are segregated at the G1 stage that a molecular validation can be completed (Fig. 1). At that stage, full characterization of animals should include sequencing of the whole region of interest, copy-counting of mobilized sequences (donor sequence and deleted interval if relevant), and sequencing of likely candidates for off-target activity.

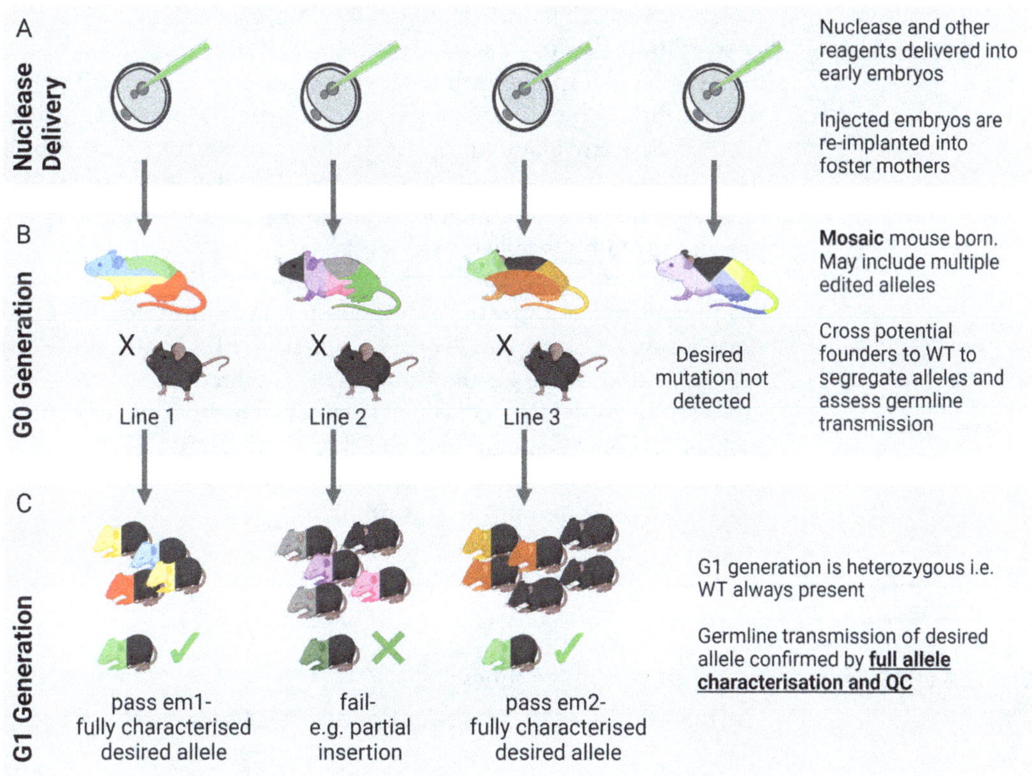

Fig. 1 Nuclease gene-editing pipeline in mice. An overview of the genetic characterization process for nuclease gene-edited mice is shown. The process is applicable to editing in other animal species. (**a**) Nuclease gene-editing reagents are delivered to one- or two-cell embryos by microinjection or electroporation. Embryos are reimplanted into a pseudopregnant dam. (**b**) The first generation of mice born from gene-edited embryos (G0) is generally mosaic. DNA lesion or nuclease activity persisting past the single-cell stage may cause multiple editing events resulting in multiple differently edited cell lineages, as indicated by different colors. These different lineages may not be represented evenly across all tissues, and so tissue biopsies taken for genotyping may not be representative of the lineages present in the germ line. G0 animals are screened to identify potential founders with evidence of the desired allele (indicated in green). These positive animals are then crossed to a WT animal in order to segregate and assess individual edited alleles transmitted through the germ line. (**c**) At the G1 generation, individual edited alleles segregate, with G1 animals having inherited one allele from their WT parent and one from their mosaic parent. Initial screening identifies animals with evidence of the desired allele (indicated in green). Animals with only a partial edit or other undesired outcome can be identified and excluded (dark green), and those with correct edits are fully characterized (bright green). Because animals from different G0 founders are created via independent nuclease-editing events, they should be used to establish independent endonuclease-mediated (em) edited colonies. (Figure created with BioRender.com)

Strategies for G0 founder screening and allele validation in the subsequent generation depend on the type of genetic modification to be produced. In this chapter, we will describe four genotyping strategies that are applicable depending on the type of mutation to be introduced: (1) indels and point mutations, (2) deletions, (3) large knock-ins (KI), and (4) floxed alleles. In all cases, it is recommended that genotyping assays are planned as part of the gene-editing design process. In this way, custom-primer binding sites can be incorporated in some instances, which can be particularly useful when the aim is to KI hard-to-prime sequences, such as LoxP sites. Early planning of the genotyping strategy also ensures time for optimization of PCR reactions. The analysis of off-target nuclease activity is described in Subheading 3.2.5 Off-Target Analysis and should be applied in all cases.

The following protocols were developed for application to ear biopsies from mice created via gene editing using CRISPR-Cas9 nuclease in embryos (*see* **Note 1**). These methods are, however, applicable to other tissue biopsies and other edited organisms. They can also be applied to genomes edited using different nucleases, such as TALENS, zinc fingers, or other Cas variants [5], including mutants made with single-stranded DNA "nickases" used in methods such as base and prime editing [6, 7].

2 Materials

2.1 DNA Extraction

1. Microcentrifuge tubes.
2. DNA Extract All Reagents Kit (Applied BiosystemsTM).

2.2 PCR and Sequencing

1. 96-well PCR plates and compatible lids or other suitable PCR tubes.
2. Tissue biopsy.
3. Desalted oligonucleotide primers.
4. Platinum™ SuperFi II PCR kit (Thermo Fisher Scientific) or equivalent.
5. UltraPureTM agarose (InvitrogenTM) or equivalent.
6. 1% w/v ethidium bromide.
7. 1× TAE buffer: 40 mM Tris, 20 mM acetic acid, 1 mM EDTA.
8. 6× gel loading dye (New England Biolabs) or equivalent.
9. 1 kb or 100 bp DNA ladders (New England Biolabs) or equivalent.
10. QIAquick® gel extraction kit (Qiagen) or equivalent.

2.3 Droplet Digital PCR (ddPCR)

1. FAM-labeled TaqMan assays against experimental target sequence and VIC-labeled TaqMan assay against a reference genomic sequence as calibrator as described below, (IDT) or equivalent.
2. PrimeTime qPCR primers (IDT).
3. ddPCR supermix for probes (no dUTP) (Bio-Rad).
4. DEPC-treated water.

5. ddPCR DG8™ cartridges, cartridge gaskets (Bio-Rad).
6. Twin-tec PCR plates 96, semi-skirted (Eppendorf®) or equivalent.
7. PCR Plate Heat Seal, foil, pierceable, for use with PX1 PCR plate sealer (Bio-Rad).
8. Droplet Generation Oil for Probes (Bio-Rad).
9. BioClean Ultra LTS 50 μl filtered pipette tips (Rainin).

2.4 Equipment

1. Plate shaker.
2. PCR thermocycler.
3. Agarose gel electrophoresis tanks.
4. Gel Doc XR+ Imager (Bio-Rad) or equivalent UV gel imaging equipment.
5. NanoDrop 1000 Spectrophotometer (Thermo Fisher Scientific) or equivalent DNA spectrophotometer.
6. Electronic E4 Multi-pipette XLS 5–50 μl LTS (Rainin).
7. QX200 droplet generator (Bio-Rad).
8. PX1 PCR plate sealer (Bio-Rad).
9. QX200 droplet reader (Bio-Rad).

2.5 Software and Online Tools

1. Primer sequences can be selected using freely available online tools such as:
   (a) NCBI Primer-Blast (https://www.ncbi.nlm.nih.gov/tools/primer-blast/).
   (b) Primer3Plus (http://primer3plus.com/).
   (c) Primer3 (https://primer3.org/).
2. Sanger sequencing chromatograms can be visualized using freely available software such as:
   (a) FinchTV (https://digitalworldbiology.com/FinchTV).
   (b) Chromatogram Explorer Lite (http://www.dnabaser.com/download/chromatogram-explorer/).
   (c) Benchling (https://www.benchling.com/).
3. DNA sequencing data can be aligned using web tools or software such as:
   (a) Gap4 (Staden Package) (https://sourceforge.net/projects/staden/files/).
   (b) SeqMan Pro (part of DNASTAR Lasergene package) (https://www.dnastar.com/software/lasergene/).
   (c) Benchling (https://www.benchling.com/).

4. DNA sequence documentation and visualization can be carried out using software such as:
   (a) SeqBuilder Pro (part of DNASTAR Lasergene package) (https://www.dnastar.com/software/lasergene/seqbuilder-pro/).
   (b) Benchling (https://www.benchling.com/).
5. CRISPR Sanger deconvolution:
   (a) DECODR v3.0 (https://decodr.org/).
   (b) ICE Analysis (https://ice.synthego.com/#/).

3 Methods

Irrespective of the category of mutation to be validated, the same types of molecular assays are employed to characterize all mutants. Here, we first describe these various assays and then show how they can be efficiently combined for the analysis of genome-edited animals according to the category of the required mutation.

3.1 DNA Extraction

Genomic DNA (gDNA) is extracted from ear clip biopsies using the DNA Extract All Reagents Kit (Applied Biosystems):

1. Collect 1–2 mm ear biopsies into in the wells of a 96-well PCR plate (*see* **Note 2**).
2. Add 50 μl lysis solution to each well, ensuring the biopsy is suspended within the liquid.
3. Seal the plate using an appropriate adhesive lid or press on caps.
   Using a PCR thermocycler, incubate the plate at 95 °C for 3 min and then hold at 16 °C until the next step.
4. Briefly centrifuge the plate to pull down condensate. Add 50 μl DNA stabilizing solution. Mix thoroughly.
5. Prepare PCR template by making a 1:3 dilution of crude lysate to ddH2O (*see* **Note 3**). Spin the plate for 1 min at ≥1520 × g to pellet tissue debris. Pipette from the top of the lysate to avoid debris which may contain potential PCR inhibitors. Undiluted crude lysate is used as template for ddPCR.
6. Store the crude lysate and dilutions at −20 °C.

3.2 PCR and Sequencing

3.2.1 Primer Design

Design primer pairs using web-based search tools (*see* Subheading 2.3). Typically, we use the NCBI Primer-BLAST. Position primers at least 200 bp either side of the region of interest. Use default setting for the NCBI Primer-BLAST with the following modifications (*see* **Note 4**):

1. PCR product size should be set relative to the size of the region to amplify.

2. Primer melting temperature (Tm): 55–65 °C. Maximum Tm difference of 3 °C.
3. Primer GC content is 40–60%.
4. Primer pair specificity Database: genomes for selected organisms.
5. Organism: *Mus musculus* (taxid: 10090) (or the organism relevant to the experiment (*see* **Note 5**)).

3.2.2 Primer Stocks

1. Centrifuge lyophilized pellet at maximum speed for 30 s.
2. Resuspend in ddH2O to a stock concentration of 100 μM according to quantity specified by the manufacturer. Vortex well and pulse centrifuge to bring solution to bottom of tube. Stocks may be kept at −20 °C.
3. Working solutions of 10 μM are made by 1/10 dilution of stock/ddH2O and stored at −20 °C.

3.2.3 PCR

1. Dilutions of crude lysate are thawed at room temperature. Once thawed, centrifuge for 1 min at top speed (≥1520 × g) to pellet tissue debris.
2. PCR reagents are thawed on ice. Once thawed, mix gently (*see* **Note 6**) and pulse centrifuge.
3. Prepare PCR master mix on ice as described in Table 1, omitting DNA, which will be added to individual reactions.
4. Mix gently (*see* **Note 6**) and pulse centrifuge.
5. Aliquot 19 μl of master mix into each PCR tube/well.
6. Add 1 μl of template DNA into each tube/well.
7. Controls to be included in every PCR:
   (a) WT control of appropriate background
   (b) A positive mutant control, where available (*see* **Note 7**)
   (c) A no-template control in which either template DNA is omitted or ddH2O is added in its place

Table 1
PCR master mix preparation

Component	Final concentration	20 μl reaction
2× Platinum™ SuperFi™ II PCR master mix	1×	10 μl
Forward primer (10 μM)	0.5 μM	1 μl
Reverse primer (10 μM)	0.5 μM	1 μl
Nuclease-free water	N/A	7 μl
Template DNA	N/A	1 μl

Table 2
Thermal cycler program for PCR

Initial denaturation	98 °C for 30 s	
Denaturation	98 °C for 10 s	25–35 cycles
Annealing	60 °C for 10 s	
Extension	72 °C for 15–30 s/kb	
Final extension	72 °C for 5 min 4 °C hold	

8. Seal tubes/plate with adhesive film or caps. Mix and pulse centrifuge to ensure the reaction mixes are pulled down and that there are no bubbles in the reactions.
9. Thermo-cycle as with program shown in Table 2.

3.2.4 Electrophoresis

PCR products are sized and visualized by agarose gel electrophoresis:

1. For a 1.5% (w/v) gel (*see* **Note 8**), 6 g of agarose is dissolved in 400 ml 1× TAE, by heating in the microwave and periodically swirling to dissolve the agarose. Take care to not vigorously boil the solution as this can alter the buffer/agarose concentration.
2. Allow the gel to cool at room temperature or in a 50 °C water bath. Once agarose is cooled enough to handle, add 20 μl (1: 20,000 volume) of 1% w/v ethidium bromide. Swirl to mix, avoiding the introduction of bubbles. Ethidium bromide is a mutagen; always handle warm solution in a fume hood (*see* **Note 9**).
3. Pour agarose into a casting tray containing appropriate well combs and leave to set in a fume hood at room temperature (*see* **Note 10**). Remove the combs and place into a gel electrophoresis tank that contains enough 1× TAE to fully submerge the gel.
4. Before loading into gel, mix 5 μl of PCR product with 1 μl of 6× gel loading dye.
5. Run DNA ladder alongside the PCR products in order to estimate the amplicon size. DNA ladders of 1 kb and/or 100 bp range are used depending on the expected amplicon size.
6. Run the gel at 4 V/cm until suitable band separation is achieved. Visualize the gel on a UV gel imager.

3.2.5 PCR Optimization

Two primer pairs are designed and tested for each genotyping assay, in order to select the pair which shows the best specificity and amplification efficiency. Where a suitable template DNA is available, the PCR design should be tested prior to gene editing so that the optimized protocol is in place when it is needed:

Template DNA

1. Template DNA for PCRPolymerase chain reaction (PCR) optimization: When assays can be tested using WT gDNA, obtain gDNA from the appropriate background strain using the method described in Subheading 3.1.
2. Mutant-specific assays in which both primers sit within the inserted (donor) sequence may be tested using donor DNA as template.
3. For assays that are specific to a novel sequence only created by successful gene editing, test both sets of primer pairs on the first litter of mice.

Assessing efficiency and specificity

4. Initial testing is carried out using the Platinum™ SuperFi II PCR kit, as described in Subheading 3.2.3.
5. Efficiency of amplification and the size of the amplicon are assessed by agarose gel electrophoresis (*see* Subheading 3.2.3). A single band of the anticipated size indicates specificity, and brighter bands indicate higher efficiency.
6. Identity of the amplicon is confirmed by Sanger sequencing (*see* Subheadings 3.2.5 and "Further Optimization" (*see* **Note 11**). The best primer pair is selected primarily for specificity and then for amplification efficiency.

Further Optimization

1. If a specific amplicon is not achieved, the annealing temperature step may be optimized by running an annealing temperature gradient of 55–65 °C. The annealing temperature that results in the most efficient and specific amplicon is selected.
2. If the optimization is still unsuccessful, alternative PCR kits and PCR additives can be tested (*see* **Note 12**).
3. If the optimization is still unsuccessful, design and test alternative primer pairs.

Purification of PCR Products for Sequencing

PCR products are purified before Sanger sequencing, using the QIAquick Gel Extraction Kit (Qiagen), following the manufacturer's instructions. All centrifugation steps are carried out at room temperature for 1 min at 17,900 × g. For a 15 μl PCR reaction:

1. Add 45 μl (3× PCR volumes) of QG buffer to the PCR and mix thoroughly.
2. Add 15 μl (or an equal PCR volume) of isopropanol and mix thoroughly before transferring to a QIAquick spin column. The column is spun and the flow-through discarded.
3. Add 500 μl of Buffer QG to the column. The column is spun and the flow-through discarded.

4. Add 750 μl of Buffer PE to the column, and incubate at room temperature for 2 min. The column is spun and the flow-through discarded.
5. Spin for a further minute to ensure spin column is dry. Transfer column into fresh collection tube.
6. To elute the DNA, add 30 μl of nuclease-free water (or elution buffer) to the column membrane and spin to elute.
7. DNA concentration is measured using by spectrophotometer and is adjusted as required for Sanger sequencing by diluting in ddH2O.
8. Purified PCR products are processed for Sanger sequencing using ABI reagents by an outsourced service (*see* **Note 13**).

3.3 Sequence Analysis

Software options for viewing Sanger chromatograms (.abi files), creating and annotating reference sequences, and aligning Sanger chromatograms to one another and to a reference sequence are suggested in Subheading 2.5.

Position primers used for Sanger sequencing a minimum of 200 bp away from the region of interest to allow sufficient run-up. When sequencing PCR amplicons, the same primers used for amplification may be used for the sequencing reaction:

1. Make a WT reference sequence. Using a suitable genome browser, download the sequence of the target locus. Include at least 1 kb on either side of the intended cut site, donor homology arms, and intended mutation.
2. Where a donor sequence has been utilized to introduce a specific mutation, a mutant reference sequence can be generated (not applicable for untemplated indels or large deletions with no donor). Make a copy of the WT reference and modify it in line with the donor used.

3.3.1 General Method of Sequence Analysis

3. The basic method involves aligning Sanger chromatograms to reference sequences and identifying regions of identity and mismatch. While this can be done by visual inspection and manual matching of the sequences, the process can be aided by using automated alignment software. Such software can be helpful in many cases, although it should not be exclusively relied upon, especially where a desirable mutant is not identified. Visual inspection can often identify patterns where the software fails, especially where traces are mixed/peak-on-peak:
   (a) Open chromatograms from forward primer.
   (b) Identify animals with evidence of modification by looking for a location (usually close to the nuclease target site) where the chromatogram loses identity to the WT reference sequence.
   (c) Where a modification has been specified by a donor, align the chromatograms to the mutant reference and look for evidence of the desired sequence.

(d) For animals identified as potentially carrying the mutation of interest, repeat alignments but with the reverse chromatogram.

(e) Identify animals with evidence of modification and analyze individually to determine its unique identity.

(f) Analysis often involves comparing the peak-on-peak/mixed peaks to the reference at the position where identity is lost. Using this approach, identify the superimposed bases at each position.

(g) Interpretation of mixed peaks from highly mosaic animals can be very challenging, and it may not always be possible to specify sequences with confidence. For low or non-mosaic animals and heterozygote G1 founders, the simple duplex peak-on-peak traces are much easier to interpret (e.g., *see* Fig. 3 panel C).

(h) For large insertions, or in regions where the sequence is repetitive or has a high GC content, it may be necessary to tile primers across the region in order to sequence the entire segment.

(i) In all cases, at the G1 generation, it is important to confirm the identity of the entire modified segment, including at least 200 bp on either side of a nuclease recognition site or the homology arms of any donor used.

3.3.2 Non-specified Indels

1. Carry out alignment to the WT reference sequence and identify superimposed bases for both the forward and the reverse chromatograms.
2. Compare the sequence of superimposed bases to the surrounding WT reference sequence to identify a shifted sequence resulting from a deletion or insertion. For highly mosaic animals, it may be very challenging to identify specific sequences.
3. Online tools for deconvolution of mixed Sanger chromatograms resulting from CRISPR editing may be helpful to identify potentially desirable indels, for example, those causing a frameshift (*see* examples in Subheading 2.5).

3.4 Non-templated Deletions

1. Align forward chromatograms to the WT reference sequence.
2. If mixed peaks are present (due to doublet or multiple PCR amplicons from two or more alleles), determine the superimposed sequence. Compare the superimposed sequence to that of the WT reference to determine whether the intended deletion is represented.
3. If a single trace is present, determine the deleted segment by identifying the position where identity is lost between the chromatogram and WT reference sequence when aligning

from each direction. A deleted segment may not be immediately obvious, as the first few nucleotides could represent a small insertion in place of the deleted segment.

4. Repeat this process with the reverse chromatogram, and check that the deleted segment identified in the forward and reverse sequence is in agreement.

3.5 Knock-ins, Floxed, and Specific Deletions

1. Align the forward chromatogram to the WT reference.
2. If the PCR amplicon contains multiple bands and mixed peaks are present, determine the position at which identity is lost between the reference sequence and the superimposed sequence. Compare this to the mutant reference sequence for evidence of targeted knock-in/deletion.
3. If the PCR amplicon is specific to the mutant allele (due to amplicon size or primer specificity), align to the mutant sequence.
4. Repeat in the reverse orientation to confirm identity.

3.6 Droplet Digital PCR Copy-Counting

3.6.1 Primer/Probe Design

1. For each ddPCR assay, design two primer/probe sets using the IDT primer quest design tool (https://eu.idtdna.com/Primerquest/Home/Index). Use the default parameters for qPCR (two primers + probe) as standard (*see* **Note 14**). Align the top five primer/probe set outputs to the relevant reference genome by BLAST search (www.ensembl.org/Multi/Tools/Blast) or similar. Avoid primers and probes with high homologies to unintended loci (*see* **Note 15**). The assay with highest specificity is selected.
2. Design and source experimental target assays as primer and probe sets, specifying standard desalted oligos, and 5′ 6-FAM dye and ZEN/Iowa Black FQ quencher probe (*see* **Note 16**).
3. Design and source a reference gene assay for the relevant species/strain with a VIC or HEX dye, which are compatible to duplex with the FAM target assay (*see* **Note 17**).

3.6.2 DNA controls

4. Use a littermate (where available) and an unrelated WT on the relevant background as controls alongside a no-template control. Run both a male and a female unrelated WT control if the region of interest resides on a sex chromosome.

3.6.3 Reaction Mix and Droplet Generation

5. Reaction mix and QX200 (manual) droplet generation:
    1. Thaw crude lysate gDNA samples on ice. Once thawed, centrifuge for 1 min at top speed (≥1520 g) to pellet tissue debris. Pipette from the top to avoid tissue debris.

Table 3
ddPCR master mix preparation

Component	Final concentration	20 μl reaction
ddPCR supermix for probes (no dUTP) 2×	1×	10 μl
Reference gene, forward primer (20 μM)	225 nM	0.225 μl
Reference gene, reverse primer (20 μM)	225 nM	0.225 μl
Reference gene, probe (5 μM)	100 nM	0.4 μl
Target gene, forward primer (15 μM)	225 nM	0.3 μl
Target gene, reverse primer (15 μM)	225 nM	0.3 μl
Target gene, probe (5 μM)	75 nM	0.3 μl
Nuclease-free water	N/A	8.25 μl
Template gDNA	N/A	2 μl

2. Thaw the ddPCR supermix and primer/probe tubes on ice. Prepare the PCR master mix as shown in Table 3, omitting DNA which will be added to individual reactions.
3. Pipette 2 μl template DNA into a 96-well plate (specified in Subheading 2.3) followed by 18 μl of master mix. Transfer reaction mix to the DG8 cartridge. Pipette 20 μl ddH2O into any empty wells.
4. Add 70 μl droplet oil into each well, and the cartridge is sealed with a DG8 gasket before loading into the QX200 droplet generator. Droplets are generated as per the manufacturers' instructions for probe-based assays.
5. Once droplet generation is complete, remove and discard the gasket and transfer the oil/reagent emulsion into a 96-well semi-skirted plate (*see* **Note 18**).

3.6.4 Thermocycling and Droplet Reading

1. Following droplet generation, seal the plate with a foil heat seal and transfer it to a PCR thermocycler and cycling, as shown in Table 4.
2. Turn on QX200 droplet reader at least 45 min before thermocycling is complete, in order to warm up.
3. Open QuantaSoft software, populate 96-well plate layout, and save the file as a new experiment. For each well specify:
   (a) Template
   (b) Reference target and dye
   (c) Experimental target and dye

Table 4
Thermocycler program for ddPCR

Initial denaturation	95 °C for 10 min	
Denaturation Annealing/extension	94 °C for 30 s 60 °C for 1 min	40 cycles. All ramp rates to be set at 3 °C/s
Enzyme deactivation	98 °C for 10 min 4 °C Hold	

4. Once PCR cycling is complete, secure the PCR plate in the QX200 plate holder, load into the droplet reader, and start the read by selecting the "run" tab on the software.
5. Once the run is complete, save the data in .qpl format.

3.6.5 ddPCR Analysis

1. Open the .qpl output file in the QuantaSoft™ Analysis Pro software.
2. Samples can be analyzed individually, or batches of samples run with the same assays can be highlighted and analyzed together (*see* **Note 19**).
3. To be confident of the result, each reaction should have a droplet count of ≥10,000. To check the droplet count, click on the Droplets tab. If any samples contain <10,000 droplets, make a note as they may need to be repeated if the copy number calling is ambiguous.
4. The threshold of fluorescence amplitude must be set for each assay (experimental and reference) to enable accurate copy number variation (CNV) calculation. The software will auto-analyze the samples; however, if the sample is flagged, thresholds should be gated manually. This can be done separately for each dye by using the 1D plot view or all together using the 2D plot view. All wells should be checked to ensure the threshold has been correctly identified, so that it separates data points on the scatter plots according to different fluorescent signal profiles. To set thresholds in the 1D plot view, select all the samples run with a given assay. Select the Threshold Multiple Wells Graph Tool, and set the threshold based on the negative droplet baseline for the no-template control (*see* **Note 20**). Pay attention to the channel amplitude of each assay, as these should be similar for all samples run with a specific assay.
5. Once the thresholds are set, the copy number for each sample will be displayed in the Copy Number tab. Table 5 shows expected CNV values for assays described in Subheading 3.3.2:

Table 5
Expected copy number values (CNV) for assays described in Subheading 3.3.2

Project type	Assay target	WT	Heterozygote	Homozygote
Small KI/KO	Mutant allele	0	1	2
Deletion	Deleted segment	2	1	0
Large KI	WT allele	2	1	0
Large KI	Inserted segment	0	1	2
Floxed allele	WT allele	2	2	2
Floxed allele	5′ LoxP	0	1	2
Floxed allele	3′ LoxP	0	1	2

(a) A tolerance of ±0.3 should be allowed, i.e., if the expected copy number is 2, then a copy number of 1.7–2.3 is within the acceptable range.

(b) Take note of the WT control to ensure that the assay calls as expected.

(c) If the assay targets a sex chromosome, the results will be different between males and females.

Strategies applicable to different allele categories are described in the following sections (*see* **Notes 1** and **21**).

3.7 Genotyping for Indel and Point Mutations

This strategy is applicable when a single nuclease recognition site is used. No donor template is included for the generation of indels, while a short donor (*see* **Note 22**) is also delivered for the introduction of a point mutation or other specific small KI or deletion (Fig. 2a, d) (*see* also **Notes 1** and **21**).

3.7.1 G0 Screening

(a) Design a PCR assay to amplify across the intended cut site/donor homologous region (Fig. 2b, e). Check for successful amplification and appropriate amplicon size by agarose gel electrophoresis (from here referred to as "on a gel") (*see* **Note 23**).

(b) For unspecified indels, where a donor is not included in the editing experiment, sequence all PCR amplicons by Sanger sequencing to assess nuclease activity. In many cases, it is possible to identify indels causing a frameshift (or other desired indel class) in the G0 mixed Sanger chromatogram, and these animals can be prioritized for breeding to G1.

(c) When a specific indel/point mutation is required and a donor is included in the editing experiment, exclude from further analysis animals for which only an unintended band shift is seen (indicative of a larger insertion or deletion) from further

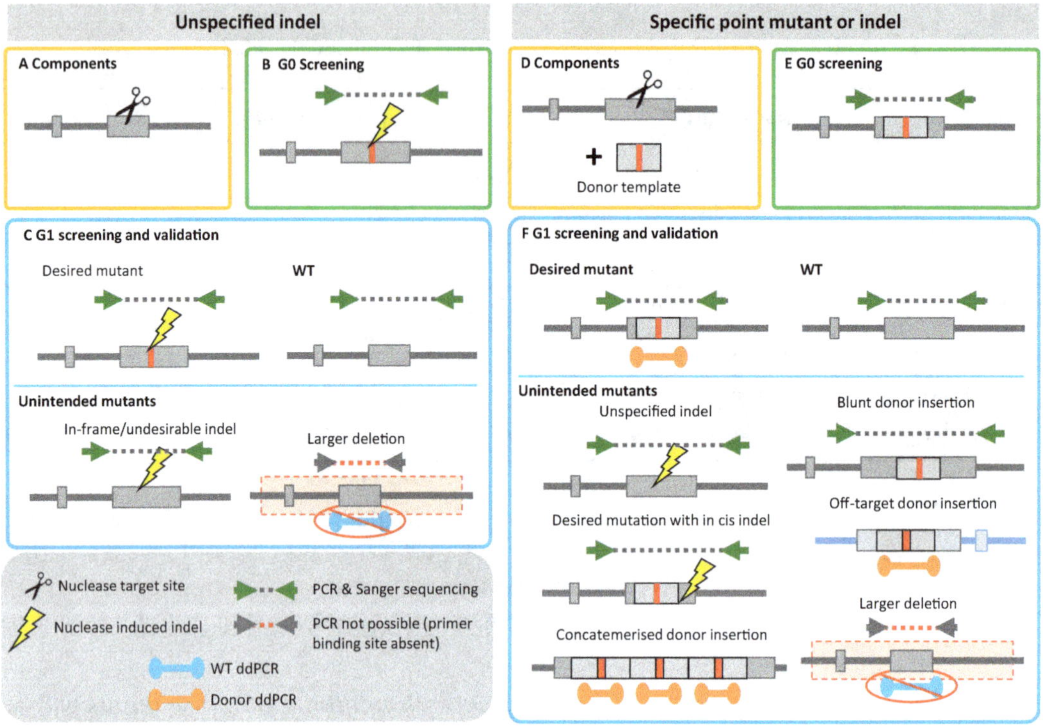

Fig. 2 Genotyping indel and point mutants. (**a**–**c**) Unspecified indels. (**a**) Molecular components: A single nuclease target is used. No donor is included. (**b**) G0 screening: The desired allele is detected using a combination of PCR and Sanger sequencing. The PCR amplifies across the nuclease cut site and is sequenced to identify evidence of the desired indel type. (**c**) G1 screening and validation: The desired allele is detected using a combination of PCR and Sanger sequencing. Undesired edits such as in-frame indels and larger deletions can be detected and excluded. (**d**–**f**) Small KI. (**d**) Molecular components: A single nuclease target is used. A short donor is included to specify the modification. This is often a short single-stranded oligonucleotide (ssODN) but may also be a double-stranded DNA donor. (**e**) G0 screening: The desired allele is detected using a combination of PCR and Sanger sequencing. The PCR amplifies across the region targeted by the donor. The amplicon is sequenced to look for evidence of targeted integration of the donor sequence. (**f**) G1 screening and validation: The desired allele is detected using a combination of PCR and Sanger sequencing. Undesired edits such as unspecified indels and blunt insertions can also be detected and excluded using PCR and Sanger sequencing. Copy-counting of the mutant allele, templated by the donor, should give a copy number of one. A copy number of more than one indicates unintended insertion events such as off-target/random insertions or concatemerized insertions of the donor

analysis. Sequence amplicons of the intended band size, and analyze for presence of the intended mutation among potentially mixed reads. In silico tools may be helpful for this analysis, as described in Subheading 3.3.2.

3.7.2 G1 Screening and Validation

(d) For G1 screening and validation of indel and point mutations, start with the same PCR assays followed by Sanger sequencing (Fig. 2, panels C and F).

(e) At G1, individual alleles present at G0 segregate. Analysis of Sanger sequencing of heterozygote amplicons will provide definitive identification of the inherited allele.

(f) When a specific indel/point mutation is required and a donor is included in the editing experiment, copy-count the donor to confirm single insertion (copy number of ~1) of the donor (*see* Subheading 3.6.5). Higher copy numbers suggest potential off-target insertions (Fig. 2, panel F).

3.8 Genotyping for Deletion Alleles

This genotyping strategy is applicable for larger deletions created using two nuclease recognition sites, one at either side of a segment of DNA to be deleted [8]. In this scenario, a donor is not included (Fig. 3, panel A) (*See* also **Notes 1** and **21**). If a specific deletion is desired, then a short donor may be included. In this case, additional screening will be required; copy-count the donor at G1 as described in Subheading 3.6.5 and Fig. 2 panel F.

3.8.1 G0 Screening for Deletion (DEL) Alleles

(a) Design a PCR assay to amplify across the intended deletion. A size shift from WT to the approximate predicted size indicates the presence of deletion alleles (Fig. 3 panel B and *see* **Note 24**). Multiple editing events could appear as multiple bands or smeared bands on the gel. If these bands are of the predicted size for the intended deletion, these animals can be bred to G1 (*see* **Note 25**). There is no need to Sanger sequence at this point; identification of the specific deletion will be carried out at G1.

3.8.2 G1 screening and Validation of Deletion (DEL) Alleles

(b) Screening at G1 uses the same PCR, with the now segregated deletion alleles detected by a size shift to the approximate predicted size. A WT (no-deletion) band may also be amplified, depending on both the size difference between the two products and the PCR conditions (Fig. 3, panel C).

(c) Sequence amplicons of the anticipated deletion size and analyze to identify the precise deletion. For cases in which only the deletion allele is amplified, the Sanger chromatogram describes the precise sequence of the deletion (Fig. 3, panel D). If both a WT and deletion amplicon are sequenced together, the Sanger trace will switch from a single trace to doublet peak-on-peak reads at the location of the deletion, describing both the WT and shifted deletion sequence (Fig. 3, panel D. *See* also **Note 26**).

(d) Copy-count the region deleted from the WT allele (*see* Subheading 3.15). The expected copy number of one confirms the presence of one copy of the WT allele. If copy number is more than one, this suggests reintegration of the deleted segment (Fig. 3, panel E). Select deletion allele founders with copy-counting results that indicate a single WT allele to establish colonies (*see* **Note 27**).

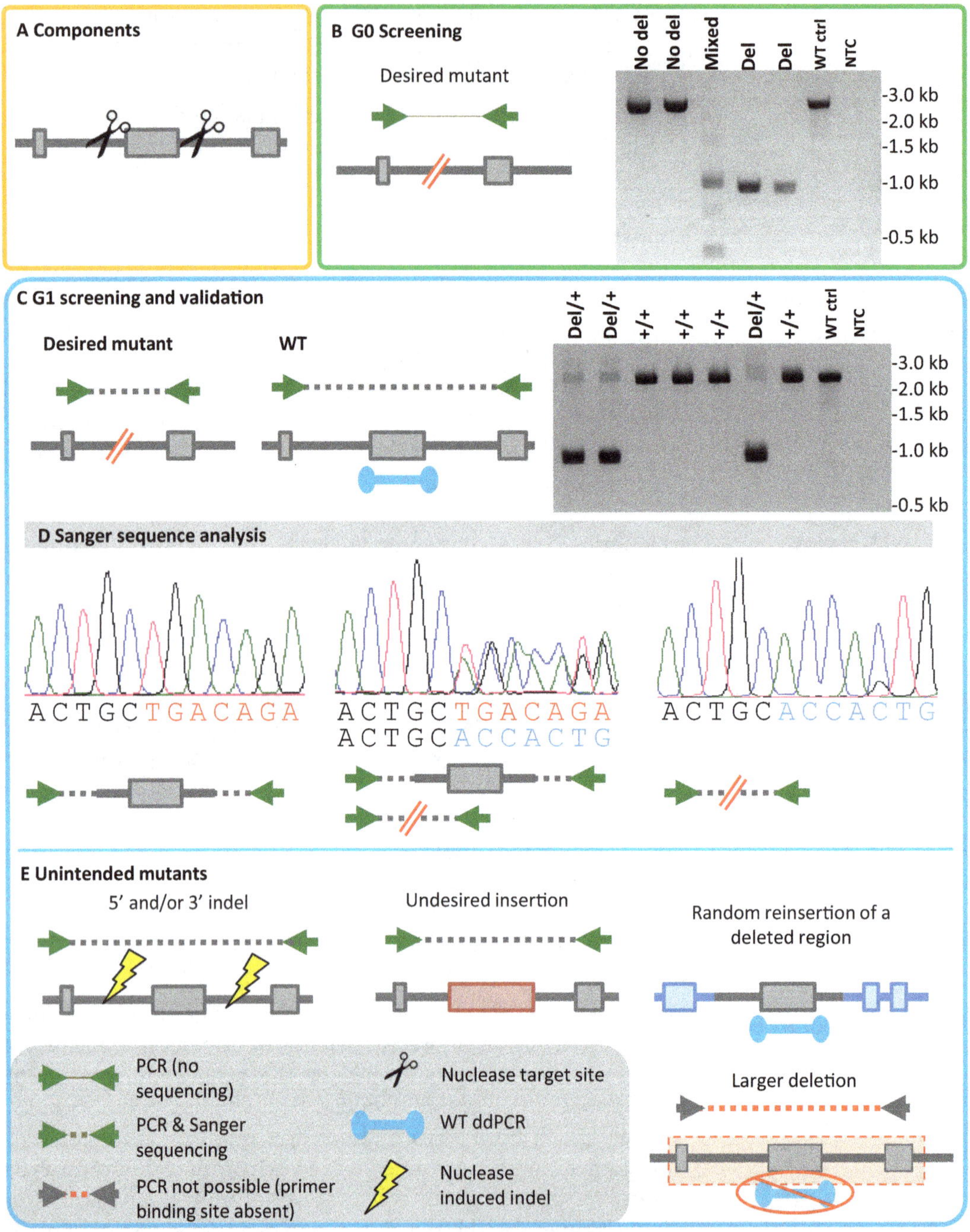

Fig. 3 Genotyping deletions. (**a**) Molecular components: Two nuclease target sites are used to delete the intervening segment of DNA. A donor is not included. (**b**) G0 screening: The desired allele is detected by PCR. The assay is designed with a pair of primers sitting on either side of the intended deletion. A size shift visible by agarose gel electrophoresis is indicative of successful deletion. In the example gel, an amplicon of ~3 kb in the WT is reduced to ~1 kb by the deletion of a 2 kb segment. Two animals, "No del," have only the 3 kb amplicon, showing no evidence of the desired deletion. One animal, "Mixed," has evidence of the desired 1 kb amplicon but appears to be mosaic with several different-sized deletions. Two animals, "Del," have the

3.9 Genotyping for Large KIs

This strategy is applicable for designs in which a longer donor with a larger KI cassette is used, for example, to knock-in tags or other coding sequences. The strategy described here is for when two nuclease recognition sites are used, for a "cut and replace" strategy (Fig. 4, panel A). If a single nuclease site is used, omit copy-counting of the WT (excised) allele at G1 (*See* also **Notes 1** and **21**).

3.9.1 G0 Screening for Large KIs

(a) Carry out two PCRs as the initial screen (Fig. 4, panel B). The small "internal" PCR A amplifies a small segment of the KI cassette. This PCR identifies any animal in which the donor has incorporated, regardless of its genomic location.

(b) The "external" PCR B amplifies across the KI, with primers positioned outside of the homology arms region. If a product of the anticipated KI size is amplified, sequence it and analyze to assess whether the desired KI cassette is present. The external PCR is potentially a very informative assay; however, it is not a robust screening assay (*see* **Note 28**) and so failure to PCR amplify an amplicon of the predicted size should not be used to screen out potential founder candidates.

(c) Carry out two junction PCRs on all animals that are positive for PCR A. PCR C straddles the left homology arm, with the forward primer positioned in the 5′ genomic sequence and the reverse primer within the KI cassette. PCR D straddles the right homology arm, with the forward primer positioned within the KI cassette and the reverse primer in the 3′ genomic sequence. Where the size and complexity of the KI allow, these two PCRs should overlap by at least 200 bp to allow for amplification and sequence confirmation of the entire KI cassette (yellow box in Fig. 4 panels B and C. *See* also **Note 29**).

Fig. 3 (continued) predicted ~1 kb amplicon, suggesting a deletion of the desired size. WT ctrl: wild-type positive control. NTC: no-template control. (**c**, **d**) G1 screening and validation: The desired allele is detected using a combination of (**c**) PCR and (**d**) Sanger sequencing. (**c**) In the example gel, four animals are genotyped as WT "+/+" as they only have an amplicon of the expected WT ~3 kb. Three animals are heterozygous for the deletion allele. Both the WT size of ~3 kb and the deletion of ~1 kb amplicon are visible, but the band corresponding to the deletion allele is much brighter due to preferential amplification of the smaller product. (**d**) Sanger sequencing is used to identify the specific deletion. If the sequenced amplicon contains both the WT allele and the deletion allele, then a switch from a single to a doublet peaks will occur at the position of the deletion. Comparison of doublets to the known WT sequence can reveal the mutant sequence. When the deleted allele is preferentially amplified, a single-trace chromatogram reveals the mutant sequence. (**e**) Unintended modifications can be identified by Sanger sequencing. Examples include indels, partial or larger deletion, or the introduction of problematic motifs, such as canonical splice acceptors or other undesirable sequence segments. Copy-counting of the deleted WT segment should result in a copy number of one for an ideal founder. A copy number higher than one indicates reintegration of the excised fragment

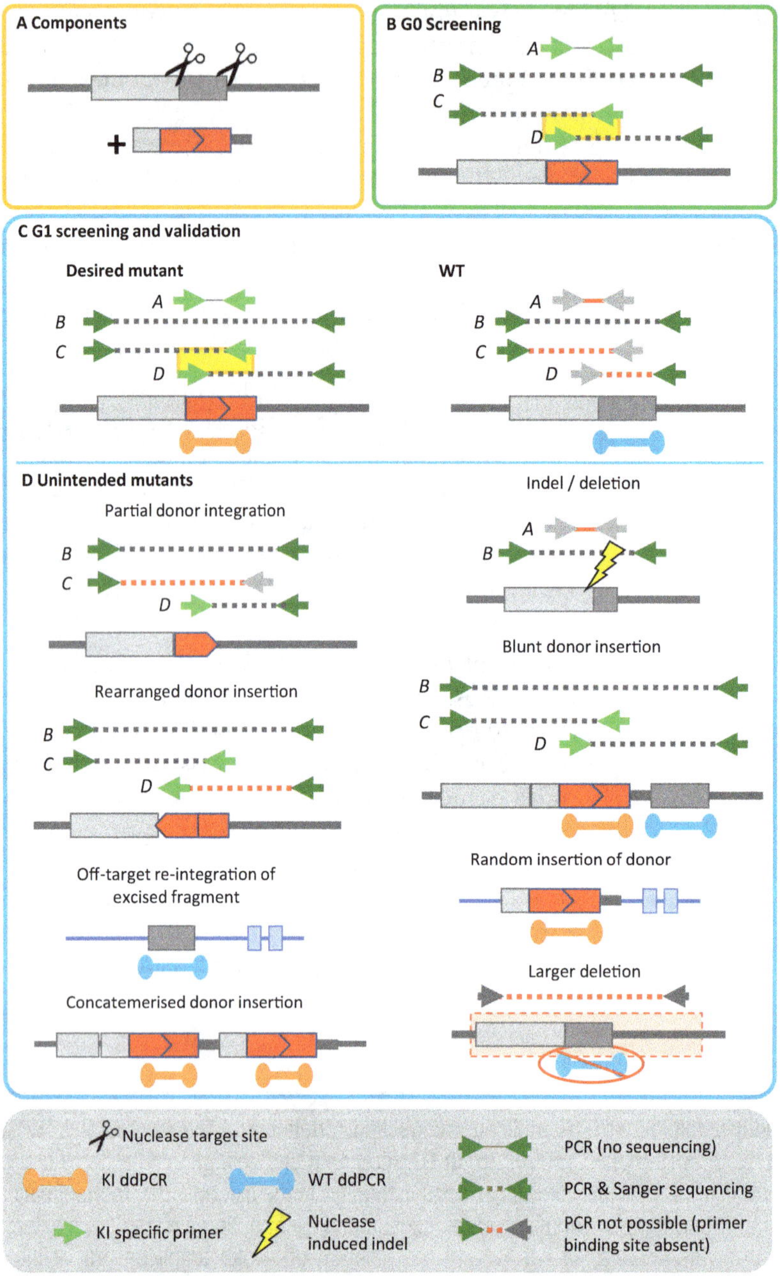

Fig. 4 Genotyping large KIs. (**a**) Molecular components: In this cut and replace strategy, two nuclease target sites are used to delete the intervening segment of DNA. A donor is also included, specifying the KI cassette that will replace the deleted segment. The donor is drawn as linear here but also can be delivered as a circular molecule. (**b**) G0 screening: The desired allele is detected by PCR and Sanger sequencing. Insertion of the donor is detected by PCR A, which amplifies a segment specific to the KI cassette. The external PCR B amplifies across the entire KI with primers positioned to straddle either side of the sequence included in the donor. Sanger sequencing can confirm the entire inserted segment if a specific amplicon is attainable, which is not guaranteed. This PCR is not essential for animals to be taken forward to G1. Junction PCRs C and D

(d) Amplification and Sanger sequencing of the two junction PCRs will identify animals with on-target integration of the 5′ (PCR C) and 3′ (PCR D) end of the donor. Unless the external PCR, B, is also successful, it is not possible at this stage to identify whether the 5′ and 3′ junctions are in cis or whether they represent independent partial *trans* integrations (Fig. 4 panel D partial donor integration).

(e) Select as potential founders animals that are positive for internal PCR A and which have confirmatory sequencing data for junction PCRs C and D. Animals with confirmatory sequencing data for PCR B may also be considered as potential founders (*see* **Note 30**).

3.9.2 G1 Screening and Validation for Large KIs

(f) Internal PCR A is an initial screen to identify all animals carrying the donor and to exclude negative animals from further analysis. External PCR B may also be run as a control for DNA extraction as a WT amplicon should be present for every animal.

(g) For animals positive for PCR A, run junction PCRs C and D to identify animals within *cis* 3′ and 5′ on-target integration of the KI cassette. PCRs C and D are then Sanger sequenced and analyzed to identify animals with a fully confirmed, on-target, KI cassette (Fig. 4, panel C and *see* **Note 31**). If a sufficient overlap of the junction PCRs is not possible, then additional PCR amplicons are required to tile across the insert in order to confirm the sequence of the full interval of the insert (*see* also **Note 32**).

(h) For all large KI animals that are fully confirmed by PCR and Sanger sequencing, copy-counting assays for the donor and the excised WT interval are required to validate the lines (Fig. 4, panels C and D and **Note 33**). The expected copy number for the donor is one (*see* Subheading 3.8.2). A higher

Fig. 4 (continued) straddle the homology arms into the KI cassette. They should overlap within the KI cassette by at least 200 bp (indicated by the yellow box) in order that the whole cassette can later be definitively confirmed at G1. At this point, it may not be possible to tell whether left and right junctions are *in cis* or *in trans*. (**c**) G1 screening and validation: The desired allele is detected using a combination of PCR and Sanger sequencing. PCRs are as described for the G0 screen. At G1, alleles have segregated, and all animals are heterozygous. Animals with positive PCR and Sanger sequence of junction PCRs C and D are confirmed KIs. (**d**) Other undesired alleles, which may be detected and excluded while screening potential founders, include partial or rearranged donors (which may be positive for just one or the other junction PCR) or blunt insertions of the donor. Copy-counting of the deleted WT segment should give a copy number of one. A copy number higher than one indicates reintegration of the excised fragment or blunt insertion of the donor. Copy-counting of the KI cassette should give a copy number of one. A copy number higher than one indicates off-target random insertion of the donor or concatemerized insertion

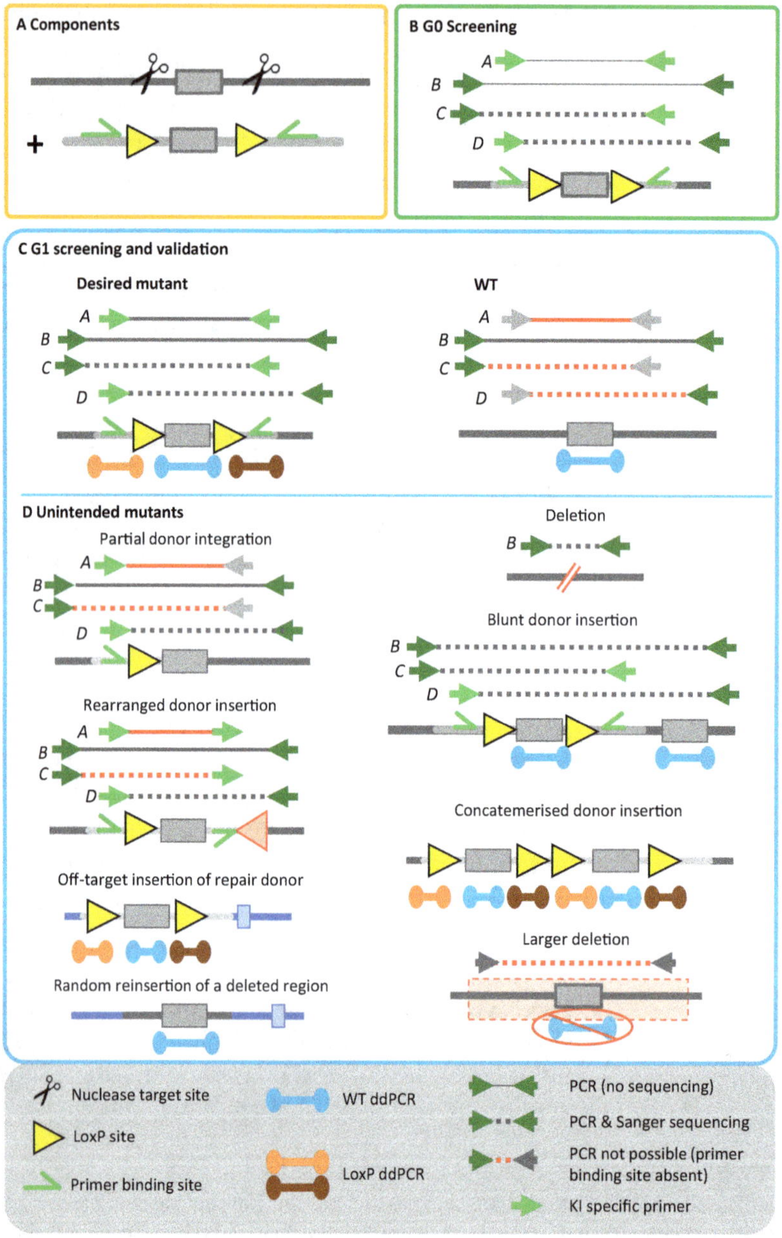

Fig. 5 Genotyping floxed alleles. (**a**) Molecular components: Two nuclease target sites are chosen, one at each site of insertion. A long donor is included, which includes the two LoxP insertions, each with its own unique primer binding site, and the intervening sequence, which is not modified. (**b**) G0 screening: The desired allele is detected by PCR and Sanger sequencing. Insertion of the donor is detected by PCR A, which amplifies across the two knock-in sites using the primer binding sites included in the donor. This PCR will amplify any insertion of the donor, irrespective of the genomic location. The external PCR B amplifies across the entire KI with primers positioned to straddle either side of the sequence included in the donor. This PCR is used as an internal control for the DNA extraction but is not the most useful G0 screening PCR because it lacks specificity for donor integration. Junction PCR C straddles the left homology arm to the 3′ LoxP primer. Junction PCR D

donor copy number indicates additional undesired insertions. The expected WT copy number is also one. A higher WT copy number indicates that the excised WT region may have reintegrated into the genome.

(i) To establish the colony, select founders with copy-counting results that indicate no additional integrations (*see* **Notes 30** and **33**).

3.10 Genotyping for Floxed Alleles

This genotyping strategy is applicable to an editing methodology that is commonly used to create a "floxed" exon (the insertion of two LoxP motifs, one at either side of an exon). Two nuclease recognition sites are used, one at each insertion location, along with a single donor that spans the entire region (as shown Fig. 5 panel A; *see* also **Notes 1**, **21** and **34**).

(a) To aid PCR genotyping, we recommend, where possible, that unique primer binding sites are included in the donor alongside the intended modification, as shown in Fig. 5 panel A. This is particularly useful when inserting sequences such as LoxP sites, which are not, themselves, an example of a good priming substrate, due to their internal repetitive sequence and their duplication within the final edited allele.

(b) For this strategy, we will follow the example of a floxed allele; however, this strategy may also be used where similar editing methodology is applied to knock-in two discrete cassettes *in cis*, leaving the intervening sequence intact. For example, this strategy can be useful when introducing two point mutations a long distance apart or a point mutation and an epitope tag at different locations within a gene. For some applications, it may not be possible to insert novel primer binding sites, in which case suitable primers should be designed as described in Subheading 3.2.

Fig. 5 (continued) straddles the right homology arm from the 5′ LoxP primer. These PCRs are specific for 5′ and 3′ on-target donor integration, respectively. Sanger sequencing can confirm the presence of both LoxPs in the same integration. At this point, it may not be possible to tell whether left and right junctions are *in cis* or *in trans*. (**c**) G1 screening and validation: The desired allele is detected using a combination of PCR and Sanger sequencing. PCRs are as described for G0. At G1 alleles have segregated and all animals are heterozygous. Animals with positive PCR and Sanger sequence for both junction PCRs C and D are confirmed KIs. (**d**) Other undesired alleles, which may be detected and excluded while screening potential founders, include partial or rearranged donors (which may amplify one or the other junction PCR) and a deletion or blunt insertions of the donor. Copy-counting of the floxed (WT) segment should give a copy number of one. A copy number higher than one indicates either off-target donor insertion, concatemerized or blunt donor insertion, or reintegration of the excised fragment. Copy-counting of the LoxP cassettes should each give a copy number of one each. A copy number higher than one indicates either off-target donor insertion or concatemerized or blunt donor insertion

3.10.1 G0 Screening for a Floxed Allele

(c) Carry out initial screening by "Lox-Lox" PCR A (*see* Fig. 5, panel B) using the two primers included in the donor. This PCR identifies any animal in which the donor has incorporated, regardless of its genomic location. External PCR B may also be run as a control for DNA extraction.

(d) For animals positive for Lox-Lox PCR A, two junction PCRs are carried out (*see* **Note 35**); PCR C amplifies across the left junction, straddling the left homology arm with the forward primer positioned in the 5′ genomic sequence and the reverse primer within the 3′ LoxP cassette. PCR D amplifies across the right junction, straddling the right homology arm with the forward primer within the 5′ LoxP site and the reverse primer in the 3′ end of the homology arm.

(e) For animals that are positive for both PCRs C and D, these amplicons are sequenced. Sanger traces are analyzed for evidence that both LoxP sites are present in both PCR amplicons and that the intervening exon is also intact.

3.10.2 G1 Screening and Validation for a Floxed Allele

(f) G1 screening for floxed alleles starts with amplification of PCRs A and B. For positive animals for PCR A, perform amplification of both junction PCRs C and D and Sanger sequence. The entire interval must be amplified, and Sanger sequence to confirm that both LoxPs and the intervening sequence are intact (Fig. 5, panel C, and *see* **Notes 34** and **35**). Incorrect insertions, such as blunt or partial insertions, are screened out at this point (*see* **Note 36**).

(g) For all floxed animals that are fully confirmed by PCR and Sanger sequencing, validation by copy-counting is also required (*see* Subheading 3.6.5). Copy-counting of each LoxP sites should indicate one copy. If more than one copy is detected, this indicates additional insertions of the donor (Fig. 5 panel D). Carry out copy-counting of the floxed exon to confirm two copies (one on the WT allele and one on the floxed allele). If more than one copy is detected, this indicates additional integrations of the donor or that the region between the two nuclease sites was excised and reinserted during editing (Fig. 5, panel D).

(h) To establish the colony, select founders with copy-counting results that indicate no additional integrations (*see* **Note 33**).

3.11 Off-Target Analysis: All Allele Types

Assay potential off-target nuclease recognition sites in G1 founder animals in order to ensure that they have not been modified. If necessary, off-target modifications can be bred out in subsequent generations, but this is only practicable when off-target sites are *in trans* of the target modification.

(a) For CRISPR-Cas9 gene editing, many of the online servers used to identify guide sites (e.g., crispor.tefor.net/ and wge.stemcell.sanger.ac.uk/) will also provide a list of potential off-target recognition sites for the chosen organism. Check off-target sites with two or fewer mismatches to protospacers of all used gRNAs by PCR amplification followed by Sanger sequencing. Employ extension times for PCRs that are sufficient to amplify potential insertions of any donor or segments excised by on-target activity. Carry out these assays on all potential founders identified at G1.

(b) In certain cases, it may also be desirable to assay off-target nuclease activity using ddPCR in order to check for large deletions which may delete primer binding sites, for example, where a target is a member of a gene family with high homology between family members, especially if these genes are *in cis* with the target, or where two off-target recognition sites are present within close proximity.

4 Notes

1. The same strategies apply whether genome-editing reagents were delivered by microinjection, electroporation, or viral vector.
2. PCR strips or 1.5 ml microfuge tubes can be used. If using microfuge tubes, the 95 °C incubation step may need to be adjusted to allow for slower thermal conductance. Biopsies may be stored frozen until ready for DNA extraction, but multiple freeze-thaw cycles should be avoided.
3. This dilution factor works well for extraction from mouse ear biopsy using the volumes described. When extracting DNA from other types of tissue biopsy, it may be necessary to test different volumes of extraction reagents and varying final dilutions in order to achieve good tissue lysis and reliable PCR amplification.
4. These are the parameters from the NCBI Primer-BLAST tool that are modified from the default settings. These and other parameters may be further modified if further PCR optimization is required.
5. Most online primer selection tools include a mispriming library, such as the "Primer pair specificity" field on the NCBI tool. This feature automatically checks potential primers for homology to other regions in the selected genome. If the relevant species is not available as a mispriming library, but there is a reference genome built on another site, manually BLAST search the primers and disqualify primer pairs which

include one or more primers with high homology to nontarget loci.

6. While it is important to thoroughly mix reagents including PCR master mixes, any reagent or solution containing enzymes should be handled with care. These reagents should be mixed by gentle flicking and not by vortexing.
7. If a previously identified correct mutant has been identified, then this DNA can be used as a mutant control. In many cases, a true mutant control gDNA will not exist. In some cases, where the primers to be tested bind within the donor used for gene editing, it may be possible to use the donor (either linear or plasmid DNA) as a mutant control either alone or spiked into WT gDNA.
8. A 1.5% agarose gel is effective for size separation of double-stranded DNA between 200 bp and 3 kb in size. When the anticipated amplicon is close to, or outside of, these limits, the percentage of the agarose should be adjusted. For smaller amplicons of 100 bp to 2 kb, a 2% gel should be used. A 1% gel is appropriate for amplicons of 0.5–10 kb. For very large amplicons, between 5 and 60 kb, the percentage of agarose can be reduced to a lower limit of 0.3% [9]. The gel will become increasingly fragile as the agarose content is reduced.
9. Ethidium bromide is a potential mutagen and is a potential hazard to both health and the environment. It should be handled in a fume cabinet and should not be added to hot agarose before it has been allowed to cool, as this may cause vaporization. Refer to local rules for guidance on handling and disposal of ethidium bromide. Alternatives to nucleic acid stains, such as SYBR Gold or SYBR Green, are also available [9].
10. Agarose gels may be made ahead of time and stored for up to 1 week in a humidified airtight container at 4 °C. This is not recommended with lower percentage gels due to their fragility.
11. This WT .abi chromatogram is also used as a control for in silico deconvolution analyses of data gathered from potential mouse lines (*see* Subheading "Further Optimization" for sequence analysis).
12. Other kits that we have found useful include Expand™ Long Range dNTPack, KAPA HiFi HotStart (Roche), and AccuPrime™ GC-Rich DNA Polymerase (Thermo Fisher Scientific). For PCR amplicons with high GC (>60%) content, the addition of dimethyl sulfoxide (DMSO) and/or betaine may be tested. For DMSO, a concentration between 2% and 10% of the total PCR reaction volume should be tested (0.4–2 μl in a 20 μl reaction). For betaine, a range of concentrations from 0.25 M to 1.25 M should be tested by addition of 5 M

molecular biology grade reagent (1–5 μl in a 20 μl reaction). In both cases, the total volume of the PCR reaction should be maintained by adjusting the ddH2O.

13. For Sanger sequencing, we outsource to a third party. Standard ABI BigDye terminator chemistry may be used.
14. In some cases, it is necessary to specify the binding location of primers and probes within the target region. In these cases, "custom design parameters" is selected. In the "Custom Target Region" section, the location of primers and probe can be specified, for example, to limit probe binding to the location of an indel/point mutant specified within a donor.
15. For many species, it is possible to run a BLAST search specifically designed to identify potential amplicons from a pair of primers in a specified reference genome/transcriptome, for example, the NCBI Primer-BLAST tool (www.ncbi.nlm.nih.gov/tools/primer-blast/). If the relevant species is represented, then this tool is recommended. Standard BLAST alignment should be used to identify unintended binding for probes.

    To use the NCBI Primer-BLAST:

    (a) PCR template is left empty.
    (b) Forward and reverse primers should be specified. Other primer parameters may be left as default.
    (c) In Section "Primer Pair Specificity Checking Parameters," Database is set to "Genome for selected organism" and the relevant organism specified.

        If the relevant species is not represented in a specific Primer-BLAST alignment tool, each primer and probe should be individually BLAST searched using an appropriate tool to identify unintended potential binding sites. Primer/probe sets with unintended binding sites located in close proximity to one another should be discounted.

16. qPCR primers and probes can also be ordered as a premixed assay specifying Dye-Quencher: 6-FAM/ZEN/IBFQ and a 3:1 primer/probe molar ratio.
17. For work in mice, we use the reference gene *Dot1l*, which is on Mmu10. Standard desalted lyophilized oligos (Dot1l-F: 5′-GCCCCAGCACGACCATT-3′ and Dot1l-R: 5′-TAGTTGGCATCCTTATGCTTCATC-3′) are used in combination with a TaqMan MGB probe (Applied Biosystems Thermo Fisher Scientific) with 5′-VIC dye and 3′ MGBNFQ quencher (5′-CCCAACAGGCCTGGATTCTCAATGC-3′).
18. Once complete, the emulsion should have two well-defined phase—a cloudy upper phase and a clear lower phase. These

two phases should still be distinct after thermocycling. The water in oil emulsion is very delicate and should be handled with care. We recommend transferring the emulsion using an automated multichannel pipette set at the slowest speed. Alternative droplet generation equipment can also be used, for example, the auto-droplet generator AutoDG (Bio-Rad). This equipment transfers droplet emulsion to a fresh PCR plate automatically.

19. Discrepancies between QuantaSoft software versions can cause problems with analysis when the plate setup and analysis are carried out using different computers. To avoid this, we recommend checking the plate layout, paying particular attention to the expected copy number for the reference gene for each sample.

20. Outliers on the 1D and 2D fluorescence plots may skew the axis making it difficult to set the florescence threshold. In this case, manually set the axis to a fixed scale by clicking on the cog in the top right, selecting fixed scale.

21. Genotyping of the founder G0 generation can be particularly complex due to their often mosaic nature. PCR amplicons are often amplified from more than one allele, and it can be hard to disentangle their Sanger chromatograms. Sub-cloning of mosaic PCR amplicons can simplify this process by separating out multiple alleles, although care should be taken when deciding how many clones to analyze [4]. Similarly, long-read sequencing such as Oxford Nanopore can be helpful to tease apart modified alleles and is particularly useful when trying to identify *in cis* modifications such as in a floxed allele [10].

    Although such methods are available, it should be stressed that it is not necessary, nor often desirable, to identify every allele represented at G0. In some experimental settings, this might be worthwhile, but it is a very labor-intensive process. Instead, the objective is more often to implement the most efficient strategy to identify potential G0 founders—those presenting evidence of the desired allele. Once backcrossed to produce the G1 generation, segregated alleles can be more easily and definitively validated.

    Once genomic validation is complete, the next step will be to functionally validate the newly edited allele. This may involve investigation of transcriptional and translational diversity and abundance or functionality of elements such as LoxP sites. This is an important step for validation of engineered alleles but is beyond the scope of this chapter.

22. This strategy may be applied to short donors that incorporate modifications (insertion, deletions, or replacements) covering a segment of up to ~200 bp. These may be in any format, for

example, short single-stranded DNA oligonucleotides (ssODN) and in vitro reverse transcribed single-stranded DNA or double-stranded DNA.

23. While a size shift may be visible if a larger indel has occurred, this is not an indication of success.

24. If no deletions are detected, consider Sanger sequencing the PCR amplicons to check for signs of nuclease activity. If no indels are detected, consider alternative nuclease sites for subsequent mutagenesis attempts.

25. If a WT allele is present, this may also amplify using this PCR. If multiple G0 animals are positive for deletion, and resources are short, prioritize mating those without a WT amplicon; the smaller deletion amplicon should be preferentially amplified over the longer WT amplicon. A bright WT band is suggestive of a lower prevalence of deleted alleles. Note, however, that very large deletions may excise primer binding sites. If only WT offspring are identified at G1, it could be that a large deletion/ WT is being misinterpreted as a +/+.

26. If a WT allele is also amplified, it is possible to gel purify the deletion band prior to sequencing. This may ease analysis; however, it is fairly straightforward to disentangle two Sanger reads when one is the known WT allele.

27. Breeding out random insertions: In some cases, all potential founders may also have copy-counting results that indicate an off-target insertion of a donor or an excised segment. In such cases, animals with undesirable off-target insertions may be bred through another generation of backcross and further ddPCR copy-counting. If the extra insertion is *in trans*, it should segregate away from the KI at Mendelian ratios. Failure to segregate away suggests possible *in cis* insertion, in which case breeding out the insertion will be exceedingly challenging and may not be possible at all. A repeat of the gene-editing experiment is probably preferable in such cases.

28. Factors inhibiting external PCR: (1) Larger KIs may be too big to be readily amplified by PCR, especially if they contain repetitive or complex sequence. (2) Because this primer pair is not specific to the KI, there may be multiple alleles competing for amplification. These could include WT alleles, deletion alleles, or partial insertions, all of which are probably smaller than the desired full KI. Because of their smaller size, they may amplify preferentially over a KI allele, obscuring the identification of a desired KI.

29. Junction PCRs for large KIs. Where a KI cassette is very large or contains repetitive or GC-rich sequence, it may be challenging to design and successfully amplify junction PCRs that also encompass the entire KI cassette. For such cases, it may be

appropriate to design shorter PCRs that do not amplify the entire KI cassette. These PCRs should straddle the entire homology arm and at least 200 bp on either side. This will allow assessment of the integration site but not the internal integrity of the cassette. At G1, amplification of the full KI cassette, followed by Sanger sequencing, is required to confirm the KI. The longer junction PCRs (C and D) which include the full insert may be more easily amplified at G1 as the KI is heterozygous, rather than potentially mosaic. If these are still challenging, then additional PCRs should be designed, tiled across the entire KI interval with >200 bp overlaps.

30. Success of only one, or other, or the junction PCRs (C and D) is suggestive of a partial donor insertion. Furthermore, as no positive control is available for these PCRs, it is not possible to properly optimize, and failure to amplify could be because of technical issues. *See* also **Note 28**. This will depend on both the likely complexity of the PCR assay (which will vary according to the precise design) and on available resources for repeating the gene-editing experiment.

31. It may be possible to amplify the external PCR B and Sanger sequence the entire segment from a single amplicon. As both WT and KI allele will be amplified, it is recommended that the amplicon is first gel purified to avoid the need to disentangle doublet peak-on-peak Sanger traces. While identifying peak-on-peak sequencing is possible, the process is more arduous for a large KI cassette.

32. It is important that at G1 the entire interval of modification is confirmed by sequencing, including a minimum of 200 bp both downstream and upstream of the insertion site, or homology arms used in a donor. Where an insertion is large, or the sequence is complex or repetitive, it may be necessary to design additional sequencing primers to fully characterize the insertion. In some cases, it may also be helpful to use alternative sequencing modalities, such as long-read nanopore [10] or next-generation sequencing [11].

33. ddPCR copy-counting of the donor integrations is of special importance for lines in which genotyping of the established colony will be anchored entirely within the KI insert. Such assays will not differentiate between on- and off-target integrations, and so these must be definitively discounted.

34. Although often less efficient, it is also possible to use two separate short donors [12], one at each insertion site. This can be particularly useful when the two insertions are a long distance from one another. Using two separate donors may increase the likelihood of only achieving insertion on one side and not the other. When two donors are used, genotyping

should be carried out as if for two independent KIs (as described in Subheading 3.2.1). Along with the assays described for small KIs, additional considerations will be required for G1 validation.

First, the entire interval must be amplified and sequenced, including at least 200 bp upstream and downstream of the donor homology arms region. If it is not possible to amplify this all in one go, PCRs should be tiled across the region.

Second, copy-counting of the WT allele (as described in Subheading 3.2.1, **step 3**) should be used to check for reintegration of the fragment excised between the two nuclease recognition sites.

35. For projects with a longer interval between insertion sites (>2.5 kb), PCR amplification across the entire region may be challenging. If this is the case, screening for each insertion site should proceed as described in Subheading 3.2.1. At G1, the entire interval must be amplified and sequenced, including at least 200 bp upstream and downstream of the donor homology arms region. If it is not possible to amplify this all in one go, or using the junction PCRs, then additional PCRs should tile across the region with at least 200 bp overlaps followed by Sanger sequencing and assembly.
36. In some cases, only animals with single-sided insertions of one or the other LoxP sites are identified. In these cases, it is possible to select a single-sided insertion and retarget these animals to add in the second LoxP site. In these cases, it may be desirable to select animals which have a LoxP on one side and an indel on the other. The indel sequence may provide novel nuclease targeting sites which will allow allele-specific retargeting in heterozygous animals.

References

1. Burgio G, Teboul L (2020) Anticipating and identifying collateral damage in genome editing. Trends Genet 36:905–914
2. Mizuno S, Dinh TTH, Kato K et al (2014) Simple generation of albino C57BL/6J mice with G291T mutation in the tyrosinase gene by the CRISPR/Cas9 system. Mamm Genome 25:327–334
3. Renaud J-B, Boix C, Charpentier M et al (2016) Improved genome editing efficiency and flexibility using modified oligonucleotides with TALEN and CRISPR-Cas9 Nucleases. Cell Rep 14:2263–2272
4. Mianné J, Codner GF, Caulder A et al (2017) Analysing the outcome of CRISPR-aided genome editing in embryos: screening, genotyping and quality control. Methods 121–122:68–76
5. Pickar-Oliver A, Gersbach CA (2019) The next generation of CRISPR-Cas technologies and applications. Nat Rev Mol Cell Biol 20:490–507
6. Komor AC, Kim YB, Packer MS et al (2016) Programmable editing of a target base in genomic DNA without double-stranded DNA cleavage. Nature 533:420–424
7. Anzalone AV, Randolph PB, Davis JR et al (2019) Search-and-replace genome editing without double-strand breaks or donor DNA. Nature 576:149–157
8. Shin HY, Wang C, Lee HK et al (2017) CRISPR/Cas9 targeting events cause complex deletions and insertions at 17 sites in the

mouse genome. Nat Commun 8:15464–15464

9. Green MR, Sambrook J (2019) Agarose gel electrophoresis. Cold Spring Harb Protoc 2019(1). https://doi.org/10.1101/pdb.prot100404
10. McCabe CV, Codner GF, Allan AJ et al (2019) Application of long-read sequencing for robust identification of correct alleles in genome edited animals. bioRxiv preprint. https://doi.org/10.1101/838193
11. Fernández A, Morín M, Muñoz-Santos D et al (2020) Simple protocol for generating and genotyping genome-edited mice with CRISPR-Cas9 reagents. Curr Protoc Mouse Biol 10:e69
12. Lanza DG, Gaspero A, Lorenzo I et al (2018) Comparative analysis of single-stranded DNA donors to generate conditional null mouse alleles. BMC Biol 16:69–69

Chapter 5

ASIS-Seq: Transgene Insertion Site Mapping by Nanopore Adaptive Sampling

Charles Yu, Roger Caothien, Anna Pham, Lucinda Tam, Tuija Alcantar, Natasha Bacarro, Juan Reyes Jr, Marques Jackson, Brian Nakao, and Merone Roose-Girma

Abstract

Generation of transgenic mice by direct microinjection of foreign DNA into fertilized ova has become a routine technique in biomedical research. It remains an essential tool for studying gene expression, developmental biology, genetic disease models, and their therapies. However, the random integration of foreign DNA into the host genome that is inherent to this technology can lead to confounding effects associated with insertional mutagenesis and transgene silencing. Locations of most transgenic lines remain unknown because the methods are often burdensome (Nicholls et al., G3: Genes Genomes Genetics 9: 1481–1486, 2019) or have limitations (Goodwin et al., Genome Research 29:494–505, 2019). Here, we present a method that we call Adaptive Sampling Insertion Site Sequencing (ASIS-Seq) to locate transgene integration sites using targeted sequencing on Oxford Nanopore Technologies' (ONT) sequencers. ASIS-Seq requires only about 3 ug of genomic DNA, 3 hours of hands-on sample preparation time, and 3 days of sequencing time to locate transgenes in a host genome.

Key words Transgenic mice, Transgene insertion site mapping, Next-generation sequencing (NGS), Oxford Nanopore Technologies (ONT), Nanopore sequencing, GridION, MinION, Adaptive sampling

1 Introduction

The first reports of transgenic mice created by the microinjection of exogenous DNA into the pronuclei of fertilized eggs appeared in 1981 [1]. The heritable capacity for one organism to express the genes of another held enormous implications for biomedical research of human diseases. Because mice and humans share many common genes, transgenic mice offered a practical vehicle for the study of human gene function or to generate models of human genetic disease.

Thomas L. Saunders (ed.), *Transgenesis: Methods and Protocols*,
Methods in Molecular Biology, vol. 2631, https://doi.org/10.1007/978-1-0716-2990-1_5,

In the years since their publication in 1981, the methods for generating transgenic mice are used routinely by hundreds of laboratories throughout the world and have increased in sophistication [2]. The addition of regulatory elements such as enhancers [3] and intronic sequences [4] was shown to improve transgene expression. However, additional transgenic elements required larger DNA constructs. The use of yeast artificial chromosomes (YACs) [5], P1-derived artificial chromosomes (PACs) [6], and bacterial artificial chromosomes (BACs) [7] later proved invaluable for generating mice with transgenes that were 100 kilobases (kb) or longer. While larger and more complex artificial chromosome-based transgenes showed improved expression levels compared to smaller and simpler constructs [8], the main limitations of micro-injecting transgenic DNA alone to generate transgenic mice remained. Low rates of transgene integration into the genome by such "passive" transgenesis approaches are often observed [9, 10]. When a transgene does integrate, it frequently integrates as tandem head-to-tail arrays containing up to several hundred copies [11] that may result in transgene silencing [12]. Recent "active" transposon-mediated transgenesis approaches using the *Sleeping Beauty* (SB) [13] or *piggyBac* (PB) [14] transposase can yield higher transgenic rates and promote the genomic insertion of a single copy transgene [15] while retaining a large DNA cargo capacity [16].

Regardless of their size, copy number, and transgenesis approach, transgenes integrate randomly within the genome. This has ramifications for transgene and endogenous gene functions. Disruption of endogenous genes or gene-regulating elements by transgene insertional mutagenesis can lead to altered phenotypes [17–19] independent of transgene function. It has been shown that the *piggyBac* transposase preferentially inserts DNA fragments into gene-containing regions and transcriptional start sites [20–22]. Such position-associated risks from random transgenesis can lead to confounding genetic events [23] that reduce the efficacy of transgenic animals as models of genetic disease. Due to limitations in tools for locating transgenes, only a small fraction of the thousands of transgenic lines in existence has been mapped and reported [23, 24]. For these reasons, novel tools to precisely map transgene integration sites are essential.

Methods such as Southern blot analysis and fluorescent in situ hybridization (FISH) [25–27] to map transgene insertion sites are common, but their low resolution limits the transgene localization to a chromosomal region and precludes sequence-level characterizations. Methods based on inverse PCR [28, 29] and restriction enzyme-PCR [30] techniques offer higher resolutions but may not yield full-length transgene sequence results. As next-generation sequencing (NGS) costs decrease, the use of whole genome sequencing (WGS) as a method to locate transgene insertion sites becomes more practical [24, 31, 32]. However, most of the data will be genomic sequences and uninformative for transgene location. Another NGS-based method called targeted locus amplification (TLA) [33, 34] can provide high coverage base-pair resolution of transgene integration sites. However, TLA has similar drawbacks as previously described methods, namely, high-input sample requirements that often require euthanizing the transgenic animal and lengthy downstream preparatory steps.

Here, we describe a method that we call Adaptive Sampling Insertion Site Sequencing (ASIS-Seq) for targeted sequencing of transgene integration sites on Oxford Nanopore Technologies' (ONT) GridION Mk1 sequencer. The technology enabling this approach is a software feature called Adaptive Sampling available on the ONT MinKNOW control software. Released in 2020 as a user-selectable option in the MinKNOW software, Adaptive Sampling is a novel and purely computational-based approach to targeted sequencing. Through real-time mapping of sequence data to a user-supplied transgene reference sequence file, the Adaptive Sampling feature enables the GridION to reverse the voltage across individual nanopores to eject DNA strands from the pores, effectively enriching or depleting sequences of interest [35, 36]. The ASIS-Seq method can precisely map transgene insertion sites starting with only 2 ug of genomic DNA (*see* **Note 10**). The method requires about 3 hours of hands-on sample preparation time and generates data for analysis after a sequencing run time of only 72 hours.

2 Materials

2.1 Equipment

1. GridION Mk1 sequencer running MinKNOW software version 21.11.7 or higher (ONT).
2. Alternatives to the GridION Mk1 sequencer are the MinION Mk1B or MinION Mk1C sequencer (ONT) connected to a host computer with the recommended hardware specifications and running MinKNOW software version 21.11.7 or higher.
3. Tabletop centrifuge capable of high speed 15,000 × *g* (Eppendorf Centrifuge 5418).
4. Qubit 3 fluorometer (Life Technologies Q33216).
5. Single-channel pipettor, up to 20 μL.
6. Single-channel pipettor, up to 200 μL.
7. Single-channel pipettor, up to 1000 μL.
8. Agarose gel electrophoresis apparatus.
9. Tabletop incubator and shaker (Eppendorf ThermoMixer).
10. 1.5 mL tube magnetic stand (Bioneer MagListo 2).

2.2 Reagents Flow Cell R9.4.1 (ONT FLO-MIN106D)

1. Ligation sequencing kit (ONT SQK-LSK109).
2. Cas9 sequencing kit (ONT SQK-CS9109).
3. Flow cell priming kit (ONT EXP-FLP002).
4. Flow cell wash kit (ONT EXP-WSH004).
5. Qubit dsDNA BR Assay Kit (Thermo Fisher Scientific Q32850).
6. g-TUBE for DNA shearing (Covaris 520079).
7. DNA LoBind 1.5 mL tubes (Eppendorf 022431021).
8. Saturated (6 M) NaCl (Ward's Science 470225-798).
9. 20% SDS (Invitrogen AM9820).
10. 5 M NaCl: dissolve 292 *g* of NaCl in 800 mL of Milli-Q water, adjust final volume to 1 L with Milli-Q water, autoclave to sterilize, and store at room temperature.
11. 1 M Tris–HCl pH 7.5: dissolve 6.05 *g* of Tris base in 30 mL of Milli-Q water, adjust pH to 7.5 with 5 M HCl, adjust final volume to 50 mL with Milli-Q water, autoclave to sterilize, and store at room temperature.
12. 0.5 M EDTA pH 8.0: dissolve 186.1 *g* of disodium EDTA•2H2O in 800 mL of Milli-Q water, adjust pH to 8.0 with NaOH, adjust final volume to 1 L with Milli-Q water, autoclave to sterilize, and store at room temperature.
13. Agarose LE (Roche 11685678001).
14. 10 mg/mL ethidium bromide (Bio-Rad 161-0433).
15. Lambda DNA Hind III digest (NEB N3012S).
16. 1 kb DNA ladder (NEB N3232S).

2.3 Buffers and Other Solutions

1. 1× TE buffer pH 7.5: add 10 mL of 1 M Tris–HCl pH 7.5 to 2 mL of 0.5 M EDTA pH 8.0, adjust final volume to 1 L with Milli-Q water, and store at room temperature.
2. 1× TBE buffer: add together 10.8 *g* of Tris base, 0.93 *g* of Na2EDTA•2H2O (EDTA, disodium, dihydrate), and 5.5 *g* of boric acid, adjust final volume to 1 L with Milli-Q water, check pH (range: 8.2 to 8.5), and store at room temperature.
3. Tail lysis buffer: mix together 12.5 mL of 1 M Tris–HCl pH 8.0, 5 mL of 5 M NaCl, 50 mL of 0.5 M EDTA, and 12.5 mL of 20% SDS, adjust final volume to 250 mL with Milli-Q water, sterilize by 0.2 μM PES filtration, and store at room temperature.
4. 70% EtOH: mix 30 mL Milli-Q water and 70 mL 200 proof ethanol, and store at room temperature.

2.4 Software and Internet Resources

1. Minimap2 sequence alignment program version 2.17-r941 or newer (https://github.com/lh3/minimap2).
2. Samtools suite of programs for interacting with high-throughput sequencing data version 1.14 or newer (http://www.htslib.org/).
3. Integrative Genomics Viewer (IGV) interactive tool for visualizing genomic data version 2.11.3 or newer (https://software.broadinstitute.org/software/igv/).
4. Basic Local Alignment Search Tool (BLAST) for finding regions of similarity between biological sequences (https://blast.ncbi.nlm.nih.gov/Blast.cgi).
5. A distribution of the Linux operating system running the Bash Unix shell.

3 Methods

3.1 Generating a Mouse Model by Random Transgenesis

A detailed description of transgene design, equipment, reagents, and methods for generating transgenic mice is beyond the scope of this chapter. The reader is referred to many excellent manuals on this topic [37, 38]. Experimental details on the piggyBac transposon [14, 39] and BAC recombineering [40, 41] as versatile tools for the genetic manipulation of the mouse are available.

3.2 Genomic DNA Isolation from Mouse Tail Biopsies

The genomic DNA isolation protocol was adapted from Beermann et al. [42]. All centrifugation steps were done in an Eppendorf 5418 centrifuge:

1. Transfer 5 mm of a mouse tail biopsy to a tube containing 750 μL of 100 mM NaCl, 50 mM Tris–Hcl (pH 8.0), 100 mM EDTA (pH 8.0), 1% SDS, and 0.5 mg/mL proteinase K.
2. Incubate the tube containing the tail biopsy in the lysis buffer overnight at 56C in an Eppendorf ThermoMixer set to shake at 500 RPM.
3. After the overnight incubation, briefly vortex the tube and add 250 μL of saturated NaCl.
4. Mix the sample for 5 minutes at room temperature in an Eppendorf ThermoMixer set to shake at 500 RPM.
5. Pellet the remaining bulk cellular proteins by centrifugation for 10 minutes at >15,000 × *g*.
6. Transfer 750 μL of the supernatant to a new tube containing 500 μL of isopropanol. Mix the sample for 5 minutes in an Eppendorf ThermoMixer set to shake at 500 RPM.

7. Pellet the DNA by centrifugation for 10 minutes at >15,000 × *g*.
8. Decant or aspirate away the supernatant without disturbing the DNA pellet. Then wash the DNA pellet with 500 μL 70% ethanol and centrifuge the tube for 5 minutes at >15,000 × *g*.
9. Decant or aspirate away the ethanol without disturbing the DNA pellet. After briefly air-drying the pellet, resuspend it in 100 μL of TE buffer (pH 7.5). To facilitate DNA resuspension, place the tube in an Eppendorf ThermoMixer with the temperature set at 42C and shaking at 500 RPM for 2 hours.
10. Determine the DNA concentration using the Life Technologies Qubit broad-range assay kit on a Qubit 3 fluorometer.
11. Store the genomic DNA in a 4C refrigerator.

3.3 Shearing Genomic DNA

The Nanopore Info Sheet on adaptive sampling (https://community.nanoporetech.com/info_sheets/adaptive-sampling/v/ads_s1016_v1_reve_12nov2020) (*see* **Note 1**) provides experimental design considerations for using adaptive sampling. Sample read lengths can affect the overall target fold enrichment compared to a sequencing run without adaptive sampling turned on. Because the adaptive sampling algorithm takes time to recognize whether a strand is on- or off-target, longer read lengths are preferred. Shorter strands may pass through the pore before the algorithm has made the decision, thereby reducing overall enrichment. In practice, we found shearing the genomic DNA to at least half the length, on average, of the transgene works. The method described in this chapter works better for longer and multicopy transgenes because the percentage of the genome represented by the transgene is higher (*see* **Note 10**). We started with approximately 2–3 ug of genomic DNA, which yielded multiple sequencing libraries of at least 50 fmol each that were loaded across several days of a sequencing run to maximize data output. Aside from the changes noted in the method below, we follow the Covaris g-TUBE user manual for shearing genomic DNA (https://covaris.com/wp-content/uploads/pn_010154.pdf):

1. Transfer approximately 2–3 ug of unsheared genomic DNA to a clean 1.5 mL Eppendorf LoBind tube.
2. Add enough TE buffer pH 7.5 to the DNA in order to bring the total volume in the tube to 150 uL. Pipet up and down to mix.
3. Remove the screw cap from the g-TUBE and load 150 uL of sample into the top of the tube.
4. Screw the cap back on tightly and load the g-TUBE (screw cap up) into an Eppendorf 5418 tabletop centrifuge. If necessary, use an extra g-TUBE to balance the rotor.

Table 1
Recommended centrifugation speeds in an Eppendorf 5424 and 5415R centrifuge for shearing DNA to a desired target size

Eppendorf® 5425 and 5415 centrifuges – 24 positions rotor – Speed (RPM)				
Targeted size Mass of DNA	**6 kbp**	**8 kbp**	**10 kbp**	**20 kbp**
=<4 μg	11,000	7200	6000	4200
8 μg	11,000	7200	6000	4200
15 μg	13,200	8600	7200	5800
30 μg	13,200	10,000	8600	7200
Processing time (seconds)	30 seconds	60 seconds	60 seconds	60 seconds

5. Table 1 is reproduced from the user manual and shows centrifugation speed recommendations for Eppendorf 5424 and 5415R centrifuges. We achieved similar results in an Eppendorf 5418 tabletop centrifuge.
6. While the user manual suggests passing the sample through the g-TUBE orifice twice, we recommend passing the sample through the orifice six times to narrow the range of sheared DNA sizes.
7. After the sixth pass through the orifice, the g-TUBE should be in an inverted position with the screw cap down. Transfer the g-TUBE from the centrifuge to the UNLOAD position of the included g-TUBE stand, keeping the screw cap down all the time.
8. Unscrew the tube body leaving the screw cap containing the sample in the stand. Transfer the sample from the screw cap to a clean 1.5 mL LoBind tube.
9. Store the sheared genomic DNA at 4C or proceed directly to visualizing the sheared DNA by agarose gel electrophoresis.

3.4 Visualize Sheared Genomic DNA by Agarose Gel Electrophoresis

1. Cast a 0.8% agarose gel using 1× TBE as the buffer and a comb with enough slots to load an unsheared DNA sample, the sheared DNA sample, and two molecular weight markers: Lambda DNA Hind III digest (NEB) and 1 kb DNA ladder (NEB).
2. Into separate comb slots, load 250 ng each of unsheared and sheared DNA, each in a total volume of 50 uL 1× TE buffer pH 7.5 containing 1× purple gel loading dye (NEB).
3. Load 500 ng each of Lambda DNA Hind III digest and 1 kb DNA ladder into separate comb slots.
4. Run the gel for 2–3 hours at 100 V.

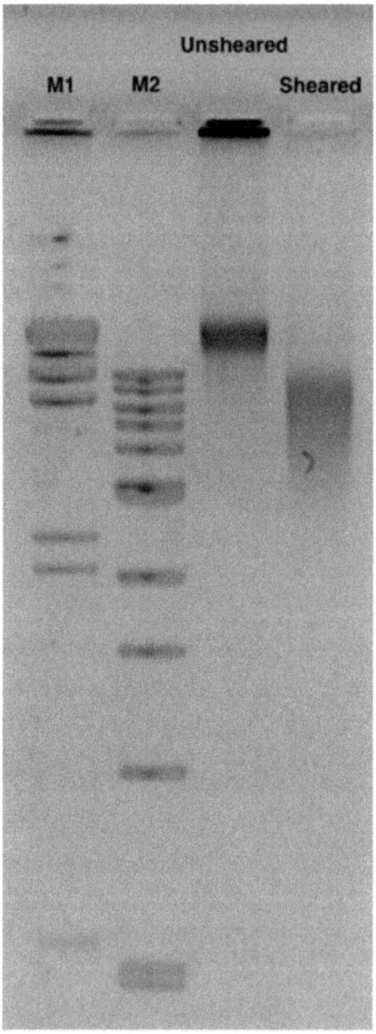

Fig. 1 Representative unsheared and sheared genomic DNA visualized on an agarose gel. Lane M1 is a NEB Lambda DNA Hind III digest ladder, and lane M2 is a NEB 1 kb DNA ladder. The sheared genomic DNA was processed in a Covaris g-TUBE centrifuged at 4218 × *g* in an Eppendorf 5418 centrifuge

5. Post-stain the gel in 500 mL of 1× TBE buffer containing 5 uL of 10 mg/mL ethidium bromide staining solution. Stain the gel for 30 minutes in a dark environment at room temperature.
6. Rinse the gel with distilled water. Please consult your local environmental health and safety guidelines for the proper handling and disposal of liquids containing ethidium bromide.
7. Visualize the DNA under UV light.
8. Figure 1 shown below is a photo of sheared and unsheared genomic DNA visualized on an agarose gel.

3.5 Sequencing Library Preparation

1. Divide the sheared genomic DNA into three equal volume aliquots of 47 uL each. Adjust the volume to 47 uL with nuclease-free water, if necessary. Each aliquot should contain approximately 600–1000 ng of sheared DNA if starting with 2–3 ug of unsheared genomic DNA.
2. For each aliquot of sheared DNA, sequencing libraries were constructed using the ONT's SQK-LSK110 ligation sequencing kit according to the manufacturer's instructions (https://community.nanoporetech.com/docs/prepare/library_prep_protocols/genomic-dna-by-ligation-sqk-lsk110/v/gde_9108_v110_revp_10nov2020) (*see* **Note 2**).
3. The final step of the library preparation process is ligating the Adapter Mix F (AMX-F). After cleanup of the ligation reaction with AMPure XP beads, the final prepared library is eluted with 15 uL of elution buffer (EB).
4. There should now be three separate tubes of library in 15 uL of EB each. Combine the three libraries into one tube and pipet gently to mix.
5. Transfer 12 uL of the combined library into a new 1.5 mL LoBind tube, and proceed to the "Priming and loading the SpotON flow cell" section of the manufacturer's instructions. A FLO-MIN106D r9.4 SpotON flow cell was used. Store the remaining library at 4 °C.

3.6 MinKNOW Control Software Settings for Adaptive Sampling

An ONT GridION Mk1 sequencer running MinKNOW control software version 21.11.7 was used. Please refer to ONT's "MinKNOW Technical Document" (https://community.nanoporetech.com/technical_documents/minknow-tech-doc/v/mitd_5000_v1_revad_16may2016) for the steps in starting an experiment. We use the default MinKNOW settings for an adaptive sampling run except for the following changes:

1. In the "Kit selection" section, choose "Ligation Sequencing Kit SQK-LSK110."
2. In the "Run options" section, toggle on the "Adaptive sampling" feature.
3. Once the "Adaptive sampling" feature is enabled, an "Adaptive sampling options" pop-up window appears. For the "Alignment reference," upload the FASTA-formatted sequence file of the transgene. Check that channels 1-512 are enabled for adaptive sampling and that "Enrich" is highlighted. Click "Save" to confirm.
4. In the "Basecalling" section, select "Super-accurate basecalling" as the basecalling option. Click "Save" to confirm.
5. In the "Output" section, uncheck the "FAST5" option while ensuring the "FASTQ" option is checked. We disable FAST5

output because FAST5 files are large and not required for data analysis. Click on the gear icon for the "FASTQ" option, and set the compression of the FASTQ to "Off." Downstream data analysis is done on uncompressed FASTQ instead of FASTQ. GZ files. Click "Save" to confirm.

6. In the "Final review" section, review the settings selected. Click "Start" to begin the experiment.

3.7 Flow Cell Wash and Reloading Libraries

A successful sequencing by adaptive sampling experiment relies on sufficient enrichment from the sample. According to the Nanopore Info Sheet on adaptive sampling (*see* Subheading 3.3), the cause for insufficient enrichment is data output decreasing quickly over time. The solution is to wash the flow cell using the Flow Cell Wash Kit to recover some channels and then reload more of your sample. Others, [36, 43], also reported improvements in enrichment by washing the flow cell in between library reloads. We recommend washing the flow cell 20–24 hours in between the library loads:

1. Twenty to 24 hours after the initial library load, on the MinKNOW control software, select your active experiment and click "Pause."
2. While the experiment is paused, follow the instructions for washing the flow cell with the EXP-WSH004 kit (https://community.nanoporetech.com/docs/prepare/library_prep_protocols/flow-cell-wash-kit-exp-wsh004/v/wfc_9120_v1_revd_08dec2020).
3. After the 60 minutes wait period with the Flow Cell Wash Mix inside the flow cell, proceed with the instructions "To run a second library on a MinION/GridION flow cell straight away."
4. Take out from the 4C storage the tube of remaining prepared library from the previous day. Place the tube on ice.
5. Transfer 12 ul of the prepared library into a new 1.5 mL LoBind tube. Follow the instructions from the SQK-LSK110 kit for "Priming and loading the SpotON flow cell." Store the remaining library at 4C.
6. After the second library is loaded into the flow cell, go to the MinKNOW control software, select the paused experiment, and click "Resume."
7. After another 20–24 hours has elapsed, repeat the steps to wash the flow cell and load the final 12 uL of the prepared library.
8. Seventy-two hours after the start of sequencing, the sequencing run will stop automatically.

3.8 Data Analysis

For the data analysis steps detailed below, the most computationally intensive step involves the minimap2 tool to align the FASTQ files to a custom reference genome sequence. We performed the data analysis separately on a high-performance computing (HPC) platform running CentOS 7.7. However, a HPC platform is not required. For GridION users, the computer hardware inside the GridION operates on the Ubuntu LTS operating system (version 16.04.7 at the time of writing) and is capable of the data analysis steps. For those using third-party laptops and desktops running different operating systems, please refer to the minimap2 github repository for support (https://github.com/lh3/minimap2/blob/master/README.md#help). The steps below that are highlighted in gray and begin with "$" indicate a Bash Unix shell prompt with a command the user types to perform a task:

1. The FASTQ files on the GridION are typically stored in a subdirectory called "fastq_pass" that is nested within the main experimental run directory located in the path "/data/" of the GridION file system. All of the commands detailed below are entered within the "fastq_pass" subdirectory. Copy the contents of the main experimental run directory to a separate storage location if performing the analysis on a computer other than the one inside the GridION (*see* **Note 3**).
2. The minimap2 alignment tool requires a reference sequence to map the FASTQ files. A custom reference sequence that combines the *Mus musculus* mouse genome (version GRCm38/mm10) and the transgene sequence needs to be created. First, copy the same FASTA-formatted transgene sequence used for adaptive sampling to the "fastq_pass" subdirectory. To the same subdirectory, download the mm10 reference sequence from the UCSC genome repository: $wget --timestamping 'ftp://hgdownload.cse.ucsc.edu/goldenPath/mm10/bigZips/mm10.fa.gz' -O mm10.fa.gz
3. Unzip the downloaded file: $gunzip mm10.fa.gz.
4. Concatenate the transgene and mm10 FASTA files to create a "custom_ref.fasta" files (*see* **Note 4**): $cat mm10.fa transgene.fasta > custom_ref.fasta.
5. The individual FASTQ files in the "fastq_pass" subdirectory consist of 4000 sequencing reads per file if the default options are used in the FASTQ settings in the MinKNOW control software. Combine all the individual FASTQ files into one merged FASTQ file named "merged.fastq": $find . -name '*.fastq' ! -name 'merged.fastq' -exec cat {} + | tee merged.fastq >/dev/null.

6. Use the minimp2 alignment tool to map the merged FASTQ files against the custom reference and output the alignments as a SAM-formatted file named "merged.sam" (*see* **Note 5**): $minimap2 -ax map-ont custom_ref.fasta merged.fastq -o merged.sam.
7. Use the samtools tool to convert the SAM file to a BAM file named "merged.bam" (*see* **Note 6**): $samtools view -uSb -o merged.bam merged.sam.
8. Use the samtools tool again to sort the alignments in the BAM file by leftmost coordinates: $samtools sort -o merged_sorted.bam merged.bam.
9. Use the samtools tool a final time to index the sorted BAM file to create a "merged_sorted.bam.bai" file: $samtools index merged_sorted.bam.
10. At this point, the "fastq_pass" subdirectory should contain the following three files: custom_ref.fasta, merged_sorted.bam, and merged_sorted.bam.bai.

    If necessary, copy these three files to the same computer where the IGV interactive tool is installed.
11. Open the IGV interactive tool. Load the "custom_ref.fasta" file by navigating in the menu: Genomes > Load Genome from File....
12. Load the "merged_sorted.bam" file by navigating in the menu: File > Load from File....
13. After the custom reference and BAM files are loaded, select the transgene reference in the "Select a chromosome to view" dropdown menu. Any sequencing reads that map to the transgene reference will appear in the IGV main window.
14. Figure 2 shows a representative alignment of ASIS-Seq sequence data to a transgene reference sequence as viewed on the IGV main window.
15. Figure 3 shows an expanded view of the genome/transgene junction on the left of Fig. 2.
16. While in the expanded view, moving the mouse cursor over an individual read and right-clicking on the read will open a pop-up window with additional menu items. Select "Copy read sequence" (*see* **Note 7**).
17. On a web browser, navigate to NCBI's online BLAST server (https://blast.ncbi.nlm.nih.gov/Blast.cgi). Select the "Nucleotide BLAST" option.
18. In the "Enter Query Sequence" box, paste the read sequence copied from IGV into this box.

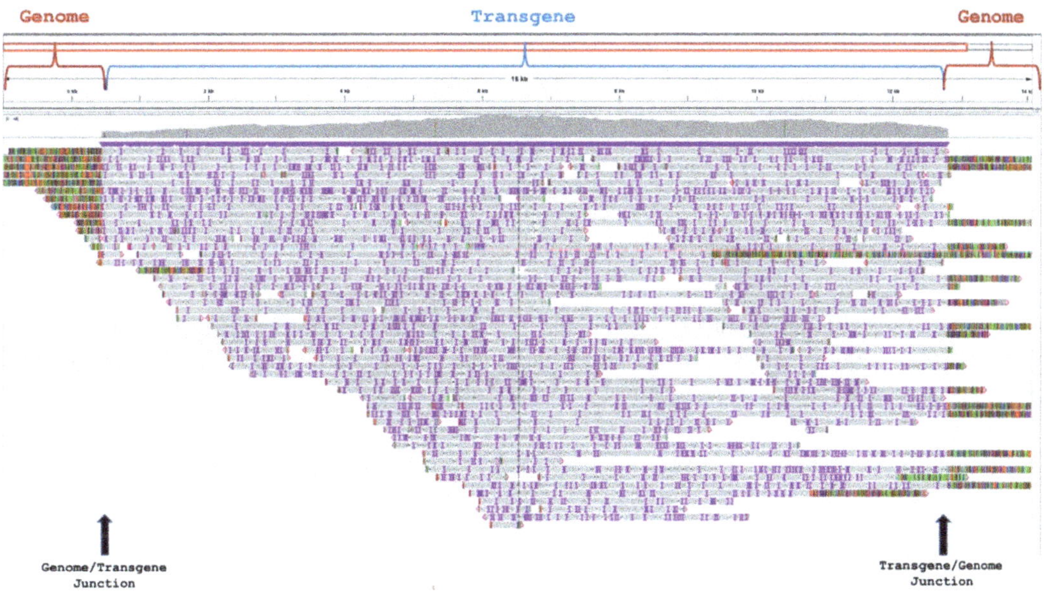

Fig. 2 An example of ASIS-Seq sequence data aligned to a transgene reference sequence as viewed on IGV. Reads with sequences in the center that align to the 12 kb piggyBac transgene sequence (highlighted by the blue curly bracket) also contain sequences on the ends that do not align to the transgene but presumably align to the mouse genome (highlighted by the red curly brackets)

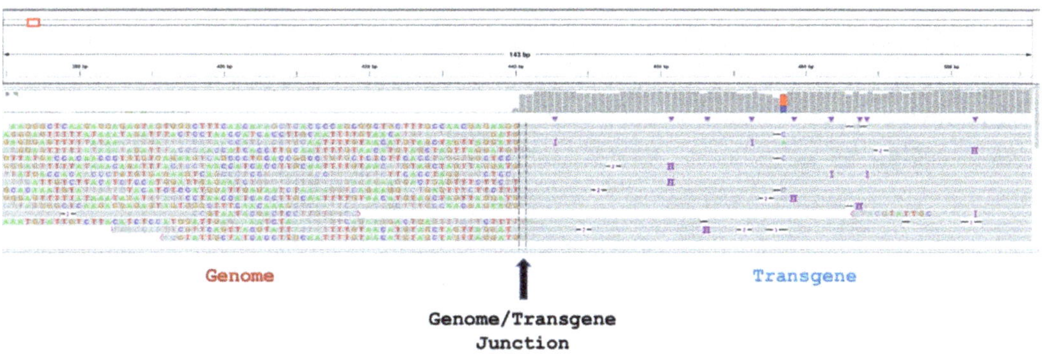

Fig. 3 An expanded view of the genome/transgene junction shown in the previous figure. In this view, the read bases that match the transgene sequence are displayed in gray (shown to the right of the genome/transgene junction), while read bases that do not match are color coded (shown to the left of the genome/transgene junction)

19. In the "Database" dropdown list, select "RefSeq Genome Database (refseq_genomes)."
20. In the "Organism" dropdown list, select "Mus musculus (taxid:10090)."
21. Click the "BLAST" button shown at the bottom of the page.

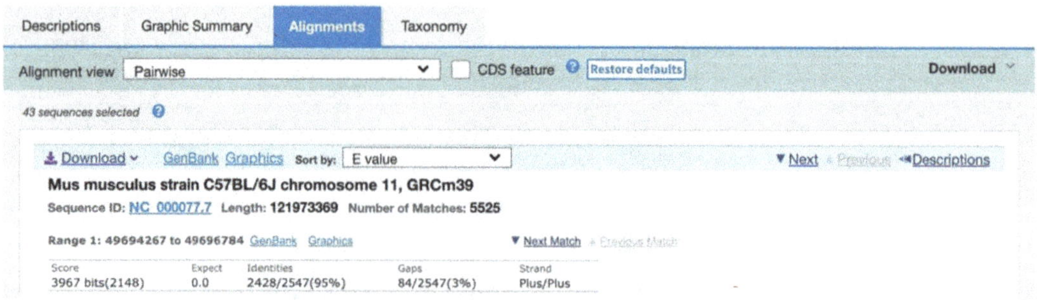

Fig. 4 A screenshot showing the BLAST search result from aligning an ASIS-Seq read sequence to the mouse reference genome (GRCm39/mm11). This read contains transgenic sequences according to the IGV alignment output and now also contains sequences matching Chr11:49694267-49696784 in the mouse according to BLAST

22. If the copied read sequence matches sequences in the mouse genome, a list of the highest scoring matches will appear showing the genomic coordinates of the matches. An example of a significant match using BLAST is shown in Fig. 4.
23. By repeating the process of BLAST-ing each read sequence that spans the transgene/genome junction on either side of the transgene, a more detailed map of the location, or locations in the case of multicopy insertions, of the transgene appears.

3.9 Confirmation of Transgene Insertion Sites by Cas9-Targeted Sequencing

The utility of the ASIS-Seq method lies in its simplicity and speed for detecting transgene/genome junctions that reveal a transgene's location in the host genome. The methods outlined in this section enable sequence confirmation of those locations as well as the conformations of or the presence of any structural variations in the transgene. By taking advantage of the RNA-programmable CRISPR/Cas9 nuclease's high specificity when cutting DNA sequences, locus-specific guide RNAs can be designed to direct the Cas9 nuclease to locations flanking the transgene and excise out the transgenic/genomic DNA fragment for sequence confirmation. ONT offers a Cas9 sequencing kit (ONT SQK-CS9109) for this purpose (*see* **Note 8**). The protocol accompanying the kit is detailed and comprehensive and will not be reprinted here. Instead, a link to ONT's protocol is provided below with additional notes:

1. Prior to starting the sequencing protocol, please read the Nanopore Info Sheet on "Targeted, amplification-free DNA sequencing using CRISPR/Cas" (https://community.nanoporetech.com/info_sheets/targeted-amplification-free-dna-sequencing-using-crispr-cas/v/eci_s1014_v1_reve_11dec2018). The Info Sheet contains important background information, recommendations for CRISPR crRNA probe design, tutorials for data analysis, and guidance on

troubleshooting. We achieved satisfactory results by following the recommendations in the Info Sheet (*see* **Note 9**).

2. The Cas9 sequencing kit (SQK-CS9109) protocol (https://community.nanoporetech.com/docs/prepare/library_prep_protocols/cas-sequencing-kit/v/cas_9106_v109_revd_16sep2020) recommends starting with 5 ug of high molecular weight genomic DNA, although 1–10 ug is an acceptable range. Target coverage scales linearly with input amount, so the input amount may be reduced accordingly if lower throughput is acceptable (*see* **Note 10**).
3. We sequenced prepared Cas9-targeted sequencing libraries in a R9.4 SpotON flow cell (FLO-MIN106D) on a GridION Mk1 sequencer running MinKNOW control software version 21.11.7. We used the default MinKNOW settings except for the following changes:
   (a) In the "Kit selection" section, choose "CAS109 Sequencing Kit SQK-CS9109" (*see* **Note 11**).
   (b) In the "Basecalling" section, select "Super-accurate basecalling" as the basecalling option. Click "Save" to confirm.
   (c) In the "Output" section, uncheck the "FAST5" option while ensuring the "FASTQ" option is checked. We disable FAST5 output because FAST5 files are large and not required for data analysis. Click on the gear icon for the "FASTQ" option and set the compression of the FASTQ to "Off." Downstream data analysis is done on uncompressed FASTQ instead of FASTQ.GZ files. Click "Save" to confirm.
   (d) In the "Final review" section, review the settings selected. Click "Start" to begin the experiment.
4. Similar to an ASIS-Seq sequencing run, preparing additional aliquots of the sequencing library in advance and washing the flow cell 20–24 hours between loading the additional libraries can yield higher enrichment. The steps for "Flow cell wash and reloading libraries" are detailed in Subheading 3.7.
5. After the sequencing run is complete, the steps for "Data analysis" detailed in Subheading 3.8 are applicable for analyzing Cas9-targeted sequencing data, except for the following changes:
   (a) Replace the transgene reference sequence with an in silico reference that represents the complete transgene sequence post-integration into the genomic locus, supported by ASIS-Seq data. The boundaries of the new reference sequence should include genomic sequences for the outermost crRNA probes used to excise the transgene

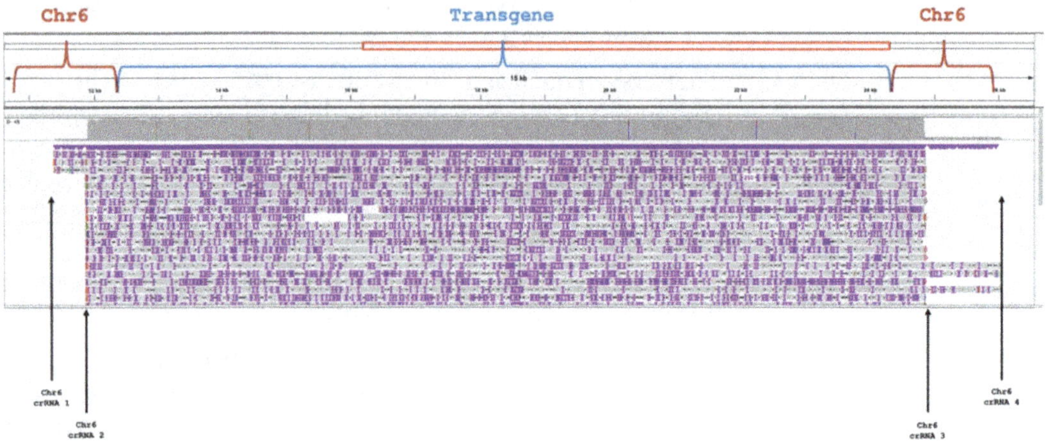

Fig. 5 A screenshot of an IGV window showing Cas9-targeted sequencing data aligning to a reference sequence predicted for a single-copy 12 kb piggyBac transgene that integrated into Chr6:47549474 as detected by ASIS-Seq. The transgene sequences are highlighted by the blue curly bracket, and the chromosome 6 sequences are highlighted by the red curly brackets. The black arrows indicate the four CRISPR/Cas crRNA probes used to excise this fragment from the genome for sequence confirmation of the transgene insertion site

fragment. A drawback for relying on a preconceived sequence as a reference during the minimap2 alignment step is the potential for missing unexpected transgene conformations such as concatemers or structural variations such as deletions, duplications, and inversions. The use of a de novo assembly tool for ONT sequencing data such as Flye (https://github.com/fenderglass/Flye) can overcome this deficiency, but the method for using this tool is beyond the scope of this publication.

(b) Figure 5 shows an example of the result from using Cas9-targeted sequencing to confirm the location of a transgene insertion site in the host genome.

(c) Instead of copying individual read sequences to paste into the online BLAST tool as detailed in **step 16** in Subheading 3.8, the user can use IGV's "Copy consensus sequence" option to copy the consequence sequence of the region in view in the IGV main window into the BLAST tool. Be aware the copied content contains additional text besides sequence bases. Only the consensus sequence bases should be pasted into the BLAST tool.

(d) The highest scoring result of the BLAST tool should display and confirm the genomic coordinates of the transgene insertion site.

4 Notes

1. URL download links to all ONT protocols and information sheets in this publication require the user to first register on ONT's website https://nanoporetech.com/.
2. The SQK-LSK110 ligation sequencing kit protocol recommends adding the included DNA Control Strand (DNACS) template when preparing a sequencing library. The template is useful for troubleshooting purposes in later steps. However, the user may omit it to maximize the number of pores available for enriching target sequences.
3. Data storage requirements depend on the amount of sequencing data generated. Ensure storage requirements are met prior to transferring data.
4. If the DNACS template was used to prepare the library, the sequence of DNACS can be downloaded here: https://assets.contentful.com/hkzaxo8a05x5/2IX56YmF5ug0kAQYoAg2Uk/159523e326b1b791e3b842c4791420a6/DNA_CS.txt.

   The command to concatenate the mm10, transgene, and DNACS sequences becomes:

   $cat mm10.fa transgene.fasta DNA_CS.txt > custom_ref.fasta
5. The minimap2 tool supports multithreading to reduce processing time. Invoke multithreading by adding the option "-t INT" to the command where INT is the number of threads to use. More information can be found at https://lh3.github.io/minimap2/minimap2.html.
6. Samtools also supports multithreading by adding the option "-@ INT" to the command.
7. While the IGV tool has a built-in BLAT tool similar to BLAST, the BLAT tool recognizes only pre-defined genomes from the UCSC BLAT server and not user-defined custom references.
8. Instead of purchasing the SQK-CS9109 Cas9 sequencing kit containing most of the necessary reagents, the user has the option to purchase the reagents separately and only follow ONT's protocol (https://community.nanoporetech.com/docs/prepare/library_prep_protocols/cas9-targeted-sequencing/v/enr_9084_v109_revs_04dec2018).
9. We purchase our CRISPR/Cas crRNA probes and the universal tracrRNA from IDT (https://idtdna.com).
10. ONT's Info Sheet on adaptive sampling recommends enriching for targets representing at least 0.1% of the genome. Transgenes, even inserted as multiple copies in a host mouse genome, often represent much less than 0.1% of the genome.

Even so, we had success using ASIS-Seq to locate a single copy of a 12 kb *piggyBac* transgene from 2 ug of genomic DNA isolated from a mouse tail biopsy. We witnessed higher enrichment for animals with high copy numbers compared to animals with low copy numbers for the transgene. Similarly, large transgenes such as those packaged in BACs, YACs, and PACs should yield higher enrichment than small transgenes. If enough starting genomic DNA is available, the ASIS-Seq method can be scaled up to process multiple sequencing libraries sequenced across multiple flow cells to maximize enrichment.

11. If only the Cas9 sequencing protocol is used with user-supplied reagents, select the "CAS109 Sequencing Protocol SQK-CAS109" as the kit.

References

1. Gordon JW, Ruddle FH (1981) Integration and stable germ line transmission of genes injected into mouse Pronuclei. Science 214: 1244–1246
2. Fielder TJ, Barrios L, Montoliu L (2010) A survey to establish performance standards for the production of transgenic mice. Transgenic Res 19:675–681
3. Hammer RE et al (1987) The rat elastase I regulatory element is an enhancer that directs correct cell specificity and developmental onset of expression in transgenic mice. Mol Cell Biol 8:2956–2967
4. Petitclerc D et al (1995) The effect of various introns and transcription terminators on the efficiency of expression vectors in various cultured cell lines and in the mammary gland of transgenic mice. Biotechnol 40:169–178
5. Schedl A et al (1993) A yeast artificial chromosome covering the tyrosinase gene confers copy number-dependent expression in transgenic mice. Nature 362:258–256
6. Ioannou PA et al (1994) A new bacteriophage P1-derived vector for the propagation of large human DNA fragments. Nat Genet 6:84–89
7. Yang XW, Model P, Heintz N (1997) Homologous recombination based modification in Escherichia coli and germline transmission in transgenic mice of a bacterial artificial chromosome. Nat Biotech 15:859–865
8. Giraldo P, Montoliu L (2001) Size matters: use of YACs BACs and PACs in transgenic animals. Transgenic Res 10:83–103
9. Brinster RL et al (1985) Factors affecting the efficiency of introducing foreign DNA into mice by microinjecting eggs. Proc Nail Acad Sci USA 82:4438–4442
10. Hirabayashi M et al (2001) A comparative study on the integration of exogenous DNA into mouse rat rabbit and pig genomes. Exp Anim 50:125–131
11. Palmiter RD, Brinster RL (1985) Transgenic mice. Cell 41:343–345
12. Garrick D et al (1998) Repeat-induced gene silencing in mammals. Nat Genet 18:56–59
13. Mátés L et al (2009) Molecular evolution of a novel hyperactive sleeping beauty transposase enables robust stable gene transfer in vertebrates. Nat Genetic 41:53–61
14. Ding S et al (2005) Efficient transposition of the piggyBac (PB) transposon in mammalian cells and mice. Cell 122:473–483
15. Katter K et al (2013) Transposon-mediated transgenesis transgenic rescue and tissue-specific gene expression in rodents and rabbits. FASEB J 27:930–941
16. Li MA et al (2011) Mobilization of giant piggyBac transposons in the mouse genome. Nucleic Acids Res 39:e148
17. Durkin ME et al (2001) Integration of a c-myc transgene results in disruption of the mouse Gtf2ird1 gene the homologue of the human GTF2IRD1 gene hemizygously deleted in Williams-Beuren Syndrome. Genomics 73: 20–27
18. Mukai HY et al (2006) Transgene insertion in proximity to the c-myb gene disrupts erythroid-megakaryocytic lineage bifurcation. Mol Cell Biol 26:7953–7965
19. Yong CSM et al (2015) Embryonic lethality in homozygous human Her-2 transgenic mice

due to disruption of the Pds5b gene. PLoS One 10:e0136817

20. Galvan DL et al (2009) Genome-wide mapping of piggybac transposon integrations in primary human T cells. J Immunother 32: 837–844
21. Li MA et al (2013) The piggyBac transposon displays local and distant reintegration preferences and can cause mutations at noncanonical integration. Sites Mol Cell Biol 33:1317–1330
22. Saha S et al (2015) Evaluating the potential for undesired genomic effects of the piggyBac transposon system in human cells. Nucleic Acids Res 43:1770–1782
23. Goodwin LO et al (2019) Large-scale discovery of mouse transgenic integration sites reveals frequent structural variation and insertional mutagenesis. Genome Res 29:494–505
24. Nicholls PK et al (2019) Locating and characterizing a transgene integration site by nanopore sequencing. G3: Genes Genomes Genetics 9:1481–1486
25. Kulnane LS et al (2002) Rapid and efficient detection of transgene homozygosity by FISH of mouse fibroblasts. Mamm Genome 13:223–226
26. Nakanishi T et al (2002) FISH analysis of 142 EGFP transgene integration sites into the mouse genome. Genomics 80:564–574
27. Ohigashi I et al (2010) Identification of the transgenic integration site in immunodeficient tgσ26 human CD3σ transgenic mice. PLoS One 5:e14391
28. Liang Z et al (2008) Identifying and genotyping transgene integration loci. Transgenic Res 17:979–983
29. Haraguchi S, Nakagawara A (2009) A simple PCR method for rapid genotype analysis of the TH-MYCN transgenic mouse. PLoS One 4: e6902
30. Bryda EC, Bauer BA (2010) A restriction enzyme-PCR-based technique to determine transgene insertion sites. Methods Mol Biol 597:287–299
31. Ji Y et al (2014) Identification of the genomic insertion site of Pmel-1 TCR α and β transgenes by next-generation sequencing. PLoS One 9:e96650
32. Dubose AJ et al (2013) Use of microarray hybrid capture and next-generation sequencing to identify the anatomy of a transgene. Nucleic Acids Res 41:e70
33. Cain-Hom C et al (2017) Efficient mapping of transgene integration sites and local structural changes in Cre transgenic mice using targeted locus amplification. Nucleic Acids Res 45:e62
34. de Vree PJP et al (2014) Targeted sequencing by proximity ligation for comprehensive variant detection and local haplotyping. Nat Biotech 32:1019–1025
35. Kovaka S et al (2021) Targeted nanopore sequencing by real-time mapping of raw electrical signal with UNCALLED. Nat Biotech 39:431–441
36. Payne A et al (2021) Readfish enables targeted nanopore sequencing of gigabase-sized genomes. Nat Biotech 39:442–450
37. Pease S, Saunders TL (eds) (2011) Advanced protocols for animal transgenesis. Springer, Berlin, Heidelberg
38. Hofker M, van Deursen J (eds) (2011) Transgenic mouse methods and protocols. Humana Press, Totowa, NJ
39. Wang W et al (2008) Chromosomal transposition of PiggyBac in mouse embryonic stem cells. Proc Natl Acad Sci U S A 105:9290–9295
40. Johansson T et al (2010) Building a zoo of mice for genetic analyses: a comprehensive protocol for the rapid generation of BAC transgenic mice. Genesis 48:264–280
41. Gong S, Kus L, Heintz N (2010) Rapid bacterial artificial chromosome modification for large-scale mouse transgenesis. Nat Protoc 5: 1678–1696
42. Beermann F et al (1993) Perinatal activation of a tyrosine aminotransferase fusion gene does not occur in albino lethal mice. Mech Dev 42: 59–65
43. Martin S et al (2022) Nanopore adaptive sampling: a tool for enrichment of low abundance species in metagenomic samples. Genome Biol 23:11

Chapter 6

High-Throughput Analysis of CRISPR-Cas9 Editing Outcomes in Cell and Animal Models Using CRIS.py

Shilpa Narina, Jon P. Connelly, and Shondra M. Pruett-Miller

Abstract

Genome editing using the CRISPR-Cas9 platform creates precise modifications in cells and whole organisms. Although knockout (KO) mutations can occur at high frequencies, determining the editing rates in a pool of cells or selecting clones that contain only KO alleles can be a challenge. User-defined knock-in (KI) modifications are achieved at much lower rates, making the identification of correctly modified clones even more challenging. The high-throughput format of targeted next-generation sequencing (NGS) provides a platform allowing sequence information to be gathered from a one to thousands of samples. However, it also poses a challenge in terms of analyzing the large amount of data that is generated. In this chapter, we present and discuss CRIS.py, a simple and highly versatile Python-based program for analyzing NGS data for genome-editing outcomes. CRIS.py can be used to analyze sequencing results for any kind of modification or multiplex modifications specified by the user. Moreover, CRIS.py runs on all fastq files found in a directory, thereby concurrently analyzing all uniquely indexed samples. CRIS.py results are consolidated into two summary files, which allows users to sort and filter results and quickly identify the clones (or animals) of greatest interest.

Key words CRISPR-Cas9, Next-generation sequencing, CRIS.py, Knock-in, Knockout, Gene modifications, Genome, Python, Genome editing, Genome engineering

1 Introduction

CRISPR-Cas9 is a versatile technology that can be used to knock out or modify genes of interest. It is a two-component system comprising a Cas9 nuclease and a single guide RNA molecule (sgRNA) which directs the Cas9 protein to a target site through complementary base pairing with genomic DNA (gDNA). The Cas9 nuclease creates a double-strand break in the genomic DNA at the target site. This break may be repaired through the nonhomologous end-joining (NHEJ) pathway or the homology-directed

Supplementary Information The online version contains supplementary material available at https://doi.org/10.1007/978-1-0716-2990-1_6.

Thomas L. Saunders (ed.), *Transgenesis: Methods and Protocols*,
Methods in Molecular Biology, vol. 2631, https://doi.org/10.1007/978-1-0716-2990-1_6,

repair (HDR) pathway. Repair by the NHEJ pathway frequently causes insertions or deletions (indels) in the genome, which leads to gene disruption. Alternatively, if a donor template that has homology to the target site is introduced into the cell with the Cas9 complex, the break can be repaired by HDR. As a result, nucleotide deletions, insertions, or substitutions present in the donor template can be copied into the genome. This targeted strategy is used to create modified cell and animal models that can be used for a wide variety of research applications.

The process of creating a modified cell line or animal generally involves introducing the CRISPR-Cas9 complex and donor template, if required, into cells or embryos and then screening the resultant cells or animals to identify those of interest. This screening process is one of the biggest challenges. Often it is necessary to single cell sort and screen several hundred clones to identify a few correctly modified clones due to low editing rates, poor clonability, and/or the need for homozygously or heterozygously modified alleles. Moreover, some clones may have one correctly edited allele as desired, but the other allele may contain an unwanted indel. Additionally, when creating transgenic animal models via direct zygote injection, the resulting animals are often mosaic. In this case, it is ideal to identify animals with the highest frequency of correctly edited alleles. Several screening methods are based on lower-throughput platforms, like Sanger sequencing; these methods require analyzing each clone individually, which is not suitable for high-throughput screening.

Next-generation sequencing (NGS) offers a method for high-throughput interrogation of CRISPR editing outcomes. Targeted NGS analysis provides the sequence identity, size of indels, and frequency of indels. However, analyzing large amounts of data generated by NGS requires efficient and accurate tools. CRIS.py is one such tool which is a Python-based program that analyzes NGS data obtained from both pools of cells and individual clones or animals [1].

In this protocol, we provide the steps of screening and analyzing clones for desired modification(s) using CRIS.py. This chapter begins at primer design and sequencing your target sites, followed by a step-by-step explanation of using CRIS.py to analyze your results.

2 Materials

2.1 Wet Lab

1. PCR primers.
2. 2× MyTaq Red Mix.
3. 2× Platinum™ SuperFi II Green PCR Master Mix.
4. gDNA from samples.

5. Crude extract (CE) buffer: 10 mM Tris pH 8.0, 2 mM EDTA, 0.2% Triton X-100, and 200 μg/mL proteinase K; final buffer pH 8.0.
6. Thermocyclers.
7. PCR plates and tubes.
8. Plate seals.
9. Forward index primers for NGS.
10. Reverse index primers for NGS.
11. Illumina PhiX Control v3.
12. Nuclease-free water.
13. Illumina sequencer.

2.2 Programming Resources

1. Anaconda environment.
2. Python 2.7 or higher.
3. CRIS.py from GitHub.

3 Methods

3.1 Overview of NGS Library Setup for Amplicon Sequencing

NGS is a massively parallel sequencing technology used for sequencing targeted regions or entire genomes. NGS offers immense scalability and an ultrahigh-throughput format with great speed [2, 3]. The resulting fastq files can be analyzed using different bioinformatic tools. CRIS.py is an NGS analysis program tailored to the interrogation of genome-edited samples, regardless of the genome-editing platform. In this protocol, we describe a two-step PCR process for NGS library preparation. In **step 1**, gene-specific primers with partial Illumina adapters are used to amplify the region of interest (*see* **Note 1**). In **step 2**, the amplicons generated in **step 1** are used as a template for indexing primers containing unique indexes to bind, amplify, and add the remaining Illumina adaptor sequence to the amplicons. This two-step PCR process allows for the multiplexing of many PCR reactions using the same set of PCR #2 primers. The PCR #2 primers can also be repeatedly used for amplifying and analyzing any number of unique target sites from different projects. There are four key steps in preparing samples for high-throughput amplicon sequencing (Fig. 1):

1. Identify the region to be sequenced, and design primers to amplify the target region with appropriate overhangs.
2. Perform PCR #1 on gDNA to amplify the target region, and add partial Illumina adaptors to each amplicon.

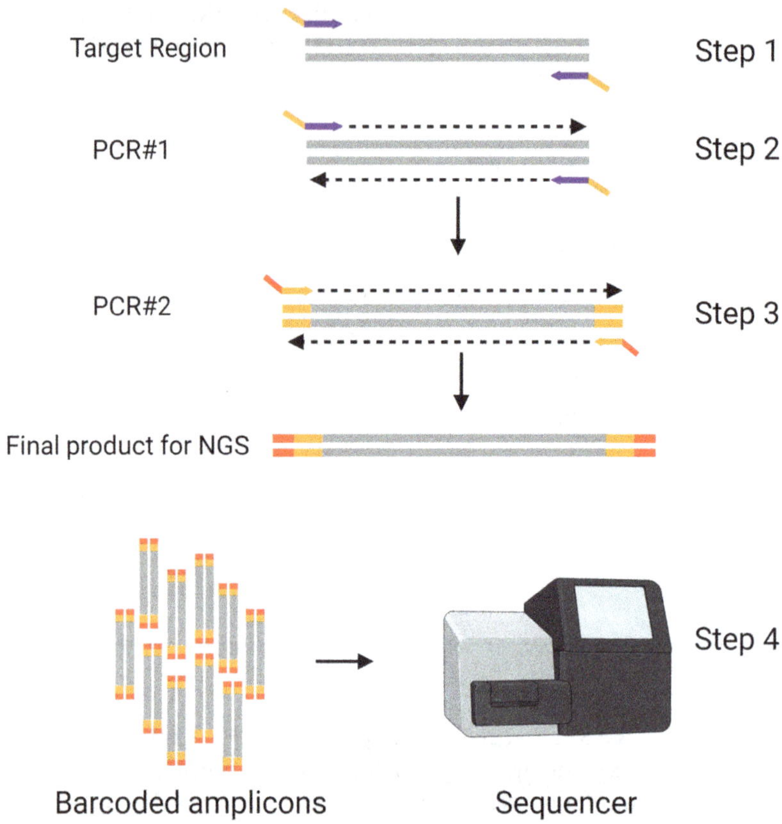

Fig. 1 Overview of PCR steps for NGS library preparation. Two rounds of PCR are used to a create an indexed pool of amplicons for targeted DS. In **steps 1** and **2**, primers are designed, and initial amplification is performed. In **step 3**, unique indexes are added to the amplicons from PCR #1, before pooling amplicons and sequencing in **step 4**. Target region in genomic DNA is shown in gray. Gene-specific primers with partial Illumina adaptors (DS tags) are shown in purple/yellow. Indexing primers are shown in yellow/red

3. Perform PCR #2 using the PCR #1 reaction as the template and primers containing unique indexes that add a distinctive identifier to each sample and the remaining Illumina adaptor sequence.
4. Sequence the PCR #2 products using an NGS platform.

The unique indexes allow all samples to be pooled together and sequenced in one NGS sequencing run. After NGS is performed, samples are demultiplexed using the unique indexes and reads assigned to each sample. In the Electronic Supplemental Materials, we have included an Excel file containing our common PCR #2 primer sequences.

3.2 Gene-Specific Primer Design

Design a primer pair to amplify the region of interest for each target for PCR#1. We recommend using PrimerBlast [4] as it also searches for and displays potential off-target amplicons:

1. Design the gene-specific primers for PCR #1 so the total length of the amplicon will be fully sequenced in the NGS run. We recommend sequencing using paired-end 250 bp reads. Design the primer pair such that the total amplicon length is less than 450 bp, with the target-modification site(s) close to the center of the amplicon and the primers binding at least 50 bp away from the cut site. At the end of this section, we will review a few basic primer design strategies for knockout (KO), knock-in (KI), and large deletion projects.
2. Add partial deep sequencing (DS) tags to the 5′ end of each primer. These tags serve two purposes: First, they function as a common annealing sequence that the indexed PCR #2 primers can bind to and amplify. Second, they are partial Illumina adaptors and are required for sequencing using Illumina platforms. The Illumina sequencing platform used is determined by the number of reads required. Partial Illumina DS tags are listed below and should be added to the 5′ end of the respective gene-specific PCR #1 primer (shown as x's below):
   (a) Forward DS tag:
   5′ CTACACGACGCTCTTCCGATCTxxxxxxxxxxxxxxxx 3′
   (b) Reverse DS tag:
   5′ CAGACGTGTGCTCTTCCGATCT xxxxxxxxxxxxxxxx 3′

The next sections describe primer design for various genome-editing strategies.

3.3 Primer Design for KO and Small Modification Projects

Design a set of primers flanking the modification site (Fig. 2). The size of the amplicon must not exceed the size of the paired sequencing read length. The modification site should be close to the center of the amplicon fragment and ideally at least 50 bp away from the primer-binding sites.

3.4 Primer Design for Large Deletion Projects

CRISPR-Cas9 may be used to create large deletions encompassing several thousand to mega base pair regions. This is achieved by designing two sgRNAs that flank the region to be deleted. Because each break point of the deletion is known, a primer pair can be designed to amplify and screen for clones containing the targeted deletion. To determine whether the deletion event has occurred, design a forward primer upstream of the 5′-break site and a reverse primer downstream of the 3′-break site (Fig. 3). If the deletion was successful, a smaller amplicon size is expected and can be detected by agarose gel or NGS. If the deletion event did not occur, the remaining sequence is often too large to be amplified with standard techniques or detected by NGS.

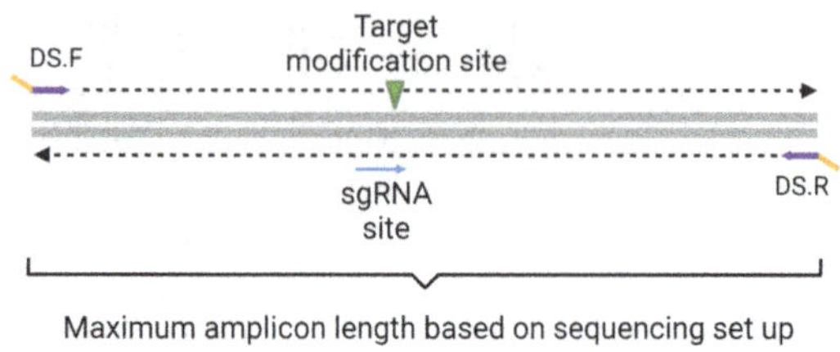

Fig. 2 Basic primer design for small insertions and KO modifications. Gene-specific forward and reverse primers (purple) with DS tags (yellow) are designed flanking the target-modification and sgRNA site. The sgRNA site is the location of the 20-nucleotide protospacer and the PAM

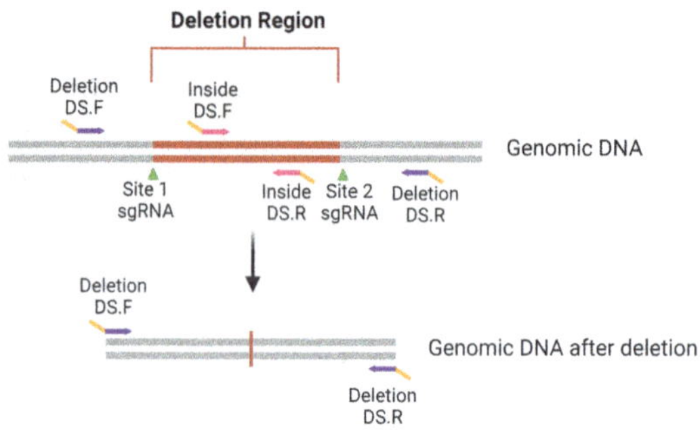

Fig. 3 Primer design and PCR strategy for large deletion projects. Gene-specific deletion primers (purple) with DS tags (yellow) are designed flanking the targeted deletion (red region flanked by sgRNA target sites) (shown by green arrows). A second set of inside primers (pink) are designed inside the targeted deletion to determine if any untargeted or inverted alleles remain

Primer design for large deletions is accomplished by first creating a sequence file in which the deletion is modeled. Next, use a primer design tool to design primers flanking the deletion that are small enough to be fully sequenced by the NGS run parameters. Often, complete-deletion clones are desired in which all alleles for a gene in a cell line contain the specified deletion. To confirm a complete deletion, use the wild-type (WT)/unedited sequence to design a second pair of primers (i.e., inside primers) that bind and amplify in the deletion region. Any clones that are not homozygous for the deletion will contain the sequence and produce an amplicon. Inversion of the intervening sequence between the two cut sites can also occur and would be positive for the inside PCR amplicon.

Because of the multiplexing capabilities of the two-step PCR method described above, both the deletion-specific and inside amplicons can be pooled, run on a single flow cell with other amplicons, demultiplexed using the unique indexes, and then analyzed using CRIS.py. A complete-deletion clone will have reads for the deletion-specific amplicon and no reads for the inside amplicon because no amplification would have occurred. The amplicons can also be run on an agarose gel, but this becomes labor intensive when screening hundreds of clones for multiple primer sets. Additionally, NGS gives the sequence identity of the deletion allele (s) and further verifies the specificity of the primers.

3.5 Primer Design for Large KI Projects

CRISPR-Cas9 technology can KI inserts ranging from one to thousands of base pairs. For smaller KIs, where the total amplicon size is within range of NGS for the KI allele, use the KO primer design strategy described in Subheading 3.3. For larger KIs, junction primers must be designed to detect targeted integration events. Donor templates contain homology arms, which are sequences that are homologous to and flank the targeted genomic region. Donor templates are generally introduced concomitantly with the sgRNA and Cas9 nuclease. Single-stranded donor templates (ssODNs or ssDNAs) have short homology arms (40–200 bp) on each side of the cut and/or modification. Double-stranded DNA (dsDNA) or plasmid donors have longer homology arms (400–800 bp) on each side of the cut and/or modification. Junction (Junc) primers are a set of primers in which one primer binds to the target gDNA outside of the donor homology arm, and another primer binds to a unique site within the sequence to be inserted. Both 5′ and 3′ junction primer sets are needed to verify complete targeted integration. It is essential to design junction primers so that they do not bind within either homology arm of the donor. Otherwise, it will not be possible to differentiate targeted integration from random integration of the donor into the genome.

3.5.1 Primer Design for Detecting KI Modifications Using ssDNA Donor Templates

The ssDNA donors used for gene targeting have small homology arms ranging from 40 bp to 200 bp. The small size allows the design of primers flanking the homology arms and producing an amplicon small enough to be sequenced using NGS. Two sets of primer pairs (PCR-A and PCR-B) are designed using the desired genomic DNA sequence after successful targeted integration as a template. These consist of a 5′ junction primer pair (PCR-A) and a 3′ junction primer pair (PCR-B) (Fig. 4). To determine if clones have any remaining untargeted alleles, perform an "out-out" PCR using the forward primer from PCR-A (5′ Gen DS.F) and the reverse primer from PCR-B (3′ Gen DS.R).

Sequence analysis will determine if the target site remains WT or if an indel has been created by Cas9 cleavage and subsequent repair by NHEJ. Here again, the NGS platform allows the pooling and sequencing of all three amplicons (5′ junction, 3′ junction, and out-out amplicon) for many clones in one sequencing run. Always ensure that you design primers that yield amplicons that are within NGS sequence read length limits.

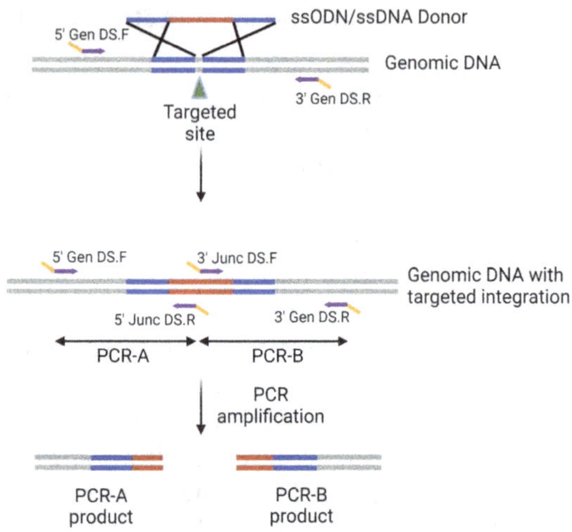

Fig. 4 Primer design and PCR strategy for large ssDNA KI projects. Two sets of primers are designed to the 5′ and 3′ regions of the targeted integration event. Each primer set consists of a gene-specific Gen primer (purple region in the Gen primer), which is outside the donor homology arms (shown in blue), and a Junc primer designed to a sequence in the unique sequence to be inserted (purple region in Junc primer). All primers contain DS tags (yellow)

3.5.2 Primer Design for KI Modifications Using Plasmid Donors

Plasmid dsDNA templates can be used for targeted KI and have larger homology arms. Traditional junction primer sets (described above) that amplify the junctions of integration events made using dsDNA donor templates generally yield amplicons that are larger than present NGS limits (>600 bp). To overcome this limitation, an additional round of nested PCR is required. The initial PCR is set up as described in the previous section with one primer binding in the gDNA outside of each homology arm and the other binding in a unique site of the integrated sequence. Because this amplicon is typically longer than NGS sequence length limits, it is referred to as long-junction PCR (Fig. 5). To generate amplicons that are within NGS limits, a nested or short-junction PCR is then performed using the long-junction PCR product as the template. For the short-junction PCR, an additional primer (5′ HA DS.F or 3′ HA DS.R) is designed that binds in the respective homology arm and is paired with the appropriate Junc DS primer. This amplicon should be within the sequence length limitations of NGS and can be used as the template for indexing PCR #2. When screening recently transfected pools of cells, it is essential to include a donor-only transfected control sample, as the donor plasmid can carry over into the gDNA preparation and result in a false-positive result. If the donor-only transfected sample yields an amplicon after the short-junction PCR, dilute the long-junction PCR product 1:10 and 1:20 and repeat the short-junction PCR. When screening clones, we do not generally see false-positive hits because the

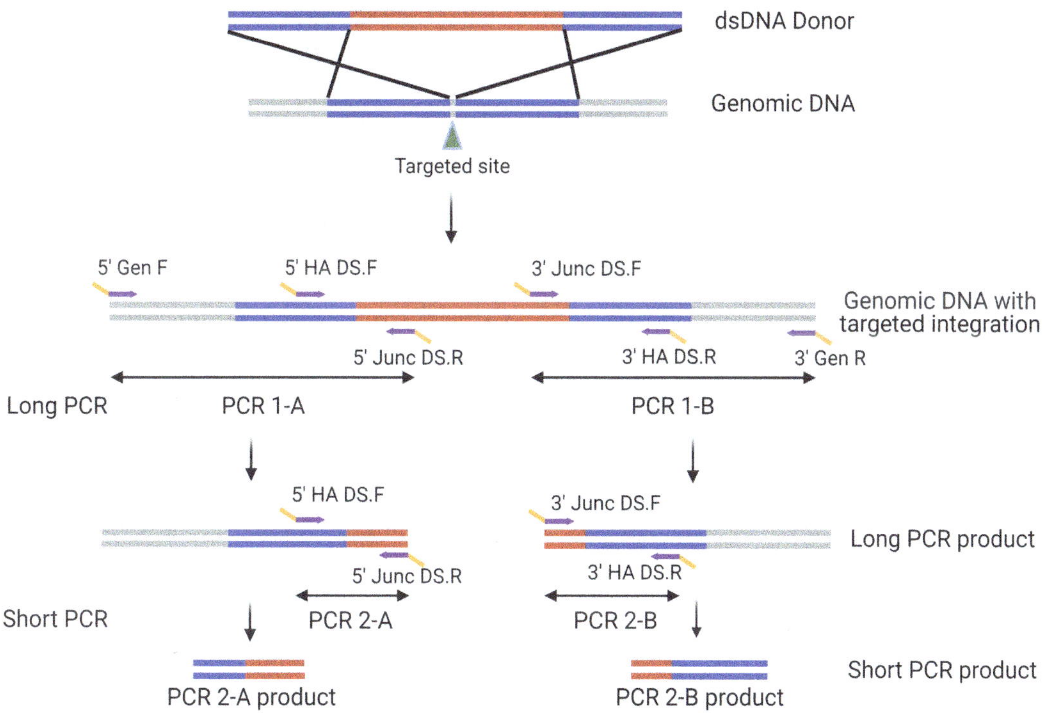

Fig. 5 Primer design and PCR strategy for KI projects using dsDNA donor. Long PCR primers are designed to the 5′ and 3′ regions of the desired targeted integration event. Each primer set contains a Gen primer designed outside of the homology arm (blue) and a Junc primer designed inside the target KI region (red). Short PCR primers are designed for nested PCR using the long PCR product as the template to create amplicons that are within the NGS sequence length limits. Short PCR primers consist of a Junc primer and a HA primer designed within the homology arm. DS tags shown in yellow

plasmid is lost over the many cell divisions required for clonal expansion. However, to ensure a clone is a true-positive hit after being identified by this nested PCR procedure, the long-junction PCR should be repeated on positive clones only, and products should be run on a gel as a final quality-control (QC) step. Although this nested PCR process requires an additional round of amplification, it enables the screening of hundreds of clones at the sequence level during one NGS run. We recommended using automation, when possible, to set up PCR reactions in the 96- or 384-well format.

3.5.3 PCR #1 to Amplify the Target Region

1. Obtain gDNA from all samples by using CE buffer as described in the next step. In general, we find it most efficient to harvest clones from 96-well tissue culture plates without consolidating clones. Additionally, PCR reactions can be minimized in volume to reduce the price per reaction which further brings down the need to make consolidation plates.

2. To harvest gDNA, pellet the cells by centrifugation at >500 × *g* for 5 minutes. Discard the supernatant and add CE buffer. The volume of CE buffer to be added depends on the approximate number of cells in the pellet. Add ~25 μL of CE buffer for every 5000 cells. As it is a crude lysis, amplification may be inhibited if not enough lysis buffer is used. When harvesting from ~80% confluent 96-well plates, we recommend using 25 μL of CE buffer per well. To lyse samples, incubate at 65 °C for 15 minutes followed by incubation at 95 °C for 5 minutes. Subjecting samples to 95 °C inactivates proteinase K in the buffer preventing further protein degradation.
3. Set up the first-round PCR master mix as shown below (Table 2) using the appropriate primer pair, as described in Subheading 3.2. Gene-specific primer design. To make the master mix, exclude the template (gDNA or long PCR product) and multiply each volume in Table 1 by the number of samples being tested. The components of the master mix are shown in blue. When possible, use automation to dispense the master mix into 96- or 384-well PCR plates, and ensure enough master mix is made to account for the dead volume of the instrument being used.
4. Add 1 μL of gDNA or long PCR product from each sample to the appropriate well containing 9 μL master mix.
5. Run the thermocycler program according to manufacturer's recommendations. We generally run 30–35 cycles of amplification for PCR #1.
6. Once PCR #1 is complete, check 2 μL of representative samples on an agarose gel to verify the presence of specific bands in appropriate wells. If the bands are present, continue to PCR #2.

Table 1
Reagents and reaction volumes for PCR#1

Reagent	Volume (μL)
2× MyTaq or Superfi II	5
Forward DS primer (10 μM)	0.5
Reverse DS primer (10 μM)	0.5
dH_20	3
gDNA or long PCR product	1
Total final reaction volume	10

Abbreviations: *dH_2O* nuclease-free water, *DS* deep sequencing, *gDNA* genomic DNA, *PCR* polymerase chain reaction

Table 2
Reagents and reaction volumes for PCR#2

Reagent	Volume (μL)
2× MyTaq or Superfi II	5
PCR2_Forward Primer (10 μM)	0.5
dH_20	2.5
Reverse index primer (10 μM)	1
PCR #1 product	1
Total final reaction volume	10

Abbreviations: *dH_2O* nuclease-free water, *PCR* polymerase chain reaction

Note: Both the long-junction PCRs and short-junction PCRs are set up as described in this section, with the exception that the long-junction PCR amplicon is used in place of gDNA for the short-junction PCR. Adjust the extension time depending on the length of the amplicon and the polymerase being used.

3.5.4 PCR #2 to Add Unique Indexes to Each Sample

1. Set up PCR #2 master mix using the table below (Table 2) as a template. Multiply the volume of each reagent according to the number of samples being tested excluding the PCR #1 template and reverse-indexing primer. The components of the master mix are shown in blue. For each sample, a uniquely indexed forward and reverse primer pair is required. We use a combination of 11 forward primers with unique 6 -bp indexes combined with 384 reverse primers with unique 10 bp indexes. For a list of functionally validated indexing primers, *see* Supplementary Tables 1 and 2.
2. Aliquot 8 μL of master mix to the appropriate number of PCR tubes or plates.
3. Add 1 μL of a 10 μM uniquely indexed reverse primer to each tube.
4. Add 1 μL of unpurified PCR #1 reaction to the corresponding tubes/plates containing PCR #2 master mix.
5. Run the thermocycler program according to the manufacturer's recommendations to add indexes. We generally run five cycles of amplification for PCR #2. Keeping the cycle number low will reduce the amplification bias.
6. After completion, pool and submit samples to the NGS facility or vendor for sequencing, demultiplexing, and merging reads.

Note: To increase sequence diversity, we recommend adding at least 10% PhiX Control v3 to the indexed amplicon pool.

Additionally, purify the amplicon to reduce primer dimer contamination. Further QC measures (e.g., Bioanalyzer or q-PCR) may be needed to determine the appropriate amount of library to load for proper clustering.

3.6 Using CRIS.py to Analyze NGS Data

CRIS.py is an analysis program that runs in a Python environment. CRIS.py enables users to analyze thousands of NGS-generated output files quickly and efficiently. The script also lets users query and quantify multiple editing events concurrently. In this section, we will discuss installation guidelines and how to modify the script to analyze multiple genome-editing outcomes in cell pools and clones.

3.6.1 Installing Python and CRIS.py

(a) CRIS.py utilizes Python 2.7 or above and the pandas module. Python can easily be installed using the Anaconda link shown here:https://www.anaconda.com/products/individual

(b) CRIS.py can be downloaded from the following link at GitHub:https://github.com/patrickc01/CRIS.py

(c) A detailed tutorial video on how to use CRIS.py is available at the link below:https://s.stjude.org/video/player.html?videoId=6000021936001

3.6.2 Setting Up a Folder with the CRIS.py Script and NGS Data

The fastq files generated from an NGS run and the corresponding CRIS.py programs should all be copied to the same directory.

3.6.3 CRIS.py Program Layout

Figure 6 shows an example CRIS.py program with user-defined inputs in single quotation marks. The example CRIS.py program shown in Fig. 6 has been referenced from GitHub [5] and modified for illustration.

```
def get_parameters():
    #Note to user- Change text inside of quote marks ('YOUR DNA SEQUENCES GO HERE') for your experiment.  Case of text does not matter.
A→  ID = 'Locus_1'
B→  ref_seq = str.upper('agggaatgccccggagggcggagaactgggacgaggccgaggtaggcgcggaggaggcaggcgtcgaagagtacggccctgaagaagacggcggggaggagtcgggcgccgaggagtccggcccggaagagtcc
C→  seq_start = str.upper('GCGGAGAACTG')
D→  seq_end = str.upper('GCCGAGGAGGA')
E→  fastq_files = '*.fastq'
F→  test_list = [
                str('g10'),   str.upper('GAGGCAGGCGTCGAAGAGTACGG'),  ┐
                str('g14'),   str.upper('CGGCCCTGAAGAAGACGGCGGGG'),  ├─ G (test_sequences)
                str('g6'),   str.upper('CCGAGGAGTCCGGCCCGGAAGAG'),   ┘
                ]

    return ID,ref_seq,seq_start,seq_end,fastq_files,test_list
```

Fig. 6 Example of a CRIS.py program. CRIS.py contains several editable variables that enable the program to be customized to the desired genome-editing outcome. *See* Subheading 3.5.4 "Variables of the CRIS.py Program" for variable definitions. Variables to be modified by the user are shown between quotation marks (")

3.6.4 Variables of the CRIS.py Program

A key feature of the CRIS.py program is that it enables users to change the script to suit their experimental needs. Script changes can be made at any time, and data can be analyzed accordingly. CRIS.py allows the parameters listed below to be changed in the script. User inputs should be entered between the quotation marks (Fig. 6). All user input variables can be entered as uppercase or lowercase letters. The letters below correspond to the lettering in Fig. 6.

(a) "ID" specifies the name of the output file that will be generated. Each program should be uniquely named.
(b) "ref_seq" defines the reference input sequence. The input sequence is the target-specific amplicon sequence obtained using the PCR#1 primers and excludes the DS tags.
(c) "seq_start" is an approximately 10–14 bp sequence downstream of the PCR #1 forward primer and upstream of the modification site. This length is generally long enough to be unique in the pool of amplicons and short enough to reduce the loss of reads due to sequencing errors.
(d) "seq_end" is an approximately 10–14 bp sequence upstream of the PCR #2 reverse primer and downstream of the target site. This length is generally long enough to be unique in the pool of amplicons and short enough to reduce the loss of reads due to sequencing errors.
(e) "fastq_files" is the location and name of the fastq files to be analyzed. If the well position of a sample to be analyzed is known, it can be inputted. Otherwise, entering *.fastq will query all the fastq files in the directory.
(f) "test_list" contains the list of sequences to be queried. The program analyzes reads for all sequences entered in the test list and reports them in a column with the corresponding name. These sequences can include full-length sgRNA-binding sites, desired HDR edits, specific indels, or even single-nucleotide polymorphisms (SNPs). The test_list sequences should be between the seq_start and seq_end, and all test_sequences (G), seq_start, and seq_end sequences must be copied from the top strand only. For example, all sequences should be found on the R1 sequence in the fastq file, as the analysis is strand specific.

3.6.5 How Does CRIS.py Work?

CRIS.py uses input variables to search and find reads within fastq files. The process CRIS.py used to analyze fastq files is described below and in Fig. 7a, b:

(a) Two flanking sequences (seq_start and seq_end) are matched to the ref_seq to determine the distance between the two sequences and establish a reference for the expected WT sequence length.

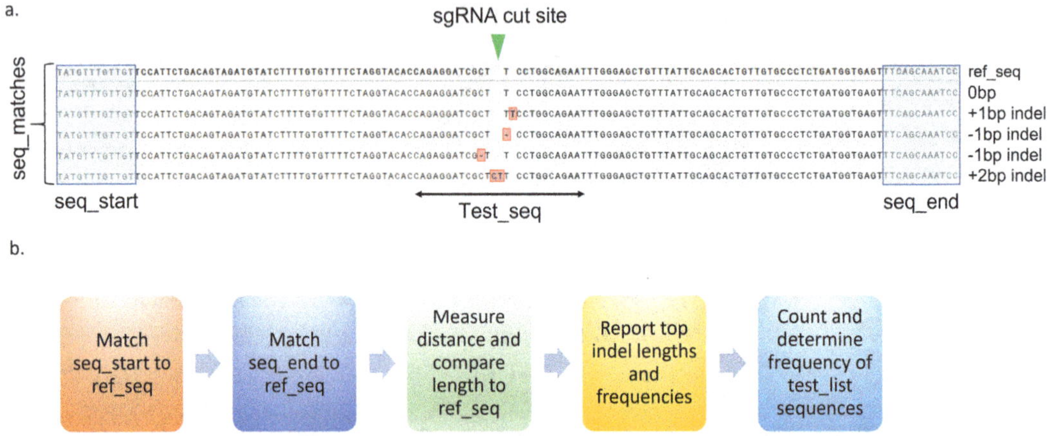

Fig. 7 Overview of CRIS.py analysis pipeline. (**a**) An example reference sequence is shown labeled with CRIS. py program variables. CRISPR-Cas9–induced indels from the .txt file were aligned to the reference sequence and highlighted in red. Indel sizes are manually labeled on the side for reference. (**b**) Process flow for CRIS.py analysis

(b) All reads in a fastq file are analyzed for both the seq_start and seq_end. If both sequences are found, they are counted (termed a "seq_match"). For every seq_match, CRIS.py calculates the length from seq_start to seq_end and compares the length to the WT reference sequence length established in **step 1**.

(c) An amplicon that has no indel and matches the WT length is reported as a 0bp indel. Amplicons that differ from the WT length are reported with their respective indel sizes. Deletions are reported as a negative number.

(d) A net zero indel is also reported as 0 bp indel. For example, if a − 2 bp and a + 2 bp indel occur in the same amplicon, the net indel size is reported as 0 bp.

(e) Indel frequencies are calculated by combining the total number of reads for each indel size and dividing that by the total number of seq_matches.

(f) CRIS.py determines the frequency of all user-defined sequences in the test_list (termed test_seq) by summing the number of times a test_seq occurs divided by the total number of seq_matches.

(g) SNP_test is calculated as part of the QC for the CRIS.py script and sequencing data. It represents a ratio between the total number of times the seq_start and seq_end sequences are counted in each fastq file. The ratio should be ~1. If a SNP occurs in either the seq_start or seq_end region, the SNP_test value varies depending on how many alleles have the SNP.

Large indels resulting in the loss of seq_start or seq_end will also affect the SNP_test score. Additionally, poor-quality NGS data can result in an aberrant SNP_test.

(h) SNPs can be present at both the seq_start and seq_end in one or more alleles. This would cause the SNP ratio to remain approximately the same. To test this potential scenario, CRIS.py includes an additional QC step that is reported as raw_wt_counter. This value represents the ratio of the number of reads found for the first test_seq of the test_list within the raw fastq file for a given index and the number of reads for the same sequence found in seq_match sequences. It should be ~1 for a WT control sample. The raw_wt_counter will be 0 if the first test_seq is not present in the samples. For example, if the sgRNA target site is the first test_seq and the sample is a complete KO clone, the test_seq will not be found in the raw fastq file, and the raw_wt_counter will be 0.

Note: It is important to align the reads to the reference sequence to know the exact indels present in a sample.

3.7 Setup and Analysis of CRIS.py-Generated Data for Various Editing Scenarios

Once the desired variables have been entered, the CRIS.py program is run. This generates a .csv file and a .txt file. The .csv file is a master summary file that provides (1) the total read count; (2) the read counts and frequencies of all sequences in the test_list; (3) the read counts, frequencies, and lengths of the top indels; and (4) the QC checks (*see* **Note 2**). The .txt file lists the test variables and provides the sequence identities and read counts of the most common reads.

The analysis of data generated by CRIS.py differs based on the scope of the genome-editing project. This section describes the setup and analysis of CRIS.py data for various types of editing outcomes.

3.7.1 KO Projects

Setting up CRIS.py script for a KO project. This section describes the process of entering variables in the CRIS.py script for analyzing a KO project. All variables should be entered between the quotation marks, as shown in Fig. 6:

(a) Enter "ID" for your project.

(b) Enter the "ref_seq" of your target PCR, as shown in Fig. 8a and below.

(c) Enter "seq_start" and "seq_end" sequences.

(d) Enter test_seqs of interest. Fig. 8a shows the sgRNA target site (i.e., 20-nucleotide protospacer plus 3 bp protospacer adjacent motif [PAM]) as the test_list sequence. The name of the sgRNA(s) used should be added within the str(''), and the sequence of the corresponding sgRNA target site should be added into str.upper('').

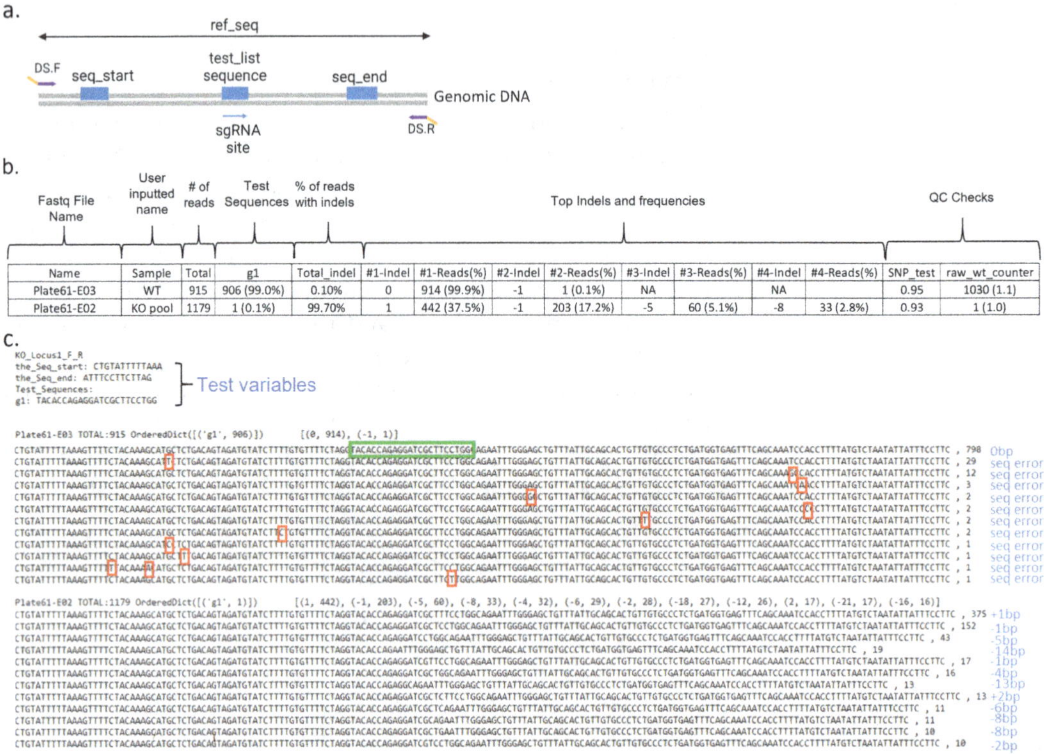

Fig. 8 Setup and analysis of CRIS.py data for a KO project. (**a**) Representation of the region of interest and CRIS.py variables in a KO project. The primers to amplify the targeted region are DS.F and DS.R. The gene-specific portion of the primers is shown in purple and the DS tags in yellow. The seq_start, seq_end, and test_list sequences to be included in the CRIS.py script are shown as blue boxes. (**b**) Example data from a .csv file generated by CRIS.py analysis of a KO cell pool with features highlighted. (**c**) Example data from a .txt file generated for CRIS.py-analyzed KO pool samples. The sgRNA (g1) sequence is shown in a green box. Indels and sequencing errors in the sample are labeled in blue text. Sequencing errors (seq error) in the WT sample are marked by red boxes

3.7.2 Analyzing and Interpreting Data for a KO Project

Interpreting the output .csv file. Figure 8b represents the output .csv file from the NGS data for a KO project (*see* **Note 3**):

(a) The first column "Name" specifies the fastq file name and plate location of the sample based on the unique indexes added during the NGS library creation.

(b) The "Sample" column is left blank intentionally, so that the user can manually assign sample names to the corresponding fastq file name. In this example, we have assigned the labels "WT" and "KO pool."

(c) The total number of reads obtained for each sample is shown in the "Total" column.

(d) All test_seq counts and frequencies will be reported in the columns between "Total" and "Total_indel." In this example, only one test_seq was queried (g1). For a KO project, the test_seq is generally the sgRNA target site including the PAM. A highly active sgRNA will result in large percentages of reads with indels at the cut site and thus low numbers of reads with the exact match to the sgRNA-specific test_seq. The lower the percentage of exact matches to the sgRNA target site test_seq, the higher the sgRNA activity. In this example, the results indicate that 0.1% of the reads in the KO pool sample have an exact match to the tested sgRNA target compared to 99% of the reads in the WT sample. This indicates a highly active sgRNA.

(e) The column "Total_indel" shows the total indel frequency for each sample. In this example, the WT sample has approximately 0.1% Total_indel. In some cases, the Total_indel for negative-control samples may be slightly higher due to sequencing or amplification errors. It is important to run a WT control sample to determine and account for amplicon-specific sequencing errors. For example, some amplicons contain long stretches of repetitive sequence that cause sequencing and/or amplification issues. Sample g1 reports 99.7% "Total_indel," which indicates that 99.7% of the reads in this sample do not have the WT sequence length and thus contain indels. The "Total_indel" is calculated by subtracting the percentage of 0 bp indels from 100%.

(f) CRIS.py reports the top 8 indels by frequency. The top 4 indels are shown in the examples for simplicity. Indel sizes relative to the ref_seq are reported in columns labeled "#-indel." An indel length of "0" is reported for reads with the same length as the ref_seq (WT length).

(g) Each "#-indel" column has a "% Reads" column adjacent to it. This column represents the total number of reads and percentage of reads for that indel length.

(h) For unedited or WT samples, the "#1-Reads (%)" value and the first test_seq (g1 in the above example) should theoretically be the same. However, the "#1-Reads (%)" value is usually higher and does not match the first test_seq because the values reported in the test_seq column are calculated based on the *exact match* to the test_seq sequence, whereas the "#1-Reads (%)" value is reported based on the *length* of the indel. The longer the test_seq is (in base pairs), the more chance for sequencing error and disparity between these two columns.

(i) A 0 bp indel indicates that the amplicon is of WT length. NHEJ events can also result in nucleotide transitions or transversions, which would not affect the amplicon length and would also be reported as 0 bp. This also contributes to disparity between the "g1" test_seq and "#1-Reads(%)" values for the WT sample (discussed above). This technical disparity is generally small and should not change the overall interpretation or clone choice.

(j) All indels of the same length are binned together and reported in the top #-Indels columns. That is, more than one distinct indel can be binned and counted together based on length. To determine if unique indels of the same length have occurred in a clonal sample, you can review the .txt file.

(k) Columns "SNP_test" and "raw_wt_counter" provide QC measures for each sample. *See* Subheading 3.6.5 ((g) and (h)) for more details (*see* **Notes 4** and **5**).

3.7.3 Interpreting the Output .txt File

The second output file generated by CRIS.py is a .txt file (Fig. 8c). The .txt file contains the test variables that the user entered into the program. This is followed by the sequence reads for each of the samples. The samples can be identified by the well and plate location corresponding to a unique barcode:

(a) For each sample, all indel sizes and their corresponding number of reads are presented in brackets next to the sample/well identifier name.

(b) For each sample, the most highly represented sequences are listed with the total number of reads for each unique sequence. Sequence reads must match *exactly* to be binned together. Although the WT sample in "Plate61-E03" contains reads of the same length, they are binned separately because of errors occurring during PCR amplification or sequencing error (*see* Fig. 8c, red boxes). All these reads would be reported as 0 bp indels because their length matches the ref_seq (*see* **Note 6**).

(c) The sequences can be used to compare the reads obtained for various samples and aligned with the ref_seq to determine the sequence identity.

(d) The .txt file can also be used to quickly identify samples in which editing has occurred by looking for a difference in the length of sequence reads as in the "Plate61-E02" sample in Fig. 8c. The differences in length are caused by NHEJ-induced indels.

3.8 Small Modification Projects

1. Setting up CRIS.py for small modification projects. This section describes the steps to set up CRIS.py for analyzing small modification projects. All variables should be entered between the quotation marks, as shown in Fig. 6.

   (a) Enter the sequences of variables, as shown in Fig. 9a.

   (b) Enter the test_list sequences that have been marked in Fig. 9a. For a small modification project, you may enter multiple sequences in the "test_list." As a default, include the sgRNA target site. The name of the sgRNA(s) used should be added within the str(''), and the sequence of the corresponding sgRNA target site should be added into str.upper('').

a.

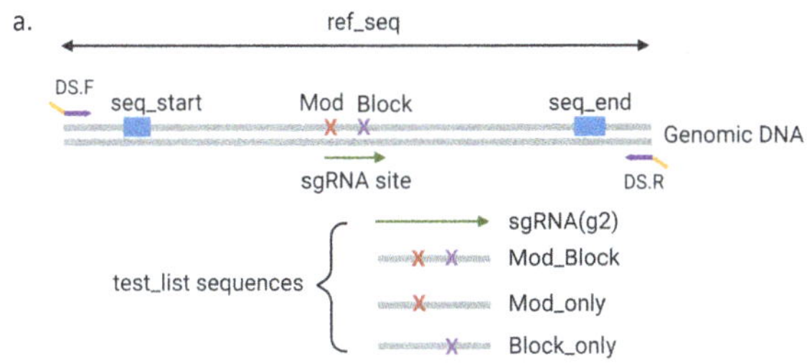

b.

Name	Sample	Total	g2	Mod_Block	Mod_only	Block_only	Total_indel	#1-Indel	#1-Reads(%)	#2-Indel	#2-Reads(%)	#3-Indel	#3-Reads(%)	#4-Indel	#4-Reads(%)	SNP_test	raw_wt_counter
Plate72-G01	WT	2346	2177 (92.8%)	0 (0.0%)	0 (0.0%)	2 (0.1%)	1.70%	0	2307 (98.3%)	-1	37 (1.6%)	-3	2 (0.1%)	NA		1	2395 (1.1)
Plate72-G02	Donor only	2528	2358 (93.3%)	0 (0.0%)	0 (0.0%)	0 (0.0%)	1.90%	0	2479 (98.1%)	-1	43 (1.7%)	-3	5 (0.2%)	-2	1 (0.0%)	1.01	2629 (1.1)
Plate72-G03	Donor + g2	2119	507 (23.9%)	277 (13.1%)	0 (0.0%)	7 (0.3%)	59.70%	0	853 (40.3%)	-1	510 (24.1%)	1	279 (13.2%)	2	75 (3.5%)	0.96	558 (1.1)

Fig. 9 Setup and analysis of CRIS.py data for a small modification project. (**a**) Example of a targeted editing site and variables used in CRIS.py for a small modification project. The gene-specific portion of the primers is shown in purple and the DS tags in yellow. The variables are shown as blue boxes, and test_list sequences represented are shown below. (**b**) Example of the .csv output file of CRIS.py analysis. The top 4 indels are shown for representation purposes only

(c) The number of sequences in the "test_list" depends on the design strategy of the project. In the example below, an ssODN was designed to create a targeted point mutation (red X). However, because a single base pair difference between the WT sequence and the edited sequence may not be enough to prevent recutting by Cas9 and subsequent repair by NHEJ after the desired integration event has occurred, we recommend incorporating a silent blocking modification (purple X). Because the goal of the project is to make the desired point mutation with or without the blocking modification, clones with either the modification only or the small modification and the blocking modification would be useful. CRIS.py facilitates the interrogation of multiple test_seqs. In this example, we recommend listing the modification only (Mod_only), the modification + blocking (Mod_Block), and the blocking only (Block_only) as test_seqs as shown in Fig. 9a.

2. Interpreting the data in the output .csv file for a small modification project.

CRIS.py analyzes and measures HDR rates in pools of cells. The desired HDR events are entered as test_seqs within the test_list portion of the Python script in the same manner as above. An example of CRIS.py analysis for small HDR events is shown below and in Fig. 9b:

(a) Columns labeled "g2," "Block_Mod," "Mod_only," and "Block_only" are the names of the test_seqs. CRIS.py allows the user to query any combination of edits within an amplicon. In this example, we are analyzing for the sgRNA target site, the desired small modification with or without the blocking modification, and the blocking modification alone.

(b) Sample "Donor only" is a control in which cells have been transfected with the donor ssODN only without the sgRNA or Cas9. The percentage of test_list sequences for this sample should be ~0, except for the "g2" test_seq as the gRNA site should be intact if no cutting has occurred.

(c) Sample "Donor + g2" shows that 23.9% of the reads have the "g2" test_seq, and the WT sample shows that 92.8% of the reads have the "g2" test_seq. The WT sample will typically have an exact match to the 23 bp target site less than 100%. This is because to be counted as a match, every base pair of the test_seq must match the sequence read exactly. Sequencing errors at any position within the test_seq will result in the read being excluded from the count. For this reason, it is important to limit the length of the test_seq and/or weigh the impact of having a long test_seq: the longer the test_seq, the lower the number of reads with an exact match to the test_seq.

(d) The percentages of the test_seqs report the percentage of reads in the sample that match exactly to the input test sequences. The sample "Donor + g2" has 13.1% of reads matching the Mod_Block sequence. The "WT" and "Donor-only" samples have 0% of the reads matching the Mod_Block sequence. Therefore, ~13% of the alleles in the pool contain the Mod_Block sequence.

(e) The "Total_indel" column reports the total number of reads that do not have the WT length (59.7% in treated sample and 1.7% in WT sample). The WT control should have very low reported rates of indels, and the top non–0 bp indel sizes are generally ±1 bp resulting from sequencing error.

(f) If you are knocking in a small insert that will yield an amplicon of different length to the ref_seq, you will see the corresponding indel length reported by frequency as with other indels. It is important to note that there may be PCR bias and a corresponding underrepresentation of the KI indel length. Amplicons that differ in size, even by 10–20 bp, will be differentially amplified with the smaller amplicon being amplified preferentially. Larger size differences will produce a bigger bias toward the smaller amplicon. This PCR bias will also be present in the frequency of test_seqs reported for small KI projects that yield amplicons that differ in length from the ref_seq.

(g) CRIS.py reports the top 8 indels by frequency. The top 4 indels are shown in the examples for simplicity. Indel sizes, relative to the ref_seq, are reported in columns labeled "#-indel." An indel length of "0" is reported for reads with the same length as the ref_seq (WT length).

(h) Each "#-indel" column has a "% Reads" column adjacent to it. This column represents the total number of reads and percentage of reads for that indel.

(i) The SNP_test and "raw_wt_counter" should be ~1, as in the other examples if the first test_seq (perfect deletion) is found in the seq_match.

3.9 Deletion Projects

1. Setting up CRIS.py for a deletion project. This section describes the process of entering variables in the CRIS.py script to analyze a deletion project.
   (a) Enter variables, as shown in Fig. 10a. The "ref_seq" for a deletion project is the expected genomic sequence after the deletion event and assuming precise rejoining of the sgRNA cut sites.
   (b) Select the region at the deletion site from the sequence obtained after successful deletion and use it as a test_seq in the test_list. This region can be named "perfect deletion" and should be ~14 to 20 bp long.
2. Interpreting the data in a .csv file of a deletion project:
   (a) The test sequence in a typical deletion project test_list is the perfect deletion sequence (Fig. 10a).
   (b) In Fig. 10b, the column "perfect deletion" represents the percentage of reads in the sample that matches the sequence obtained if a perfect deletion event has occurred. Deletion 1 and Deletion 2 samples have 25.2% and 20.3% of the reads, respectively, that have the perfect deletion sequence.

a.

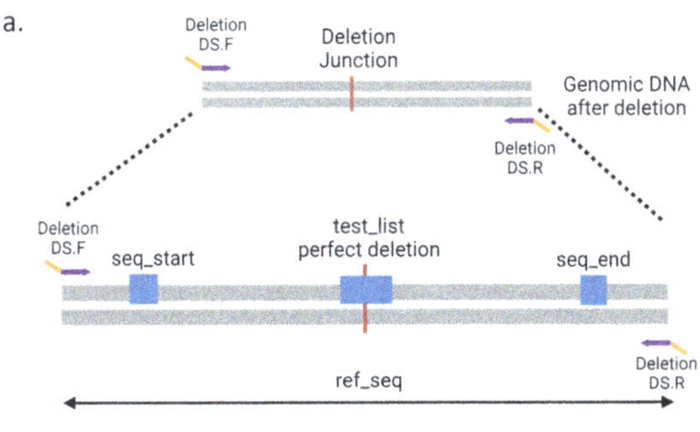

b.

Name	Sample	Total	perfect deletion	Total_indel	#1-Indel	#1-Reads(%)	#2-Indel	#2-Reads(%)	#3-Indel	#3-Reads(%)	#4-Indel	#4-Reads(%)	SNP_test	raw_wt_counter
Plate57-F11	Deletion 1	1484	374 (25.2%)	75.00%	-6	573 (38.6%)	0	371 (25.0%)	-7	124 (8.4%)	-15	71 (4.8%)	0.99	1583(1.1)
Plate57-G11	Deletion 2	1666	338 (20.3%)	79.70%	-6	700 (42.0%)	0	339 (20.3%)	-7	132 (7.9%)	-25	65 (3.9%)	0.99	1687(1.1)

Fig. 10 Setup and analysis of CRIS.py data for a deletion project. (**a**) Example of a deletion product and the variables used in the CRIS.py program for deletion analysis. The red line indicates the joining after the deletion event occurred. The gene-specific portion of the primers is shown in purple, and the DS tags in yellow. The variables in the script are shown as blue boxes. (**b**) Example of a .csv output file for a deletion project analysis by CRIS.py. Two representative deletion pool samples are shown

(c) For deletion projects, a 0 bp indel also represents the reads with a perfect deletion, as the ref_seq length is the expected length after a perfect rejoining of the sgRNA cut sites. If a precise deletion is not needed for the downstream application, reads with other indels may also be useful.

(d) As previously described, there may be small differences between the 0 bp indel and the "perfect deletion" test_seq frequencies because of the difference in how each is calculated: the "perfect deletion" column will report back *exact* matches to the test_seq, whereas the 0 bp indel (#2-indel in Fig. 10b) reports the frequency of the reads with the ref_seq length regardless of sequence.

(e) The SNP_test and "raw_wt_counter" should be ~1, as in the other examples if the first test_seq (perfect deletion) is found in the seq_match.

Notes: Writing a CRIS.py program for the inside PCR (*see* Fig. 3) and including the inside PCR amplicon in the NGS run can be done to determine the zygosity of clones in a deletion project. If a given sample has reads for both the deletion PCR and the inside PCR, that indicates the clone is heterozygous for the deletion. This can also be confirmed by analyzing the PCR products by gel electrophoresis. A clone with no remaining unedited alleles would have no sequence reads for the inside PCR.

3.10 Large KI Projects

1. Setting up CRIS.py for a large KI project. As described in the "Primer Design for Large KI Projects" (Subheading 3.5), the analysis for large KI projects involves multiple PCR amplicons (Figs. 4 and 5). Thus, there will be multiple CRIS.py programs: one for each amplicon. The three key analysis scripts for a KI project are (1) FR, (2) 5Junc, and (3) 3Junc programs. The 5Junc and 3Junc programs are used to detect the integration of the large KI at the 5′ and 3′ ends, respectively. The FR program uses primers that are designed in the gDNA region flanking the targeted integration site; this program is used to determine if any unedited or indel alleles remain. This section describes the process of entering variables in the CRIS.py scripts for analyzing a large KI project.

   (a) Enter variables for the 5Junc and 3Junc CRIS.py programs, as shown in Fig. 11a (for ssDNA KI) or Fig. 11b (for plasmid dsDNA donor KI).

   (b) If you are analyzing clones and want to know if they are heterozygous or homozygous for the KI, perform a PCR using the Gen DS.F/Gen DS.R primer set for an ssDNA KI and using the HA DS.F/HA DS.R primer set for a

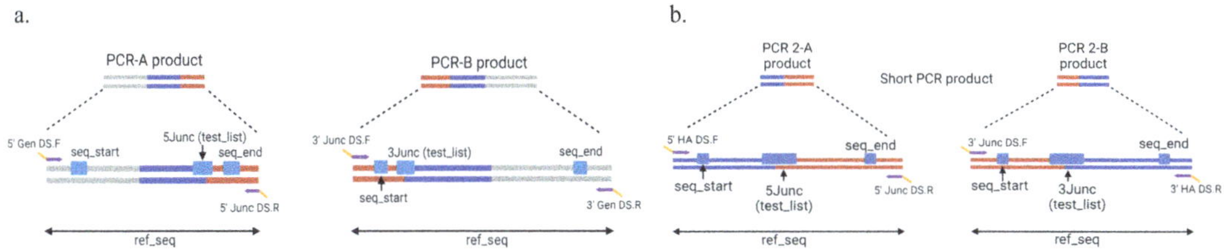

c.

Name	Sample	Total	5Junc	Total_indel	#1-Indel	#1-Reads(%)	#2-Indel	#2-Reads(%)	#3-Indel	#3-Reads(%)	#4-Indel	#4-Reads(%)	SNP_test	raw_wt_counter
Plate65-A02	KI Pool	1378	1238 (89.8%)	4.50%	0	1316 (95.5%)	-1	61 (4.4%)	-2	1 (0.1%)	NA		1.01	1487 (1.1)

d.

Name	Sample	Total	3Junc	Total_indel	#1-Indel	#1-Reads(%)	#2-Indel	#2-Reads(%)	#3-Indel	#3-Reads(%)	#4-Indel	#4-Reads(%)	SNP_test	raw_wt_counter
Plate65-B02	KI Pool	1325	1169 (88.2%)	2.30%	0	1295 (97.7%)	-1	27 (2.0%)	1	2 (0.2%)	-2	1 (0.1%)	0.79	1679 (1.4)

Fig. 11 Setup and analysis of CRIS.py data for a large KI project. Example of the KI products obtained by junction PCRs and the variables used in the CRIS.py program for a KI project using ssODN/ssDNA (**a**) or dsDNA (**b**). The PCR products shown in Figures **a** and **b** can be referenced to the final PCR products in Figs. 4 and 5, respectively. The gene-specific portion of the primers is shown in purple and the DS tags in yellow. The variables in the script are shown as blue boxes. (**c**) Example of the .csv output file for 5junc PCR analysis by CRIS.py. (**d**) Example of the .csv output file for 3junc PCR analysis by CRIS.py

dsDNA KI project. Write a CRIS.py script using the WT genome sequence as the "ref_seq" (*see* Subheading 3.6.1 and Fig. 8a).

2. Interpreting the data in a .csv files of a KI project:
   (a) A separate .csv file will be generated for each CRIS.py program/amplicon: one for each junction program (5Junc and 3Junc).
   (b) The respective "5Junc" and "3Junc" test_seq columns will report the percentage of reads in the samples that have the precise junction.
   (c) In example Fig. 11c, the KI pool sample has 89.8% of the reads with the 5Junc test_seq.
   (d) In example Fig. 11d, the KI pool sample has 88.2% of the reads with the 3Junc test_seq.
   (e) The total indels for the 5Junc program and 3Junc program are 4.5% and 2.3%, respectively. This indicates that the remaining 95.5% and 97.7% of the sequences match the ref_seq length, which is reported in the "#1-Reads (%)" column. In this example, ref_seq is the sequence with the precisely targeted integration incorporated at either the 5′ or 3′ junction.

Notes: Depending on the size of the insert, it is likely that the out-out primer sets will yield no reads in homozygous KI clones as the resulting amplicon size would be beyond the read length limit for NGS. If a clone is heterozygous for the KI, NGS data is obtained for both the junction PCR sets and the out-out primer PCR sets. The modified allele(s) would be positive for the junction PCRs, and the nonintegrated allele(s) would be positive for the out-out PCR.

3.11 Interpreting the CRIS.py .csv Output File for Clonal Data

The above examples show how to use CRIS.py and interpret the output files to determine editing frequencies within a pool of edited cells. CRIS.py is also highly useful for screening and identifying desired edited clones. The output .csv file is searchable and sortable, allowing for quick and efficient clone identification. In general, the same program written for determining the rates of editing in the pool can be used for identifying correctly edited clones. However, the analysis is different.

Below we discuss how to interpret the output .csv file (Fig. 12) when screening for clones with a desired point-mutation modification. The example shown below is the single-cell screening data for the same point mutation project described in Subheading 3.6.2:

1. Finding a correctly edited KI clone can require screening hundreds of individual clones. To quickly narrow the candidates that have the desired modification, sort the .csv file by "%Mod_Block." Because the column labeled "Mod_Block" lists both the number and percentage of reads in each clone with the given test_seq, you cannot use this column for sorting. The frequency of all test_seqs is listed separately in the corresponding column with a percent sign "%" before the test_seq name. These columns are listed after the raw_wt_counter to facilitate sorting of the .csv file on a given variable. In the example in Fig. 12, only "%Mod_Block" is shown for simplicity.

Name	Sample	Total	g2	Mod_Block	Mod_only	Block_only	Total_indel	#1-Indel	#1-Reads(%)	#2-Indel	#2-Reads(%)	SNP_test	raw_wt_counter	%Mod_Block
Plate47-H08	Low Reads	107	0 (0.0%)	103 (96.3%)	0 (0.0%)	0 (0.0%)	0.065	0	100 (93.5%)	-1	6 (5.6%)	1.08	0	96.3
Plate45-C04	Hom Clone	3565	0 (0.0%)	3393 (95.2%)	2 (0.1%)	0 (0.0%)	0.029	0	3460 (97.1%)	-1	98 (2.7%)	1.1	0	95.2
Plate48-H08	Hom Clone	2056	0 (0.0%)	1955 (95.1%)	2 (0.1%)	0 (0.0%)	0.02	0	2014 (98.0%)	-1	40 (1.9%)	1.07	0	95.1
Plate45-E02	Hom Clone	3035	0 (0.0%)	2873 (94.7%)	2 (0.1%)	0 (0.0%)	0.024	0	2962 (97.6%)	-1	60 (2.0%)	1.08	0	94.7
Plate44-F07	Clean Het	3574	1680 (47.0%)	1611 (45.1%)	4 (0.1%)	0 (0.0%)	3.40%	0	3452 (96.6%)	-1	109 (3.0%)	1.1	1937 (1.2)	45.1
Plate41-C06	Het Clone + Indels	3642	6 (0.2%)	1639 (45.0%)	2 (0.1%)	1 (0.0%)	51.70%	1	1829 (50.2%)	0	1759 (48.3%)	1.09	8 (1.3)	45
Plate48-F08	Het Clone + Indels	2354	0 (0.0%)	1058 (44.9%)	1 (0.0%)	1 (0.0%)	54.00%	-1	1229 (52.2%)	0	1084 (46.0%)	1.07	0	44.9
Plate41-B07	Clean Het	3295	1548 (47.0%)	1472 (44.7%)	2 (0.1%)	2 (0.1%)	3.20%	0	3190 (96.8%)	-1	87 (2.6%)	1.09	1742 (1.1)	44.7
Plate47-G09	KO	3289	5 (0.2%)	1461 (44.4%)	3 (0.1%)	1 (0.0%)	98.40%	1	1675 (50.9%)	11	1491 (45.3%)	1.07	5 (1.0)	44.4

Fig. 12 Analysis of CRIS.py data for screening single-cell clones. Example of a .csv output file for screening single-cell-derived clones for a point mutation project. The different clone types identified in the screen are labeled in the sample column. Clones with low read counts are highlighted in red. Homozygous clones (Hom Clone) are highlighted in green. Heterozygous clones (Clean Het) with the desired modification (Mod_Block) on one allele and the WT sequence on the other allele are highlighted in blue. Heterozygous clones with indels (Het Clone + Indels) on one allele and the desired modification (Mod_Block) on the other allele are shown in yellow. Clones with all out-of-frame indels (KO) are highlighted in purple. The top 2 indels are shown for representation purposes only

2. Clones with low reads may have fewer cells in the harvested well and may have skewed allelic frequencies. In general, we recommend picking clones with at least 100 reads in the "Total" column.
3. For a point mutation project, you should identify clones with the desired modification and ensure that there are no indels in the desired clone by examining the "Total_indel" column. The "Total_indel" should be ~0, the "#1-indel" should be 0 bp, and the "%Mod_Block" should be ~100% for a homozygously edited clone (Fig. 12, green rows). A diploid clone, in which one allele contains the desired edit and the other contains an NHEJ event, should have a "%Mod_Block" of ~50% and a "Total_indel" of ~50%, and the top two indels would be the 0 bp indel (KI) and the indel length of the NHEJ allele.
4. A clean heterozygously edited clone (Fig. 12, blue rows) has the desired modification on one allele and remains WT on the other allele. In this case, "Total_indel" remains ~0%, but the "% Mod_Block" column will have ~50%, and the "%g2" column (not shown) will have ~50%. This is because only one allele has the desired modification and the other allele has a WT sequence.
5. Indel frequencies (#-Reads(%)) with low percentages can generally be attributed to sequencing errors and typically account for less than 3% of the reads per specific indel. However, before choosing a clone for expansion, the user should align the sequence reads from the .txt file with the desired sequence to confirm that no indels align with the sgRNA site, which might be indicative of a nuclease-induced indel.
6. It is often possible to identify KO clones when creating a targeted KI clone if the resulting indels lead to a premature stop codon. KO clones will have only out-of-frame indels in the "#-Reads(%)" columns with adequate read counts and near 100% in the "Total_indel" column (Fig. 12, purple row). Note that the KO clone in Fig. 12 contained the desired "Mod_Block" modification. However, the top two indels were not 0 bp, which indicates that additional insertions or deletions have occurred on the edited allele.
7. The SNP_test and the raw_wt_counter columns are reported for each clone. The SNP_test should always be considered, but the raw_wt_counter will be 0 if the first test sequence is not found in the clone.
8. Align the sequence data of the selected clones in the .txt file with the reference file to confirm the desired edit and precise KI of the sequence.

3.12 Summary

NGS is a powerful tool for determining genome-editing outcomes. However, it is not without limitations, including read-length limits, the inability to detect larger chromosomal aberrations, and PCR bias for smaller amplicons. Nevertheless, targeted NGS enables high-throughput screening because many individual samples can be indexed, pooled, and sequenced and the resulting data demultiplexed to provide the sequence identity of edited samples. The large amounts of data generated by NGS can be efficiently analyzed using CRIS.py, which allows users to simultaneously analyze many clones at great speed and with high accuracy. In this chapter, we detailed NGS library preparation, primer design, and setup and analysis of CRIS.py for a variety of genome-editing scenarios. The flexibility in modifying the CRIS.py script, per the user's needs, makes it a great tool for analyzing any number of editing outcomes in cell pools, single clones, and animal models. The output files generated by CRIS.py are editable and in a format that enables users to quickly search, rank, and identify clones of interest. The ease of running CRIS.py locally on most computers and the speed of analysis make it a powerful tool for analyzing NGS data for genome-editing outcomes.

4 Notes

1. Unable to amplify the target regions:
   1. Check the concentration of DNA in the sample. Too little or too much gDNA can inhibit amplification.
   2. Optimize PCR conditions for a given amplicon.
   3. Check GC content of the desired amplicon. If it is higher than 65%, we recommend using an additive (e.g., DMSO, betaine, or GC enhancer) to improve amplification rates for GC-rich amplicons.
2. Low read count:
   1. Check the QC metrics of the NGS quality via the Illumina software.
   2. Check amplicon size and relative quantity on an agarose gel.
   3. To increase sequence diversity, it is recommended to add at least 10% PhiX Control v3 (Illumina Cat# FC-110-3001) to the indexed amplicon pool. Low sequence diversity can impair correct base calling on the Illumina sequencer.
   4. Reduce primer dimer contamination by amplicon purification. This will reduce primer dimers which can consume reads in the sequencing run.

3. No data in .csv file:
   1. Check the CRIS.py program for any errors.
   2. Verify that the script is being executed in the same directory where the fastq files are located.
   3. Check if each test_list sequence is separated by a comma (,).
   4. Verify that the sample location specified in the variable "fastq_files" is correct.
   5. The most common errors are the "ref_seq," "seq_start," "seq_end," and sequences in the "test_list." Check each one to ensure the correct sequences from top strand have been provided.
   6. Check the size of the final amplicon to ensure it is within the NGS run length limits.
   7. Move the "seq_start" and "seq_end" sequences, as one or both may be landing on a SNP or a repetitive region, which would result in no reads.
   8. Confirm the amplification of the PCR product by gel electrophoresis.
   9. Download and test analysis using data from the CRIS.py GitHub site.
4. SNP_test ratio is not near 1.0:
   1. A SNP may be present in the seq_start and/or seq_end in one or more alleles. Pick a different seq_start and/or seq_end sequence inside the primer sequences and rerun.
5. The raw_wt_counter is not as expected:
   1. Check to see if your first test_seq should be found in the raw reads or seq_matches. The raw_wt_counter value would be "~0" for total KO clones or deletion clones if the first test_seq is the sgRNA target site that was used for the editing event. This is because the test_seq will not be found in the fastq file if editing has occurred.
6. Total indel value is high in the WT/negative control:
   1. Check if there is a repetitive stretch inside the seq_start and/or seq_end sites. If one is present, move the seq_start and/or seq_end sequences to avoid the region and rerun the program.

Acknowledgments

We would like to thank our colleagues from the Center for Advanced Genome Engineering at St. Jude Children's Research Hospital for their support, stimulating discussions, and helpful comments on this manuscript. We thank Angela J. McArthur, PhD, ELS, for editing the manuscript. We thank the American Lebanese Syrian Associated Charities, St. Jude Children's Research Hospital, and the Comprehensive Cancer Center Grant (P30 CA021765) for financial support. Figures 1, 2, 3, 4, 5, 8a, 9a, 10a, and 11a, b have been created using BioRender (https://biorender.com).

References

1. Connelly JP, Pruett-Miller SM (2019) CRIS.py: a versatile and high-throughput analysis program for CRISPR-based genome editing. Sci Rep 9:4194
2. Behjati S, Tarpey PS (2013) What is next generation sequencing? A Arch Dis Child Educ Pract Ed 98:236–238
3. https://www.illumina.com/science/technology/next-generation-sequencing.html
4. https://www.ncbi.nlm.nih.gov/tools/primer-blast
5. https://github.com/patrickc01/CRIS.py

Chapter 7

Advanced Technologies and Automation in mES Cell Workflow

Charles Yu, Roger Caothien, Marques Jackson, Brian Nakao, Anna Pham, Lucinda Tam, and Merone Roose-Girma

Abstract

Gene targeting in mouse ES cells replaces or modifies genes of interest; conditional alleles, reporter knock-ins, and amino acid changes are common examples of how gene targeting is used. To streamline and increase the efficiency in our ES cell pipeline and decrease the timeline for mouse models produced via ES cells, automation is introduced in the pipeline. Below, we describe a novel and effective approach utilizing ddPCR, dPCR, automated DNA purification, MultiMACS, and adenovirus recombinase combined screening workflow that reduces the time between therapeutic target identification and experimental validation.

Key words Mouse embryonic stem cells (ESC), Genotyping, MultiMACS (magnetic-activated cell sorting), ddPCR (droplet digital PCR), dPCR (digital PCR), Biomek i5

1 Introduction

Although CRISPR technology has enabled fast and efficient generation of mice with knockout and some types of knock-in alleles, ES cell technology is still a preferred route for more complex alleles. Automation is introduced to streamline and increase the efficiency in the ES cell pipeline and decrease the timeline for mouse models produced through ES cells in many places in the pipeline as possible. A significant boost in the number of 96-well ES cell plates is handled per unit time by incorporating automated DNA purification, MultiMACS [1], adenovirus recombinase, droplet digital PCR [2, 3], and multiplex digital PCR [4]. The Biomek i5 liquid handler (B87583/Biomek i5 Multichannel with enclosure) automates liquid handling to complete DNA purification [5, 6] steps for up to five 96-well or 24-well plates at one time. Additionally, improvement in the ES cell DNA quality is achieved with a MultiMACS™ M96 Separator to remove the contaminating feeder cells in ES cell culture. The elimination of feeder cell DNA from ES cell

Thomas L. Saunders (ed.), *Transgenesis: Methods and Protocols*,
Methods in Molecular Biology, vol. 2631, https://doi.org/10.1007/978-1-0716-2990-1_7,

DNA subjected to multiplex digital PCR analysis reduces background readings and improves the reliability of gene copy number analysis. Multiplex amplification of the critical elements of gene-targeted ES cells such as expected copy number of the drug selection cassette, insertion of elements to confer conditional gene expression, and correct copy number of the targeted gene, along with internal controls, is quickly achieved by droplet digital PCR and multiplex digital PCR. The number of ES cell clones that can be simultaneously analyzed per unit time is significantly higher by droplet digital PCR and multiplex digital PCR rather than by conventional assays that process a single genetic feature at a time. After high-throughput multiplex screening, a handful of ES cell calls that pass the first screen are thawed, expanded, and analyzed more completely with long-range PCR assays for final validation. Utility of this pipeline accelerates the identity of genetically engineered ES cell lines [7], improves ergonomic issues in the workplace, and enhances the reproducibility of genetic screening results.

2 Materials

2.1 MultiMACS Materials

1. MultiMACS EQ buffer: DPBS (Dulbecco's phosphate-buffered saline), minus calcium chloride and magnesium chloride with 0.5% bovine serum albumin and 2 mM EDTA.
2. MultiMACS M96 Separator benchtop instrument (MSE) Cat#130-091-937.
3. Feeder removal microbeads, Miltenyi Biotec Cat# 130-095-531.
4. MultiMACS 96w column block, Miltenyi Biotec Cat#130-092-445.
5. Integra Viaflo 96 channel electronic pipette Cat#6001.
6. ES media: Gibco Knockout Media, 15% fetal bovine serum (FBS), 10 mM Gibco nonessential amino acids (NEAA), 10 mM 200 mM L-glutamine, 2-mercaptoethanol, and 1000 IU leukemia inhibitory factor (LIF) per mL.
7. Phase-contrast microscope with 40×, 100×, and 200× magnification.
8. A 96-well plate shaker.
9. A 96-well plate centrifuge.
10. Incubator suitable for mES cell culture. 37 °C, 5% CO_2, and 95% relative humidity.

2.2 Biomek i5 Materials

1. Biomek i5 Multichannel with enclosure Cat# B87583.
2. Mobile Workstation, i5 (if countertop not available) Cat# C02612.

3. HEPA Kit, i5, 115 VAC Cat# B98217.
4. Pipetting head, MC96, 5–1200 uL Cat# B87589.
5. Wash station, MC96 pipette tips Cat# B87689.
6. 5 1 × 1 TIP LOAD ALP Cat# C02867.
7. 7 ALP, 1 × 1 static Cat# B87477.
8. 2 ALP, 1 × 3 static Cat# B87478.
9. ALP, 1 × 5 static Cat# B87479.
10. ALP, mounting plate Cat# B87485.
11. CIRC.RESERVOIR TIPBOX ALP110 VAC Cat# 394586.
12. ALP, VIBRATION ISOLATION PLATE Cat# C02750.
13. ORBITAL SHAKER ALP Cat# 379448.
14. Biomek FX Device Controller I/O Box Cat# 719366.
15. CABLE, CAN INTERFACE, 160 CM (63 INCHES) LENGTH Cat# B14088.
16. Greiner Bio-One UV-Star® Microplates Cat# EK-25801.
17. Agilent polypropylene reservoir microplates Cat# EK-2035.
18. Abgene™ 96- well 2.2 mL Polypropylene Deepwell Storage Plate Cat# AB0661.
19. Agencourt® DNAdvance,™ 100 × 96 (4 plate prep) Cat# A48706.
20. Biomek 1025 uL Filtered Pipette Tips Cat# B85955.
21. 20 L purified laboratory water.
22. 70% ethanol.
23. 10% bleach.
24. Costar 96-well round bottom Cat# 3799.
25. Costar 24-well flat bottom Cat# 3524.
26. Bradley lysis buffer: 10 mM TRIS pH 7.5 (10 mL of 1 M), 10 mM EDTA pH 8.0 (40 mL of 0.25 M), 10 mM NaCl (2 mL of 51 M), 0.5% N-lauroylsarcosine or sarcosyl (5 *g*), and sterile purified laboratory water → 500 mL.
27. Roche proteinase K Cat# 03115828001.
28. Titer Tops® sealing film for microplates Cat# Z688630.
29. Oven set at 50 °C.

2.3 Materials for Two-Channel Droplet Digital PCR

Materials reflect a two-channel microfluidic system with the capacity to run two probes:

1. FAM-labeled target probe, Integrated DNA Technologies.
2. HEX-labeled endogenous control probe, Integrated DNA Technologies.

3. Primers for target and endogenous control, Integrated DNA Technologies.
4. ddPCR Supermix for Probes (No dUTP) Cat #1863025.
5. ddPCR™ 96-Well Plates Cat#12001925.
6. DG32 Automated Droplet Generator Cartridges Cat#1864108.
7. Automated Droplet Generator Cat#1864101.
8. Automated Droplet Generation Oil for Probes Cat#1864110.
9. Applied Biosystems™ Veriti™ 96-Well Thermal Cycler Cat# 4375786.
10. ddPCR™ Droplet Reader Oil Cat#1863004.
11. QX200 Droplet Digital PCR System #1864001.

2.4 Materials for Five-Channel Digital PCR

Materials reflect a five-channel microfluidic system with the capacity to run five probes:

1. FAM-labeled target probe, Integrated DNA Technologies.
2. ROX-labeled target probe, Integrated DNA Technologies.
3. Cy5-labeled target probe, Integrated DNA Technologies.
4. ABY-labeled target probe, Integrated DNA Technologies.
5. HEX-labeled endogenous control probe, Integrated DNA Technologies.
6. Primers for targets and endogenous control, Integrated DNA Technologies.
7. QIAcuity Eight Platform System PRM-3 Cat# 911059.
8. QIAcuity Nanoplate 8.5 k 96-well Cat#250021.
9. QIAcuity Nanoplate Tray Cat#250098.
10. QIAcuity Nanoplate Seals Cat#250099.
11. QIAcuity Probe PCR Kit Cat#250101.

3 Methods

3.1 Plate Preparation on Day of Sorting

1. Aspirate culture media in a 96-well ES cell DNA plate.
2. Wash a 96-well ES cell DNA plate with 150 uL DPBS minus CaCl/MgCl (*see* **Note 1**).
3. Aspirate off 150 uL 1 DPBS.
4. Apply 50 uL/well trypsin to your 96-well ES cell DNA plate.
5. Incubate 15–20 minutes until ES cells dissociate from well.
6. Break up cells by pipetting up and down (ten times) with a 12-channel multichannel pipette (*see* **Note 2**).

7. Observe ES cells on phase-contrast microscope and make sure cells are broken up into single cells.
8. Add 150 uL ES media to neutralize the trypsin.
9. Add 10 uL of feeder removal microbeads (Cat#130-095-531).
10. Incubate at room temp for 10 minutes with 300 rpm shaking.

3.2 Prepare MultiMACS Block While Cells Are Mixed with Microbeads

1. Equilibrate column block (Cat#130-092-445) with 300 uL EQ buffer.
2. Use Viaflo 96 to transfer the EQ buffer the MultiMACS block.
3. Spin 300 rpm for 2 minutes. Set deceleration speed to 2.
4. After equilibration, go to "MSE" and use "GNE protocol."
5. MSE will prompt you for each step.
6. Apply your ES cell/bead mixture to the column block (*see* **Note 3**).
7. Wait for ES cells to pass through column block (10–15 minutes).
8. Wash with 100 uL EQ buffer.
9. After collecting all flow-through, transfer flow-through fraction from deep well plate to 96w U-bottom plate using the Viaflo 96.
10. Perform high-speed spin at 3000 rpm for 20 minutes (set deceleration to 2).
11. After spin, carefully aspirate off media and avoid disrupting your ES cell pellet.
12. Add 100 uL/well Bradley lysis buffer with 1 mg/mL proteinase K freshly added.
13. Overlay with sealing film for microplates.
14. Lyse overnight at 50 °C.
15. Place plates in −70 °C freezer for DNA analysis.

3.3 Biomek i5 Method for DNA Preparation in 96-Well Plates of mES Cells

1. Aspirate culture media from a 96-well plate.
2. Add 1 mg/mL proteinase K (Cat# 0311582800) to Bradley lysis buffer. To calculate how much proteinase K to add to the buffer, determine buffer amount required (given the number of DNA plates), and divide that volume by the concentration of the proteinase K listed on the bottle (18 mg/mL).
3. Add 50 uL Bradley lysis buffer to each well for a 96-well plate.
4. Lyse overnight in an incubator at 35 °C.
5. The next morning, take the plate out of incubator, spin at 3000 rpm for 1 minute, and set aside while setting up the Biomek i5 instrumentation (Cat# B87583).

6. Check the Biomek i5 instrument prior to running method:
   (a) Ensure there is 20 L purified laboratory water present to isolate gDNA from a 96-well plate with appropriate tube lining from the purified laboratory water carboy to the Biomek i5 unit.
   (b) Ensure there is 300 mL of 70% ethanol present in the circulating reservoir.
7. Open the Biomek i5 software
8. Click on home all axes icon. Be sure the gripper is not facing the plexiglass.
9. The 96-well on-site method implementation by Beckman Coulter will follow this specific method.
10. Click File → Open → Method → DNAdvance i5 96MC 5-plate_PartI_Bind1Addition_AB0661_Final.
11. Click Finish to ensure the script does not contain any glitches. If yes, a red bar will flag at the bottom of the window.
12. If no glitches, Click Start → Click Run icon (green arrow).
13. A new window will open. This is the Guided Laboratory Setup (GLS). From here, select the number of plates.
14. Follow all steps from GLS from the method provided by Beckman Coulter.
15. Deck setup for Bind 1 addition is as follows (Fig. 1).
16. Click Finish and Run will begin.
17. Click File → Close Method.

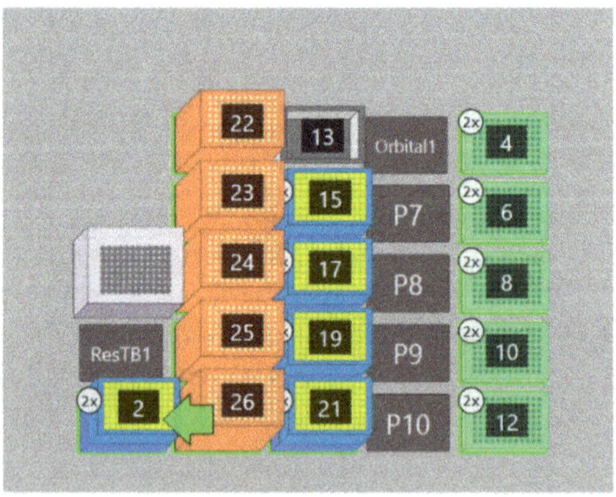

Fig. 1 Positions of materials on Biomek deck. 22–26 refer to Biomek 1025 uL Filtered Pipette Tips Cat# B85955. 2, 15, 17, 19, and 21 refer to Abgene™ 96-well plate 2.2 mL Polypropylene Deepwell Storage Plate Cat# AB0661. 4, 6, 8, 10, and 12 refer to Costar 96-well round bottom Cat# 3799 containing lysed ES cells with Bradley lysis buffer

18. Click File → Open → Method → DNAdvance i5 96MC 5-plate_PartII_Bind1Addition_AB0661_Final.
19. Click *Finish* to ensure the script does not contain any glitches. If yes, a red bar will flag at the bottom of the window.
20. If no glitches, Click *Start* → Click *Run* icon (green arrow).
21. A new window will open. This is the GLS, Guided Laboratory Setup. From here, select the number of plates to be prepped.
22. Remove Bind1 reservoir.
23. Follow all steps (see following pages for proper labware and reagent setup).
24. Deck setup for Bind 2 addition, and wash steps are seen in Fig. 2.
25. Click Finish.
26. Run will begin.
27. After the run ends, remove bleach and ethanol as hazardous waste material in separate containers for proper disposal.
28. Thoroughly rinse reservoirs that had bleach and ethanol with purified laboratory water.
29. Use sealing film for microplates to seal finished plates and place in the −70 freezer until needed for screening.

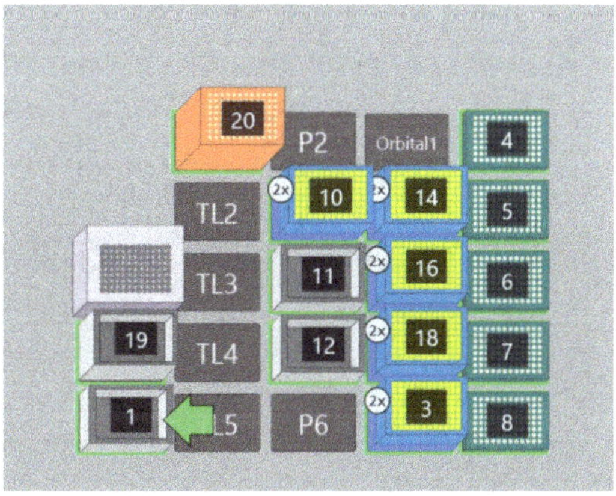

Fig. 2 Positions of materials on Biomek deck. 20 refers to Biomek 1025 uL Filtered Pipette Tips Cat# B85955. 1 refers to elution buffer provided by Agencourt® DNAdvance, ™ 100 × 96 (4 plate prep) Cat# A48706. 11 refers to Bind2 solution provided by Agencourt® DNAdvance, ™ 100 × 96 (4 plate prep) Cat# A48706. 12 refers to 70 mL of 10% bleach. 3, 10, 14, 16, and 18 refer to Abgene™ 96-well plate 2.2 mL Polypropylene Deepwell Storage Plate Cat# AB0661 containing lysed ES cells with Bradley lysis buffer. 4, 5, 6, 7, and 8 refer to final Greiner Bio-One UV-Star® Microplates Cat# EK-25801

3.4 Biomek i5 Method for DNA Preparation in 24-Well Plates of mES Cells

The method is the same for 96-well and 24-well plates, but an added advanced tip loading method accommodates a 24-well plate gDNA isolation (*see* **Note 4**).

3.5 Droplet Digital PCR Method for a Two-Channel ddPCR Instrument

The PCR reaction setup for a 96-well plate (total volume 25 uL per well) combines the following reagents:

1. 20 ng gDNA
2. 10 uL ddPCR Supermix for Probes (No dUTP) Cat #1863025
3. 900 nmol Primer 1
4. 900 nmol Primer 2
5. 900 nmol Primer 3
6. 900 nmol Primer 4
7. 250 nmol FAM-labeled probe
8. 250 nmol HEX-labeled probe

After reaction setup, follow directions based on the droplet digital PCR instrument in your laboratory.

3.6 Droplet Digital PCR Method for a Five-Channel ddPCR Instrument

1. The PCR reaction setup for a 96-well plate (total volume 12 uL per well) combines the following reagents:
2. 20 ng gDNA
3. 3.75 uL 4× Probe Master Mix from QIAcuity Probe PCR Kit Cat#250101
4. 800 nm Primers 1–10
5. 400 nm FAM-labeled probe
6. 400 nm ROX-labeled probe
7. 400 nm Cy5-labeled probe
8. 400 nm ABY-labeled probe
9. 400 nm HEX-labeled probe
10. After reaction setup, follow directions based on the droplet digital PCR instrument in your laboratory.

4 Notes

1. 96-well plates containing mouse ES cells and feeder cells were washed once with 200 uL of PBS. PBS is aspirated and replaced with 40 uL of 0.25% trypsin-EDTA. Cells are placed in a 37 °C, 5% CO2 humidified incubator for 20 minutes.
2. Trypsinized cells are pipetted 10–20 times to produce a single cell suspension. After adding 150 uL of ES cell medium, 10 uL of feeder removal microbeads is added and resuspended by

gentle pipetting. The mixture is then incubated for 20 minutes at room temperature on an orbital shaker (300 rpm).

3. Following incubation, the suspension is loaded onto a 96-well column. The ES cell-containing flow-through is collected, while the feeder cells remain in the column. The collected ES cells are transferred to a U-bottom 96-well plate and spun down in a centrifuge at 3000 rp. Genomic DNA in a 96-well plate format was prepared as described above.
4. Advanced tip loading method.

   On the deck of the Biomek i5, place one empty tip box in TL1 and one full tip box in TL3 as shown in Fig. 3.

   Then open the Biomek i5 software. Click on home all axes icon. Be sure the gripper is not facing the plexiglass. Under tab "Setup & Device Steps," select Instrument Setup (Fig. 4).

   Once Instrument Setup has been selected, click on the tab "Liquid Handling Steps" and select "Select Tips." Then select "Advanced Load Tips" with the following specifications: Tips BC1025F; Tips Location TL3; A1 Row 5; A1 Column 7 (Fig. 5). The gray area will denote where the pod head is located and what tips will be selected to move.

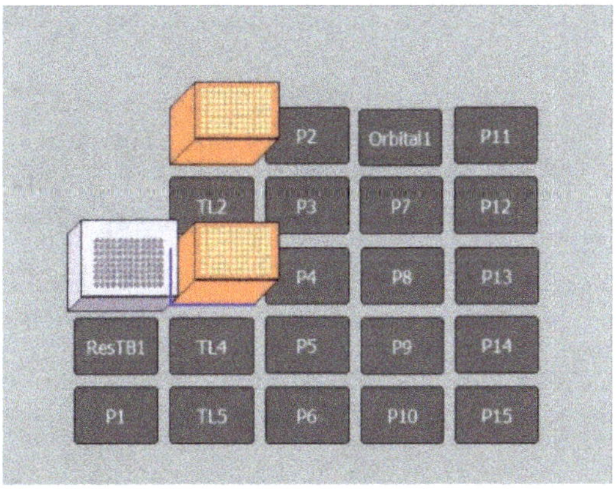

Fig. 3 The Biomek i5 deck configuration for the advanced tip loading method

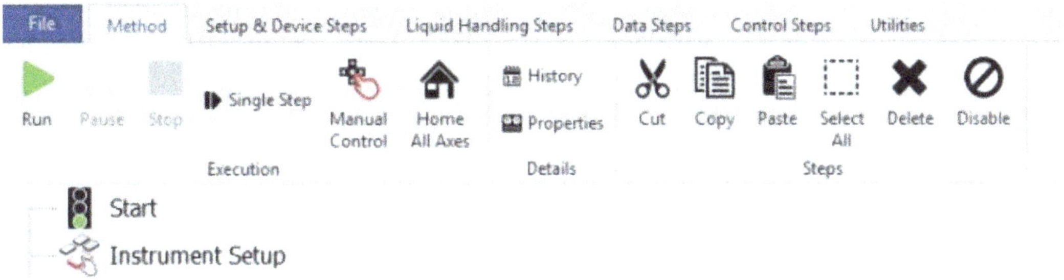

Fig. 4 Elements of Biomek i5 software taskbar

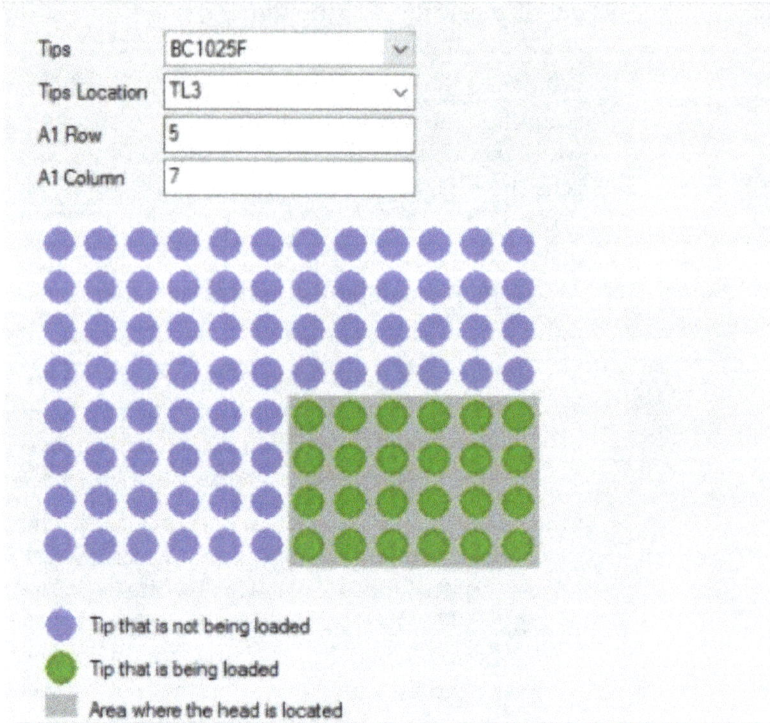

Fig. 5 Biomek i5 software screen for loading specifications for "Advanced Load Tips"

Then select "Advanced Unload Tips" with the following specifications: Tips Location TL1; A1 Row 5: A1 Column 7. The unload specifications will dispense tips picked up from first command to TL1 position (Fig. 6).

From TL1 position, the tips will need to be repositioned for a 24-well plate by programming the following commands. Select "Advanced Load Tips" with the following specifications: Tips BC1025F; Tips Location TL1; A1 Row 8; A1 Column 1 (Fig. 7).

Then select "Advanced Unload Tips" with the following specifications: Tips Location TL1; A1 Row 1; A1 Column 1 (Fig. 8).

Then select "Advanced Load Tips" with the following specifications: Tips BC1025F; Tips Location TL1; A1 Row 7; A1 Column 1 (Fig. 9).

Then select "Advanced Unload Tips" with the following specifications: Tips Location TL1; A1 Row 3; A1 Column 1 (Fig. 10).

Then select "Advanced Load Tips" with the following specifications: Tips BC1025F; Tips Location TL1; A1 Row 6; A1 Column 1 (Fig. 11).

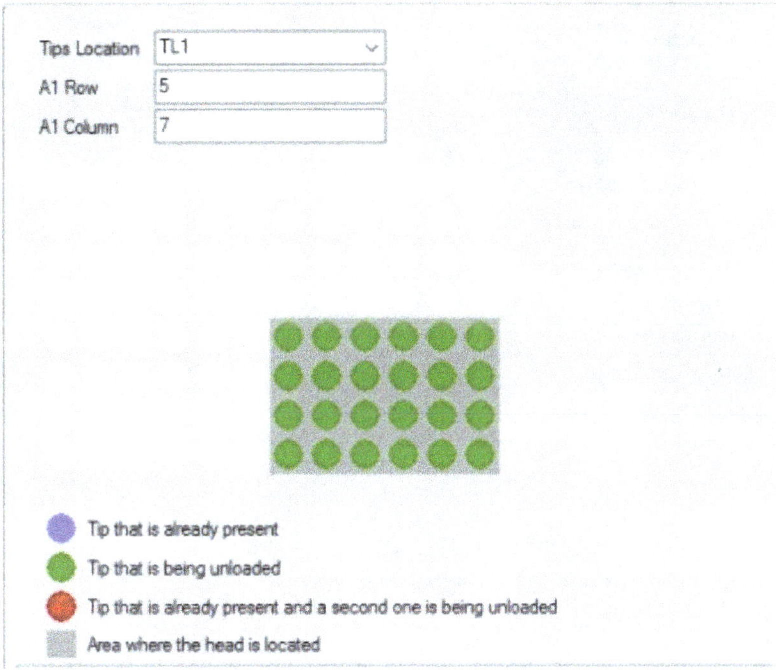

Fig. 6 Biomek i5 software screen for loading specifications for "Advanced Unload Tips"

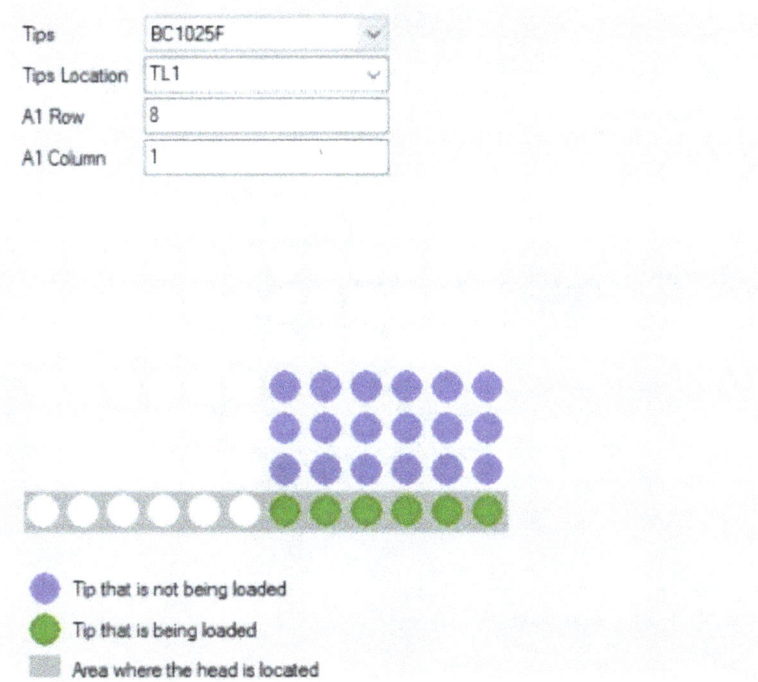

Fig. 7 Biomek i5 software screen for loading specifications for "Advanced Unload Tips" for a 24-well tip positioning

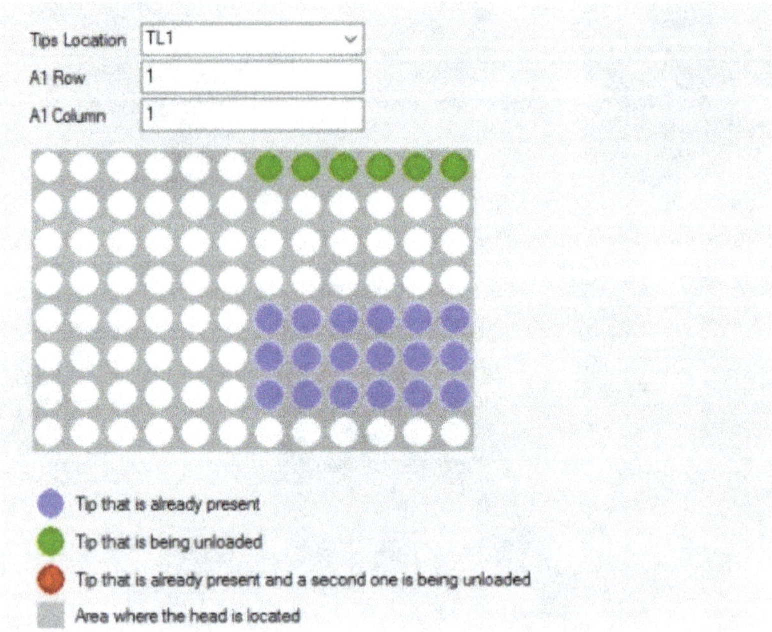

Fig. 8 Continuation of Biomek i5 software screen for loading specifications for "Advanced Unload Tips" for a 24-well tip positioning

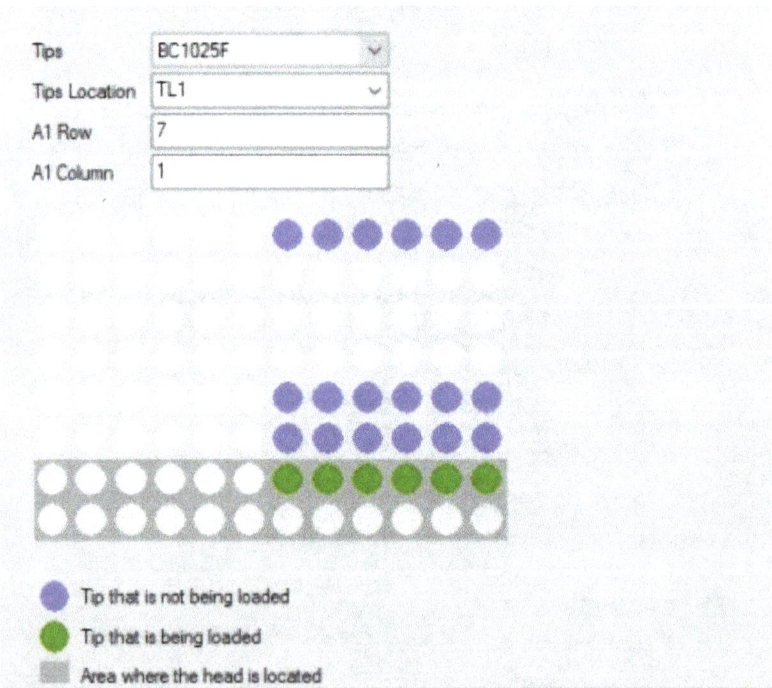

Fig. 9 Continuation of Biomek i5 software screen for loading specifications for "Advanced Load Tips" for a 24-well tip positioning

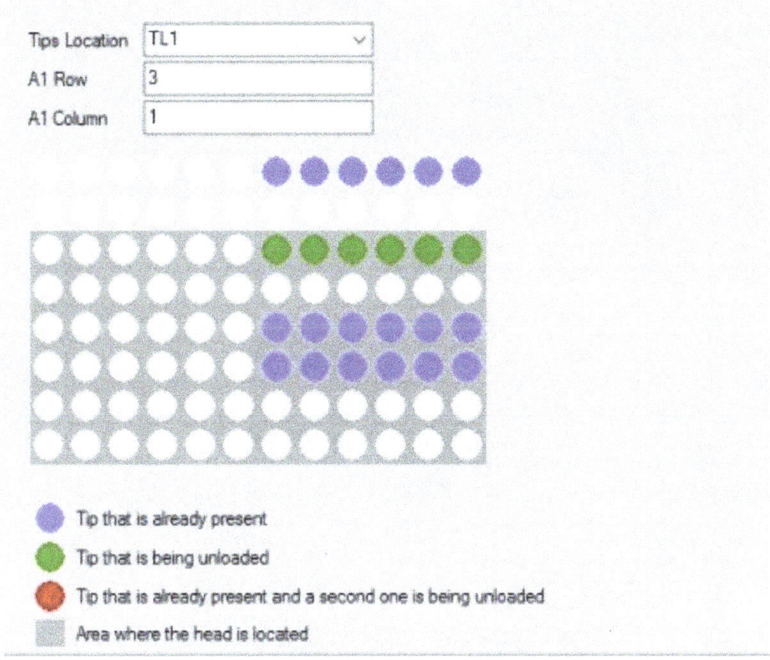

Fig. 10 Continuation of Biomek i5 software screen for loading specifications for "Advanced Load Tips" for a 24-well tip positioning

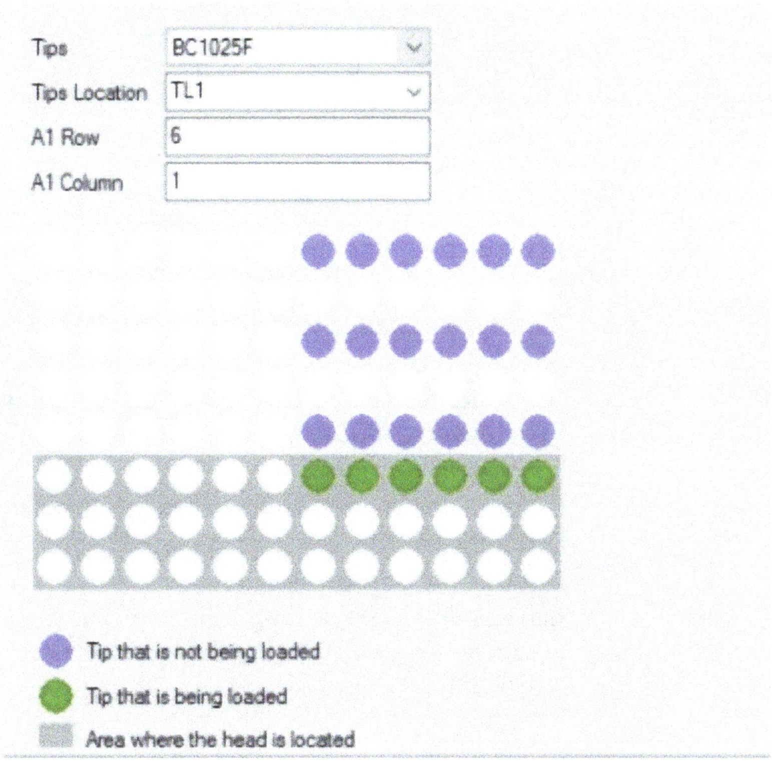

Fig. 11 Continuation of Biomek i5 software screen for loading specifications for "Advanced Load Tips" for a 24-well tip positioning

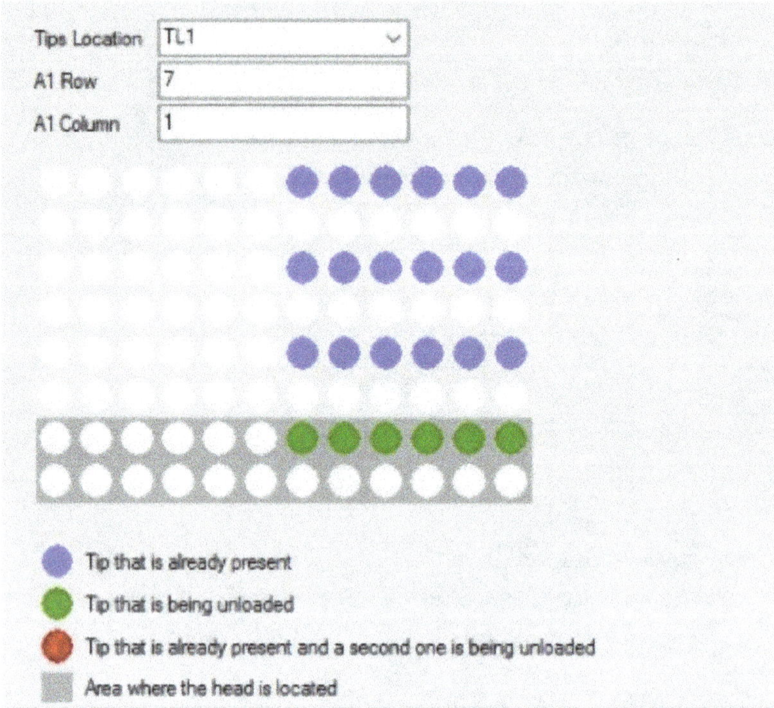

Fig. 12 Continuation of Biomek i5 software screen for loading specifications for "Advanced Unload Tips" for a 24-well tip positioning

Then select "Advanced Unload Tips" with the following specifications: Tips Location TL1; A1 Row 7; A1 Column 1 (Fig. 12).

Then select "Advanced Load Tips" with the following specifications: Tips BC1025F; Tips Location TL1; A1 Row 1; A1 Column 12 (Fig. 13).

Then select "Advanced Unload Tips" with the following specifications: Tips Location TL1; A1 Row 1; A1 Column 1 (Fig. 14).

Then select "Advanced Load Tips" with the following specifications: Tips BC1025F; Tips Location TL1; A1 Row 1; A1 Column 11 (Fig. 15).

Then select "Advanced Unload Tips" with the following specifications: Tips Location TL1; A1 Row 1; A1 Column 3 (Fig. 16).

Then select "Advanced Load Tips" with the following specifications: Tips BC1025F; Tips Location TL1; A1 Row 1; A1 Column 10 (Fig. 17).

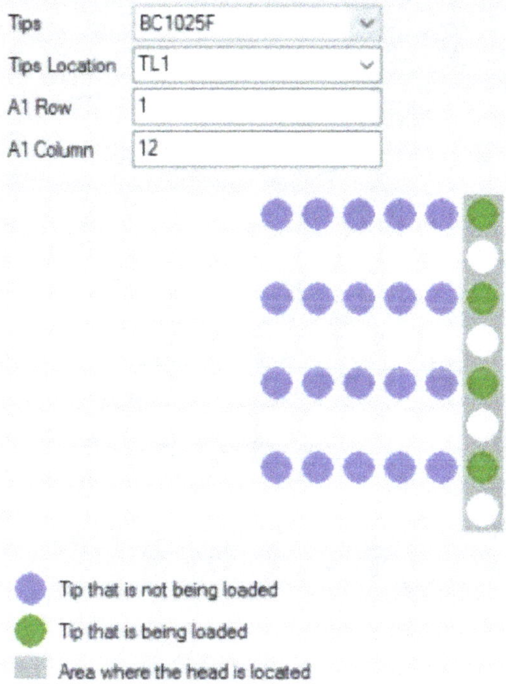

Fig. 13 Continuation of Biomek i5 software screen for loading specifications for "Advanced Unload Tips" for a 24-well tip positioning

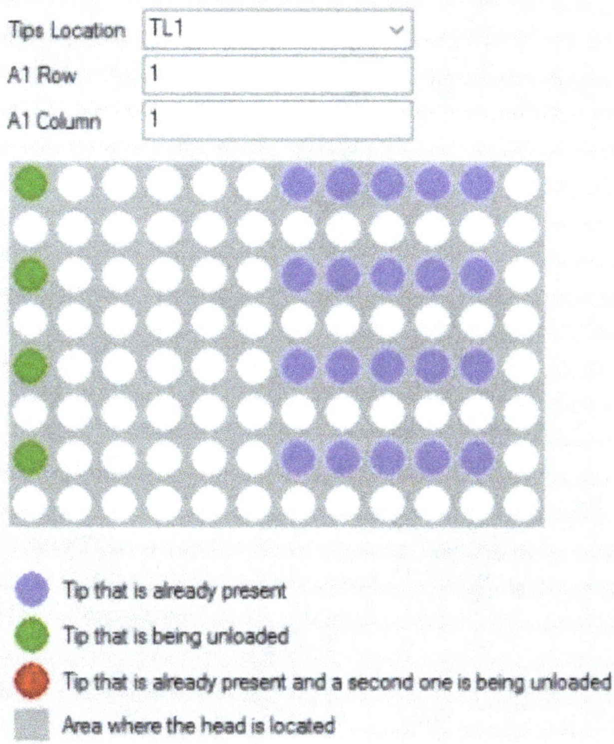

Fig. 14 Continuation of Biomek i5 software screen for loading specifications for "Advanced Unload Tips" for a 24-well tip positioning

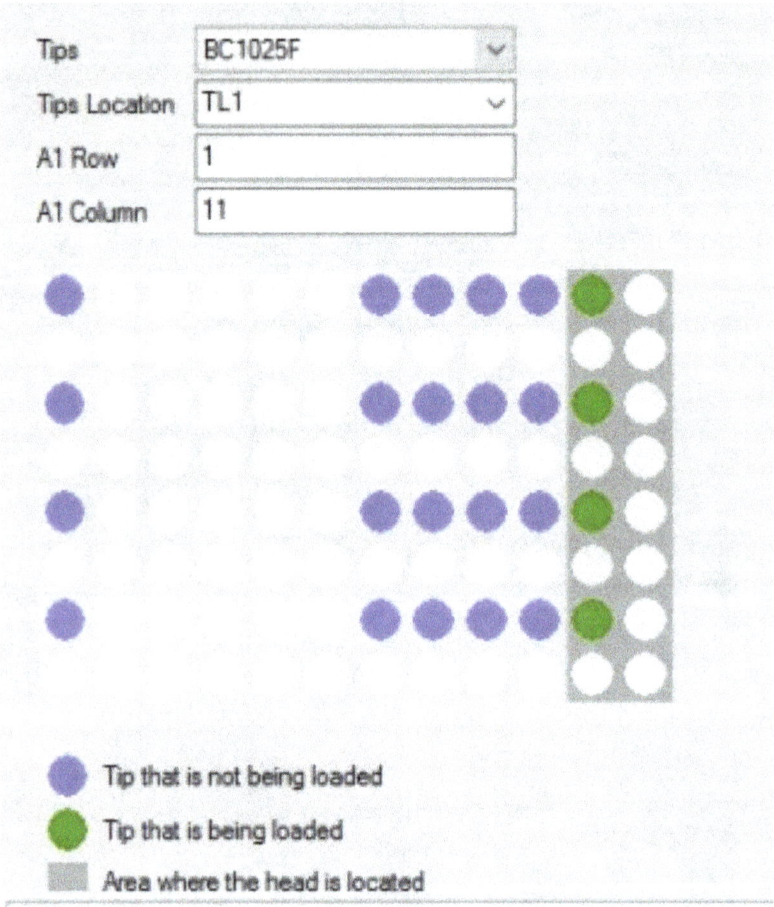

Fig. 15 Continuation of Biomek i5 software screen for loading specifications for "Advanced Load Tips" for a 24-well tip positioning

Then select "Advanced Unload Tips" with the following specifications: Tips Location TL1; A1 Row 1; A1 Column 5 (Fig. 18).

Then select "Advanced Load Tips" with the following specifications: Tips BC1025F; Tips Location TL1; A1 Row 1; A1 Column 8 (Fig. 19).

Select "Advanced Unload Tips" with the following specifications: Tips Location TL1; A1 Row 1; A1 Column 9 (Fig. 20).

Then select "Advanced Load Tips" with the following specifications: Tips BC1025F; Tips Location TL1; A1 Row 1; A1 Column 10 (Fig. 21).

Then select "Advanced Unload Tips" with the following specifications: Tips Location TL1; A1 Row 1; A1 Column 11 (Fig. 22).

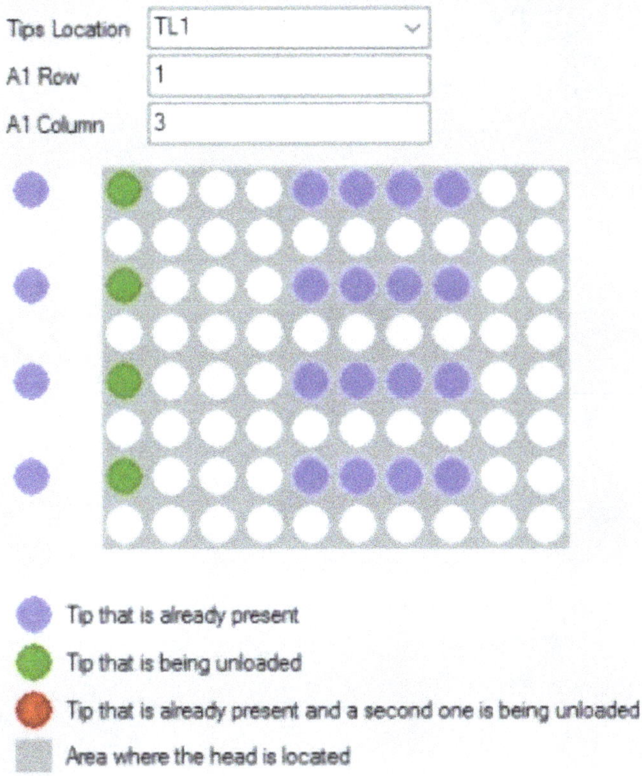

Fig. 16 Continuation of Biomek i5 software screen for loading specifications for "Advanced Unload Tips" for a 24-well tip positioning

Once completed, click on "Liquid Handling Steps" and select "End Using Select Tips" in the Biomek i5 software taskbar. The command should end under Instrument Setup with a Finish Icon as seen in Fig. 23. Methods can continue under Bind1 and Bind 2 with alterations to amount aspirated and dispensed based on needs.

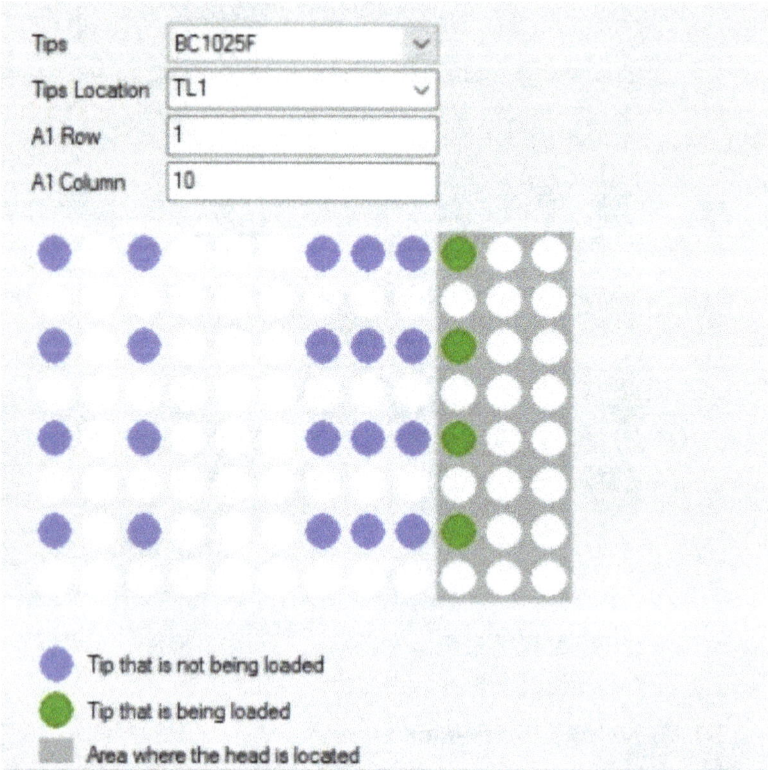

Fig. 17 Continuation of Biomek i5 software screen for loading specifications for "Advanced Load Tips" for a 24-well tip positioning

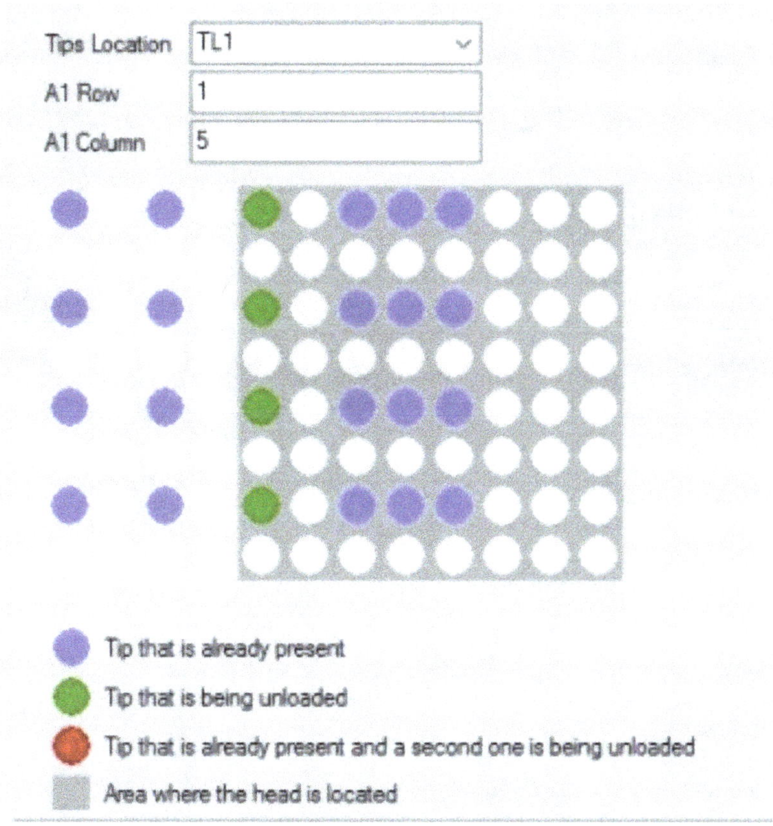

Fig. 18 Continuation of Biomek i5 software screen for loading specifications for "Advanced Unload Tips" for a 24-well tip positioning

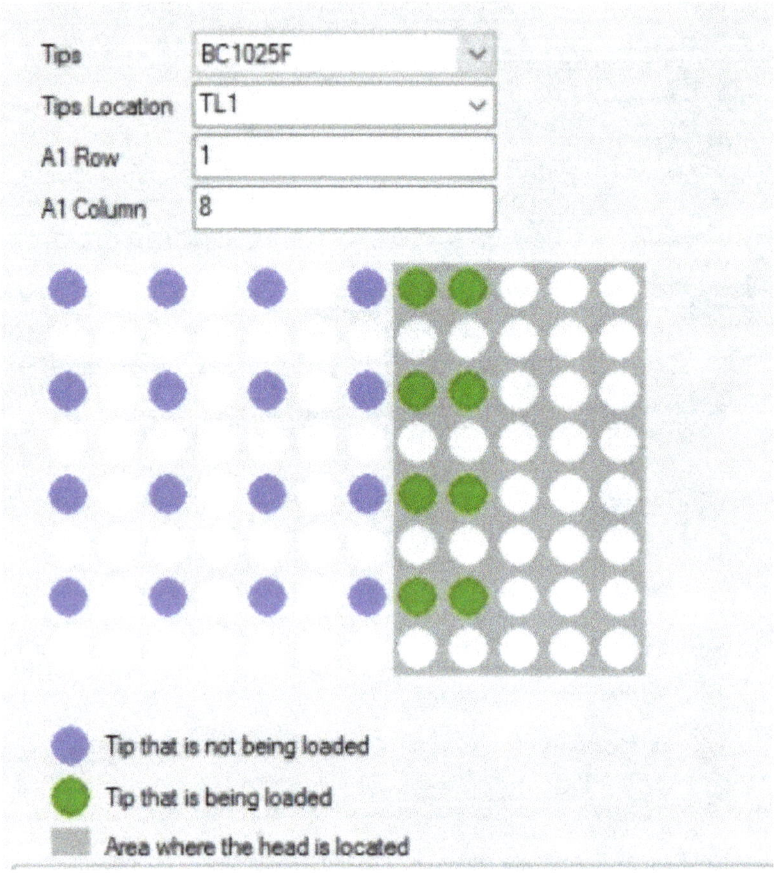

Fig. 19 Continuation of Biomek i5 software screen for loading specifications for "Advanced Load Tips" for a 24-well tip positioning

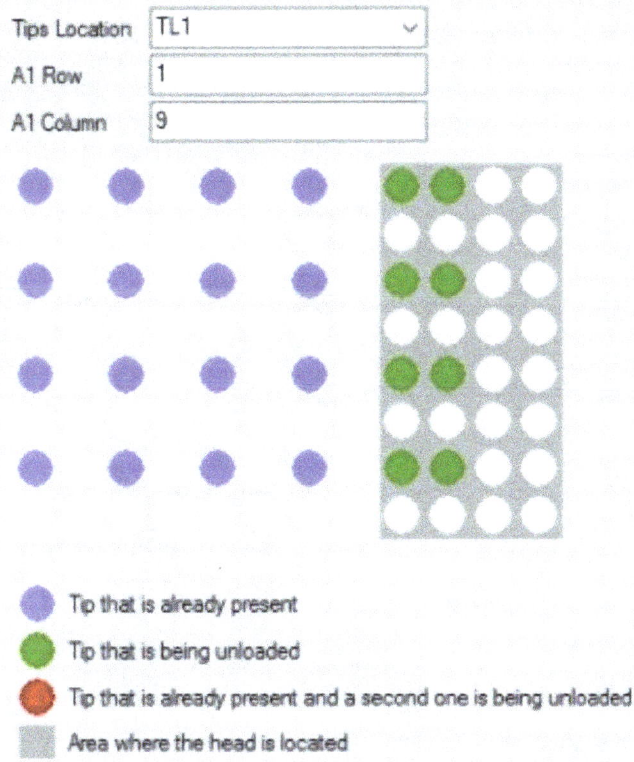

Fig. 20 Continuation of Biomek i5 software screen for loading specifications for "Advanced Unload Tips" for a 24-well tip positioning

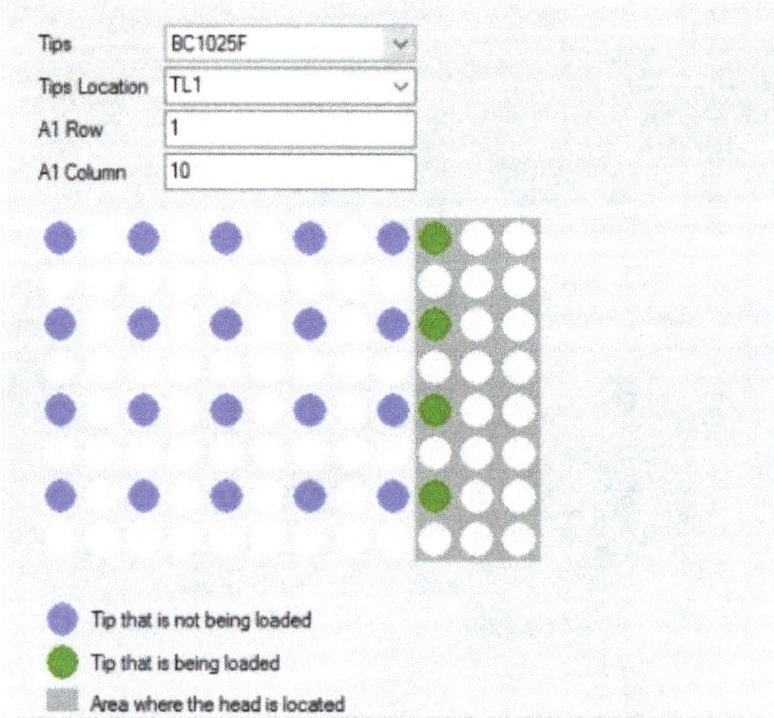

Fig. 21 Continuation of Biomek i5 software screen for loading specifications for "Advanced Load Tips" for a 24-well tip positioning

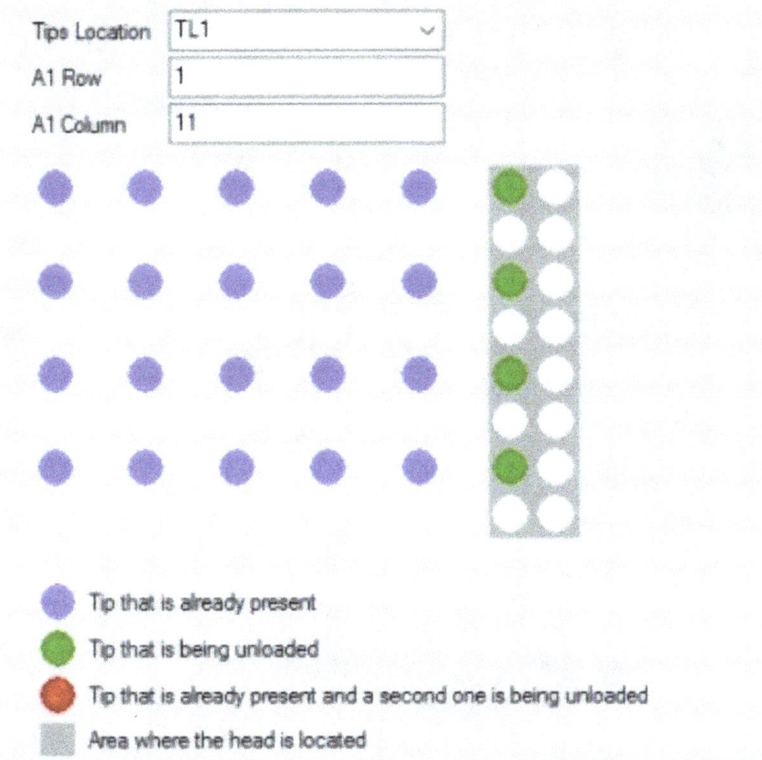

Fig. 22 Continuation of Biomek i5 software screen for loading specifications for "Advanced Unload Tips" for a 24-well tip positioning

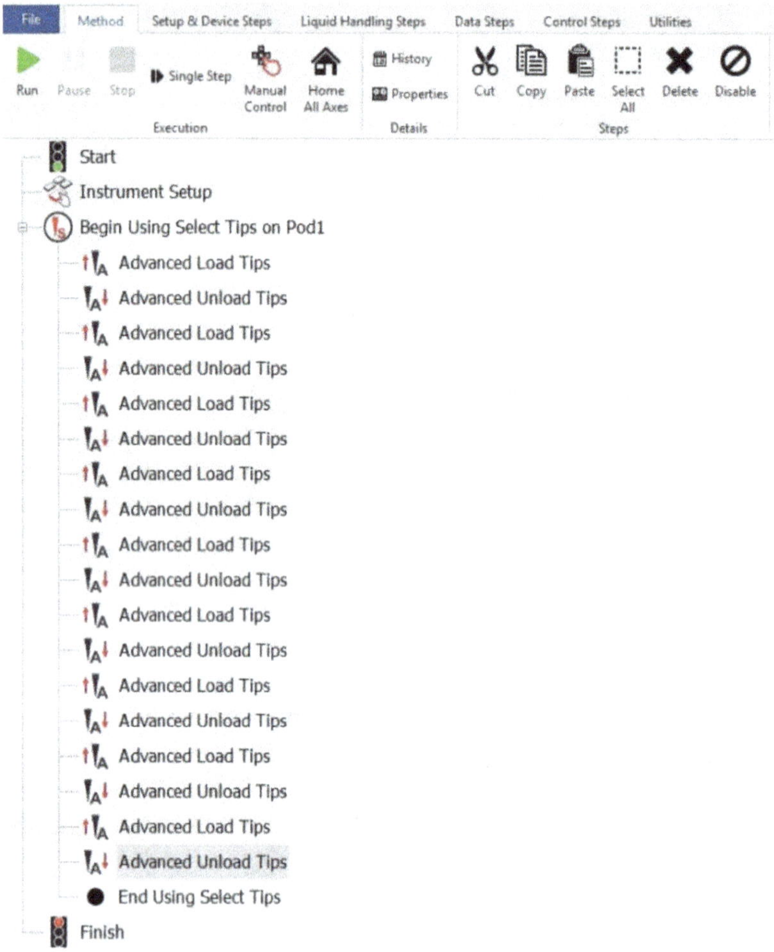

Fig. 23 Image of Biomek i5 software taskbar at conclusion of reprogramming for a 24-well plate tip configuration

References

1. Baskale D et al (2012) Use of magnetically enriched pluripotent stem cells increases chimerism rate after blastocyst injection and enables the use of inbred ES cell lines for tetraploid complementation. MACS&more 14(2):8–10
2. Pinheiro LB et al (2012) Evaluation of a droplet digital polymerase chain reaction format for DNA copy number quantification. Anal Chem 84:1003–1011
3. Zhong Q et al (2011) Multiplex digital PCR: breaking the one target per color barrier of quantitative PCR. Lab Chip 11:2167–2174
4. Karlen Y et al (2007) Statistical significance of quantitative PCR. BMC Bioinform 8:131
5. Gaudernack G et al (1986) Isolation of pure functionally active $CD8^{+}$ T cells. Positive selection with monoclonal antibodies directly conjugated to monosized magnetic microspheres. J Immunol Methods 90:179–187
6. Trickett A, Kwan YL (2003) T cell stimulation and expansion using anti-CD3/CD28 beads. J Immunol Methods 275:251–255
7. Caothien R et al (2022) Accelerated embryonic stem cell screening with a highly efficient genotyping pipeline. Mol Biol Rep 49:3281–3288

Chapter 8

Gene Editing in Mouse Zygotes Using the CRISPR/Cas9 System

Benedikt Wefers, Wolfgang Wurst, and Ralf Kühn

Abstract

Engineering of the mouse germline is a key technology in biomedical research for studying the function of genes in health and disease. Since the first knockout mouse was described in 1989, gene targeting was based on recombination of vector encoded sequences in mouse embryonic stem cell lines and their introduction into preimplantation embryos to obtain germline chimeric mice. This approach has been replaced in 2013 by the application of the RNA-guided CRISPR/Cas9 nuclease system, which is introduced into zygotes and directly creates targeted modifications in the mouse genome. Upon the introduction of Cas9 nuclease and guide RNAs into one-cell embryos, sequence-specific double-strand breaks are created that are highly recombinogenic and processed by DNA repair enzymes. Gene editing commonly refers to the diversity of DSB repair products that include imprecise deletions or precise sequence modifications copied from repair template molecules. Since gene editing can now be easily applied directly in mouse zygotes, it has rapidly become the standard procedure for generating genetically engineered mice. This article covers the design of guide RNAs, knockout and knockin alleles, options for donor delivery, preparation of reagents, microinjection or electroporation of zygotes, and the genotyping of pups derived from gene editing projects.

Key words CRISPR, Cas9, Mouse, Zygotes, Gene editing, HDR, NHEJ, Knockin, Knockout, Genome engineering

1 Introduction

1.1 Gene Editing in Mouse Zygotes

In 2013, the CRISPR/Cas9 nuclease system was adapted from *Streptococcus pyogenes* for creating targeted double-strand breaks (DSBs) in the genome of mammalian cells, enabling targeted genome editing at high efficiency [1, 2]. For the induction of DSBs, two components must be introduced into cells: the Cas9 protein and a short guide RNA (gRNA) that associates with Cas9 to form a ribonucleoprotein (RNP) complex. The first 20 nucleotides of gRNAs direct the RNP complex to a specific complementary DNA target sequence via RNA-DNA hybridization. If the target sequence is located upstream of the invariant sequence NGG (PAM—protospacer adjacent motif), Cas9 nuclease is

Thomas L. Saunders (ed.), *Transgenesis: Methods and Protocols*,
Methods in Molecular Biology, vol. 2631, https://doi.org/10.1007/978-1-0716-2990-1_8,

activated and creates a blunt-ended DSB located three base pairs upstream of the PAM site. Using this system, Cas9-induced DSBs can be directed to any unique PAM-containing DNA sequence by recoding of the first 20 nucleotides of the gRNA which correspond to the new target DNA sequence. Using CRISPR/Cas9, new target sites can be easily addressed by synthesis of a short RNA sequence whereas the Cas9 protein is an invariant component.

Numerous studies confirmed the utility and efficacy of CRISPR/Cas9-mediated gene editing in mouse zygotes, enabling the direct production of knockout and knockin mutants in a single step (*see* Fig. 1) [3, 4]. Therefore this technology replaced previously established mutagenesis techniques based on embryonic stem cells or earlier generations of nucleases and revolutionized mouse genetics [5, 6]. Since 2013, the protocols for use of CRISPR/Cas9 in zygotes have evolved and have further streamlined the production of genetically modified mouse lines in shorter time with less resources. Initially, Cas9 mRNA and gRNAs were generated by in vitro transcription and delivered into zygotes by pronuclear or cytoplasmic microinjection. Both delivery routes lead to comparable results and the protocol of choice depends on trained skills and the available equipment. Later, the delivery of Cas9 was further improved and simplified by the use of recombinant protein and of chemically synthesized RNAs that are readily available from commercial providers. Since 2016, protocols for the electroporation of zygotes were described that provide an alternative to the time-consuming and tedious microinjection procedure [7, 8].

Upon the introduction of Cas9 and gRNA into zygotes, gene editing occurs by the targeted induction of DSBs, followed by endogenous DSB repair. In most cases, DSBs are repaired by the non-homologous end-joining (NHEJ) pathway that religates open DNA ends by DNA ligase IV without the use of a repair template [9]. DSB sites that underwent processing by NHEJ repair frequently exhibit the random deletion and/or insertion of nucleotides (InDels). If these InDels are located within coding regions and generate a shift in the reading frame (i.e. the size of the InDel is not a multiple of three), a knockout mutation occurs. Using pairs of gRNAs, knockout alleles can be also created by the deletion of gene segments. Alternatively to NHEJ, DSBs can be repaired in cycling cells by homology-directed repair (HDR), an endogenous pathway that utilizes DNA molecules as repair template [10]. To achieve gene editing, HDR is utilized by the addition of an artificial DNA template molecule that includes sequences as homology regions that are identical to the genomic sequence located up- and downstream of the DSB. In the targeting molecule or vector, these homology regions are flanking the desired sequence modification or insertion. In the repair process, sequence conversion extends from the template's homology regions into the heterologous sequence and copies the genetic modification into the target gene

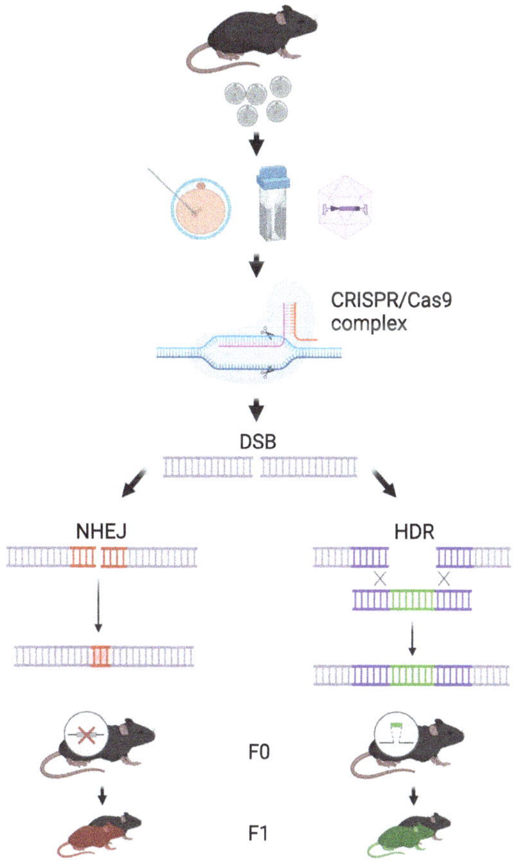

Fig. 1 Gene editing in mouse zygotes using CRISPR/Cas9. Mouse zygotes are treated by pronuclear microinjection (PNI) or electroporation (EPO) with Cas9 protein, guide RNA, and (optionally) a repair template. The double-strand break (DSB) in the target gene can be either repaired by NHEJ, leading to small sequence deletions and gene knockout alleles. Precise sequence modifications (knockin alleles) are introduced by HDR with the homology regions of DNA template molecules (oligonucleotide, plasmid vector, AAV). Mice derived from microinjected zygotes (F0 generation) can harbor a variety of mutant founder alleles. Mating of the F0 founder mutants with wild-type mates leads to the transmission of mutant alleles into individual F1 pups. Figure created with BioRender.com

(knockin), enabling the introduction of preplanned mutations such as codon replacements or the insertion of reporter genes. Large sequence insertions require the construction of plasmid-based gene targeting vectors which include homology regions of several thousand base pairs, whereas small sequence modifications can be introduced using synthetic single-stranded DNA oligonucleotides (ssODN) with lengths of 100–200 nucleotides. In mouse zygotes, both repair mechanisms occur side by side, but DSBs are mostly repaired by the highly active NHEJ pathway whereas HDR occurs less frequently. Therefore, the induction of DSBs in the pronuclei of zygotes leads to a variety of individually repaired alleles at the target locus and to a diverse collection of founder mutants (F0 generation; *see* Fig. 1), which needs to be carefully characterized by genotyping. For increasing the frequency of HDR in zygotes, various modifications regarding the structure or application of the repair donor molecules were described. These include donor delivery as linear double-stranded DNA (HMEJ, TILD) [11, 12], long single-stranded DNA donors (lssDNA) (Easi-CRISPR) [13–15], or by the delivery of donors up to a size of 4.5 kb through packaging into AAV viral particles (CRISPR-READI) [16–18]. Besides current standard applications in C57BL/6 zygotes, the CRISPR/Cas9 system enables also the introduction of mutant alleles into other wild-type strains, or strains that already carry genetic mutations, such as NSG or NOD mice [19, 20]. Furthermore, chromosome engineering for the deletion, inversion, or duplication of megabase segments, initially performed using CRISPR/Cas9 in ES cells, has also been achieved directly in zygotes [21, 22]. In addition to Cas9, nuclease DSB induction by other methods has been validated for use in mouse zygotes but is not yet widely used. These include Cas9 orthologues from other bacterial species [23–26], base editors that combine Cas9 nickase with an enzyme domain able to deaminate A or C nucleotides leading to A to G or C to T replacements [27–30], and prime editors that combine Cas9 nickase with reverse transcriptase and an extended gRNA that includes an RNA-based donor sequence [31].

1.2 Timing of CRISPR/Cas9 Gene Editing in Zygotes

Overall CRISPR/Cas9 projects require four steps represented by protocols in the methods section: (1) target sequence selection and the design of mutant alleles, (2) preparation of reagents, (3) microinjection or electroporation of zygotes and embryo transfer, and (4) the genotyping of F0 pups. The first two steps are required for preparation and need only a few days for ordering gRNA but more time if a HDR template vector needs to be constructed. Standard times apply for **steps 3** and **4**. Female mice are treated with hormones for superovulation starting 3 days before zygote manipulation (*see* Fig. 2a). Upon the mating of superovulated females to males, zygotes are collected on the next day, microinjected or electroporated with CRISPR/Cas9 reagents, and transferred into foster females (*see* Fig. 2b), which give rise to birth after 3 weeks. Biopsies can be taken from the pups derived from transferred zygotes at the age of 4 weeks and used for genotyping 7 weeks after embryo treatment (*see* Fig. 2c). Mutant F0 founders reach fertility at 7–8 weeks and can be further mated to obtain F1 progeny which can be genotyped 11 weeks after mating (*see* Fig. 2d). In many cases, a single or a few days of embryo manipulation result into desired mutant founders such that mutant F1

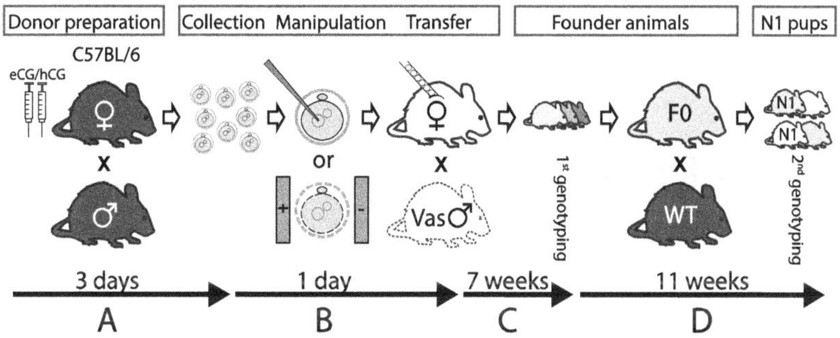

Fig. 2 Timing of CRISPR/Cas9 gene editing projects. (**a**) Three days before microinjection, donor females are treated with hormones and mated. (**b**) On the next day, zygotes are collected, microinjected or electroporated with CRISPR/Cas9 reagents, and transferred to pseudopregnant foster mothers that were mated to vasectomized (Vas) males. (**c**) After 7 weeks, biopsies for genotyping are obtained from 4 week-old pups of the F0 generation derived from treated zygotes. (**d**) 11 weeks later, N1 pups can be genotyped (4 week-age) that were obtained from mating F0 founder mutants to wild-type mice

progeny is available within 5 months. However, the total number of F0 pups recovered from manipulated C57BL/6 zygotes is variable and sometimes insufficient, requiring time-consuming repeat experiments. For the delivery of CRISPR/Cas9 reagents into zygotes, various options with regard to the nature, concentration, and delivery are available. The protocols described in this article are based on the pronuclear zygote microinjection using Cas9 protein with synthetic gRNA and optionally a DNA oligonucleotide or plasmid vector as HDR template. Alternatively, zygotes can be electroporated with Cas9 protein, synthetic gRNA, and optionally a DNA oligonucleotide or incubated with AAV particles harboring a donor template. Using this approach, we routinely result in 20–70% of F0 pups alleles harboring small deletions and HDR alleles in 5–30% of the pups. Note that all animal experimentation is regulated by national guidelines and needs approval by local authorities.

1.3 Design of Gene Editing Experiments

1.3.1 Targeting Strategies

Each experimental question requires a customized edited allele that includes a sequence modification obtained by the deletion, insertion, or replacement of nucleotides in the genomic target site. These demands must be translated into practice by induction of DSBs with Cas9 and gRNAs at the site of interest and forecasting the products obtained by the repair through the NHEJ or HDR pathway. Small deletions and insertions (InDels) of variable length can be obtained by NHEJ repair induced with Cas9 and a single gRNA, whereas gRNA pairs can be used for the deletion of larger DNA segments (*see* Fig. 3a). The outcome of InDel approaches can be predicted by in silico tools [32, 33]. NHEJ can also be utilized to generate sequence insertions (HITI) and replacements (Replace) in mammalian cells [34, 35] but are not included in Fig. 3 since these approaches are not validated in mouse zygotes. In contrast to

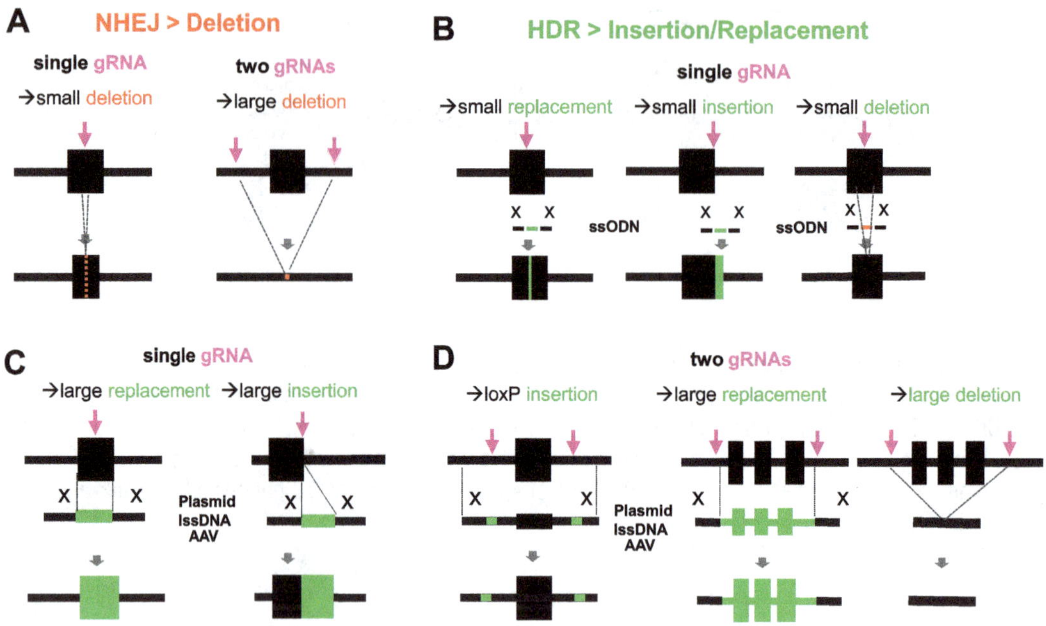

Fig. 3 Design of gene editing experiments. (**a**) Small or large deletions with imprecise ends are obtained by using single or pairs of gRNAs and repair of DSBs by the NHEJ pathway. (**b**) Small precise sequence modifications (<40 bp, green line) can be introduced by DSB repair and ssODNs as HDR repair donor templates that include homology sequences flanking the DSB site. (**c**) Large insertions or replacements can be introduced with single DSBs and plasmid, lssDNA, or AAV particles as large HDR donor templates. (**d**) Two gRNAs can be used for large replacements including the insertion of two loxP sites or large precise deletions using plasmid, lssDNA, or AAV particles as large HDR donor templates

imprecise products obtained by NHEJ repair, edited alleles that require precise sequence modifications over the entire sequence need to be generated by HDR repair, requiring a repair template that contains the desired donor sequence. Using single gRNAs and ssODN templates enables the introduction of short sequence replacements, insertions or deletions up to a size of approximately 40 nucleotides as required for the insertion of a loxP site, or a Flag Tag (*see* Fig. 3b). Larger replacements or insertions of, e.g., GFP or CreERT2 coding sequences require larger template molecules, either cloned into a plasmid backbone or prepared as lssDNA, together with a single or two gRNAs (*see* Fig. 3c, d). For the insertion of expression cassettes into the Rosa26 locus, premade donor vectors are available [36]. Alternatively to CRISPR/Cas9, an BxbI integrase-based approach for Rosa26 integrations has been described recently [37]. To generate precise large deletions, the insertion of two distant loxP sites, or the replacement of genes or gene segments, two gRNAs can be used together with a large donor vector (*see* Fig. 3d). For the generation of loxP flanked alleles, it is more efficient to use a large lssDNA or plasmid vector comprising both sites as compared to a pair of independent

Table 1
Selection guideline for type of delivery method and targeting molecule

Planned modification	DSB repair	gRNAs	Delivery	Donor template
Deletion (imprecise)	NHEJ	1	*EPO* or PNI	None
Small replacement, insertion, deletion (precise)	HDR	1	*EPO* or PNI	ssODN
Large replacement or Insertion (<3 kb)	HDR	1 or 2	*PNI* or EPO + AAV	*lssDNA* or plasmid or AAV
Large replacement or Insertion (>3 kb)	HDR	2	PNI	Plasmid
Large deletion (precise)	HDR	2	*PNI* or EPO + AAV	*lssDNA* or plasmid or AAV

Preferred options italicized

ssODNs [14, 38]. The established donor templates for HDR repair differ by size, nature, and the delivery routes. Synthetic ssODNs are small and cheap templates (up to 200 nucleotides) that enable short sequence modifications and can be delivered by electroporation or microinjection. To introduce larger sequence alterations, plasmid-based donor vectors allow the cloning of inserts up to 10 kb that together with homology regions reach the cargo limit of high copy number plasmids. Plasmid inserts can be either generated by classical restriction cloning procedures or by gene synthesis. High rates of HDR are obtained using lssDNA as donor that can be prepared from plasmid inserts, by in vitro transcription-based DNA synthesis, or are available from commercial providers. Due to their size, plasmids and lssDNAs (>1 kb) do not pass the zona pellucida and can be introduced into zygotes only by microinjection. Plasmid inserts up to 4.5 kb in between ITR motifs of AAV can be packaged into viral particles. These particles can be added to the embryo culture medium prior to electroporation with Cas9 RNPs and lead to high rates of HDR. Table 1 shows the different classes of edited alleles following the requirement for a donor template and its size. The advantages and limitations of delivery techniques into zygote are compared in Table 2.

1.3.2 Design of Guide RNAs

Planning a desired sequence modification requires first to analyze the sequence region of interest for Cas9 target sites. These sites can be easily identified by inspection of sequences for NGG motifs, but online analysis tools are preferable that also predict the gRNA off-target profile and on-target activity. We use the CRISPOR webpage that ranks target sites along with specificity and includes the results of efficiency scoring tools in a single table (http://crispor.tefor.net/) [39]. From this table, we focus on highly specific guide RNAs and use these with the off-target-reduced Cas9 HiFi protein that is

Table 2
Comparison of different delivery methods for RNPs and donor molecules

	Advantages	Limitations
PNI	Robust and common technique Applicable to all types of mutations	Sophisticated training needed Expensive equipment Low throughput Difficult in some genetic strains
EPO	Easy and fast Cheap equipment High throughput possible Applicable to all genetic strains	Size limit of targeting molecule 200 nt (for ssODN) or ~1 kb (for lssDNA)
EPO plus AAV	Extends targeting possibilities of electroporation	Size limit of targeting molecule 4.5 kb Extra work for AAV production/purification

sensitive to a single mismatch and can be commercially purchased as recombinant protein. To avoid failures due to a single, inefficient gRNA, we use whenever possible two alternative gRNAs for each target region that are introduced each into half of the number of zygotes. In addition, gRNAs may be verified for on-target activity upon introduction with Cas9 into cultured mouse cells or zygotes followed by PCR amplification and T7 endonuclease I assays (*see* Subheading 3.5). The integrity of synthetic gRNA batches can be further confirmed by an in vitro Cas9 nuclease assay using a PCR product or plasmid that includes the target sequence. For the generation of mutant mice identified, off-target mutations are not that troublesome since unlinked mutations can be easily segregated through breeding. Moreover, whole-genome sequencing of CRISPR/Cas9-generated mutant mice confirmed that unexpected genome modifications, besides the predicted off-target sites, do not occur [40–42]. For the rare case that highly specific gRNAs in the desired region are not present, targeting must be performed using medium- or low-specific guide RNAs. In this case, use guide RNAs with off-targets located on a different chromosome than the on-target locus, so that off-target mutations can be segregated by backcrossing of founder mutants to wild-type mice. As routine off-target screen, we verify the sequence integrity of five of the best matching off-target sites by Sanger sequencing of PCR products amplified from F1 pups of a newly generated mouse line.

1.3.3 Preparation of Donor Template Molecules

ssODNs are synthetic single-stranded oligonucleotides that are used as HDR donor template for the introduction of short sequence modifications up to 40 nucleotides that are recombined with the DSB site by the Fanconi anemia pathway [43]. The use of purchased ssODNs as donor templates up to 200 nucleotides is simple and saves own hands-on work. In mouse zygotes, we use the symmetric design of homology sequences each of 50–70 nucleotides flanking the sequence modification although in human cell lines an asymmetric design is preferred [44]. Within the gRNA and PAM target site, silent nucleotide replacements should be included in the ssODN sequence to prevent recutting of the recombined sequence by Cas9. We recommend to use full length ssODNs purified by PAGE such as Alt-R™ HDR oligos (IDT). ssODNs pass the zona pellucida and can be introduced into zygotes by electroporation or microinjection.

lssDNA are long single-stranded DNA (>1000 nucleotides) donor molecules that enable the insertion of larger sequence regions (e.g., GFP reporter) or two loxP sites by HDR at enhanced frequency (EASI-CRISPR) [13, 14, 38]. Due their size, lssDNA molecules do not pass the zona pellucida and can be introduced into zygotes only by microinjection. We conveniently use synthesized lssDNA (Megamer, IDT) with homology regions of 100–300 nucleotides each and a maximum size of 3000 nucleotides. Various protocols based on phage lambda exonuclease (Strandase) or polymerases can also be used for lssDNA preparation in molecular biology lab settings [45–48].

Plasmids are the classical double-stranded DNA donor templates for HDR that can be introduced into zygotes only by microinjection. Since high copy plasmids have a backbone of 2–3 kb and a size limit of ~20 kb, inserts including homology regions of 1–3 kb and up to ~10 kb donor sequence can be cloned. We recommend the careful preparation of plasmid DNA to avoid embryo toxicity. Plasmid inserts can be cloned by using restriction enzymes, Gibson assembly of, e.g., PCR products, or can be derived by gene synthesis from service providers. Supercoil plasmid DNA is amenable for HDR but isolated linear inserts may be also used [11]. A convenient positive control vector for HDR in zygotes is targeting the β-actin gene with RFP [12] and is available from Addgene (www.addgene.org, ID 97317). Using supercoiled plasmid DNA, the expected knockin frequency of ~20% can be detected by fluorescence microscopy of blastocysts derived from microinjected embryos.

AAV donors: plasmid inserts up to 4.5 kb that are flanked by ITR sequence motifs can be packaged into AAV particles. AAV particles with the coat of serotypes 1 and 6 can infect zygotes and serve as superior donor template for HDR at DSBs introduced by electroporation of Cas9 RNPs (CRISPR-READI) [17]. Zygotes are precultured in 20 μL KSOM with 10^8–10^9 AAV genome copies. Viral particles are produced upon the transfection of HEK293 cells with helper and capsid plasmids and require several days for purification and titering. Due to the additional workload, we use AAV donors only as a backup strategy for projects that failed by using a standard plasmid vector.

1.3.4 CRISPR/Cas9 Reagents

Initially Cas9 and gRNAs were delivered into zygotes as RNA molecules prepared by in vitro transcription from plasmid templates. Meanwhile the use of Cas9 protein and of chemically synthesized RNAs is the preferred option since these reagents save own hands-on time and are cost-effective. We routinely use high-fidelity Cas9 nuclease protein together with the dual or single RNA system (IDT). The dual system is more economic but needs handling of two components. For each day of embryo manipulation, aliquots are prepared by mixing gRNA with Cas9 protein from stocks, followed by a short incubation to form RNPs. Besides Cas9, related nucleases and base and prime editors were shown to work in mouse zygotes; however, most of these reagents are not available as recombinant proteins and need to be prepared and microinjected as mRNAs.

1.3.5 Genotyping Strategies

Genotyping is of critical importance for the identification of the desired founder mutants. Before starting embryo manipulation, it is possible to validate the on-target activity of gRNAs by the introduction of RNPs into zygotes followed by PCR amplification of the target sequence using genomic DNA from cultured blastocysts. For the detection of InDels, primers flanking the target site are required, and amplicons can be analyzed by T7 endonuclease I assay or Sanger sequencing (*see* Fig. 4a). The same type of genotyping applies for the detection of InDels in genomic DNA from F0 mice obtained from manipulated embryos (*see* Fig. 4b left). In case of short sequence replacements leading to, e.g., point mutations, PCR products can be conveniently analyzed by digestion and gel electrophoresis if a new restriction enzyme site can be introduced through silent nucleotide replacements (*see* Fig. 4b center). In case of large sequence insertions, we prefer in the first PCR screen of F0 mice to use PCR primers that are specific and both located within the inserted new sequence, e.g., within a GFP coding sequence, and therefore a positive control is available (*see* Fig. 4b right). By this means, F0 mice that contain donor vector copies in their genome can be quickly identified, and further genotyping efforts can be focused on a small number of potential founder mutants. In the next step, candidate founders should be analyzed in secondary genotyping screens using one genomic primer located outside of the donor vector homology regions and a second primer located within a vector specific sequence (*see* Fig. 4c In-Out-PCRs). By Sanger sequencing, these PCR products can be verified for presence of the expected recombination products. Genomic primers both located outside of the donor vector homology regions can be used to amplify and sequence the entire edited allele (*see* Fig. 4c Out--Out-PCR). Furthermore, additional quality controls can be performed by quantitative PCR to test for the presence of additional randomly integrated vector copies. Once F0 founder mutants are identified, they are mated with wild-type mice to obtain F1

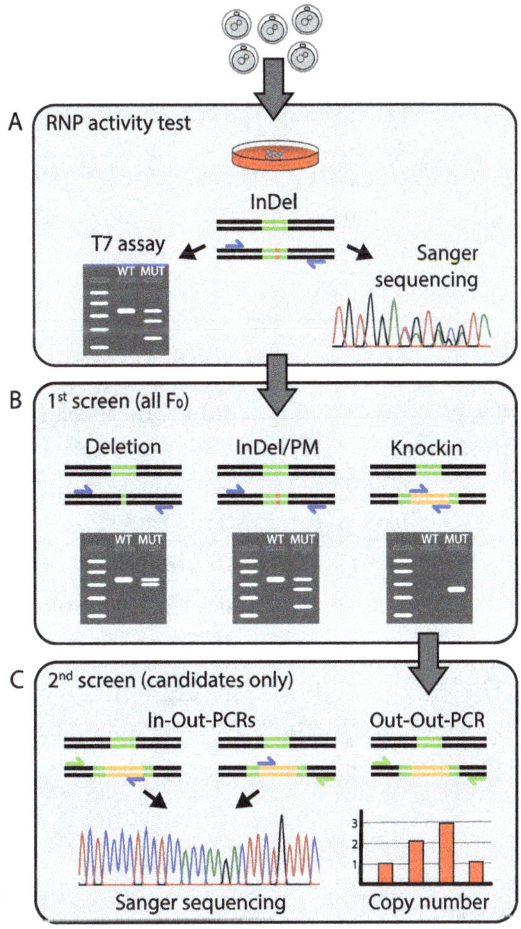

Fig. 4 Overview of genotyping workflow to identify correct founder mutants. (**a**) Efficiency of CRISPR/Cas9-mediated gene targeting is initially assessed in an in vitro RNP activity test. Manipulated one-cell embryos are cultivated until the blastocyst stage and the targeted locus is subsequently PCR amplified and analyzed by T7 endonuclease I assay or Sanger sequencing of amplicons. (**b**) All F0 founder animals undergo a primary genetic screen using suitable PCR strategies. Mutants are identified by band size shifts (deletions), restriction fragment length polymorphism (point mutations, PM), or presence of a transgene-specific amplicon using internal primers (knockin). (**c**) Candidate founders of complex knockins should be further analyzed in a secondary screen by In-Out-PCRs using an internal and an external (outside the homology regions) primer as well as whole-transgene by Out-Out-PCRs and Sanger sequencing to validate sequence integrity. In addition, copy number assays by qPCRs can be performed to exclude additional, concatemeric or random integrations of the donor vector

offspring. These pups should be reconfirmed for the presence and integrity of the edited allele before establishment of a breeding colony since F0 mice are often mosaic mutants. If genotyping is based on the analysis of DNA from ear tissue of such a founder, the germline of this mouse may include other edited alleles that may not be identified in ear DNA. In addition, F1 pups should be analyzed for the presence of modifications at predicted off-target sites.

2 Materials

2.1 Preparation of Fertilized Embryos

1. eCG (equine chorionic gonadotropin) stock solution, 50 IU/mL in sterile water (*see* **Note 1**).
2. hCG (human chorionic gonadotropin) stock solution, 50 IU/mL in sterile water.
3. Female donor mice (3–6 weeks old).
4. Stud males (2–6 months old) (*see* **Note 2**).
5. M2 medium.
6. Hya/M2 medium: 0.3 mg/mL hyaluronidase in M2 medium.
7. Embryo-tested mineral oil (*see* **Note 3**).
8. Transfer pipettes.
9. Surgical instruments.
10. Stereomicroscope.
11. Humidified incubator (37 °C, 6% CO_2, 5% O_2).

2.2 Targeting Constructs

1. SpCas9 protein (*see* **Note 4**).
2. Locus-specific crRNA and tracrRNA guide RNA (*see* **Note 5**).
3. Embryo-tested water.
4. Optional: targeting molecule (*see* **Note 6**).
5. Microinjection buffer: 10 mM Tris–HCl, 0.1 mM EDTA (*see* **Note 7**, for microinjection only).
6. OptiMEM reduced serum medium (for electroporation only).
7. Dialysis filter disc membrane (0.025 μm pore size).
8. Centrifugal filter column (PVDF, 0.1 μm pore size).

2.3 Pronuclear Injection of Fertilized Embryos

1. Inverted microscope (*see* **Note 8**).
2. Micromanipulators for holding and microinjection devices.
3. Microinjection device for holding embryos (e.g., Eppendorf CellTram).
4. Microinjection device for pronuclear injection (e.g., Eppendorf Femtojet).
5. Microinjection chamber (*see* **Note 9**).

6. Transfer pipettes.
7. Holding capillaries (inner diameter 20–25 μm) (*see* **Note 10**).
8. Microinjection capillaries (inner diameter 1.6 μm) (*see* **Note 10**).
9. Capillary loading tips.

2.4 Electroporation of Fertilized Embryos

1. Electroporation device (e.g., Biorad Gene Pulser X-Cell, Nepagene NEPA21, or BTX ECM 830).
2. Electroporation slide or 1 mm cuvette (*see* **Note 11**).
3. Transfer pipettes.
4. KSOM-AA medium.
5. M2 medium.
6. KSOM/BSA: 1 mg/mL BSA in KSOM-AA medium.
7. Optional: AAV particle preparation (*see* **Note 12**).

2.5 In Vitro Blastocyst Analysis

1. DNA extraction solution for blastocysts (e.g., QuickExtract DNA Extraction Solution (Lucigen), *see* **Note 13**).
2. Thermocycler.
3. PCR polymerase or master mix.
4. Locus-specific PCR primers.
5. PCR purification kit.
6. T7 endonuclease I.
7. Proteinase K: 20 mg/mL in distilled water.
8. Agarose gel electrophoresis setup.

2.6 Embryo Transfer

1. Pseudo-pregnant females.
2. Vasectomized males (2–6 months old).
3. Anesthesia and analgesia (in accordance with local regulations).
4. M2 medium.
5. Surgical instruments.
6. Suture material (USP 6/0 and 5/0).
7. Stereomicroscope.

2.7 Genotyping

1. Tail lysis buffer: 50 mM Tris–HCl pH 8.0, 100 mM NaCl, 100 mM EDTA pH 8.0, 1% SDS; alternatively use commercial genomic DNA purification kit (*see* **Note 14**).
2. Proteinase K: 20 mg/mL in distilled water.
3. Saturated NaCl solution: 6 M in distilled water.
4. Isopropanol.
5. Ethanol p.a.
6. DNA storage buffer: 10 mM Tris–HCl pH 8.0, 1 mM EDTA.

7. PCR polymerase or master mix.
8. Locus-specific PCR primers.
9. Gene specific restriction enzyme(s) for restriction fragment length polymorphism analysis.
10. Agarose gel electrophoresis setup.

3 Methods

3.1 Zygote Preparation

1. At day −3, inject donor females with 5 IU eCG at 9 am to 10 am (*see* **Note 15**).
2. 48 hours later (day −1), inject donor females with 5 IU hCG at 9 am to 10 am.
3. After hCG injection, mate each donor female with a stud male overnight.
4. On the day of zygote treatment (day 0), identify plug-positive donor females and sacrifice them at 8 am using an approved method.
5. Dissect out the oviducts and place each oviduct into a drop of 50 μL of Hya/M2 in a 6 cm petri dish.
6. To dissect the cumulus complex with the fertilized oocytes, tear the ampullae with a pair of fine forceps and incubate for 3–5 minutes until the cumulus cells fall off.
7. Transfer the one-cell embryos to a fresh drop of 50 μL prewarmed M2 medium (without hyaluronidase) for removal of cumulus cells and pool all embryos in a prewarmed 200 μL drop of M2 medium.
8. Keep embryos in a humidified incubator (37 °C, 6% CO_2, 5% O_2) until injection/electroporation.

3.2 Zygote Treatment

1. For each day of microinjection or for each electroporation of up to 40 zygotes, prepare one single-use aliquot following Table 3. Scale up for multiple injection days/electroporations.
2. For PNI aliquots: prepare a 10 μM working stock of heteroduplex RNA (cr:tracrRNA) for PNI: mix 1 μL 100 μM crRNA, 1 μL 100 μM tracrRNA, and 8 μL microinjection buffer.
3. Heat to 95 °C for 5 minutes, then let cool to RT and store on ice (*see* **Note 16**).
4. Mix 0.3 μL 1 μg/μL Cas9 protein, 0.6 μL 10 μM cr:tracrRNA, and (9.1-x) μL microinjection buffer (x being the amount of targeting molecule).
5. Incubate 10 minutes at RT for the assembly of the RNP complex.

Table 3
Composition of microinjection and electroporation mixes

	Microinjection mix	Electroporation mix[a]
SpCas9	0.18 μM (≈30 ng/μL)	1.2 μM (≈200 ng/μL)
crRNA:tracrRNA or sgRNA	0.6 μM (≈20 ng/μL)	6.0 μM (≈200 ng/μL)
ssODN (opt.)	25 ng/μL	300 ng/μL
lssDNA (opt.)	10 ng/μL	100 ng/μL
plasmid (opt.)	5 ng/μL	–
Microinjection buffer	Filled up to 10 μL	–
OptiMEM	–	Filled up to 5 μL

[a]If using cuvettes: prepare 2× mix as it will be mixed with 5 μL embryo-containing OptiMEM

6. If applicable, add x μL targeting molecule.
7. For EPO aliquots, mix 0.6 μL 100 μM crRNA and 0.6 μL 100 μM tracrRNA with (8.6-x) μL OptiMEM.
8. Heat to 95 °C for 5 minutes, then let cool to RT and store on ice (*see* **Note 16**).
9. Add 0.2 μL 10 μg/μL Cas9 protein to the cr:tracrRNA and incubate 10 minutes at RT for the assembly of the RNP complex.
10. If applicable, add x μL targeting molecule.
11. Purify the injection/electroporation mix by filtering through a centrifugal filter column for 2 minutes @ 12.000 × *g* (*see* **Note 17**).
12. Aliquot in fresh tubes and store at −80 °C until the day of treatment.
13. On the day of treatment, thaw one aliquot, centrifuge at 16.000 × *g* for 2 minutes, and transfer the supernatant to a fresh tube.
14. Keep the aliquot on ice until zygote treatment.

3.3 Option A: Microinjection

1. Load the pronucleus injection capillary with 2 μL of injection solution using a capillary loading tip.
2. Connect the capillary to the microinjector and attach the injector to the micromanipulator.
3. Set the injection conditions to 240 hPa as injection pressure and to 140 hPa as compensation pressure (*see* **Note 18**).
4. Mount the injection chamber, pipette 100 μL M2 medium in the middle, and cover the drop with 200 μL of mineral oil (*see* **Note 3**).

5. Transfer the embryos to be injected into the upper region of the 100 μL M2 drop within the injection chamber.
6. Fix the first fertilized embryo using the holding pipette and place it into the center of the M2 drop.
7. Push the injection capillary to penetrate the zona pellucida and guide the capillary into the larger (male) pronucleus.
8. Remain the capillary in the pronucleus for 0.5–2 seconds until a swelling of the pronucleus becomes visible.
9. Carefully withdraw the capillary and place the injected embryo in the lower part of the M2 drop.
10. Repeat the procedure until all embryos are injected.
11. Identify and remove lysed embryos and transfer the surviving embryos into a prewarmed 200 μL M2 drop in a 6 cm petri dish covered with mineral oil.
12. Keep embryos in a humidified incubator (37 °C, 6% CO_2, 5% O_2) until the implantation.

3.4 Option B: Electroporation

1. Equilibrate KSOM/BSA medium (3–4 hours before electroporation) in a humidified incubator (37 °C, 6% CO_2, 5% O_2) and pre-warm the cuvette/slide electrode.
2. Wash the zygotes five times in M2 and keep them in the incubator until electroporation.
3. Prepare a 6 cm petri dish with a 200 μL drop of OptiMEM.
4. For each electroporation, wash 30–40 zygotes once in OptiMEM.

3.4.1 For Electroporation in Cuvettes

1. Using a P20 pipette, transfer 10 μL OptiMEM containing all zygotes to a dry region of the petri dish (*see* **Note 19**).
2. Using a P20 pipette, add 10 μL of the electroporation mix.
3. Mix 5–10 times by careful pipetting (*see* **Note 20**).
4. Using a P20 pipette, transfer the mix into a 1 mm cuvette.
5. Insert the cuvette into the device and apply four square wave pulses (30 V amplitude, 3 ms duration) with a 100 ms interval (30 V/3 ms—100 ms—30 V/3 ms, etc.).
6. Recover embryos from the cuvette by flushing with 200 μL KSOM/BSA and pipette on petri dish.
7. Repeat recovery with additional 100 μL KSOM/BSA.
8. Collect and count all surviving embryos into a prewarmed 200 μL KSOM/BSA drop in a 6 cm petri dish covered with mineral oil.
9. Keep embryos in a humidified incubator (37 °C, 6% CO_2, 5% O_2) until the implantation.

3.4.2 For Electroporation in Slide Electrodes

1. Using a P10 or P20 pipette, add 5 µL of the electroporation mix to the chamber (*see* **Note 19**).
2. Using a transfer pipette, place the zygotes in the chamber (*see* **Note 21**).
3. Measure the impedance (*see* **Note 22**).
4. Apply electroporation pulses: poring pulse: 40 V, 3.5 ms length, 50 ms interval, 4 pulses, 10% decay rate, + polarity; transfer pulse: 5 V, 50 ms length, 50 ms interval, 5 pulses, 40% decay rate, +/− polarity.
5. Recover embryos from the slide electrode to a prewarmed 200 µL KSOM/BSA drop in a 6 cm petri dish covered with mineral oil.
6. Keep embryos in a humidified incubator (37 °C, 6% CO_2, 5% O_2) until the implantation.

3.4.3 Option C: Electroporation Combined with AAV Incubation

1. Transfer all zygotes into a 25 µL drop of KSOM/BSA in a 3.5 cm petri dish.
2. Add up to 2.5 µL of AAV particle preparation (serotype 1 or 6) containing 10^8–10^9 genome copies.
3. Cover with mineral oil and incubate before electroporation for 4 hours in a humidified incubator (37 °C, 6% CO_2, 5% O_2).
4. Collect zygotes from KSOM/BSA drop and electroporate as described above in option B.
5. Return electroporated zygotes into KSOM/BSA drop with AAV (from **steps 1** to **3**) for 18–24 hours.
6. Wash incubated embryos 5–10 times in 100 µL drops of M2 medium before the transfer of 2 cell embryos into foster females at day 1.5 after plug.

3.5 Optional: In Vitro Blastocyst Analysis

Especially for sophisticated projects, it is advised to initially analyze the efficiency of CRISPR/Cas9-mediated gene targeting in blastocysts without the efforts of embryo transfers and longsome genotyping of living animals. In a PCR-based genotyping assay on cultivated blastocysts, issues with the targeting reagents or strategy can be identified and solved in a timely manner.

1. After injection/electroporation, grow zygotes in KSOM/BSA medium in a humidified incubator (37 °C, 6% CO_2, 5% O_2) for 3 days until they reach blastocyst stage.
2. Pick each blastocyst and transfer it into 10 µL DNA extraction solution in PCR tubes (*see* **Note 23**).
3. Incubate the PCR strips in a thermocycler at 68 °C for 15 minutes followed by 95 °C for 15 minutes.
4. Use 2 µL DNA solution for a 10 µL PCR reaction using locus-specific primers to amplify the targeted genomic region.

5. Keep DNA extracts on ice until further proceeding.
6. After PCR, purify the PCR product using a PCR purification kit according to the manufacturer's instructions.
7. For the T7 endonuclease I assay, add 250 ng purified PCR product and 2 μL 10× restriction buffer to a fresh PCR tube and fill with H_2O to 19 μL.
8. To form the heteroduplex, denature and anneal the mix in a thermocycler: 95 °C for 5 minutes, cool to 85 °C with a ramp of −2 °C/minutes, cool to 25 °C with a ramp of −0.1 °C/minutes, cool to 4 °C.
9. Add 1 μL T7 endonuclease I and incubate at 37 °C for 1 hour.
10. Stop digestion reaction with 1 μL proteinase K at 37 °C for 5 minutes.
11. Visualize the digested samples by gel electrophoresis and check for the expected fragment sizes. Use wild-type samples for comparison (*see* **Note 24**).
12. Instead of T7 endonuclease I assay, the purified PCR product can also be analyzed by Sanger sequencing using the forward or reverse primer. Single or continuous mixed sequencing peak indicate CRISPR/Cas9-mediated mutations.

3.6 Embryo Transfer of Manipulated Zygotes

1. The day before the embryo transfer, the foster females are mated with vasectomized males.
2. The next day, identify plug-positive foster females.
3. For the embryo transfer, anesthetize a plug-positive foster female and make a 5 mm skin incision parallel to the dorsal midline above the position of the left or right oviduct. The ovarian fat pad should become visible.
4. Pull out the ovary and fix the fat pad using a vessel clamp.
5. Locate the infundibulum and tear the bursa using two fine forceps.
6. Transfer the treated embryos to a drop of 200 μL prewarmed M2 medium (without mineral oil to avoid coating of the transfer capillary).
7. Load the capillary with 20–30 embryos (*see* **Note 25**).
8. Insert the tip of the transfer capillary into the infundibulum and carefully release the first group of embryos. Relocate the ovary and fat pad into the abdomen and sew the peritoneum (USP 6/0) and skin (USP 5/0).
9. For bilateral transfer, repeat the surgery for the second ovary and the second group of 15 embryos. Repeat the same procedure until all embryos are transferred.

3.7 Genotyping

The selection of an appropriate genotyping strategy depends on the anticipated mutation (*see* Fig. 4). We strongly suggest to establish the genotyping protocol well in advance since founder animals can be mosaic with multiple mutant alleles, hence identifying correctly targeted animals might be challenging. Here, we present the most common genotyping strategy by PCR amplification, analysis of amplicons or restriction digest fragments by agarose gel electrophoresis, and sequence validation by Sanger sequencing.

3.7.1 Isolation of Genomic DNA from Tissue Biopsies

1. At the age of 3 weeks, take ear-punch, tail-tip, or toe-clip biopsies from F0 founder animals (*see* **Note 26**).
2. Place each biopsy in a fresh 1.5 mL tube for extraction of genomic DNA (*see* **Note 14**).
3. Add 600 μL tail lysis buffer and 35 μL proteinase K to the biopsy and incubate 3–12 hours at 55 °C until sample is digested completely.
4. Cool to room temperature and add 200 μL saturated NaCl solution.
5. Vortex for 30 seconds and incubate on ice for 5 minutes.
6. Centrifuge at 16.000 × *g* for 4 minutes.
7. Transfer supernatant to a fresh tube containing 600 μL isopropanol.
8. Mix by inverting tube several times.
9. Centrifuge at 16.000 × *g* for 1 minutes.
10. Discard the supernatant and wash the pellet with 400 μL 70% EtOH.
11. Centrifuge at 16.000 × *g* for 1 minutes.
12. Discard the supernatant and let the pellet air-dry for approx. 15 minutes.
13. Resuspend the DNA pellet in DNA storage buffer.

3.7.2 PCR Amplification of Targeted Region

1. Use 0.3–2.0 μL genomic DNA solution for a 25 μL PCR reaction using locus-specific primers to amplify the targeted genomic region.
2. Check 5 μL of the PCR product on an agarose gel to confirm successful and specific amplification.
3. Purify the PCR product using a PCR purification kit according to the manufacturer's instructions.

3.7.3 Analysis by Restriction Enzyme Digest or Sanger Sequencing

1. Digest 500 ng of purified PCR product with 5 U restriction enzyme and incubate restriction assay for 1 hour at the enzyme's recommended temperature.
2. Visualize the digested samples by gel electrophoresis and check for the expected fragment sizes. If possible, use wild-type or mutant samples for comparison.
3. Instead of restriction digest, the purified PCR product can also be analyzed by Sanger sequencing using the forward or reverse primer (*see* **Note 27**).

4 Notes

1. Ultra-superovulation reagent ("Hyperova" from Cosmobio) can be used to increase embryo yield per female donor. While superovulation usually results in 20–30 zygotes per donor, ultra-superovulation can yield up to 100 oocytes [49]. However, the reagent is costly and of use mainly for the collection of oocytes for in vitro fertilization since in our experience the mating of ultra-superovulated C57BL/6 females to C57BL/6 males leads only to a fertilization rate of ~30%.
2. Stud males should ideally be from the same genetic background as donor females. Mate males only once per 1 or 2 weeks and replace males every 3–4 months.
3. Mineral oil prevents evaporation of medium; however, mineral oil often contains embryo toxic contaminants. Pre-test every lot of mineral oil for embryo toxicity, use special IVF-certified mineral oil, or use no oil overlay at all. In the latter case, increase the volume of M2 and work quickly to reduce excessive evaporation effects.
4. We routinely use wild-type and high-fidelity SpCas9 protein from IDT, but any commercial or self-made Cas9 protein will work. Caution: the presence of at least one NLS is obligatory. For preparation of PNI aliquots, Cas9 protein should be pre-diluted to 1 μg/μL with, e.g., NEB enzyme diluent buffer B and can be stored at −20 °C for several weeks.
5. We routinely use synthetic gRNA from IDT. crRNA, tracrRNA, and sgRNA should be resuspended in embryo-tested microinjection buffer.
6. ssODNs and lssDNA should be dialyzed against embryo-tested water to remove impurities and salts. Plasmid templates should be EtOH precipitated, washed twice with 70% EtOH (prepared with embryo-tested water), and resolved in microinjection buffer.
7. For the preparation of microinjection buffer, use highest quality reagents and embryo-tested water to prevent embryo toxicity.

8. Microscope optics should be equipped with 5×, 20×, and 40× objective with phase-contrast and differential interference contrast (DIC, only for 40×) for the visualization of pronuclei.
9. Common plastic petri dishes can be used as microinjection chambers; however, many plastics are incompatible with DIC optics and prevent visualization of pronuclei. Use special petri dishes or custom-made chambers with glass-bottom as alternative.
10. Holding and injection capillaries can be purchased from commercial suppliers (e.g., BioMedical Instruments or Eppendorf) or self-made using borosilicate glass capillary tubes and a micropipette puller.
11. Electroporation can be performed either in cuvettes or in slide electrodes. We successfully used both kind of electrodes; however, since handling and recovery of embryos is easier with slide electrodes, we recommend these.
12. Reagents for AAV cloning and production can be purchased from Cell Biolabs Inc. (www.cellbiolabs.com) or customized ready-to-use AAV preparations can be ordered from commercial suppliers.
13. Some DNA extraction solutions might be incompatible with the low input amounts of blastocyst tissue and subsequent PCR amplification might fail.
14. Many different protocols and commercial kits are available for the isolation of genomic DNA. While crude extracts or alkaline lysis methods might work for many projects, we recommend using more elaborate protocols that result in very clean, high molecular weight genomic DNA preparations, like the proteinase K method by Beermann et al. [50]. Thereby, problems during genotyping of founder animals, where usually no positive controls are available, can be avoided.
15. Time points of eCG injection, hCG injection, and embryo collection vary depending on genetic background and light-dark cycle and therefore have to be adjusted empirically.
16. For a controlled cooldown, you can use a thermocycler with slow cooling rate and 1 minutes incubation step every −5 °C.
17. Filtering the mix is crucial for PNI, since it removes dust particles that could clog the injection capillary.
18. Parameters are dependent on the microinjection device and capillaries and have to be adjusted empirically.
19. Process next steps quickly to minimize evaporation.
20. Be careful not to harm the embryos by harsh pipetting.
21. Transfer as little OptiMEM as possible to minimize dilution of electroporation mix.

22. Ideal range for impedance in 1 mm electrode: 0.18–0.22 kΩ. If impedance is <0.18, remove 1–2 μL of solution; if impedance is >0.22, add 1–2 μL of OptiMEM.
23. Under a stereomicroscope, a P10 micropipette or an embryo transfer pipette can be used for this purpose. Transfer as little medium as possible.
24. The estimated percentage of modification can be calculated using the formula: % modification = $100 \times (1 - (1 - \text{fraction cleaved})^{1/2})$.
25. Usually, 30 embryos are transferred bilaterally per foster mother (15 per oviduct). When loading into the capillary, leave a small air bubble between groups to identify the embryos. If unilateral transfer is preferred, use 17–23 embryos per foster.
26. Biopsy procedures, especially tail-tip and toe-clip biopsies, might be subject to institutional and governmental regulations.
27. Mixed sequencing peaks (indicating heterozygous substitutions by HDR) or continuous mixed peaks (indicating InDel mutations caused by NHEJ) can be deconvoluted by online tools, e.g., Tide (http://tide.nki.nl/) [51] or ICE (https://ice.synthego.com/).

References

1. Cong L, Ran FA, Cox D et al (2013) Multiplex genome engineering using CRISPR/Cas systems. Science 339:819–823
2. Mali P, Aach J, Stranges PB et al (2013) CAS9 transcriptional activators for target specificity screening and paired nickases for cooperative genome engineering. Nat Biotechnol 31:833–838
3. Wang H, Yang H, Shivalila CS et al (2013) One-step generation of mice carrying mutations in multiple genes by CRISPR/Cas-mediated genome engineering. Cell 153:910–918
4. Yang H, Wang H, Shivalila CS et al (2013) One-step generation of mice carrying reporter and conditional alleles by CRISPR/Cas-mediated genome engineering. Cell 154:1370–1379
5. Kühn R (2021) Genome engineering in rodents – status quo and perspectives. Lab Anim 56:83–87
6. Tröder SE, Zevnik B (2021) History of genome editing: from meganucleases to CRISPR. Lab Anim 56:60–68
7. Qin W, Dion SL, Kutny PM et al (2015) Efficient CRISPR/Cas9-mediated genome editing in mice by zygote electroporation of nuclease. Genetics 200:423–430
8. Chen S, Lee B, Lee AY-F et al (2016) Highly efficient mouse genome editing by CRISPR ribonucleoprotein electroporation of zygotes. J Biol Chem 291:14457–14467
9. Lieber MR (2010) The mechanism of double-strand DNA break repair by the nonhomologous DNA end-joining pathway. Annu Rev Biochem 79:181–211
10. Heyer W-D, Ehmsen KT, Liu J (2010) Regulation of homologous recombination in eukaryotes. Annu Rev Genet 44:113–139
11. Yao X, Zhang M, Wang X et al (2018) Tild-CRISPR allows for efficient and precise gene Knockin in mouse and human cells. Dev Cell 45:526–536
12. Yao X, Wang X, Hu X et al (2017) Homology-mediated end joining-based targeted integration using CRISPR/Cas9. Cell Res 27:801–814
13. Quadros RM, Miura H, Harms DW et al (2017) Easi-CRISPR: a robust method for one-step generation of mice carrying conditional and insertion alleles using long ssDNA

donors and CRISPR ribonucleoproteins. Genome Biol 18:92
14. Codner GF, Mianné J, Caulder A et al (2018) Application of long single-stranded DNA donors in genome editing: generation and validation of mouse mutants. BMC Biol 16:70
15. Lanza DG, Gaspero A, Lorenzo I et al (2018) Comparative analysis of single-stranded DNA donors to generate conditional null mouse alleles. BMC Biol 16:69
16. Yoon Y, Wang D, Tai PWL et al (2018) Streamlined ex vivo and in vivo genome editing in mouse embryos using recombinant adeno-associated viruses. Nat Commun 9:412
17. Chen S, Sun S, Moonen D et al (2019) CRISPR-READI: efficient generation of knockin mice by CRISPR RNP electroporation and AAV donor infection. Cell Rep 27:3780–3789
18. Mizuno N, Mizutani E, Sato H et al (2018) Intra-embryo gene cassette Knockin by CRISPR/Cas9-mediated genome editing with adeno-associated viral vector. iScience 9: 286–297
19. Lin X, Pelletier S, Gingras S et al (2016) CRISPR-Cas9-mediated modification of the NOD mouse genome with Ptpn22R619W mutation increases autoimmune diabetes. Diabetes 65:2134–2138
20. Tirado-Gonzalez I, Czlonka E, Nevmerzhitskaya A et al (2018) CRISPR/Cas9-edited NSG mice as PDX models of human leukemia to address the role of niche-derived SPARC. Leukemia 32:1049–1052
21. Boroviak K, Doe B, Banerjee R et al (2016) Chromosome engineering in zygotes with CRISPR/Cas9. Genesis 54:78–85
22. Birling M-C, Schaeffer L, André P et al (2017) Efficient and rapid generation of large genomic variants in rats and mice using CRISMERE. Sci Rep 7:43331
23. Fujii W, Kakuta S, Yoshioka S et al (2016) Zygote-mediated generation of genome-modified mice using Streptococcus thermophilus 1-derived CRISPR/Cas system. Biochem Biophys Res Commun 477:473–476
24. Kim Y, Cheong S-A, Lee JG et al (2016) Generation of knockout mice by Cpf1-mediated gene targeting. Nat Biotechnol 34:808–810
25. Zhang X, Liang P, Ding C et al (2016) Efficient production of gene-modified mice using Staphylococcus aureus Cas9. Sci Rep 6:32565
26. Hur JK, Kim K, Been KW et al (2016) Targeted mutagenesis in mice by electroporation of Cpf1 ribonucleoproteins. Nat Biotechnol 34:807–808
27. Kim K, Ryu S-M, Kim S-T et al (2017) Highly efficient RNA-guided base editing in mouse embryos. Nat Biotechnol 35:435–437
28. Zuo E, Sun Y, Wei W et al (2019) Cytosine base editor generates substantial off-target single-nucleotide variants in mouse embryos. Science 364:289–292
29. Lee HK, Smith HE, Liu C et al (2020) Cytosine base editor 4 but not adenine base editor generates off-target mutations in mouse embryos. Commun Biol 3:19
30. Caso F, Davies B (2021) Base editing and prime editing in laboratory animals. Lab Anim 56:35–49
31. Liu Y, Li X, He S et al (2020) Efficient generation of mouse models with the prime editing system. Cell Discov 6:27
32. Shen MW, Arbab M, Hsu JY et al (2018) Predictable and precise template-free CRISPR editing of pathogenic variants. Nature 563: 646–651
33. Allen F, Crepaldi L, Alsinet C et al (2018) Predicting the mutations generated by repair of Cas9-induced double-strand breaks. Nat Biotechnol 10:1038
34. Suzuki K, Tsunekawa Y, Hernandez-Benitez R et al (2016) In vivo genome editing via CRISPR/Cas9 mediated homology-independent targeted integration. Nature 540:144–149
35. Danner E, Lebedin M, de la Rosa K, Kühn R (2021) A homology independent sequence replacement strategy in human cells using a CRISPR nuclease. Open Biol 11:200283
36. Chu VT, Weber T, Graf R et al (2016) Efficient generation of Rosa26 knock-in mice using CRISPR/Cas9 in C57BL/6 zygotes. BMC Biotechnol 16:4
37. Low BE, Hosur V, Lesbirel S, Wiles MV (2022) Efficient targeted transgenesis of large donor DNA into multiple mouse genetic backgrounds using bacteriophage Bxb1 integrase. Sci Reports 12:5424
38. Gurumurthy CB, O'Brien AR, Quadros RM et al (2019) Reproducibility of CRISPR-Cas9 methods for generation of conditional mouse alleles: a multi-center evaluation. Genome Biol 20:171
39. Haeussler M, Schönig K, Eckert H et al (2016) Evaluation of off-target and on-target scoring algorithms and integration into the guide RNA selection tool CRISPOR. Genome Biol 17:148
40. Dong Y, Li H, Zhao L et al (2019) Genome-wide off-target analysis in CRISPR-Cas9 modified mice and their offspring. G3 (Bethesda) 9: 3645–3651

41. Iyer V, Shen B, Zhang W et al (2015) Off-target mutations are rare in Cas9-modified mice. Nat Methods 12:479
42. Anderson KR, Haeussler M, Watanabe C et al (2018) CRISPR off-target analysis in genetically engineered rats and mice. Nat Methods 15:512–514
43. Richardson CD, Kazane KR, Feng SJ et al (2018) CRISPR-Cas9 genome editing in human cells occurs via the Fanconi anemia pathway. Nat Genet 50:1132–1139
44. Richardson CD, Ray GJ, DeWitt MA et al (2016) Enhancing homology-directed genome editing by catalytically active and inactive CRISPR-Cas9 using asymmetric donor DNA. Nat Biotechnol 34:339–344
45. Miura H, Quadros RM, Gurumurthy CB, Ohtsuka M (2018) Easi-CRISPR for creating knock-in and conditional knockout mouse models using long ssDNA donors. Nat Protoc 13:195–215
46. Minev D, Guerra R, Kishi JY et al (2019) Rapid in vitro production of single-stranded DNA. Nucleic Acids Res 47:11956–11962
47. Hao M, Huang H, Hu Y, Qi H (2020) Construction of a system for single-stranded DNA isolation. Biotechnol Lett 42:1663–1671
48. Liang C, Li D, Zhang G et al (2015) Comparison of the methods for generating single-stranded DNA in SELEX. Analyst 140:3439–3444
49. Takeo T, Nakagata N (2015) Superovulation using the combined administration of inhibin antiserum and equine chorionic gonadotropin increases the number of ovulated oocytes in C57BL/6 female mice. PLoS One 10:e0128330
50. Beermann F, Hummler E, Schmid E, Schütz G (1993) Perinatal activation of a tyrosine aminotransferase fusion gene does not occur in albino lethal mice. Mech Dev 42:59–65
51. Brinkman EK, Kousholt AN, Harmsen T et al (2018) Easy quantification of template-directed CRISPR/Cas9 editing. Nucleic Acids Res 46:e58

Chapter 9

Floxing by Electroporating Single-Cell Embryos with Two CRISPR RNPs and Two ssODNs

Mia Wallace, J. Michael White, Evgenea Kouranova, Zi Teng Wang, and Xiaoxia Cui

Abstract

Floxed alleles and Cre drivers are two components of most conditional knockout mouse models, which are not only important for studying a given gene in a tissue-specific manner, but also useful for functional analysis of various sized genomic regions. With the increased demand for floxed mouse models in biomedical research, reliable and economical creation of floxed alleles is clearly highly valuable yet remains challenging. Here we provide technical details on the method consisting of electroporating single-cell embryos with CRISPR RNPs and ssODNs, next-generation sequencing (NGS)-based genotyping, an in vitro Cre assay (recombination followed by PCR) for loxP phasing determination, and optional second round targeting of an indel in cis with one loxP insertion in embryos obtained via in vitro fertilization (IVF). As importantly, we present protocols for validation of gRNAs and ssODNs before electroporation of embryos, to confirm phasing of loxP and the indel to be retargeted in individual blastocysts and an alternative strategy to insert loxP sites sequentially. Together, we hope to help researchers reliably obtain floxed alleles in a predictable and timely manner.

Key words Conditional knockout, Electroporation, CRISPR, NGS, In vitro fertilization, NHEJ

1 Introduction

Floxed mouse models in combination with Cre driver lines allow spatial and/or temporal excision of floxed sequences from the genome and have now been used far beyond the initial goal of circumventing embryonic lethality [1]. We have observed a steady increase in requests for floxed models from investigators on the Washington University campus over the past 5 years.

Like other types of modifications of the mouse genome, floxed alleles can be generated by using reprogrammable nucleases, such as CRISPR, ZFNs, and TALENs, by direct manipulation of single-cell embryos, bypassing the need for using mouse embryonic stem cells [2].

Thomas L. Saunders (ed.), *Transgenesis: Methods and Protocols*,
Methods in Molecular Biology, vol. 2631, https://doi.org/10.1007/978-1-0716-2990-1_9,

Creation of a floxed allele involves insertion of loxP sites into two loci on the same chromosome, either by using two individual donors [3–8] or one donor harboring both loxP sites [9–11].

The two RNPs/two ssODNs strategy covered in this chapter uses one gRNA and one ssODN carrying the loxP site with short flanking sequences for the insertion of each loxP site and has the following advantages:

1. Flexibility in the size of the region to be floxed. Given that one set of RNP/ssODN is designed for insertion of each loxP site, the two sets of reagents can be designed to be of any distance from each other, unlike single donor strategies where the size of the donor is limiting.
2. Short turnaround and relatively low cost of donors. ssODNs can be obtained with reasonable cost and usually within 2 weeks.
3. Ease of validation in cells and genotyping by NGS. The short homology arms required for an ssODN-mediated insertion allow convenient genotyping by using NGS on short PCR amplicons for both validation and founder identification.
4. Compatibility with embryo electroporation. Instead of conventional microinjection, CRISPR RNPs and ssODNs can be delivered to single-cell embryos by electroporation that often leads to significantly higher birth rates and founder rates [12–14]. Electroporation, however, does not deliver large DNA molecules effectively, such as plasmid donors or long single-stranded DNA.

On the other hand, by its nature, the 2 RNPs/2 ssODNs floxing strategy is not going to be a highly efficient event given the various competing outcomes that can result from exposing the genome to nucleases targeting two loci on the same chromosome. Each target site can have wild type, an indel, or loxP insertion, and a given allele can have any combinations of these modifications at each target site. Additionally, deletions between the two target sites form at different frequencies for some target genes, although the resulting global knockout sometimes is a useful byproduct. Even in animals where both loxP sites are detected by NGS, the loxP sites can be in trans in the same cells or even in different cells, which can be identified and excluded using the in vitro Cre assay. Other animals may only have one loxP site or none. In short, floxed alleles are only one of the multiple possible genotypes.

In this chapter, we will focus on the floxing strategy as reported in Sentmanat et al. [8] with some additional information, as we continue to optimize the conditions to obtain floxed alleles with higher confidence. The critical points of the method are the following:

1. To maximize efficiency, we make every effort to minimize human errors, such as defective or mislabeled reagents, by validating gRNAs and ssODNs in cultured cells. Although at first we validated each RNP/ssODN pair in separate reactions, we found that co-nucleofecting both sets of reagents yielded results that are more reflective of potential outcomes in vivo and makes it possible to adjust the dosage of each reagent pair according to relative activity, if needed.
2. Genotyping with NGS allows higher resolution analysis of editing outcomes, critical to the design of second round retargeting as well as identifying low percentage loxP insertions that are potentially in cis (an example in Sentmanat et al., *see* Fig. 1 [8]).
3. The use of the Cre assay allows us to determine phasing of loxP sites and identify founders with not only both loxP sites but a floxed allele, confirming floxing in founder generation and avoiding unnecessary breeding of animals with loxP sites in trans.
4. When floxed founders are not identified, we specifically retarget an indel with a cis loxP insertion in oocytes obtained by IVF. Resulting animals with both loxP sites are guaranteed to carry a floxed allele.

One caveat to the retargeting strategy is that larger-than-expected indels [15], mostly deletions, can occur at individual target sites and escape amplification by NGS PCRs (*see* Fig. 1), misleading the phasing of the loxP and indel in a given animal. These larger than expected deletions are rather infrequent. We did not notice any among the first 60 targets we reported in Sentmanat et al. To avoid wasting time and resources on the wrong male founders, we recommend genotyping individual blastocysts from IVF as a quality control test for the sperm for retargeting. Only sperm carrying both a loxP site and an indel can produce blastocysts with both.

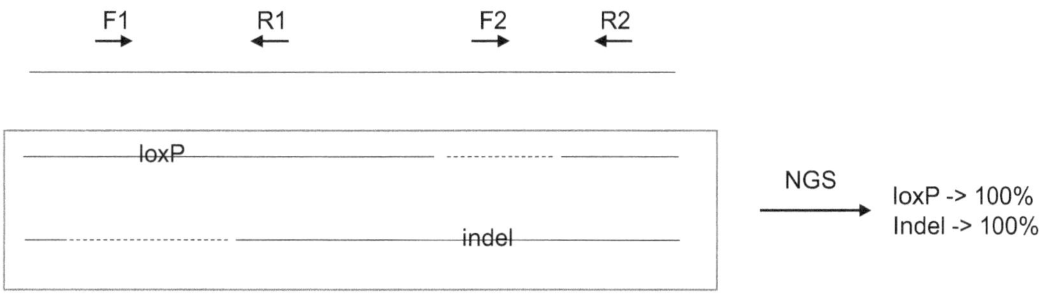

Fig. 1 Schematic showing larger than expected indels at both target sites on separate chromosomes lead to misinterpretation of NGS results

One solution to circumvent the impact of larger deletions between target sites or from one target site is to sequentially insert the loxP sites. One RNP/ssODN pair will be used to electroporate embryos first, and sperm from a founder with a loxP allele will be used in IVF of wild-type oocytes to obtain single-cell embryos for targeting at the second site. The in vitro Cre assay can be used to determine phasing and identify floxed animals. Just as in retargeting, IVF is critical for sequential targeting to shorten the cycle and avoid extensive breeding.

In essence, both retargeting and sequential targeting break down low-efficiency single-step floxing into two more efficient single loxP insertions with fewer competing outcomes. In our hands, most projects with single-step targeting have resulted in at least one floxed founder, whereas sequential targeting is likely more of a guarantee. Without trying to suggest one method over the other, we aim to provide assays at various steps to help maximize efficiency and minimize failure. The workflows of one-step targeting with possible retargeting and sequential targeting are shown in Fig. 2.

In this chapter, we provide details on design and validation of gRNAs and ssODNs, zygote electroporation, founder genotyping, and IVF. Although these methods can be found scattered across various methods papers, we have compiled them here for ease of reference. Between the two mouse cores that contributed to the

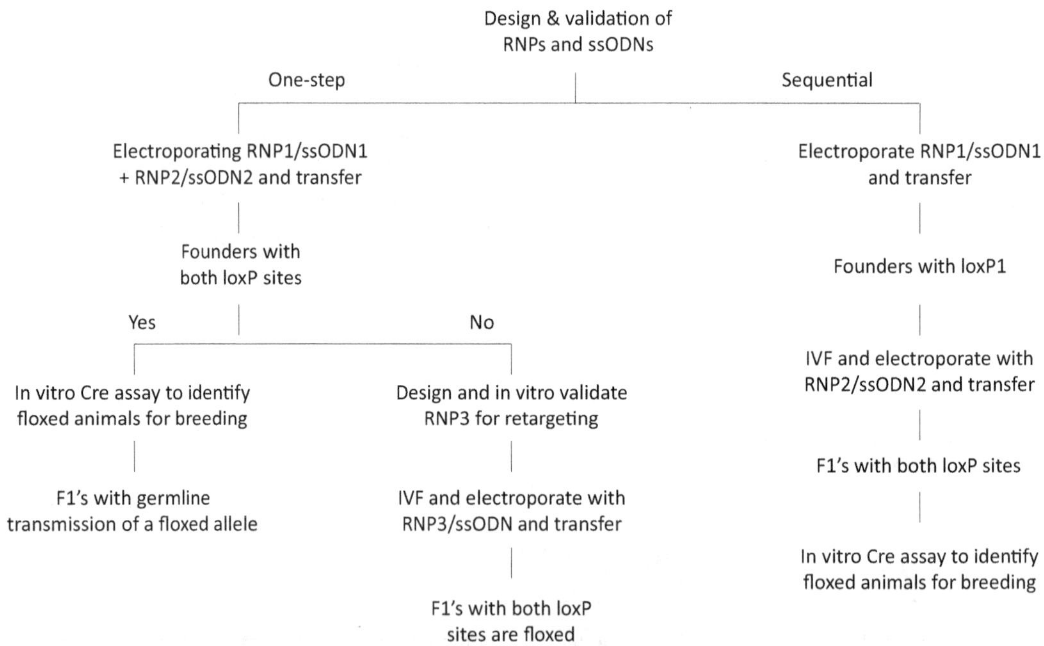

Fig. 2 Comparison of one-step and sequential targeting strategies

work, if protocols are different, we present both so that you can decide what is best for you. With the protocols described below, one should be able to obtain a confirmed floxed model in as short as 20 weeks of time. In the meantime, we hope that the methods on floxing will continue to be improved and optimized with more people working on them.

2 Materials

2.1 Validation of gRNAs and ssODNs in Cultured Cells

1. Equipment: Biosafety cabinet, CO_2 incubator, desktop centrifuge, Lonza 4D-Nucleofector.
2. SF Cell Line Nucleofector Kit.
3. CRISPR reagents: synthetic gRNAs (at 10 nmol scale) and ssODNs (as 4 μmol Ultramers) are purchased from IDT and kept in 100 μM stocks upon reconstitution. DTT-free and glycerol-free Cas9 recombinant protein (10 μg/μL) is also purchased from IDT, and recombinant Cas9 protein at 6.4 μg/μL is obtained from QB3 Macrolab at University of California, Berkeley.
4. Neuro-2a (CCL-131, ATCC) or BV2 cells (CVCL_0182, Cellosaurus).
5. DMEM, 4.5 *g*/L D-glucose, L-glutamine, 10% FBS, and 1× Penicillin-Streptomycin.
6. Lysis buffer: 10 mM Tris–HCl pH 8.0, 2 mM EDTA, 0.2% Triton X-100, 200 μg/mL proteinase K, or QuickExtract DNA Extraction Solution: Lucigen, QE09050.

2.2 In Vitro Validation of RNPs

1. Synthetic gRNA and Cas9 protein.
2. Buffer 3.1 for restriction enzymes and 10× BSA: New England Biolabs.
3. PCR cleanup kit.
4. Agarose gel setup.

2.3 Embryo Electroporation

2.3.1 Sample Prep for Electroporation

(a) Buffer 1: 5 mM Tris–HCl, 0.1 mM EDTA, pH 7.4, RNase-free or Buffer 2: 10 mM Tris–HCl, 0.1 mM EDTA, pH 7.6, RNase-free (*see* **Note 1**).

(b) Synthetic gRNAs at 100 μM, ssODNs at 100 μM, Cas9 protein at 10 μg/μL.

2.3.2 Embryo Electroporation With or Without Acid Treatment of Zona

(a) Opti-MEM (1×): Gibco, 31985-062.

(b) M2 with HEPES: Sigma, M7167.

(c) EmbryoMax Acidic Tyrode's Solution: Millipore, MR-00/D.

(d) KSOM: Millipore, MR-106-D.

(e) Paraffin Oil: Life Global, LGPO-500 or Vitrolife Ovoil, 10,029.
(f) Gene Pulser XL electroporator (Bio-Rad, 1,652,660) and Gene Pulser MicroPulser Cuvette, 0.1 cm gap (Bio-Rad, 1,652,089).
(g) Embryo handling pipette.
(h) Culture dishes, 35 mm and 60 mm.
(i) Stereomicroscope with 20× and 40× lens.

2.4 Sperm Cryopreservation

1. Paraffin Oil: Life Global, LGPO-500.
2. Modified Human Tubal Fluid: Millipore, MR-070-D.
3. Sperm cryoprotectant agent: raffinose (Sigma, R7630), skim milk (Becton Dickinson, 232,100), L-glutamine (Sigma, G8540), embryo-grade water (Sigma, W1503).
4. 0.25 mL Sperm Storage Straws: IMV Technologies, 5565.
5. Straw Storage Cassettes: Minitube, 16,980.
6. Two pairs of #5 forceps, one blunt straight, and the other blunt curved.
7. Fine scissors.
8. 25 gauge needles.
9. Stereomicroscope.
10. Culture Dishes, 35 mm and 60 mm.
11. 1.5 mL Eppendorf tubes.
12. Makler Sperm Counter: Sefi-Medical Instruments.
13. Cooling Chamber: Fisher Scientific, 11-676-13 (Sonoco 304).
14. Cooling Raft: Home Depot, Owens Corning Rigid Foam Insulation R-7.5.
15. Push Rod: 1/16 in. stainless steel TIG wire that can be found in welding supply stores.

2.5 In Vitro Fertilization (IVF) with Frozen Sperm

1. Modified Human Tubal Fluid: Millipore, MR-070-D.
2. GSH: L-glutathione reduced, Sigma G6013.
3. KSOM: Millipore, MR-106-D.
4. Fertiup Mouse Sperm Preincubation Medium [PM]: Fisher, KYD-002-EX.
5. Paraffin Oil: Life Global, LGPO-500.
6. PMS: Lee Bio Solutions, 493-10.
7. hCG: Sigma; CG10-1VL.
8. Female C57BL6/J mice, JAX, 000664, 3–4 or 8–20 weeks of age.

9. Forceps: two #5 and one fine pair.
10. Fine scissors.
11. Culture dishes, 35 mm and 60 mm.
12. 15 mL Falcon tubes.
13. Syringe filters, 0.22 μm.
14. Stereomicroscope with 20× or 40× lens.

2.6 IVF with Fresh Sperm

1. K-RVFE 50 medium: Cook Medical K-RVF-50.
2. 100 mM GSH stock solution: Dissolve 0.1845 *g* GSH (L-glutathione reduced, Sigma G6013) in 6 mL K-RVFE 50, pass through a 0.2 μm syringe filter and freeze in 100 μL aliquots in −80 °C.
3. High calcium K-RVFE 50: dissolve 0.02275 *g* of $CaCl_2 \cdot 2H_2O$ (Sigma 7902) in 5 mL of K-RVFE 50 and pass the solution through 0.2 μm syringe filter to 45 mL of K-RVFE 50 to reach final calcium concentration of 5.14 mM.
4. 10× Salt Solution: per 100 mL, dissolve 6.97 *g* of NaCl, 0.356 *g* of KCl, 0.162 *g* of KH_2PO_4, 0.293 *g* of $MgSO_4 \cdot 7H_2O$, 1 *g* of glucose (Sigma G6152) in embryo-grade water (Sigma W1503) to final volume and pass through a 0.22 μm filter. Store at room temperature.
5. 100× $CaCl_2$ solution: per 10 mL, dissolve 0.2510 *g* $CaCl_2 \cdot 2H_2O$ in 5 mL of embryo-grade water. Store at room temperature.
6. MBCD Medium: per 100 mL, add 10 mL 10× Salt Solution, 1 mL 100× $CaCl_2$ Solution, 0.2106 *g* of $NaHCO_3$, 5.5 mg of Sodium Pyruvate, 7.5 mg of Penicillin G, 5 mg of Streptomycin Sulfate, 0.1 *g* of Methyl-B-Cyclodextrin (Sigma C4555), 0.1 *g* of Polyvinyl alcohol (Sigma P8136), and embryo-grade water to final volume and pass through a 0.22 μm filter.

2.7 NGS-Based Genotyping

1. QuickExtract DNA Extraction Solution (QE buffer): Lucigen, QE09500.
2. NGS primers with adaptor sequences for indexing to amplify a 250–450 bp genomic region flanking each loxP insertion site:

   Forward: 5′- CACTCTTTCCCTACACGACGCTCTTCC GATCT-target-specific sequence.

   Reverse: 5′- GTGACTGGAGTTCAGACGTGTGCTCTTCC GATCT-target-specific sequence

   (a) Barcoded index primers.

   (b) DNA polymerase mixes for PCR, such as Platinum SuperFi II Green Master Mix (ThermoFisher) or Jump-Start Redtaq ReadyMix (Sigma).

2.8 In Vitro Cre Assay

1. Qiagen's DNeasy Blood & Tissues kit.
2. Cre recombinant protein (NEB MO298S).
3. NGS primers from Subheading 2.7 above.
4. Agarose gel running setup.

3 Methods

3.1 Criteria for Designing and Choosing Insertion Sites

1. If the critical domain(s) for protein function are known, part or all exons coding for the domain(s) should be floxed. When possible, aim at floxing one or more exons with coding sequences that are not in multiple of three in base pairs, e.g., $3n + 1$ or $3n + 2$ bp, so that Cre-mediated excision will lead to frameshift and formation of premature stop codons before the last exon.
2. If critical domains are not known, one should flox one or more earlier exons so that all isoforms lose $3n + 1$ or $3n + 2$ bp coding sequences upon Cre-mediated excision. It is important to confirm that premature stop codons form first in downstream exons that are not the last exon, in order to trigger nonsense-mediated decay of mRNA. Always avoid disrupting splicing signals if insertion sites are chosen in introns.
3. For genes with only one or two coding exons or with all exons encoding $3n$ bp, the design can be trickier. We follow these rationales:
   (a) Avoid inserting into 5′ UTRs without knowing the consequences. The introns before the first coding exon, if available, are good candidates for 5′ loxP insertion.
   (b) If it is necessary to insert one loxP site into the 3′ UTR, we insert right after the stop codon.
   (c) If necessary, one can flox the whole gene region. If loxP is inserted into the 3′ UTR, one might want to consider inserting the 5′ loxP upstream of the promoter.
   (d) If there are annotated noncoding RNAs or other regulatory elements in introns, we try to avoid them when possible.

3.2 gRNA Design

We use an in-house script to design gRNAs and primers for genotyping, which is based on Zhang Lab's algorithm [16], with results very similar to those obtained from the CRISPOR site: http://crispor.org/.

3.3 ssODN Design

ssODNs in general are 60 bases on each side of the loxP sequences and ordered as ultramers from IDT with two phosphorothioate bonds at both ends (*see* **Note 2**).

3.4 Choice of Founders for Retargeting

Based on NGS data, the best candidates for retargeting are males with 100% loxP insertion at one site and 100% indel at the other site, with a PAM site available nearby so that the spacer sequence for the new gRNA contains the indel, and the wild-type allele will not be recognized. Otherwise, males with the highest percentage of loxP insertion at one site and an indel with comparable percentage at the other site should be used. A female with the above genotype can be bred for retargeting. Male offspring with 50% loxP insertion and 50% indel should be used to collect sperm for IVF. Using NGS to genotype individual embryos at blastocyst or morula stages is an additional confirmational step to demonstrate that the loxP site and the indel to be retargeted are in phase.

3.5 Validation in Cell Culture

1. Combine 6 μg of Cas9, 0.75 μL each gRNA and 0.75 μL of each ssODN.
2. Collect all content to bottom of tube by a low-speed, quick spin.
3. Incubate at room temperature for 10 minutes and leave on ice until use.
4. BV2 or Neuro-2a cells are passaged 1:2 to 1:5 every 48–72 hours as needed (*see* **Note 3**).
5. Trypsinize cells at 70–80% confluence and pellet 100,000 cells for each reaction, including a GFP transfection control.
6. Wash cells once with equal volume of PBS as growth medium to remove nucleases in the serum.
7. Resuspend cells in Solution F at 100,000 cells/20 μL.
8. Add 20 μL of cells to each tube from **step 3**, transfer the mixture to a well in a Lonza 4D-nucleofector 16-well Nucleocuvette strip and nucleofect immediately (*see* **Note 4**).
9. Transfer each nucleofection to 500 μL of growth media in a 24-well plate and return to the CO_2 incubator.
10. 24–72 hours after nucleofection, transfection efficiency is estimated by percentage of GFP-positive cells in the GFP control sample, and cells are trypsinized and pelleted in 0.2 mL PCR tubes.
11. Add 50–100 μL lysis buffer to cell pellets, mix and incubate in a PCR machine using program: 65 °C for 15′ and 98 °C for 5′.
12. One microliter of extracted lysate is used as template for genotyping as described in **NGS-based genotyping**. GFP transfected sample is used as a negative control for each target.

3.6 Validation In Vitro

Retargeting requires a new gRNA against a specific indel in a founder, which is not present in available cell lines. The nuclease activity of gRNA/Cas9 RNP is validated by in vitro cleavage of a PCR product that is amplified from the genomic region containing the indel to be targeted. Amplicons should be 300–500 bp in length for ease of the assay.

1. PCR amplify ROI in a reaction ≥30 μL.
2. Run a small amount (1–4 μL) on an agarose gel to confirm the presence of desired amplicon.
3. Purify the rest of PCR reactions using a PCR cleanup kit (*see* **Note 5**).
4. Measure the concentration of purified PCR product using a spectrophotometer, such as Nanodrop, or by another assay, such as Qubit.
5. Assemble RNP by mixing 1 μL of gRNA (~3.6 μg) with 0.5 μL of Cas9 protein (5 μg) and incubate at RT for 10 minutes. Use a non-target gRNA as a negative control. The control mix should be assembled the same way.
6. Add 300 ng of PCR product, 2 μL of NEB buffer 3.1, 2 μL of BSA to each RNP mix and bring the reaction volume to 20 μL with water.
7. Incubate the reactions for a maximum of 60 minutes at 37 °C (*see* **Note 6**).
8. Stop the reactions by heating at 75 °C for 10 minutes to denature the Cas9 protein.
9. Run the digests on a 2% agarose gel together with 300 ng of uncut PCR product as a control (*see* Fig. 3).

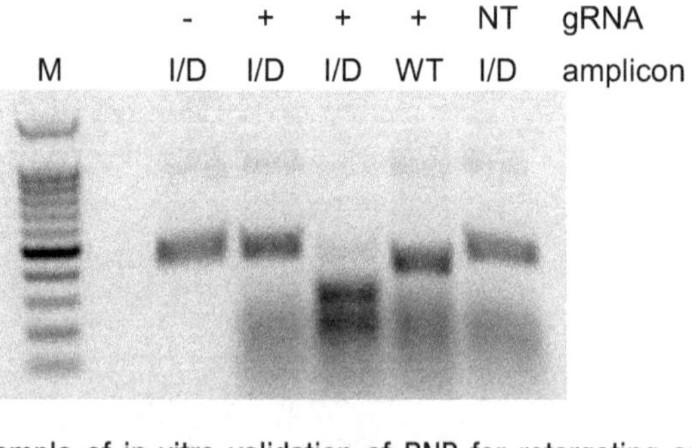

Fig. 3 An example of in vitro validation of RNP for retargeting an indel. M is 100 bp DNA ladder. I/D: PCR product amplified from the allele with an indel to be retargeted. WT: wild-type amplicon without the indel. NT stands for non-targeting gRNA

Table 1
Electroporation sample preparation (50 μL)

	Stock (μM)	Final concentration (μM)	Volume of stock to add (μL)
Cas9	62.5	2	1.5 × 2
gRNA1	100	3	1.5
ssODN1	100	10	5
gRNA2	100	3	1.5
ssODN2	100	10	5
Buffer	–	–	34
Total	–	–	50

3.7 Sample Prep for Embryo Electroporation

1. Mix each gRNA with Cas9 separately and incubate at room temperature for 10 minutes.
2. Combine RNPs and add ssODNs, then add buffer to the final volume, see the example of a 50 μL mixture in Table 1, typically for one electroporation session. We prepare for 4–6 electroporation sessions at one time and freeze aliquots of 50 μL in single-use tubes at −20 °C or −80 °C for up to a week.

3.8 Embryo Electroporation with Acid Treatment of Zona

This protocol was adapted from Wang et al. [14].

1. Prepare your dishes of media prior to starting with zygote electroporation. For every 40 zygotes, prepare the following:
   (a) "Pre Tyrode's dish": one 35 mm dish with three 50 μL drops of M2 each covered in paraffin oil in benchtop incubator (37 °C, no added CO_2)
   (b) "Acidic Tyrode's dish": One 60 mm dish with one 150 μL drop of Acidic Tyrode's solution covered in paraffin oil—keep this dish at room temp
   (c) "Post Tyrode's dish": one 60 mm dish with six 50 μL drops of M2 covered in paraffin oil in benchtop incubator (37 °C, no added CO_2)
   (d) One Eppendorf tube with 1 mL KSOM (to flush cuvette)—kept in incubator (37 °C, 5% CO_2), with lid uncapped.
   (e) "Post-Shock dish": one 35 mm dish with three 50 μL drops of M2 covered in paraffin oil in benchtop incubator (37 °C, no added CO_2)
   (f) "Post Dish": one 35 mm dish—five 50 μL drops of KSOM covered in paraffin oil in benchtop incubator (37 °C, 5% CO_2)

2. In a 35 mm dish, make one 45 μL drop and one 10 μL drop of Opti-MEM, respectively, and leave at room temperature.
3. Thaw an aliquot of electroporation sample mix on ice.
4. Bring isolated zygotes from the incubator and wash 45 zygotes through 3 drops M2 in **step 1a**.
5. Place Acidic Tyrode's dish (**step 1b**) and Post-Tyrode dish (**step 1c**) right next to the stereomicroscope to minimize the time it takes to move the zygotes from the Tyrode's treatment into the Post-Tyrode dish.
6. Under the scope, move the 45 zygotes into the drop of Acidic Tyrode's solution. Spread them out and then quickly pick them right back up (*see* **Note 7**).
7. Wash the acid-treated zygotes through 3 drops of M2 in the "Post Tyrode's" wash dish (**step 1c**) and discard zygotes without zona.
8. Move the zygotes from the Post-Tyrode's wash dish with as little medium carryover as possible to the 45 μL drop in the OPTI-MEM dish (**step 2**).
9. Move 40 of the zygotes with thinned but intact zonas (wipe off any paraffin oil on the pipette with a Kimwipe) to the 10 μL drop of Opti-MEM in the same dish.
10. Slowly pipette up and down once before picking up 10 μL of electroporation sample mix and add to the 10 μL Opti-MEM drop with zygotes (**step 9**).
11. Load the 20 μL of zygotes and sample mixture to the electroporation cuvette using a regular pipette tip. Be as gentle as possible, but make sure you pick up all the zygotes.
12. Put the lid on cuvette, and place the cuvette in the Electroporator shock pod.
13. Electroporate using ten pulses of 1 ms duration of square wave at 30 V with 100 ms intervals.
14. Remove the cuvette from the electroporator.
15. Place the lid of a 60 mm dish on the benchtop and flush the cuvette by pipetting 100 μL of KSOM from the Eppendorf tube in the incubator (**step 1d**), into the cuvette, then drawing the 100 μL back up and placing it in the empty 60 mm dish lid.
16. Add a second 100 μL aliquot of KSOM to the cuvette and draw it back out, placing it in a separate drop in the 60 mm dish lid.
17. Repeat the flushing step two more times.
18. Retrieve the zygotes from all four drops.
19. Wash the zygotes through three drops in the "Post-Shock" wash dish (**step 1e**).

20. Move the zygotes to the "Post Dish" (**step 1f**) in incubator (37 °C, 5% CO_2), washing once through the KSOM wash drop.
21. Allow embryos to sit overnight to move to the two-cell stage. Transfer healthy two cells into recipient females for gestation.

3.9 Embryo Electroporation Without Acid Treatment of Zona

This protocol was adapted and modified from Chen et al. [13].

1. Set up plates containing several ~30 μL drops of KSOM overlaid with quality paraffin oil the afternoon before electroporation and equilibrate in the CO_2 incubator overnight.
2. On the morning of electroporation, isolate cumulus oocyte complexes (COCs) from hormone-primed and mated donor females per standard procedure and treat with hyaluronidase to isolate single embryos. Wash through two 500 μL drops of M2+ P/S then return embryos to the equilibrated KSOM drop in the CO_2 incubator.
3. After an hour in KSOM, remove embryos from KSOM drop and place on hanging drop slide with 75–100 μL of M2. Place on inverted microscope and sort for fertile embryos as determined by the presence of two pronuclei. Return fertile embryos to a new KSOM drop in the incubator.
4. Count the embryos and determine the number of cuvettes to electroporate. We group about 40–60 embryos per cuvette.
5. Deposit a 50 μL drop of M2 for each 40–60 embryos and form a horizontal line across the top half of a 10 cm dish.
6. Deposit two 100 μL Opti MEM drops for every 40 60 embryos below the M2 drops in the same 10 cm dish as above.
7. Wash 40–60 embryos sequentially in one M2 drop and two Opti-MEM drops in the 10 cm dish from **step 6**.
8. When embryos are in the second Opti-MEM wash drop, make a 10 μL drop of Opti-MEM and transfer the 40–60 embryos for each cuvette to this 10 μL drop.
9. Immediately add 10 μL of thawed electroporation sample mix to each 10 μL Opti-MEM drop with embryos and load the whole 20 μL into an electroporation cuvette.
10. Repeat **steps 8–9** until all embryos are loaded into cuvettes.
11. Electroporate using square wave with two pulses of 3 ms duration at 30 V with an interval of 0.1 second.
12. Flush the embryos from the cuvette using a P1000 micropipette with 500 μL of M2 into a watch glass and repeat. Use a P200 and gently recover as much medium remaining from the cuvette as possible and add to the flushed volume. Take care to avoid introducing air bubbles in the medium.

13. Identify and count the embryos flushed from the cuvette and pool into a 50 μL drop of M2 in a watch glass. Repeat **steps 11–12** for the remaining cuvettes.
14. Transfer the embryos back into an equilibrated KSOM culture drop in incubator for a couple hours until same day embryo transfer into d0.5 pseudopregnant recipient females (*see* **Note 8**).

3.10 Mouse Sperm Cryopreservation: CPA Preparation

This protocol was adapted from Takeo et al. [17] and Ostermeier et al. [18].

1. Dissolve 3.6 *g* raffinose, 0.6 *g* skim milk, and 0.292 *g* L-glutamine into 20 mL distilled water at 60 °C in a 50 mL centrifuge tube.
2. Incubate in a 60 °C water bath for 90 minutes and vortex for 3 minutes every 30 minutes.
3. Divide solution into 1.0 mL aliquots and pipette into 1.5 mL Eppendorf tubes.
4. Centrifuge the 20 samples at 10,000 × *g* for 60 minutes.
5. Carefully collect ~0.7 mL of the supernatant.

   CRITICAL STEP: if the supernatant is not clear, continue to centrifuge until the solution becomes clear
6. Filter the supernatant with a 0.22 μm disposable filter and store the Eppendorf tubes sealed with parafilm. Use within 3 months.

3.11 Mouse Sperm Cryopreservation: Sperm Freezing

1. Label 10 cryo straws per male and mark lines at 2.3 cm and 4.5 cm from the open end.
2. Fill a Thermosafe cooling chamber to 9 cm with LN_2 and float a cooling raft on top.
3. Add 27 μL of CPA to an Eppendorf tube. Place in a benchtop incubator (37 °C without CO_2).
4. Add a 120 μL drop of CPA to a 35 mm dish and cover with ~3 mL of paraffin oil. Once covered, inflate the drop with an additional 120 μL of CPA, for a final volume of 240 μL per two males (use 200 μL total if you are only freezing 1 male). Cover and place in a benchtop incubator (37 °C without CO_2) or on a hotplate set at 37 °C.
5. Sacrifice two males, dissect, and remove the cauda epididymides, taking care to remove the testicular artery. Blood will contaminate the sperm. Transfer all four cauda epididymides with a pair of forceps to the 240 μL CPA droplet.
6. Hold down, one at a time to anchor the tissue, and with a 25-gauge needle, poke or tear the tissue open to release the sperm. Place dish in a benchtop incubator (37 °C without CO_2) or a hot plate set at 37 °C and swirl gently every minute for 3 minutes to aid dispersion.

7. As the sperm equilibrates into the CPA, use a syringe to fill all straws to the second line with an HTF spacer and then draw air up to first line.
8. At the end of 3 minutes, use a micropipette to make five 15 μL drops of the sperm in CPA, each in the top and bottom halves of a 60 mm dish (10 total drops), making sure to wipe the pipette tip with a piece of Kimwipe after each dip, to prevent oil from getting into the droplets.
9. Draw up one droplet into each of the ten straws, then aspirate air until the plunger is fully extended.
10. Remove the syringe and heat seal the open end.
11. Load the straws into the cassettes and place on a cooling chamber raft for 15 minutes.
12. While waiting, add 3 μL of sperm to the micro centrifuge tube containing 27 μL of CPA and perform a sperm count and morphology check using a Makler sperm counter.
13. After 15 minutes, tip the raft to plunge the cassettes into liquid nitrogen.
14. Carefully and quickly transfer the cassettes to a cryo storage unit.

3.12 In Vitro Fertilization with Frozen Sperm

This protocol was adapted from Takeo et al. [17].

1. Female donor mice are superovulated using 5 IU of PMS on day 1 between 4:00 and 6:00 PM, and 5 IU of hCG on day 3 between 4:00 and 6:00 PM.
2. Prepare the following reagents on day 3 (the day before the IVF):
   (a) A wash dish with four 150 μL drops of HTF in a 60 mm dish to be kept in a CO_2 incubator until use.
   (b) An Eppendorf tube with 30.7 mg GSH to be kept in a 4 ° C fridge.
   (c) A 15 mL Falcon tube with 5 mL of HTF with loosened cap to be kept in a CO_2 incubator until use.
   (d) A 15 mL Falcon tube with 2 mL of HTF with loosened cap to be kept in a CO_2 incubator until use.
   (e) An empty 15 mL Falcon tube with loosened cap to be kept in a CO_2 incubator until use.
   (f) 50 mL tubes of H_2O; two tubes for each straw and kept in a 37 °C incubator
3. On day 4 (the day of IVF), prepare the following:
   (a) In each 35 mm dish, add 90 μL of CARD Fertiup pre-incubation medium [PM], cover with paraffin oil and equilibrate in a CO_2 incubator—capacitation dishes (one for each straw of frozen sperm).

(b) GSH medium: dissolve 30.7 mg of GSH powder in 1 mL of HTF medium and then add 50 μL of the GSH solution into 5 mL of HTF medium to reach a final concentration of 1 mM of GSH. Filter sterilize using a 0.22 μm syringe filter.

(c) In each 60 mm dish, add one 90 μL drop of GSH medium for fertilization and one 250 μL drop of GSH medium for oocyte isolation covered with paraffin oil and incubate in a CO_2 incubator—fertilization dishes (one for each wash dish).

4. Remove a straw of cryopreserved sperm from LN_2 and move to a 37 °C water bath. Put the straw directly into a 50 mL tube H_2O bath, making sure the frozen sample is below the water. As soon as the ice disappears (~5–10 seconds), move the straw to a second 50 mL tube water bath and move the water bath to a 37 °C incubator for 10 minutes to warm to 37 °C.
5. Remove the straw from the H_2O bath and wipe with a paper towel to dry.
6. Cut the straw in the center of the clay plug, then cut off the heat sealed end of the straw.
7. Use the metal plunger to push from the clay end of straw to expel the sperm sample directly into the 90 μL drop of Fertiup PM in the capacitation dish. Do not dispense the HTF spacer.
8. Move the capacitation dish into the incubator (37 °C, 5% CO_2) and allow the sperm to capacitate for 30 minutes.
9. Thirteen to fifteen hours after the hCG injection, euthanize the superovulated mice, five females at a time, and quickly remove the oviducts and place them into the 250 μL isolation drop of the 60 mm dish of HTF in **step 3c**.
10. Remove the oocytes from the swollen oviducts and move the cumulus–oocyte clusters with a #5 forceps to the 90 μL fertilization drop.
11. Add 10 μL of sperm sample from the outer edge of the drop from your capacitation dish to the fertilization drop using a wide bore pipette tip.
12. Place the fertilization dish back into the incubator (37 °C, 5% CO_2) for 3–4 hours.
13. After incubation, collect eggs from the fertilization dish and move them to the first drop of the wash dish.
14. Wash through three drops of the wash dish, then choose healthy-looking round oocytes to move to the fourth (last) drop. Proceed to the electroporation protocol.
15. Leave electroporated zygotes to incubate overnight (37 °C, 5% CO_2).
16. The following day, transfer 2-cell embryos to pseudopregnant females.

3.13 In Vitro Fertilization with Fresh Sperm

This protocol was adapted from Mouse IVF Using Frozen Sperm [20].

1. Preparations on the day before IVF.
   (a) Place 1–2 mL of high calcium K-RVFE 50 and 40 mL of oil per 10 donor females in falcon tubes, respectively, with caps loosened in the CO_2 incubator to equilibrate overnight.
   (b) A 250 μL drop of MBCD overlaid with oil in a 60 mm petri dish is placed in CO_2 incubator overnight.
   (c) Equilibrate one 30 mm petri dish with 3 mL of K-RVFE 50 for each group of five donor females in the incubator overnight.
2. Preparations in morning of IVF:
   (a) Add 2.5 μL of stock of reduced glutathione (GSH) to each mL of the tube of high calcium K-RVFE 50 medium in **step 1a** at 2.5 μL/mL of equilibrated medium.
   (b) IVF drops: make two drops of 200 μL of the above medium in a 60 mm dish for every 10 donor females and overlay with pre-equilibrated mineral oil from **step 1a.**
   (c) Wash drops: in a 60 mm dish make 4–6 100 μL drops of K-RVFE 50 medium and overlay with oil and return to the incubator.
3. Euthanize the male mouse, isolate epididymis, and place one or both epididymis into the pre-equilibrated drop of MBCD in **step 1b**.
4. Using a 30 *g* needle attached to a syringe, slice open the epididymis tissue without tearing into bits to allow the sperm to swim out. Return dish to the incubator for ~15 minutes to capacitate the sperm. In our hands, the fresh sperm does not require as much time to capacitate as frozen sperm.
5. While sperm is capacitating, euthanize the donor female mice in groups of five or fewer such that all the COC clutches can be isolated and transferred to the IVF dishes within a few minutes. Place the oviducts isolated from the females into the dishes in **step 1c**.
6. Working quickly, "Pop" the ampulla containing the COC clutches from the set of oviducts.
7. Use a P20 set to 20 μL with wide barrel tip to remove all COC clutches, place them into one IVF drop in **step 2b** and return the 60 mm dish to the incubator. **Critical timing:** COCs should be placed into the IVF drops within 3–4 minutes from the time of euthanasia of the females.
8. Repeat **steps 6** and 7 until COCs from all females have been placed into the IVF drops.

9. Remove the IVF plates and the sperm from the incubator and add 5 μL of sperm from the MBCD drop in **step 3** directly onto the COC clutches in each IVF drop containing COCs from **step** 7. Return the dishes to the incubator.
10. After 4 hours of incubation, recover all the oocytes from each drop and wash through 2–3 of the K-RVFE 50 wash drops made in **step 2c** to remove as much sperm and debris as possible. At this point, pronuclei can be visualized using an inverted microscope (optional) to estimate fertilization rate. We regularly obtain >90% fertilization.
11. Oocytes are electroporated with RNP/ssODN mix and placed into a pre-equilibrated drop of KSOM overlaid with oil for culturing overnight in the CO_2 incubator.
12. The two-cell embryos are transferred to d0.5 pseudopregnant females or frozen down the next day.

3.14 NGS-Based Genotyping

3.14.1 NGS-Based Genotyping of Founder

1. Tail clips from 7 to 10 day-old founder animals (*see* **Note 9**) are extracted in 100 μL of QuickExtract buffer using the following program on a thermal cycler: 50 °C, 30′; 65 °C, 15′; and 98 °C, 5′.
2. Assemble PCR reaction as following: 2 μL of lysate in **step 1**, 10 μL of 2× PCR mix, 1 μL of 10 μM of each NGS primer, add H_2O to 20 μL final volume.
3. Most targets at this size can be efficiently amplified by the following program (SuperFi II): 98 °C, 2′, 30 cycles of 15″ 98 °C, 15″ at 60 °C and 30″ at 72 °C.
4. Assemble indexing PCR (PCR2) reactions as following: 1 μL of PCR reaction in **step 1**, 5 μL of 2× PCR mix, 1 μL of 10 μM of each indexing primer, add H_2O to 10 μL final volume.
5. PCR2 reactions are then combined into a library, purified, and submitted for 2 × 250 MiSeq (Illumina) NGS sequencing.
6. Extracted reads from generated FASTQ files are analyzed by using CRIS.py [19] to identify genotype at each target site: indels, loxP insertion, or wild type (*see* **Note 10**).

3.14.2 Genotyping Blastocysts

(a) Mouse blastocysts (or morulae) are collected individually into 10 μL QE buffer and extracted on a thermocycler using the following program: 50 °C, 30′; 65 °C, 15′; 98 °C, 5′.

(b) Assemble PCR reaction as following: 3 μL of lysate in **step 1**, 10 μL of 2× SuperFi II mix, 2 μL of 10 μM of each NGS primer, add H2O to 20 μL final volume.

(c) Run the following program on a thermal cycler: 95 °C, 2′, 37 cycles of 15″ at 98 °C, 15″ at 60 °C, and 30″ at 72 °C.

(d) PCR reactions are then indexed for NGS as in Subheading 3.5 of **step 1**.

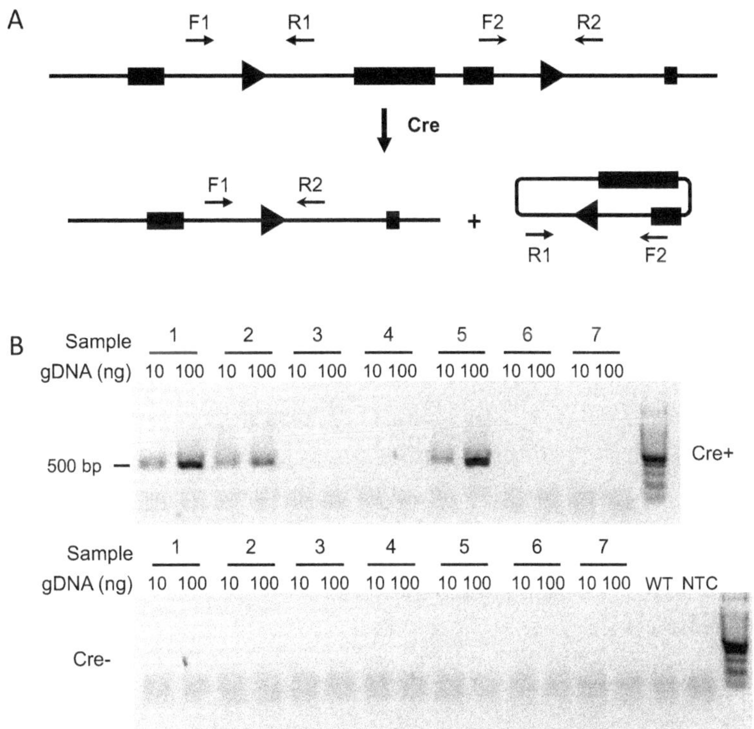

Fig. 4 In vitro Cre assay. (**a**) Schematic of how the in vitro Cre assay works. F1/R1 and F2/R2 are two sets of primers flanking the two target sites, respectively. Cre mediates recombination between the two loxP sites in a floxed allele and resulting in an allele missing the floxed sequences and a circular product where primers R1 and F2 are facing toward each other, flanking the remaining loxP site. The amplicon by R1 and F2 is unique to the circular product. (**b**) An example of a positive Cre assay. Samples 1–6 are six founders with both loxP sites. Sample 7 is a mixture of two animals each with one loxP site. The top panel is with Cre incubation, and the bottom, without Cre treatment. WT wild-type mouse DNA, NTC non-template control where water was added instead of gDNA

3.15 In Vitro Cre Assay for Phase Determination

Samples with both loxP sites are tested using in vitro Cre assay to determine whether the insertions are on the same allele. High quality double-stranded genomic DNA is needed for the assay, instead of lysate used in NGS genotyping. The assay detects an amplicon that is unique to the circular product excised by Cre-mediated recombination (*see* Fig. 4a).

1. Genomic DNA is purified from tissue samples, preferably ear punches using Qiagen's DNeasy Blood & Tissues kit and quantified with NanoDrop.
2. To assemble a reaction, add 10 ng or 100 ng gDNA (*see* **Note 11**), 2 μL of 10× Cre buffer, and 1 μL of Cre recombinase and add H_2O to a final volume of 20 μL.

3. Set up a second set of reactions as in **step 2** without adding Cre recombinase. This set serves as negative controls.
4. Set up an additional set of negative controls by combining DNA from two animals, one only with 5′ loxP insertion, the other with only 3′ loxP insertion. Use the same total gDNA amount as in the experimental samples (10 ng and 100 ng).
5. Incubate all reactions at 37 °C for 30 minutes.
6. Assemble the following PCR reactions: 5 μL of Cre assay, 12.5 μL of 2× SuperFi II Green Master Mix, 500 μM of NGS primers F2 and R1 as in Fig. 4a, +/− GC Enhancer (*see* **Note 12**), add H_2O to 25 μL final volume.
7. Run PCR for 35 cycles, and resolve on 1.5% agarose gel.
8. An example is in Fig. 4b. Desirable results: clean PCR bands (≤500 bp) in +Cre samples (none in negative controls), and no PCR product in the "no Cre" samples.

4 Notes

1. Buffers are prepared in nuclease-free water, filtered with 0.2 μm filters, frozen in single-use aliquots. Each prep is tested for RNase activity by incubating with an in vitro transcribed (IVT) RNA at 37 °C for 45 minutes. The treated samples are then run on an agarose gel alongside of untreated IVT RNA and should look identical to the untreated sample. The two buffers work equally well.
2. End protection in ssODNs greatly improves KI efficiency in cultured cells over ssODNs without phosphorothioate bonds. We have not observed a consistent correlation in embryos. However, including the protection does not seem to harm efficiencies in embryos and gives us more reliable readouts in validation in cultured cells.
3. We initially performed validations in Neuro-2a cells, but after running into issues of genotyping the target region during validation for certain genes due to strain-specific variations, we recently switched to BV2 cells derived from the B6 background.
4. Nucleofection should be carried out as soon as possible upon mixing of RNP cell suspension to minimize the degradation of gRNAs by residual nucleases in the cell suspension.
5. If the PCR product has nonspecific bands, the correct-sized band should be band isolated and gel extracted in **step 3**.
6. Longer digestion time may cause non-specific cleavage.

7. It is best to use a large orifice pipette to increase the speed at which you can pick the zygotes back up. Make sure you move as quickly as possible to get the zygotes from the Tyrode's treatment into your Post dish or the zygotes can be damaged. The speed at which you perform this step will affect survival rates.
8. We experience a higher birth rate doing same day embryo transfer at one-cell stage into d0.5 pseudopregnant recipients. Transfer at the two-cell stage can be performed the following day into d0.5 pseudopregnant recipients. We use CD1 mice from Charles River labs as our embryo recipients.
9. It is critical that the tail clips are under 2 mm in length. Too much material is the most common reason for failed PCR reactions.
10. Founder animals (F0) are often mosaic, and more than two alleles in one animal is common with percentage of reads for a given allele ranging from 10% to 99%.
11. In the in vitro Cre assay, low concentration of genomic DNA is used to promote intramolecular recombination between loxP sites. Two different amounts of genomic DNA are used because of the difficulty to accurately measure genomic DNA concentrations. Occasionally, only one of the genomic DNA concentration results in positive amplification.
12. Depending on the sequence context, the in vitro Cre assay may require the GC enhancer that comes with the SuperFi mix to work efficiently. We routinely include reactions with and without the GC enhancer to maximize the success rate.

Refcrences

1. Gu H, Marth JD, Orban PC, Mossmann H, Rajewsky K (1994) Deletion of a DNA polymerase beta gene segment in T cells using cell type-specific gene targeting. Science 265:103–106
2. Carroll D (2014) Genome engineering with targetable nucleases. Annu Rev Biochem 83:409–439
3. Yang H, Wang H, Shivalila CS, Cheng AW, Shi L, Jaenisch R (2013) One-step generation of mice carrying reporter and conditional alleles by CRISPR/Cas-mediated genome engineering. Cell 154:1370–1379
4. Bishop KA, Harrington A, Kouranova E et al (2016) CRISPR/Cas9-mediated insertion of loxP sites in the mouse dock7 gene provides an effective alternative to use of targeted embryonic stem cells. G3 6:2051–2061
5. Pritchard CEJ, Kroese LJ, Huijbers IJ (2017) Direct generation of conditional alleles using CRISPR/Cas9 in mouse zygotes. Methods Mol Biol 1642:21–35
6. Hai L, Szwarc MM, Lanza DG, Heaney JD, Lydon JP (2019) Using CRISPR/Cas9 engineering to generate a mouse with a conditional knockout allele for the promyelocytic leukemia zinc finger transcription factor. Genesis 57:e23281
7. Gurumurthy CB, O'Brien AR, Quadros RM, Adams J, Alcaide P, Ayabe S (2019) Reproducibility of CRISPR-Cas9 methods for generation of conditional mouse alleles: a multicenter evaluation. Genome Biol 20:171
8. Sentmanat MF, White JM, Kouranova E, Cui X (2022) Highly reliable creation of floxed alleles by electroporating single-cell embryos. BMC Biol 20:31
9. Quadros RM, Miura H, Harms DW et al (2017) Easi-CRISPR: a robust method for one- step generation of mice carrying conditional and insertion alleles using long ssDNA donors and CRISPR ribonucleoproteins. Genome Biol 18:92

10. Miura H, Gurumurthy CB, Sato T, Sato M, Ohtsuka M (2015) CRISPR/Cas9-based generation of knockdown mice by intronic insertion of artificial microRNA using longer single-stranded DNA. Sci Rep 5:12799
11. Bennett H, Aguilar-Martinez E, Adamson AD (2021) CRISPR-mediated knock-in in the mouse embryo using long single stranded DNA donors synthesised by biotinylated PCR. Methods 191:3–14
12. Qin W, Dion SL, Kutny PM et al (2015) Efficient CRISPR/Cas9-mediated genome editing in mice by zygote electroporation of nuclease. Genetics 200:423–430
13. Chen S, Lee B, Lee AY-F, Modzelewski AJ, He L (2016) Highly efficient mouse genome editing by CRISPR ribonucleoprotein electroporation of zygotes. J Biol Chem 291:14457–14467
14. Wang W, Kutny PM, Byers SL et al (2016) Delivery of Cas9 protein into mouse zygotes through a series of electroporation dramatically increases the efficiency of model creation. J Genet Genomics 43:319–327
15. Kosicki M, Tomberg K, Bradley A (2018) Repair of double-strand breaks induced by CRISPR-Cas9 leads to large deletions and complex rearrangements. Nat Biotechnol 36: 765–771
16. Hsu PD, Scott DA, Weinstein JA, Ran FA, Konermann S, Agarwala V et al (2013) DNA targeting specificity of RNA-guided Cas9 nucleases. Nat Biotechnol 31:827–832
17. Takeo T, Sztein J, Nakagata N (2019) The CARD method for mouse sperm cryopreservation and in vitro fertilization using frozen-thawed sperm. In: Liu C, Du Y (eds) Microinjection. Methods in molecular biology, vol 1874. Humana Press, New York, NY
18. Ostermeier GC, Wiles MV, Farley JS, Taft RA (2008) Conserving, distributing and managing genetically modified mouse lines by sperm cryopreservation. PLoS One 3(7):e2792
19. Connelly JP, Pruett-Miller SM (2019) CRIS.py: a versatile and high-throughput analysis program for CRISPR-based genome editing. Sci Rep 9(1):4194
20. UC Davis Mouse Biology Program (2018) Mouse rescue IVF using 70–100 μL of cryopreserved sperm. Available via https://mmrrc.ucdavis.edu/files/protocols/Cryo-IVF-protocol.pdf. Accessed 24 Mar 2022

Chapter 10

CRISPR/Cas9 Endonuclease-Mediated Mouse Genome Editing of One-Cell and/or Two-Cell Embryos by Electroporation, and the Use of Rad51 to Enhance Knock-In Allele Homozygosity via Interhomolog Repair Mechanism

Selika Garza and Raehum Paik

Abstract

Electroporation of mouse embryos with CRISPR/Cas9 endonuclease tool is a facile and efficient method to edit endogenous genome sequences for generating genetically engineered mouse models (GEMMs). Common genome engineering projects, such as knock-out (KO), conditional knock-out (cKO), point mutation, and small foreign DNA (<1 Kb) knock-in (KI) alleles, can be effectively accomplished with a simple electroporation procedure. The use of electroporation in sequential gene editing at the one-cell (0.7 days post-coitum (dpc)) and at two-cell (1.5 dpc) embryonic stages provides a fast and compelling protocol to safely introduce multiple gene modifications on the same chromosome by limiting chromosomal fractures. In addition, the co-electroporation of the ribonucleoprotein (RNP) complex and single-stranded oligodeoxynucleotide (ssODN) donor DNA with the strand exchange protein Rad51 can significantly increase the number of homozygous founders. Here we describe a comprehensive guideline for mouse embryo electroporation to generate GEMMs and the implementation of Rad51 in RNP/ssODN complex EP medium protocol.

Key words CRISPR/Cas9, Electroporation (EP), Genetically engineered mouse model (GEMM), Knock-out (KO), Point mutation, Knock-in (KI), Ribonucleoprotein (RNP), Single-stranded oligodeoxynucleotide (ssODN), Rad51

1 Introduction

Since the publication of Jinek et al. (2012) science article [1] describing the CRISPR/Cas9 system of two RNA structures, namely, crRNA and tracrRNA, directing Cas9 endonuclease to cause double strand breaks (DSB) in targeted DNA, numerous articles have been published delineating efficient methods to introduce gene modifications in various species, including mus musculus. Although mouse pronuclear microinjection remains a reliable and effective method to deliver RNP (and donor DNA) complex

Thomas L. Saunders (ed.), *Transgenesis: Methods and Protocols*,
Methods in Molecular Biology, vol. 2631, https://doi.org/10.1007/978-1-0716-2990-1_10,

into targeted endogenous DNA sequences, it requires high-cost and sophisticated microinjection instrument rigs and skilled personnel to perform the procedures. The electroporation of mouse embryos represents a cost-effective and simple avenue to dispatch CRISPR/Cas9 materials into a selected DNA locus for mouse genome editing [2–7].

Moreover, the ability of the strand exchange protein Rad51 to convert the heterozygous alleles into homozygous alleles via interhomolog repair mechanism is a powerful application that significantly increases homozygosity of desired gene modification when ssODN is used as the donor DNA [8]. While electroporation "only" method can be applied to generate indel based frameshift (early stop codons) knock-outs, and knock-ins of foreign DNA with limited length [9, 10], in combination with AAV packaged donor DNA, a larger foreign DNA knock-in can be also achieved [11–13].

Between 2018 and 2021, we applied electroporation method to successfully deliver 5 KO, 8 point mutation, 20 cKO, and 2 large foreign DNA KI GEMM projects. In this chapter, we describe in detail the protocols we have finalized to obtain the best successful outcomes for each project and what we have learned from our experiences.

2 Materials

2.1 Embryo Preparation and Culture

1. Embryo Collection: M2 (Millipore Sigma, M7167) and Hyaluronidase (Millipore Sigma, H4272).
2. Embryo Culture: Global Media (Cooper Surgical, LGGG) and Paraffin Liquid (Nacalai Tesque, 26137-85).

2.2 RNP Components (sgRNA and Cas9)

1. sgRNA: Synthetic sgRNA, with chemical modification (Synthego, custom order, *see* **Note 1**).

   Reconstitution Solution = 3 ug/uL in 10 mM Tris-NaCl and 0.5 mM EDTA, stored in −80 °C.

   Stock Solution = 300 ng/uL in Opti-MEM (1:10 dilution from reconstitution solution), stored in −80 °C (final working solution = 30 ng/uL)
2. Cas9: eSpCas9 (Millipore Sigma, ESPCAS9PRO, *see* **Note 2**).

   Reconstitution Solution = 5 ug/uL in Reconstitution Solution provided from the manufacturer.

   Stock Solution = 500 ng/uL in Opti-MEM (1:10 dilution from Storage Solution), aliquoted with 2 uL in 15 mL Eppendorf tube, and stored in −20 °C (final working solution = 50 ng/uL).

2.3 Donor DNA (ssODN and lssDNA) and Rad51

1. ssODN: (IDT-DNA, Eurofins..., custom order).

Reconstitution Solution = 5 ug/uL in 10 mM Tris-NaCl and 0.5 mM EDTA, stored in −20 °C.

Stock Solution = 1 ug/uL in Opti-MEM (1:5 dilution from reconstitution solution), stored in −20 °C (final working solution = 100 ng/uL).

2. lssDNA: (IDT-DNA, GeneWiz, GenScript, or other vendor, custom order).

Reconstitution Solution = 1ug/uL in 10 mM Tris-NaCl and 0.5 mM EDTA, stored in −20 °C.

Stock Solution = 250 ng/uL in Opti-MEM (1:4 dilution from reconstitution solution), stored in −20 °C. (final working solution = 25 ng/uL)

3. In-House synthesized lssDNA: Guide-it Long ssDNA Production System (Takara, 632666).

Stock Solution = the concentration varies from preparation to preparation, depending on the concentration of purified PCR amplicon required to synthesize lssDNA using Strandase exonuclease. Purified lssDNA is eluted in 10 mM Tris-NaCl and 0.5 mM EDTA, stored in −20 °C (final working solution = 25 ng/uL)

4. dsDNA: purified PCR Amplicon.

Stock Solution = the concentration varies from preparation to preparation, depending on the concentration of purified PCR amplicon. PCR amplicon is gel extracted, and purified, in 10 mM Tris-NaCl and 0.5 mM EDTA, stored in −20 °C (final working solution = 25 ng/uL)

5. Rad51 (Abcam, AB63808).

Stock Solution = 200 ng/uL in Opti-MEM, aliquoted with 1.2uL in 200 uL tube strip, stored in −80 °C (final working solution = 10 ng/uL).

2.4 Electroporation

Two apparatuses employed to perform mouse embryo EP are (1) electrode (made of high electrical conductivity element, such as platinum) and (2) electroporator (capable of generating accurate and low voltage square wave pulses and desired pulse length and pulse interval). There are several commercially available electrodes and electroporators. Most popular commercial electrodes for mouse embryo EP are Bio-Rad Gene Pulser cuvette electrode (165-2081), Harvard Apparatus platinum plated 10 mm oocyte electrode (LF501PT110), and Protech International Inc. 15 mm solid platinum plate electrode (PD1.15), and as for the electroporators, BTX ECM830, BEX Genome Editor, and Nepa Gene CUY21EDIT are widely used for such purpose. Our lab uses custom-made electrodes and a modified electroporator.

1. Electroporation buffer: Opti-MEM: (Gibco, 31985062).
2. Electroporator: A-M Systems Model 4100, (modified): Model 4100 Isolated High Power Stimulator was modified to increase the voltage and amperage output to match BEX and Nepa Gene instruments. As modified, the power is increased to 40 volts and 350 mA.
3. Electrode: custom made (15 mm channel length platinum plate, *see* **Note 3**).

3 Methods

3.1 Embryo Collection and Culture

1. Embryos are collected around 20 hours post hCG injection, at 09:00 AM (*see* **Note 4**): Collect two oviducts in 100 uL M2/Hyaluronidase media drop (from five female mice, per collection session), extract two ampullas from two oviducts, and place in a new 100 uL M2/Hyaluronidase media drop. Tear the ampullas apart to release the embryos and let the cumulus masses come loose from the embryos. Further wash the embryos in 200 uL M2 media without Hyaluronidase.
2. Transfer the collected embryos from M2 to pre-equilibrated Global media (300 uL Global media, immersed in Paraffin liquid, equilibrated for at least 30 minutes in 37 °C, 95% $C0_2$ incubator), and culture them in the incubator (37°, 95% $C0_2$) until the electroporation is applied.

3.2 Electroporation (at One-Cell or Two-Cell Embryonic Developmental Stage)

1. Set the electroporation stage: place the electrode on the inverted microscope stage and connect it to the electroporator (Fig. 1).
2. Program the electroporator parameter settings (*see* **Notes 5** and **6**):

   30 mV (3 msec ON and 97 msec OFF, bi-directional 7 pulses) (*see* Fig. 2).
3. Prepare RNP mix by adding 2 uL sgRNA (300 ng/uL) into 2 uL Cas9 (500 ng/uL) aliquot tube. Add donor DNA if needed (*see* **Notes** 7 and **8**). Fill up the master mix to 20 uL final volume with Opti-MEM (*see* **Note 9**). Final concentration: eSpCas9 = 50 ng/ul, and sgRNA = 30 ng/uL.
4. Store the master mix at RT (25 °C) for 10 to 30 minutes (*see* **Note 10**).

   Optional: thaw Rad51 by holding the aliquot tube with fingers for 1 minute, and add 1 uL (200 ng/uL) in the master mix right before the EP. Final concentration: 10 ng/uL.
5. Wash embryos 3× times in 100 uL Opti-MEM drops.

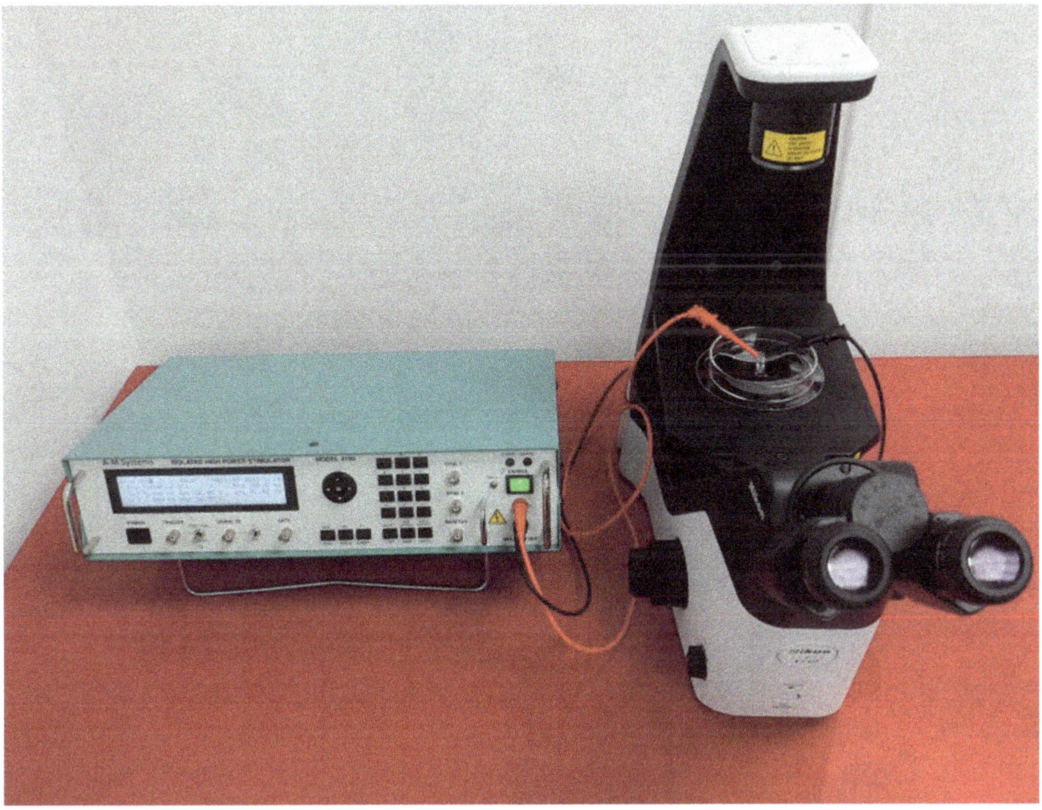

Fig. 1 Electroporation workstation ready for use. The three principal components are the electroporator, the electroporation chamber, and a suitable inverted microscope with optics for brightfield 40× and 100× magnification. Note the electrical leads going from the electroporator to the chamber

6. Fill the electrode channel with master mix. The solution should reach to the top of the electrodes without overflowing (*see* Fig. 3) and should not extend beyond the ends of the electrodes (*see* Fig. 4).
7. Align embryos along the electrode channel as shown (*see* Fig. 5).
8. Electroporate embryos with pre-set parameters (*see* **Note 11**). A sign of successful delivery of the electrical pulse is the formation of gas bubbles on the electrodes due to the hydrolysis of water (*see* Fig. 6).
9. Wash 3 times in pre-equilibrated 100 uL Global Media drops in paraffin liquid.
10. Incubate washed embryos O/N at 37 °C (95% CO_2) in the third Global Media drop.
11. Transfer two-cell embryos (fertilized eggs) into pseudo pregnant female (*see* **Note 12**).

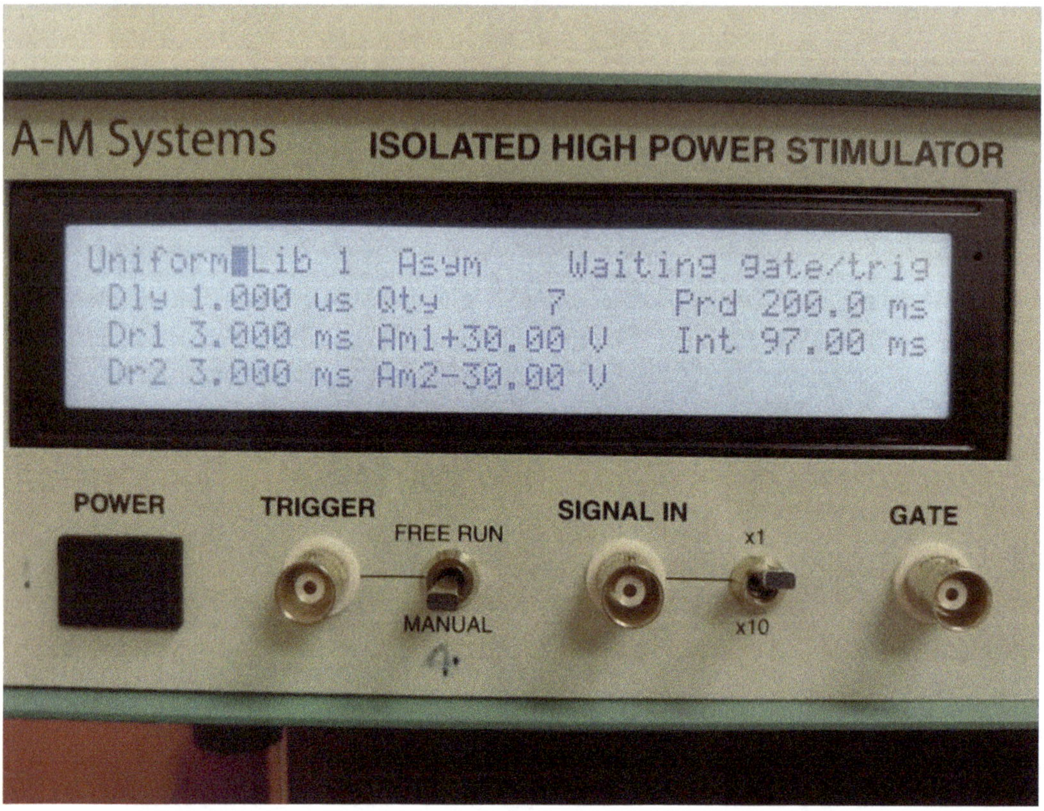

Fig. 2 Example of optimized electroporation parameters displayed on our modified A-M Systems Model 4100 Isolated High Power Stimulator

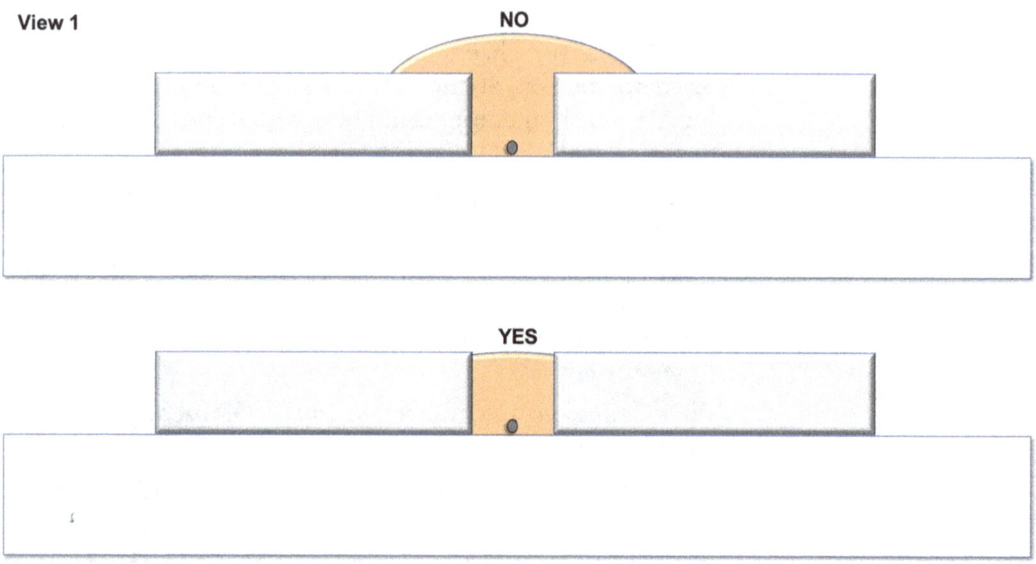

Fig. 3 Electrode gap filled with electroporation solution is shown. The solution should completely fill the space between the electrodes without overtopping the electrodes

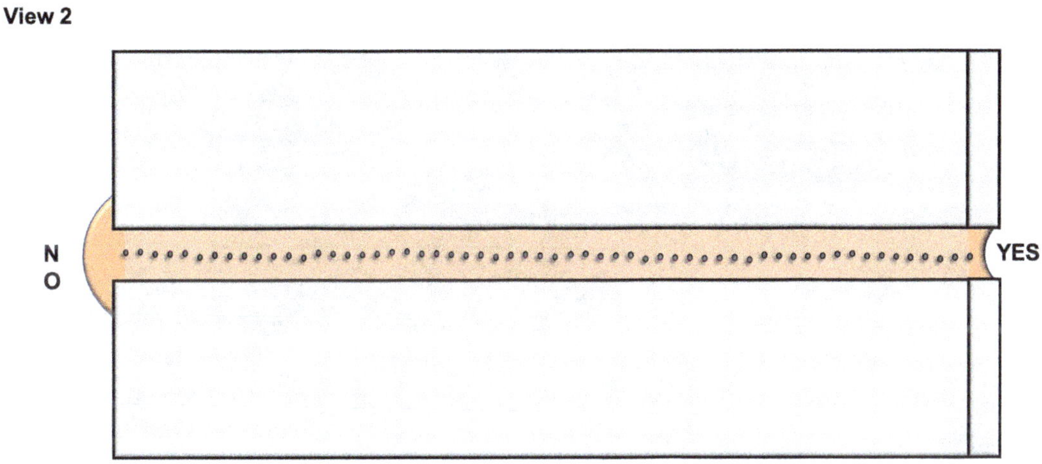

Fig. 4 Electrode gap filled with electroporation solution is shown. The solution should completely fill the space between the electrodes without overflowing the ends of the electrodes onto the glass surface

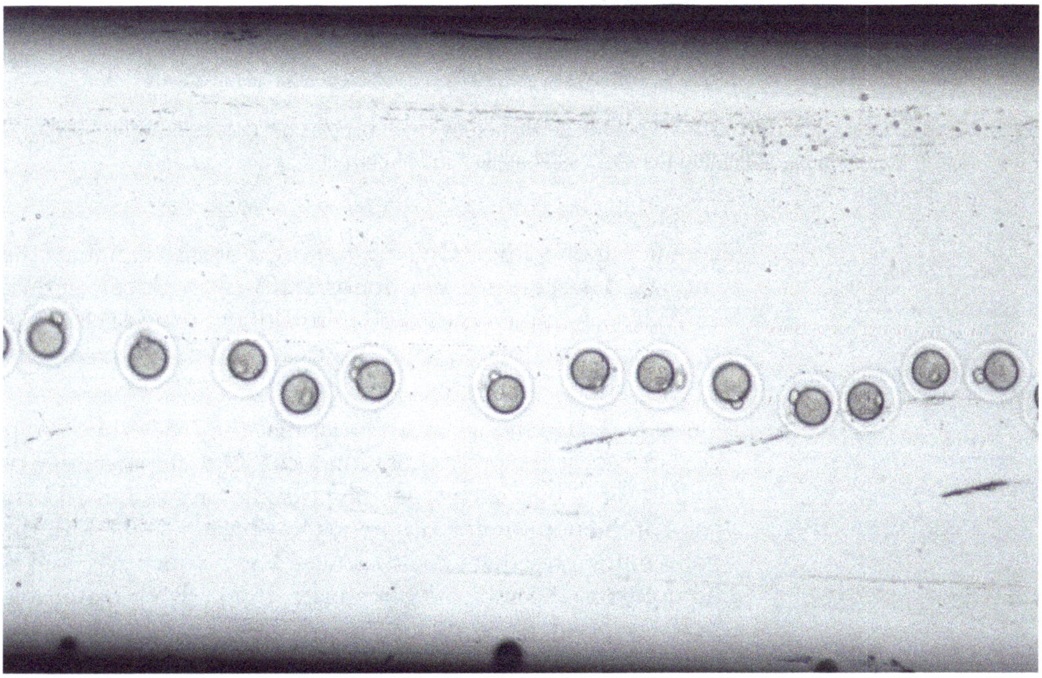

Fig. 5 Mouse embryos lined up in the electroporation chamber. The electrodes can be seen as the dark line/shadow above and below the mouse embryos

3.3 GEMM Individual Project Plan Details

1. *KO model*: use one sgRNA to induce indels in the endogenous coding sequence, or two sgRNAs to generate a large deletion. Try to target the first coding exon to produce an early stop codon from an indel originated frameshift. We usually target coding sequences with two sgRNAs flanking multiple coding exons; instead of two distinct indels, the sequence between two sgRNA will be deleted most of the time, eliminating the

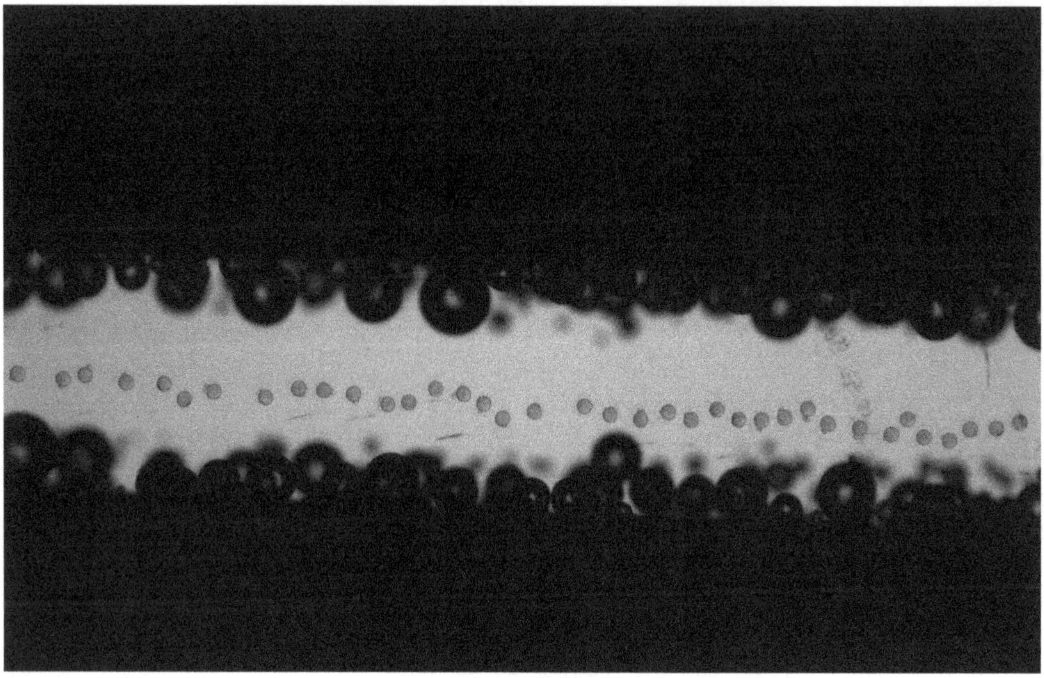

Fig. 6 After electroporation, the embryos are intact. Numerous small bubbles generated by water hydrolysis are evident in the chamber, indicating the successful application of current

coding region within. One can remove a precise region of the endogenous sequence if a donor DNA is provided: ssODN containing homology arms (60 nt to 100 nt each side) flanking the desired deletion sequence (upstream and downstream of 5′ and 3′ crRNAs DSB sites).

2. *Point mutation* model: it is crucial that the DSB site of your crRNA is near the point mutation locus. Our empirical rule of thumb is to find the endonuclease cleavage site within 15 base pairs of the point mutation site. ssODN donor DNA with the point mutation sequence and silenced PAM sequence is used as homologous recombination template. If no crRNA is available near the point mutation site, then two sgRNAs flanking the point mutation site can be also employed. Such donor DNA will contain middle sequence with the point mutation and two homology arms flanking two crRNA DSB sites with PAM sequence modifications. Homology arm size can vary from 60 nt to 100 nt in each arm.
3. *cKO* (floxed) model: our protocol for cKO GEMM generation is a hybrid method between the application of the sequential introduction of loxP [14] and Rad51 [8]. Electroporate zygotes in master mix containing 5′ (or 3′) loxP ssODN

donor DNA and corresponding RNP complex with Rad51. Incubate overnight. Electroporate two-cell embryos in master mix containing 3′ (or 5′) loxP ssODN donor DNA and corresponding RNP complex with Rad51. Incubate for an hour to screen fused embryos from the electroporation at two-cell developmental stage. Discard fused embryos and transfer the intact ones to female pseudopregnant mice. In contrast to Wilde et al. [8], we do not observe any increased homozygosity when RNP-ssODN complex is introduced via EP to the zygotes with Rad51, but the effect of Rad51 to convert heterozygous alleles into homozygous alleles is significantly high when the reagents are electroporated at two-cell embryos.

4. KI model: Although we were able to electroporate donor lssDNA up to 2.3Kb, we do not no longer use lssDNA EP when the DNA donor is longer than 800 nucleotides. Despite its high KI efficiency, the founder mice often have small deletions within the foreign DNA sequence when longer lssDNA donors are used. In our lab, we find pronuclear injection of plasmid donor DNA a superior method to introduce donor DNA segments longer than 800 base pairs, due to complete insertion of foreign DNA sequences in founder mice and extremely low frequency of random donor DNA integration. We synthesize lssDNA in-house with the Guide-it lssDNA Synthesis Kit from Takara.

3.4 Conclusion

Although, most of KO gene editing can be efficiently achieved by one-cell zygote electroporation, the efficiency of foreign DNA KI integration is greatly enhanced when the electroporation is performed at two-cell due to longer S-phase cell cycle [15, 16]. During the last 3 years, we delivered over 25 projects with 2-cell EP, and only 1 project resulted in a mosaic editing outcome.

Despite published articles on the successful use of two distinct ssODNs (each one containing loxP and HAs flanking critical exon-s) or lssDNA (with the whole sequence encompassing two loxPs and critical exon-s) as donor DNA for the conditional KO gene editing [10, 17], we found the simultaneous delivery of two sgRNA was not a productive method. The prevalent outcome of such a protocol, in our hands, resulted in deletion of the sequence between two crRNA DSB sites. Additional observations we observed after the concurrent delivery of two sgRNAs include single loxP insertion, trans loxP integration, and small deletions within the donor lssDNA sequence.

4 Notes

1. Chemical modifications, 2′-O-methyl at first three and last bases and 3′ phosphorothioate bonds between first three and last two bases, provide increased stability and protection from exonucleases and deliver improved gene editing efficiency [18].
2. We use "enhanced specificity" SpCas9 (eSpCas9) sold from Millipore Sigma. eSpCas9 is a Cas9 variant where positive charges in the nt-groove are neutralized. Such modification reduces off-target effects and maintains robust on-target cleavage [19].
3. Our electrode is made of a pair of 1 mm thick × 3 mm wide × 25 mm long bent platinum plates (*see* Fig. 7). Full 15 mm (current channel length) of both plates are glued to the 10 cm glass Petri dish using dental glue, and the remaining 10 mm is bended to 90° with 1.5 mm diameter hole for clamp cable connection (*see* Fig. 8). To attach platinum plate electrode to the glass Petri dish, Krazy Glue® also worked well (~150 EP uses and washes before it came loose), but it was not as strong as acrylic resin dental glue (~250 EP uses and washes before it came loose). Two electrode plates are placed at the center of the Petri dish with 1 mm gap between them.
4. Our superovulation timing is as follows: PMSG (5 IU) at 02: 00 PM (mouse age = P25), hCG (5 IU) at 01:00 PM (47 hours post PMSG). Super-ovulated female mice are placed with stud male mice (1:1) right after hCG injection.

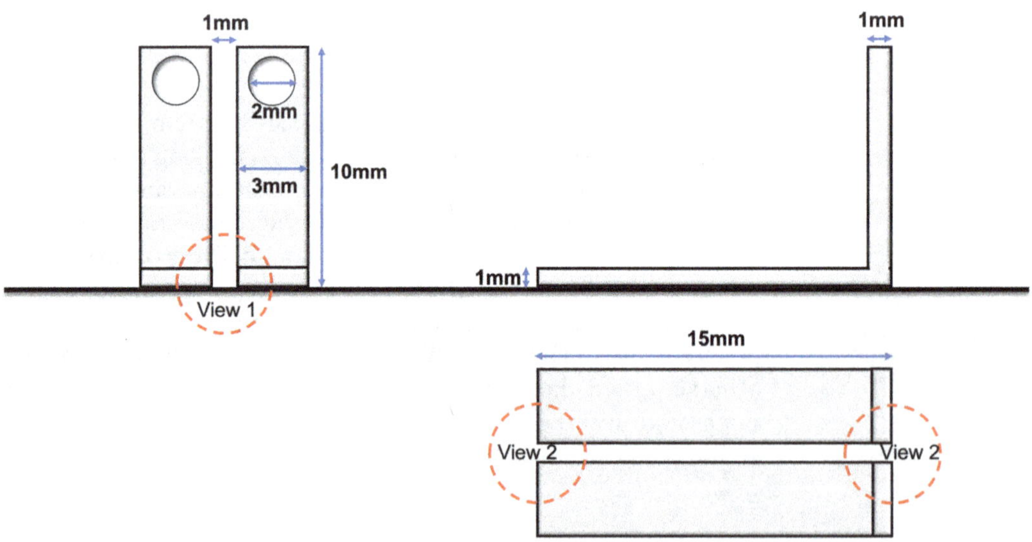

Fig. 7 Schematic of electrodes for the construction of a customized electrode chamber, as described in the text

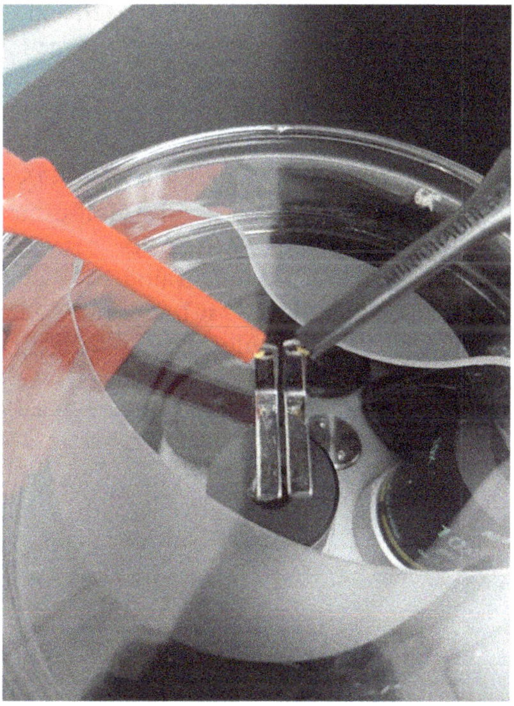

Fig. 8 Final assembled electroporation chamber connected to electrodes, positioned on the microscope stage, and ready for current application

5. It is very important that you find best electroporator parameters specific to your settings. Different electrodes (shape and nature) and electroporation media volume in the electrode channel will generate different electric fields, hence different electric resistances. We have tested 25 V, 27 V, and 30 V electric output paired with 5, 7, 9, and 11 bi-directional pulses (note that 1 bi-directional pulse equals to 2 total pulses, 1 opposite direction of the other), while maintaining 3 msec ON/97 msec OFF electroporation triggering interval times. In our specific setting, we found 30 V applied in seven bi-directional pulses to be most efficient and safe parameter for mouse embryo CRISPR/Cas9 gene editing electroporation applications, using our electrode.
6. The oscillation in resistance value results in current value alteration, and as a byproduct, it affects experimental reliability. It is an imperative that one finds optimal electroporation parameters, specific to each setting, in order to deliver the optimal current that is safe and effective for the mouse embryo gene editing.

7. ssODN donor (<200 nt): we use ssODN for projects where the foreign DNA sequence falls below ~100 nt (point mutation, small tag KI, loxP, and other lox variants). We have observed only one project (point mutation, two-cell EP) that exhibited mosaicism in founder mouse. Founders from another 28 projects where ssODN EP was applied did not produce any mosaic founder mouse lines. During three distinct electroporation sessions for three different projects, after initial PCR screening, founder mice were sequenced to detect point mutations. One nucleotide insertion within the 34 nt loxP sequence, a single nucleotide point mutation within the loxP central spacer sequence, and two nucleotide point mutations within the lox2272 sequence were identified after Sanger sequencing. No other projects presented with mutations in ssODN donor DNA.

   Pros: High success rate in relatively short time frame, low-cost donor DNA production, easy PCR outside HAs, and subsequent Sanger sequencing due to relatively small donor DNA size.

   Cons: high non-specific random integration rate.

8. lssDNA donor (200<): We synthesize in-house lssDNA with Guide-it Long ssDNA Production System kit from Takara. We have tried Nickase-based lssDNA synthesis method but abandoned it due to its labor-intensive nature required for extraction and relatively low final lssDNA yield. lssDNA can be purchased from commercial companies, such as IDT-DNA and Eurofins. Although the homology directed repair (HDR) rate of a lssDNA donor is far higher than that of a plasmid donor DNA, when the foreign DNA size is over ~800 nt, we always choose to use plasmid donor DNA delivered via pronuclear microinjection (PNI). For lssDNA >800 nt, we often find a deletion within the foreign DNA sequence. In our case, we observed deletions varying from 17 bp to 187 bp, which cannot always be identified by visual examination of UV excited amplicon band after PCR genotyping. On one occasion, we had generated one correct founder mouse using lssDNA as long as ~2300 nt; it had an EGFP Conditional by Inversion (COIN) module with ~250 nt long homology arms. Among eight initial founders thought to carry the correct locus insertion, we found only one founder with the proper intact sequence, while the other seven had deletions of various sizes. This is not a viable method given amount of time and resources one must invest. Hence, we use plasmid donor DNA and PNI method to deliver foreign DNA donors bigger than 800 bp. Nonetheless, when HDR is a consistent obstacle with the plasmid donor DNA KI, lssDNA donor is an alternative option to generate desired GEMM.

9. We prepare 20 uL of electroporation master mix. eSpCas9 aliquot (500 ng/uL, 2 uL per tube) in 1.5 mL Eppendorf tube is used as master mix vessel where all the components are added. From 10 pregnant mice, we collect an average of 200 to 300 embryos. Our electrode channel length can easily accommodate 120 to 150 embryos per electroporation session. In our setting, the theoretical volume inside the electrode measures 0.1 cm × 0.1 cm × 1.5 cm = 0.015 cubic centimeter = 15 uL. Due the superficial tension of Opti-MEM, and voluntary placement of the master mix within the electrode length (to avoid electric field deviation at the edge of the electrodes), we find 9 uL of RNP/Opti-MEM to be the ideal volume for electroporation in our electrode system. Below 9 uL, we observe big bubbles generated after applying electroporation, causing embryos to lyse more promptly. One should test the specific electrode system they are using to find the ideal parameters to execute efficient EP-based gene editing. Embryo survival rates and gene editing efficiency should be analyzed by using same RNP complex, while titrating various parameters, such as voltage, pulse number, and buffer volume.
10. We do not observe any editing efficiency differences when RNP mix is stored at RT for 10 minutes or longer, up to 3 hours. We tested EP after 10 minutes, 30 minutes, 1 hour, 2 hours, or 3 hours RT incubation of RNP.
11. Despite Opti-MEM being an ideal EP medium, it is an unsuitable culture medium for mouse embryos: embryos will not survive if left for a long time in Opti-MEM. Perform the wash and EP as quickly as you can do it. In our hands, it takes less than 5 minutes from the first Opti-MEM wash to EP and transfer to the Global Media drop. There are several factors that contribute to the electroporation outcome including standardized voltage, time interval between pulses, and number of pulses. Assuming the EP buffer is composed with at least 70% Opti-MEM, the distance between electrodes and the buffer volume that occupies the space between electrodes will influence the electric field required for RNP materials to penetrate the nuclear envelope of each embryo.
12. We perform bi-lateral transfer of two-cell embryos in pseudopregnant CD1 female mice (age: P45 to P75). 12–13 embryos are transferred into each oviduct.

Acknowledgements

We thank Dr. Thomas Saunders for kindly providing his feedback on this manuscript

References

1. Jinek M et al (2012) A programmable dual-RNA-guided DNA endonuclease in adaptive bacterial immunity. Science 337(6096): 816–821
2. Hashimoto M, Takemoto T (2015) Electroporation enables the efficient mRNA delivery into the mouse zygotes and facilitates CRISPR/Cas9-based genome editing. Sci Rep 5:11315
3. Qin W et al (2015) Efficient CRISPR/Cas9-mediated genome editing in mice by zygote electroporation of nuclease. Genetics 200: 423–430
4. Chen S et al (2016) Highly efficient mouse genome editing by CRISPR ribonucleoprotein electroporation of zygotes. J Biol Chem 291: 14457–14467
5. Sato M et al (2016) Nucleic acids delivery methods for genome editing in zygotes and embryos: the old the new and the old-new. Biol Direct 11:16
6. Wang W et al (2016) Delivery of Cas9 protein into mouse zygotes through a series of electroporation dramatically increases the efficiency of model creation. J Genet Genomics 43:319–327
7. Modzelewski AJ et al (2018) Efficient mouse genome engineering by CRISPR-EZ technology. Nat Protoc 13:1253–1274
8. Wilde JJ et al (2021) Efficient embryonic homozygous gene conversion via RAD51-enhanced interhomolog repair. Cell 184: 3267–3280 e18
9. Miura H et al (2018) Easi-CRISPR for creating knock-in and conditional knockout mouse models using long ssDNA donors. Nat Protoc 13:195–215
10. Miyasaka Y et al (2018) CLICK: one-step generation of conditional knockout mice. BMC Genomics 19:318
11. Yoon Y et al (2018) Streamlined ex vivo and in vivo genome editing in mouse embryos using recombinant adeno-associated viruses. Nat Commun 9:412
12. Chen S et al (2019) CRISPR-READI: efficient generation of knockin mice by CRISPR RNP electroporation and AAV donor infection. Cell Rep 27(13):3780–3789. e4
13. Mizuno N et al (2018) Intra-embryo gene cassette knockin by CRISPR/Cas9-mediated genome editing with adeno-associated viral vector. iScience 9:286–297
14. Horii T et al (2017) Efficient generation of conditional knockout mice via sequential introduction of lox sites. Sci Rep 7:7891
15. Gu B, Posfai E, Rossant J (2018) Efficient generation of targeted large insertions by microinjection into two-cell-stage mouse embryos. Nat Biotechnol 36:632–637
16. Abe T et al (2020) Pronuclear microinjection during S-phase increases the efficiency of CRISPR-Cas9-assisted knockin of large DNA donors in mouse zygotes. Cell Rep 31:107653
17. Lanza DG et al (2018) Comparative analysis of single-stranded DNA donors to generate conditional null mouse alleles. BMC Biol 16:69
18. Hendel A et al (2015) Chemically modified guide RNAs enhance CRISPR-Cas genome editing in human primary cells. Nat Biotechnol 33:985–989
19. Slaymaker IM et al (2016) Rationally engineered Cas9 nucleases with improved specificity. Science 351:84–88

Chapter 11

Electroporation-Mediated CRISPR/Cas9 Genome Editing in Rat Zygotes

Daniel J. Davis, Hongsheng Men, and Elizabeth C. Bryda

Abstract

Genetic engineering in the rat has been revolutionized by the development of CRISPR-based genome editing tools. Conventional methods for inserting genome editing elements such as CRISPR/Cas9 reagents into rat zygotes include cytoplasmic or pronuclear microinjections. These techniques are labor-intensive, require specialized micromanipulator equipment, and are technically challenging. Here, we describe a simple and effective method for zygote electroporation in which CRISPR/Cas9 reagents are introduced into rat zygotes via pores produced by precise electrical pulses applied to the cells. Zygote electroporation allows for high-throughput efficient genome editing in rat embryos.

Key words CRISPR/Cas9, Genome editing, Rat, Electroporation, Zygote, Embryo

1 Introduction

CRISPR-based genome editing tools have greatly facilitated the production of genetically engineered rat models. Ribonucleoprotein complexes (RNPs) comprise a single-guide RNA, and Cas9 nuclease can be programmed to target a sequence of interest within the genome and create a double-stranded DNA break. Aberrant repair of this DNA break can result in the introduction of insertion or deletion (INDEL) mutations within the DNA sequence which can disrupt the target gene's function. Alternatively, through the co-delivery of a DNA repair template, this DNA break can be precisely restored along with the addition of any desired engineered DNA sequence through a process called homology-directed repair. Introduction of CRISPR/Cas9 reagents into rat zygotes has initially been achieved through a microinjection process [1]. This low-throughput process is labor-intensive, requires specialized micromanipulator equipment, and is technically challenging.

Thomas L. Saunders (ed.), *Transgenesis: Methods and Protocols*,
Methods in Molecular Biology, vol. 2631, https://doi.org/10.1007/978-1-0716-2990-1_11,

Electroporation offers an alternative method of delivering CRISPR/Cas9 components into zygotes. Although electroporation has traditionally been used to transfer reagents into cultured cell lines, this technique has been proven effective at introducing genome editing reagents directly into embryos. Zygote electroporation was first demonstrated in rat embryos [2] and since has been shown to be an effective delivery method for mouse [3–8], pig [9, 10], and other animal model organisms [11–13]. For this process, zygotes are placed into a cuvette or onto a slide between two electrodes while suspended in a medium containing CRISPR/Cas9 reagents. An electroporator directs pulses of electrical currents from one electrode to the other through the zygotes in order to create temporary holes in the zona pellucida and plasma membrane allowing the transfer of CRISPR/Cas9 components into the zygotes. Typically, mid-level electrical pulses are used as "poring" pulses to create the temporary holes, while subsequent lower level electrical "transfer pulses" are used to facilitate movement of the negatively charged genome editing components. The zygote electroporation approach offers a high-throughput and efficient system for introducing CRISPR/Cas9 reagents for the production of genetically engineered rat models.

2 Materials

2.1 Superovulation and Zygote Production

1. Immature embryo donor rats and mature stud males: 4–5-week-old female Sprague Dawley rats (*see* **Note 1**) and sexually mature Sprague Dawley stud males (proven breeders preferred).
2. PG600 stock solution (*see* **Note 2**): Dilute the 2000 IU bottle of PG600 (Valley Vet 16,820) with 2 mL of diluent (DPBS w/o CaCl2, MgCl2 Gibco 14190-144) for a concentration of 100 IU/100 μL. Aliquot in 80 μL volumes to generate 25 tubes of PG600 stock solution and store at −80 °C.
3. Human chorionic gonadotropin (hCG) stock solution: Carefully add 2 mL of DPBS at neutral pH to dilute the 1 mg, 3100 IU bottle of hCG (Calbiochem 230,734). Remove reconstituted hCG and bring the total volume to 15.5 mL in a 50 mL conical tube. Keeping the hormone on ice, aliquot in 1 mL volumes resulting in 200 IU/mL hCG stock solutions and store at −80 °C.

2.2 Zygote Collection

1. Rat stock solution (10×): Dissolve the reagents in Table 1 and bring to 500 mL total volume with Lonza's water for cell culture application.
2. R1CC (high NaCl) stock solution: Dissolve the reagents in Table 2 and bring to 250 mL total volume with Lonza's water for cell culture application. After all components have

Table 1
Reagents for 10× rat stock solution

Company	Catalog	Reagent	FW	Concentration	500 mL
Sigma	S5886	NaCl	58.44	80 mM	23.376 g
Sigma	P5405	KCl	74.55	3.2 mM	1.1928 g
Sigma	G6152	D-Glucose	180.2	7.5 mM	6.7576 g
Sigma	P7794	Penicillin G K Salt	372.2	100 μg/mL	0.375 g
Sigma	S1277	Streptomycin Sulfate	1457	50 μg/mL	0.25 g
Sigma	L7900	Sodium Lactate (60% syrup)	186.8	13.53 mM	12.637 g
Sigma	C7902	$CaCl_2$-$2H_2O$	147.02	2 mM	1.4702 g
Sigma	M2393	$MgCl_2$-$6H_2O$	203.31	0.5 mM	0.5083 g

Table 2
Reagents for R1CC (high NaCl) stock solution

Company	Catalog	Reagent	FW	Concentration	250 mL
In-house	N/A	Rat stock solution	N/A	N/A	25 mL
Sigma	S5886	NaCl	58.44	30 mM	0.4383 g
Sigma	S5761	$NaHCO_3$	84.007	25 mM	0.5251 g
Sigma	P4562	Sodium Pyruvate	110.04	0.5 mM	0.0138 g
Invitrogen	11140-050	MEM NEAA 100×	N/A	N/A	2.5 mL
Invitrogen	11130-051	MEM EAA 50×	N/A	N/A	5 mL
Invitrogen	35050-061	GlutaMAX 1	N/A	0.1 mM	0.125 mL

been dissolved, transfer the contents to the appropriate volumetric flask and gently gas the solution with 5% CO_2, using a sterile Pasteur pipette, for 15–30 min. Check to ensure the pH of the solution is approximately 7.4 and the osmolality is between 280 and 300 mOsm. Filter-sterilize and store at 4 ° C for up to 4 weeks.

Add 4 mg/1 mL (0.04 g/10 mL) Fatty Acid Free BSA (Sigma A7638) to make the R1CC working solution immediately prior to use.

3. R1CH stock solution: Dissolve the reagents in Table 3 and bring to 500 mL total volume with Lonza's water for cell culture application. After all components check to ensure the pH of the solution is approximately 7.4 and the osmolality is between 290 and 320 mOsm. Filter-sterilize and store at 4 °C for up to 4 weeks.

Table 3
Reagents for R1CH stock solution

Company	Catalog	Reagent	FW	Concentration	500 mL
In-house	N/A	Rat stock solution	N/A	N/A	50 mL
Sigma	S5886	NaCl	58.44	30 mM	0.8766 g
Sigma	S5761	$NaHCO_3$	84.007	25 mM	0.21 g
Sigma	P4562	Sodium Pyruvate	110.04	0.5 mM	0.0276 g
Invitrogen	11140-050	MEM NEAA 100×	N/A	N/A	5 mL
Invitrogen	11130-051	MEM EAA 50×	N/A	N/A	10 mL
Sigma	H6147	HEPES	238.31	22 mM	2.6214 g
Invitrogen	35050-061	GlutaMAX 1	N/A	0.1 mM	0.25 mL

4. Add 4 mg/1 mL (0.04 g/10 mL) Fatty Acid Free BSA (Sigma A7638) to make the R1CH working solution immediately prior to use.
5. Hyaluronidase stock solution: Reconstitute Hyaluronidase (Sigma H4272-30 mg) by carefully adding 30 mL of R1CH working solution (containing 4 mg/mL BSA) and store 1 mL aliquots at −20 °C.

2.3 sgRNA/Cas9 RNP Formation

1. Single-guide RNA (sgRNA): We use commercially produced synthetic chemically modified sgRNAs. Chemical modifications are 2′-O-methyl analogs and 3′ phosphorothioate internucleotide linkages at the first three 5′ and 3′ terminal RNA residues. The sgRNAs are designed to target the desired region of the rat genome and can be ordered from a variety of commercial sources (*see* **Note 3**).
2. Cas9 protein: Enhanced-specificity Cas9 variants can be used to reduce potential off-target activity.
3. DNA repair template (when applicable for knock-in rat models): Single-stranded oligodeoxynucleotides (ssODNs) are designed to harbor the desired knock-in DNA sequence and ordered as a 200 bp Ultramer™ DNA oligonucleotides from IDT (*see* **Note 4**).

2.4 Zygote Electroporation

1. NepaGene NEPA21 Super Electroporator equipped with alligator clips and a glass slide electrode (platinum plate electrodes with 1.5 mm gap) (*see* **Note 5**).
2. Dissection microscope equipped with 37 °C heated stage.

2.5 Surgical Embryo Transfer

1. Pseudopregnant Sprague Dawley recipient female rats at appropriate stage for manipulated embryos (day 0.5 recipient for one-cell zygotes): Recipient females are mated with vasectomized males the day prior to embryo collections and zygote electroporation.
2. Ketamine/Xylazine working stock: Add 2 mL of XylaMed (xylazine hydrochloride at 100 mg/mL) to a 10 mL vial of Ketathesia (ketamine hydrochloride at 100 mg/mL) and mix thoroughly.
3. Banamine working stock: Add 0.4 mL of Banamine (flunixin meglumine at 50 mg/mL) to 9.6 mL of sterile saline and mix thoroughly.

3 Methods

3.1 Zygote Superovulation

1. On day 1 (see procedure timing in Table 4), thaw 80 μL aliquot of PG600 stock solution and add 720 μL of saline to a final volume of 800 μL. This will give a working concentration of 80 IU/0.8 mL or 20 IU/0.2 mL. Each immature female rat should be injected intraperitoneally (IP) with 0.2 mL diluted PG600 between 9:00 and 10:00 am (our light cycle is 12:12 where the lights come on at 6:30 am and turn off at 6:30 pm). Each aliquot should then be enough for three rats (due to losses in the syringe/needle).
2. On day 3, thaw a 1 mL aliquot of hCG and administer 0.2 cc/animal (40 IU) intraperitoneally between 11:00 am and 12:00 pm.

3.2 Zygote Collection

1. The next morning, check for copulation plugs and euthanize embryo donor female rats via CO_2 (*see* **Note 6**).
2. Place animal on benchtop and spray abdominal area with 70% ethanol. Grasp the abdominal skin with serrated forceps and make a lateral incision at the midline using scissors (Fig. 1a, b). Push the intestines away to reach the uterine horns (they are positioned laterally) to collect the oviducts (Fig. 1c–e). Wash twice and place oviducts in R1CH working solution.

Table 4
Superovulation and embryo transfer schedule

Day	Procedure	Time
1	Superovulation (PG600 administration)	9:00 am–10:00 am
2		
3	Inducing ovulation (hCG administration) and mating	11:00 am–12:00 pm
4	Zygote collection, electroporation, and embryo transfers	9:00 am–5:00 pm

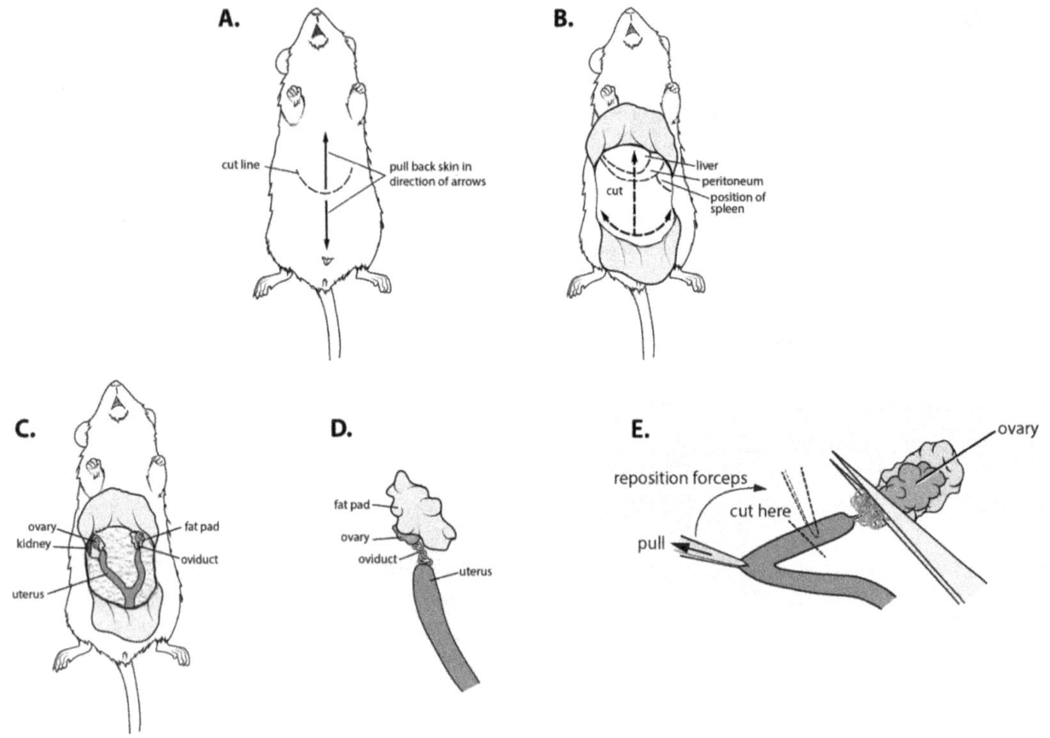

Fig. 1 Reproductive organ dissection from a female embryo donor rat. Cut the skin with scissors as indicated by dashed line and then pull the skin apart in the direction of solid arrows (**a**). Cut the body wall (peritoneum) as indicated by dashed line in (**b**). Layout of internal reproductive organs of a female rat (**c**). An enlarged view of one side of the internal reproductive tract of a female rat (**d**). Collect oviduct by cutting away uterus and ovary (**e**)

3. Move oviducts to a prewarmed (37 °C) petri dish of R1CH working solution containing 1 mg/mL Hyaluronidase. Hold oviduct coil with a serrated forceps and release zygotes and cumulus mass complexes by tearing the swollen ampulla with 27-gauge needle (Fig. 2).
4. Let the cumulus zygote complexes sit for about 5 min (or watch them until the cumulus cells fall apart). Using a pipette, pick up the embryos and gently wash them twice in R1CH working solution. Transfer embryos to 35 mm culture plates containing 30 μL drops of R1CC working solution covered in mineral oil and culture in 37 °C, 5% CO_2 incubator.

3.3 Zygote Electroporation of CRISPR Reagents

1. Create a mixture containing a final concentration of 100 ng/μL sgRNA, 100 ng/μL enhanced-specificity Cas9 protein, and 100 ng/μL ssODN immediately prior to electroporation: Form sgRNA/Cas9 RNP complexes by gently mixing the sgRNA and Cas9 protein together and incubating at room temperature for 10 min (*see* **Note 7**). After RNP formation, add the ssODN repair template to the mixture. Reaction mixture should be made in Opti-Mem™ reduced serum media.

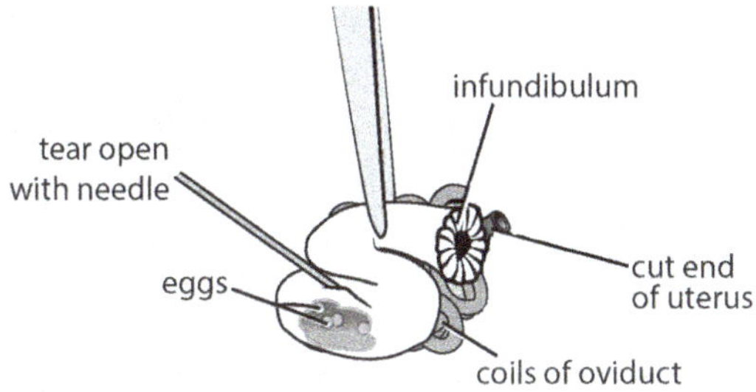

Fig. 2 Hold oviduct coil with a serrated forceps and release zygotes and cumulus mass by tearing the swollen ampulla with a 27-gauge needle as indicated

Fig. 3 Place zygotes in 5–7 μL of the CRISPR electroporation mixture in between the electrodes taking care not to introduce bubbles or allowing zygotes to touch electrodes

2. Using a NepaGene21 electroporator along with a 1.5 mm gap glass slide electrode, gently pipette 5–7 μL of the CRISPR electroporation mixture created above between the two electrodes (taking care not to introduce bubbles). Line up zygotes in the mixture without allowing them to touch the side of the electrodes (Fig. 3). Check to ensure impedance is between 0.100 and 0.300 Ω. The reaction volume will alter the impedance (add volume to decrease impedance and subtract volume to increase impedance).
3. Electroporate under the following conditions: Poring pulse: 40 V, 3.5 ms length, 50 ms interval, 10% decay rate, positive polarity (×4 pulses). Transfer pulse: 5 V, 50 ms length, 50 ms interval, 40% decay rate, alternating polarity (×5 pulses) (*see* **Note 8**). Electroporated zygotes can be cultured overnight or surgically transferred to pseudopregnant surrogate females the same day as the electroporation.

3.4 Surgical Embryo Transfer

1. Mature Sprague Dawley female rats are mated with vasectomized males the day prior to embryo collections and zygote electroporation. The following morning, these recipient females are checked for a copulation plug. Any females with a visible copulation plug can be used as surrogate female to receive manipulated embryos.
2. Recipient females are anesthetized by intramuscular (IM) administration of 5.5 mg ketamine and 1.0 mg xylazine per 100 g body weight. Lubricating ointment should be placed on both eyes to prevent corneal ulceration.

3. To manage surgical pain and inflammation, the recipients are then given Banamine (flunixin meglumine) subcutaneously: 0.2 mg for rats under 250 g and 0.3 mg for rats over 251 g.
4. Shave the back of the rat using clippers and wipe the shaved area with 70% ethanol to clean and remove loose hairs. Scrub the surgical area using betadine-soaked cotton swabs (3 times, alternating between 70% ethanol and betadine). Gently pick up the rat and place on the lid of a 150 mm petri dish.
5. Make a 1-cm-long transverse incision in the skin using fine dissection scissors about 1 cm to the left of the spinal cord, at the level of the last rib. Slide the skin to one side and undermine it using scissors until the incision is over the ovary or fat pad. They are both visible through the body wall.
6. Using serrated forceps, pick up the body wall (dorsolateral) over the fat pad and make a small cut to create a hole (paralumbar fossa). Gently grab the fat pad using serrated/curved blunt forceps to exteriorize the oviduct, ovary, and part of uterine horn. Grip the fat pad with a bulldog clamp being careful to avoid the reproductive organs. Lay the fat pad on top of gauze to avoid retraction of the reproductive tract back into the abdominal cavity; the bulldog clamp will act as an additional weight.
7. Pick up the petri dish lid with the rat, place on a dissecting microscope stage, and position a fiber-optic lamp to illuminate the tissues. Apply a couple of drops of 10% epinephrine to the ovarian bursa to minimize bleeding. Using sharp forceps, tear the ovarian bursa to expose the infundibulum. Be careful to avoid large blood vessels.
8. Retrieve embryos, washing them in R1CH working solution. Load the embryos into a 1.5 inch pulled glass pipette or 125 μL Stripper Tip with as little media as possible. We typically transfer 20–30 embryos per recipient female (10–15 embryos per side of uterine horn).
9. Under the microscope, insert the pipette with embryos into the opening of the infundibulum. Use fine tweezers to gently apply pressure on either side of the pipette and expel the embryos into the oviduct. As soon as you see the air bubbles enter the infundibulum (or see the oviduct swell up), carefully pull the pipette out and crush the infundibulum (for 5–10 s) with the fine tweezers. This should prevent back-flow of the fluid into the surgical field or abdominal cavity. Apply sterile saline on the tissues and gently replace them into the abdominal cavity.
10. Repeat the same procedure on the opposite side of the animal. Close the skin with an autoclip and return the animal to a warm cage located in a silent place. Monitor the animals during the post-operative period to ensure a safe recovery from anesthesia.

4 Notes

1. This protocol is optimized for Sprague Dawley rats but may be used on other rat strains. However, it is important to note that superovulation efficiency can be strain dependent.
2. Immature rats can be superovulated by either PG600 or PMSG. We primarily use PG600 and use PMSG as a substitute when PG600 is unavailable.
3. We use chemically modified synthetic sgRNAs from Synthego. However, in vitro transcribed or non-modified sgRNAs from other commercial sources can also be used.
4. Knock-ins that require a DNA repair template larger than 200 bp may not be efficiently introduced through an electroporation approach and may require techniques such as pronuclear microinjection.
5. Other square wave electroporators may be used but parameters may need adjusted.
6. We used a CO_2 flow rate of 30% chamber volume per minute. Length of time exposed to CO_2 could affect embryo quality.
7. If using two or more sgRNAs, mix them prior to adding Cas9 protein for RNP complex formation.
8. Timing of electroporation process should be as quick as possible (less than 1–2 min) due to quick evaporation of the reagent mixture altering impedance.

Acknowledgments

This work was done in collaboration with the Rat Resource and Research Center (RRRC) funded by NIH grant OD011062 and University of Missouri Animal Modeling Core (AMC).

References

1. Li D et al (2013) Heritable gene targeting in the mouse and rat using a CRISPR-Cas system. Nat Biotechnol 31(8):681–683
2. Kaneko T et al (2014) Simple knockout by electroporation of engineered endonucleases into intact rat embryos. Sci Rep 4:6382
3. Chen S et al (2016) Highly efficient mouse genome editing by CRISPR ribonucleoprotein electroporation of zygotes. J Biol Chem 291(28):14457–14467
4. Hashimoto M, Yamashita Y, Takemoto T (2016) Electroporation of Cas9 protein/sgRNA into early pronuclear zygotes generates non-mosaic mutants in the mouse. Dev Biol 418(1):1–9
5. Wang W et al (2016) Delivery of Cas9 protein into mouse zygotes through a series of electroporation dramatically increases the efficiency of model creation. J Genet Genomics 43(5):319–327
6. Tröder SE et al (2018) An optimized electroporation approach for efficient CRISPR/Cas9 genome editing in murine zygotes. PLoS One 13(5):e0196891
7. Teixeira M et al (2018) Electroporation of mice zygotes with dual guide RNA/Cas9 complexes

for simple and efficient cloning-free genome editing. Sci Rep 8(1):474

8. Remy S et al (2017) Generation of gene-edited rats by delivery of CRISPR/Cas9 protein and donor DNA into intact zygotes using electroporation. Sci Rep 7(1):16554
9. Tanihara F et al (2020) Efficient generation of GGTA1-deficient pigs by electroporation of the CRISPR/Cas9 system into in vitro-fertilized zygotes. BMC Biotechnol 20(1):40
10. Tanihara F et al (2021) Generation of CD163-edited pig via electroporation of the CRISPR/Cas9 system into porcine in vitro-fertilized zygotes. Anim Biotechnol 32(2):147–154
11. Camargo LSA et al (2020) Efficient one-step knockout by electroporation of ribonucleoproteins into zona-intact bovine embryos. Front Genet 11:570069
12. Lin JC, Van Eenennaam AL (2021) Electroporation-mediated genome editing of livestock zygotes. Front Genet 12:648482
13. Betters E, Charney RM, García-Castro MI (2018) Electroporation and in vitro culture of early rabbit embryos. Data Brief 21:316–320

Chapter 12

CRISMERE Chromosome Engineering in Mouse and Rat

Laurence Schaeffer, Loic Lindner, Guillaume Pavlovic, Yann Hérault, and Marie-Christine Birling

Abstract

CRISPR/Cas9 technology is a versatile tool for engineering biology that has dramatically transformed our ability to manipulate genomes. In this protocol, we use its capacity to generate two double-strand breaks simultaneously, at precise positions in the genome, to generate mouse or rat lines with deletion, inversion, and duplication of a specific genomic segment. The technic is called CRISMERE for CRISpr-MEdiated REarrangement. This protocol describes the different steps to generate and validate the different chromosomal rearrangements that can be obtained with the technology. These new genetic configurations can be useful to model rare diseases with copy number variation, understand the genomic organization, or provide genetic tools (like balancer chromosome) to keep lethal mutations.

Key words CRISPR/Cas9, PCR, Sanger sequencing, Droplet digital PCR, Structural variants, Deletion, Duplication, Inversion

1 Introduction

Before CRISPR/Cas9 genome editing, modelling segmental duplication or deletion was achieved, only in the mouse, by using the Cre-lox technology [1–3] in combination with ES cells or alternative strategies [4–6]. The overall strategy necessitated at least 3 years to generate a model but demonstrated the power of a mapping panel of duplications to better understand the contribution of structural variants (SV) [1, 3]. CRISPR/Cas9 is now used to manipulate large genomic regions in vitro and in vivo (for review [7]) . For example, gene clusters of up to 800 kb can be deleted, duplicated, and inversed in HEK293T cells [8]. In mice, the Wu lab [9] demonstrated that DNA elements of up to 29 kb can be manipulated directly using the CRISPR/Cas9 technology. Zheng and Li's lab showed that deletion of up to 95 kb can be obtained [10, 11]. Similarly, CRISPR/Cas9 was used for the generation of up to 1 Mb structural deletion and inversion around the Tyrosinase

Thomas L. Saunders (ed.), *Transgenesis: Methods and Protocols*,
Methods in Molecular Biology, vol. 2631, https://doi.org/10.1007/978-1-0716-2990-1_12,

locus [8] in mouse zygotes. Duplications appeared less frequently and did not pass the germline. Deletion, duplication, and inversion were obtained for the contactin-6 gene [12] but no proof of the complete duplication or inversion was shown as only junction PCRs were performed.

We showed that, using CRISMERE, mouse or rat segmental duplications, and other large-scale rearrangements, can be easily generated using the CRISPR/Cas9 genome editing in zygotes. This approach is effective from single genes up to large (hundreds of kb) or very large (in Mb) genomic region rearrangements [13]. We also showed that a proper validation of these lines needs more than single-junction PCRs and has to be coupled with the counting of copy numbers of SV alleles as larger deletion or insertions can occur at the junction points [13, 14]. Here, we describe a detailed protocol to generate and characterize SV alleles in rodents. These models and their analyses will allow us to better understand the complexity of human diseases associated with SV.

2 Materials

Diligently follow all waste disposal regulations when disposing of waste materials.

2.1 RNA Synthesis

1. MEGAshortscript Kit (Life Technologies; AM1354).
2. MEGAclear Kit (Life Technologies) or RNA Clean & Concentrator-25 (Zymo Research; R1018).
3. TE buffer (10 mM Tris-HCl, 0.1 mM EDTA, InVitrogen).
4. PCR machine.

2.2 RNA

1. Trans-activating crRNA and tracrRNA can be ordered from different suppliers (IDT, Synthego, Dharmacon, or other vendor).
2. Ultrafree-MC Centrifugal filter – Millipore UFC30LG25.

2.3 Cas9 Protein

We use the recombinant *Streptococcus pyogenes* Cas9 protein from IDT (Alt-R SpCas9 Nuclease V3; ref 1081059). Other versions of the *Streptococcus pyogenes* Cas9 (SpCas9) protein can be used and ordered from other companies.

2.4 Agarose Gels

2.5% agarose gels in TAE are used for the migration of the PCR products from RNA synthesis.

2.5 DNA Ladder

GeneRular DNA Ladder Mix (100bps to 10 kb) SM0333.

2.6 Sample Lysis and PCR

1. DNA Extract All Reagents including Lysis Solution and DNA stabilizing solution (Applied-4403319).
2. Phusion Taq HS (Fermentas-F549L).
3. PCR machine.

2.7 Droplet Digital PCR

1. QX200™ ddPCR Supermix for Probes (Bio-Rad).
2. 20× primer/probe mix (IDT).
3. Reference 20× primer/probe mix (IDT).
4. DG8 Cartridges for QX200™ (Bio-Rad).
5. DG8 Gaskets for QX200 Droplet Generator (Bio-Rad).
6. QX200™ Droplet Generation Oil for probe (Bio-Rad).
7. QX200™ droplet generator (Bio-Rad).
8. QX200™ droplet reader (Bio-Rad).
9. ddPCR Droplet Reader Oil (Bio-Rad).
10. QuantaSoft analysis software (Bio-Rad).

3 Methods

3.1 Strategies for Generating the SV

1. The first step is to clearly define the genomic region to be duplicated, deleted, or inverted and especially its two borders. Guide RNAs (gRNAs) are selected accordingly in the two borders (*see* Fig. 1). To increase the likelihood of generating double-strand breaks (DSBs) simultaneously at both borders, we design two close gRNAs per extremity (*see* Fig. 1). We assume that it is very unlikely that the two gRNAs per extremity are inefficient and so increase our chance to obtain at least one DSB at each extremity of the region of interest. Ideally, the two gRNAs of the pair are selected in close proximity, as small indel events can also occur and complicate the interpretation of the new junction sequences.
2. For segmental duplication models, it is important that the target region includes all the regulatory elements of the genes located at each border (or of the target gene when only one gene is to be duplicated).

3.2 CRISPR/Cas9 RNA Component

The definition of the term guide RNA may vary according to different authors. Here, for sake of clarity, we defined the gRNA as the RNA component of the CRISPR/Cas9 system irrespective of

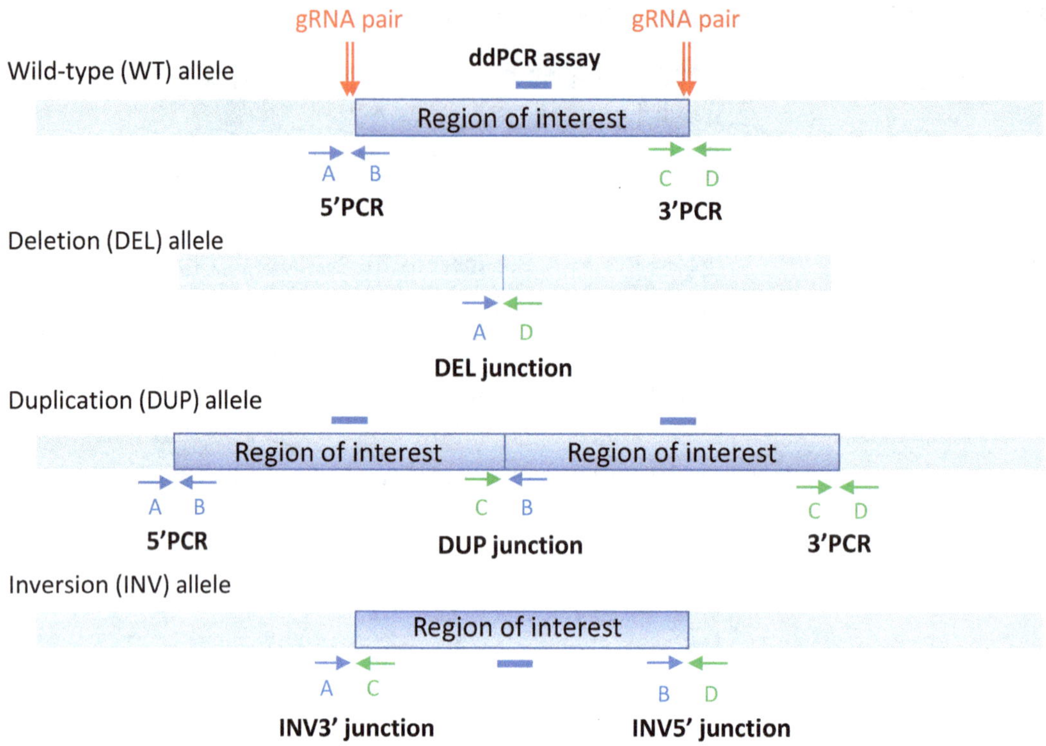

Fig. 1 Schematic drawing of the CRISPR approach and the resulting alleles that can be obtained. Two pairs of gRNAs (red arrows) are selected at both extremities of the genomic sequence to be manipulated. Each gRNA of a pair is selected to be near the other one or can even be overlapping. A ddPCR probe is designed in the second third of the genomic fragment of interest (large blue line). The position of the primers is indicated with letters and arrows. They are positioned at 200 to 300 bp from the expected DSB. The size of the amplified PCR fragment should be between 400 and 600 bp but can be different as it depends on the DSB cellular DNA repair pathway choice

its format: single guide RNA (sgRNA—the use of a single chimeric RNA) or crRNA:tracrRNA duplex (the use of the dual RNAs). Guide sequence is defined as the 20 nucleotides that are complementary to the genome target sequence when using SpCas9.

1. Guide RNA sequence selection. Guide sequences are selected using the CRISPOR software (http://crispor.tefor.net/) [16]. Each gRNA name refers to the specificity score given by the software. Other websites can be used and will give approximately the same guide scores. From our experience, almost all of the gRNAs selected on the CRISPOR website with a specificity score higher than 50 showed a good efficiency (data not shown). We chose the pair of gRNAs according to the desired position of the DSB and make sure their expected DSBs are

Table 1
Primers to produce sgRNA templates with the T7 promoter

gRxxfwT7	GAAATTAATACGACTCACTATAGnnnnnnnnnnnnnnnnnnnGTTTTAGAGCTAG AAATAGCAAG
gRrevcommon	AAAAGCACCGACTCGGTGCCACTTTTTCAAGTTGATAACGGACTAGCCTTA TTTTAACTTGCTATTTCTAGCTCTAAAAC

close. We do not necessarily select sgRNA with the highest score but make sure that the DNA sequence surrounding the expected DSBs is not surrounded with repeats or repeated genomic region that would render PCR amplification difficult. We are using two gRNA formats for the CRISPR/Cas9 RNA component: single guide RNA (sgRNA) or crRNA:tracrRNA duplex. Both formats give efficient SV mutations and can be used indifferently (data not shown, *see* **Note 1**).

2. sgRNA synthesis.
   (a) sgRNA templates with T7 promoter are obtained by PCR amplification on any plasmid containing the gRNA scaffold (example pX330 Plasmid #342230 from Addgene) [17] with primers detailed in Table 1.
   (b) The cycling conditions are the following: one cycle of 30 seconds at 98 °C, 30 cycles: including 8 seconds at 98 °C, 10 seconds at 60 °C, and 30 seconds at 72 °C and finally one cycle of 5 minutes at 72 °C.
   (c) The MEGAshortscript Kit (Life Technologies; AM1354) is used to transcribe sgRNAs from the PCR product template.
   (d) sgRNAs are then purified with the MEGAclear Kit (Life Technologies) or alternatively with the RNA Clean & Concentrator-25 Kit (Zymo Research; R1018) and eluted in TE buffer (10 mM Tris-HCl, 0.1 mM EDTA, Invitrogen).
3. crRNA and tracrRNA synthesis. Both crRNA and tracrRNA are directly ordered from IDT (https://eu.idtdna.com/pages/products/crispr-genome-editing/alt-r-crispr-cas9-system) (*see* **Note 2**).
4. Cas9 mRNA
   (a) The Cas9 vector (T7-hSpCas9, the humanized SpCas9 having been amplified on pX330 Plasmid #342230 from Addgene and cloned in pUC57) is first linearized with AccI (just after the stop codon of SpCas9 cDNA) for use as a template for in vitro transcription with T7 polymerase.

(b) Cas9 mRNA is transcribed using mMESSAGEmMACHINE T7 Ultra Kit according to the manufacturer's manual (Life Technologies AM1345).

(c) Cas9 mRNA is purified with the MEGAclear Kit (Life Technologies) or RNA Clean & Concentrator-25 (Zymo Research; R1018) and eluted in TE buffer (10 mM Tris-HCl, 0.1 mM EDTA, Invitrogen) for microinjections.

5. Cas9 protein. The Cas9 protein (Alt-R SpCas9 Nuclease V3; ref 1081059) is ordered from IDT.

3.3 Preparation of the Mix for Microinjection or Electroporation

We describe here CRISPR/Cas9 mixes and conditions that were successful in our hands.

1. sgRNAs and Cas9 mRNA for microinjection. For most experiments, 50 ng/μL Cas9 mRNA were mixed with 25 ng/μL of each sgRNAs in injection buffer (*see* **Note 3**).
2. Cas9 RNP with dual crRNA:tracrRNA duplex for electroporation.

   (a) For one electroporation slide, mix 2 μL of 100 μM crRNA and 2 μL of 100 μM tracrRNA to form the dual RNA (each crRNA to be prepared separately).

   (b) Incubate for 5 minutes at 95 °C, and then reduce temperature to room temperature for 15 minutes.

   (c) Mix in a tube: 5.18 μL Opti-MEM (Gibco™ Opti-MEM™ 31985062), 0.42 μL of each dual gRNAs, and add 0.14 μL of Cas9 protein (Alt-R SpCas9 Nuclease V3, IDT). The final concentration of crRNA/tracrRNA duplex should be 6 μM and 1.2 μM for Alt-R SpCas9 Nuclease V3 in a final volume of 7 μL.

   (d) The preparation is then filtered on an Ultrafree-MC-filter to avoid any precipitate.

   (e) Load the 7 μL mix on the filter, centrifuge 3 minutes at 12000 g. Recover the filtered solution and incubate for 10 minutes at room temperature.

3.4 Preparation of Mouse One-Cell Embryos

All mice are housed at 21 °C on a 14/12 hours light-dark cycle (5: 00 am–07:00 pm) in an SPF facility. Sexually immature C57BL/6NCrl females (4–5 weeks old) are superovulated by intraperitoneal injection of 5 IU eCG followed by 2.5 IU hCG at an interval of 48 hours and mated overnight with C57BL/6NCrl male mice that are >10 weeks old. Zygotes are collected after 20 hours of hCG injection by oviductal flashing, and pronuclei-formed zygotes are put into the M2 medium (Sigma M-7167).

3.5 Preparation of Rat One-Cell Embryos

All rats are housed at 21 °C on a 14/12 hours light-dark cycle (5:00 am–07:00 pm) in an SPF facility. Sexually immature Sprague Dawley females (4–5 weeks old) are superovulated by intraperitoneal injection of 20 IU eCG followed by 30 IU hCG at an interval of 48 hours and mated overnight with Sprague Dawley male rats that are >10 weeks old. Zygotes are collected after 20 hours of hCG injection by oviductal flashing, and pronuclei-formed zygotes are put into the M2 medium.

3.6 Delivery of CRISPR/Cas9 in One-Cell Embryos

Both microinjections or electroporations in one-cell embryos were used to generate SV models. Both approaches give a high rate of mutations but electroporation does not require microinjection skills, and is faster as up to 30 embryos can be electroporated simultaneously. Also a better survival of the eggs is observed (as described in [18–20]).

3.6.1 Microinjection

(a) Microinjections are performed using a microinjector (Eppendorf Femtojet 4i) equipped microscope.

(b) RNA solution is injected into the cytoplasm and the pronucleus of each zygote using continuous pneumatic pressure.

(c) After injection, embryos are cultured in vitro in the KSOM medium (MERCK MR-151-D) at 37 °C in a 5% CO2 incubator.

(d) The surviving injected embryos are implanted on the same day into the oviducts of pseudo-pregnant B6CBAF1 mice for mouse models and Sprague Dawley rats for rat models.

3.6.2 Electroporation

Electroporation is performed using a NEPA21 type II machine (NEPAGENE) with an electroporation slide (CUY501P1-1.5). Only zygotes with two visible pronuclei are selected for electroporation. The same protocol is used for mouse and rat zygotes.

(a) Wash zygotes in two drops (100 μL) of M2 medium (to remove oil) and in two drops (50 μL) of OptiMEM media (to remove FBS) before electroporation.

(b) Pipette 5 μL of RNP complex into the chamber of the electrode.

(c) Press the Ω button of the NEPA electroporator to measure impedance.

(d) The impedance should be in the range of 0.20–0.24 kΩ. If needed, adjust the impedance (if the impedance is below 0.20 kΩ, remove some of the cell solution from the chamber to increase the impedance; if the impedance is above 0.24 kΩ, add Opti-MEM to the chamber to decrease the impedance).

(e) Prepare and warm culture media for use after electroporation.

(f) Wash embryos with Opti-MEM (no need to weaken the zona pellucida with acidic Tyrode's solution).

(g) Place embryos in parallel of the electrode in the chamber. We use the CUY501P1-1.5 electrode to perform electroporation of 20–30 embryos simultaneously.

(h) Press the Ω button of the NEPA21 electroporator and record the impedance. The impedance should be in the range of 0.18–0.22 kΩ. If needed, adjust the impedance as described above.

(i) Set up these parameters for zygote electroporation on NEPA21 electroporator:

Poring pulses: 40 V, 3.5 ms, 50 ms interval, 4 pulses, decay rate 10%, polarity +.

Transfer pulses: 7 V, 50 ms, 50 ms interval, 5 pulses, decay rate 40%, polarity +/−.

(j) Press the Start button to execute the electroporation program, and record the values of current and joules displayed in the Measurements frame.

(k) Remove the embryos from the chamber.

(l) After the electroporation, the zygotes are cultured for 2 hours, then transferred into the oviduct of surrogate mothers (mouse or rat) the same day.

3.7 Characterization of Founders

3.7.1 DNA Extraction

Animals born after reimplantation (potential founders) are biopsied (tail or ear clipping) following the protocol described in the tissue sampling methods and procedure section in [21] at approximately 2 weeks of age. The lysates are then prepared using DNA Extract All Reagents from Applied Biosystems. These lysates are, without any additional purification, used for both junction PCRs and ddPCR analyses. We tested different lysis buffers and found that only the DNA Extract All Reagents from Applied Biosystems allowed us to obtain interpretable ddPCR results without performing further DNA purification.

3.7.2 Junction PCRs for the Identification of the New Junctions

The different primer sets are used to identify inversions, duplications, and deletions junctions (A–D for a deletion; C–B for a duplication; A–C and B–D for an inversion). The primer positions are shown in Fig. 1. Primers are designed at 200–300 bps from the expected CRISPR/Cas9-mediated DSB. The primers are thus far enough from the DSB to allow the identification of each junction in the majority of founders.

Junction PCRs are performed with Phusion Taq HS (Fermentas) in a final volume of 20 μL. The PCR conditions are:

(a) Pre-denaturing at 94 °C for 30 seconds,

(b) Followed by 30 cycles.

(c) 94 °C denaturation for 8 seconds.
(d) 60 °C annealing for 10 seconds.
(e) 72 °C extension for 30 seconds.
(f) Followed by a final extension at 72 °C for 5 minutes.

3.7.3 Junction PCR Products Are Sent for Sanger Sequencing

If PCR products are obtained for new junctions, their sequence is carefully analyzed after Sanger sequencing. The precise junction is defined as the point of junction of two sequences that are not located one close to the other in the wild-type animal but corresponds to the fusion of two DNA fragments expected if a rearrangement occurs. A theoretical map can be made for the DEL, DUP, and INV (INV5′ and INV3′) alleles with a theoretical junction at the expected DSB of one guide of the pair. The chromatogram of a newly sequenced junction can then be aligned to the theoretical map and a map corresponding to the real new junction is saved (*see* **Note 4**).

3.7.4 Droplet Digital PCR for Copy Number Counting

(a) To confirm deletions or duplications events, ddPCR quantification is performed on DNA Extract All Reagents founder lysates as described in [15]. Details on how to perform droplet digital PCR (ddPCR) experiments are described in [13, 15].
(b) Regarding primers and probes design, the assay is preferentially chosen in the middle third of the genomic region to delete or duplicate (*see* Fig. 1).
(c) Because most founders are mosaics [13], ddPCR gives only indicative information allowing the selection of the most relevant founders for the establishment of the SV line (*see* [13]; Figs. 2a, 3, and 4). Indeed, the ddPCR assay quantifies all possible alleles (WT, DUP, and DEL) excluding inversions (INV) and is therefore not specific to one type of rearrangement (*see* Fig. 1). Another limitation of ddPCR is observed in a founder in which one allele is deleted and the other allele is duplicated. In this case, the ddPCR copy counting will be, like a WT animal, of 2 (*see* Fig. 3a).

3.8 Interpretations of the F0 Results

Among the different rearrangements that can be obtained via CRISMERE, the deletion of the entire genomic fragment of interest is the most frequent outcome (>60% of SV observed), but duplications and inversions are also present (Table 1 in [13] and *see* Fig. 6 for an illustration). The following paragraphs give a good overview of the different cases encountered. However, they do not describe all of them, as the range of possible rearrangements appears to be very large.

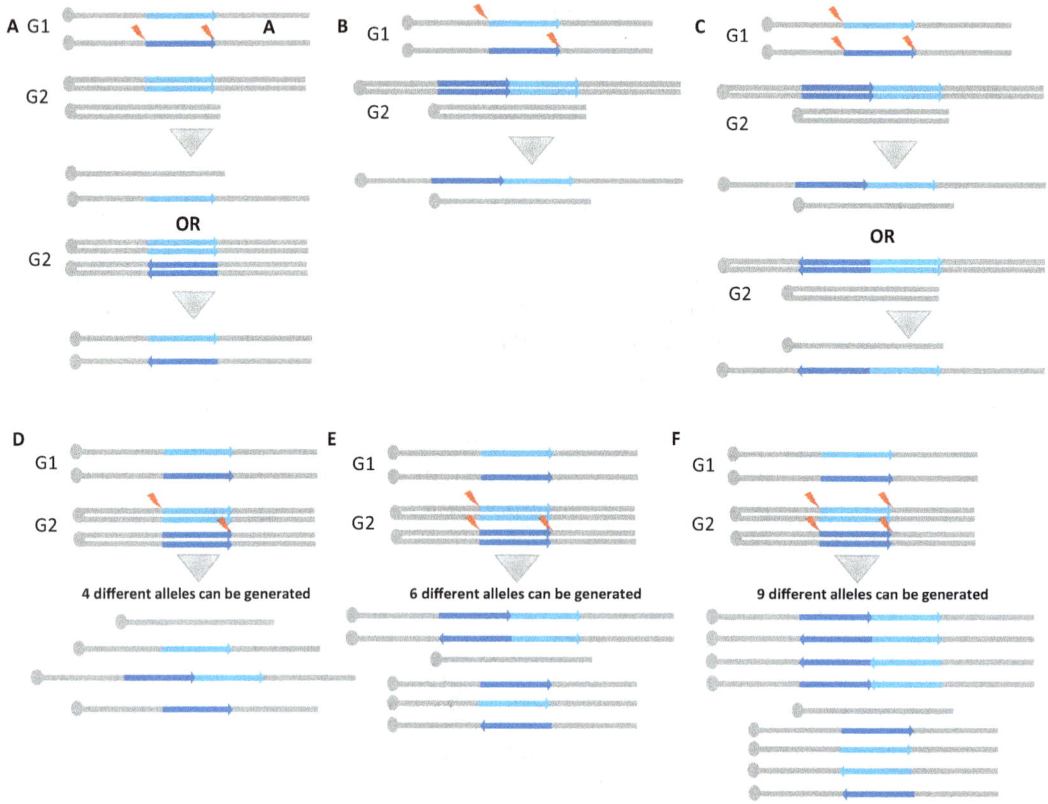

Fig. 2 CRISpr-MEdiated Rearrangement (CRISMERE) mechanisms illustrating the variety of possible outcomes depending on the time and position of the occurring DSBs. (**a**–**c**) Standard chromosomic recombination when Cas9 edits the genome in G1 phase. After mitosis, two alleles distinct from the initial WT allele will be obtained. (**a**) Intra-chromosomal recombination between two DSBs on a single chromosome. (**b**) Trans-allelic recombination between two DSBs each on one of the two chromosomes. (**c**) Trans-allelic recombination between three (or four DSBs) on the two chromosomes ending with head to head, tail to tail duplication. (**d**–**f**) Standard chromosomic recombination when Cas9 edits the genome in G2. (**d**) Schematic of the event that should take place in the eggs were Cas9 edits the genome in a *cis* configuration in G2 leading to monosomic and trisomic daughter cells after mitosis. Trans-allelic recombination between two DSBs on the two chromatids in G2. (**e**) Trans-allelic recombination between three DSBs when two breaks are located on only one chromatid and the third DSB is located on another chromatid. (**f**) Trans-allelic recombination between four DSBs when two breaks are located on one chromatid and the two others are located on another chromatid. As many as nine possible alleles can be obtained

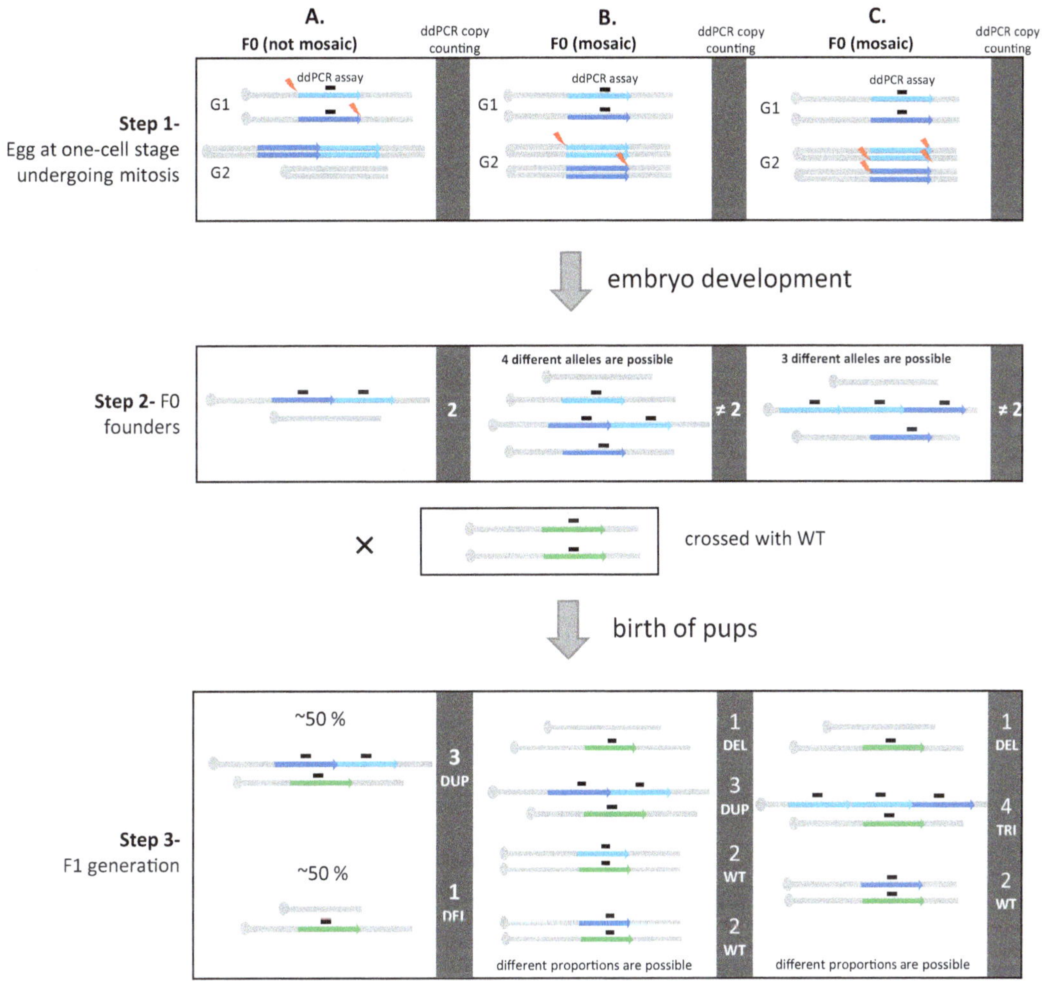

Fig. 3 Scheme for three possible CRISMERE events and their outcome (alleles and copy numbers). The region of interest is in blue (light and dark blue arrows to illustrate both alleles) in the egg and the F0 founder. The WT allele transmitted after breeding founders with WT animals is represented in green. The ddPCR assay is represented by a black line. Step 1 is representing the fertilized eggs undergoing mitosis, the position of the DSB generated by the gRNA pair is illustrated by a red lightning. Case A symbolizes the expected rearrangements if the DSB takes place during the G1 interphase. Cases B and C illustrate the expected rearrangements if the DSB takes place during the G2 interphase with, respectively, two or four generated DSBs. Step 2 illustrates the alleles that should be observed in an F0 founder animal. The columns in gray indicate the copy number that should be detected by the ddPCR assay. Step 3 illustrates all the possible combinations that can be obtained in the first generation after crossing with WT animals (always one WT allele and one allele version from the F0 founder). This figure only illustrates the rearrangements that are possible if the DSB repair takes place at the one-cell stage

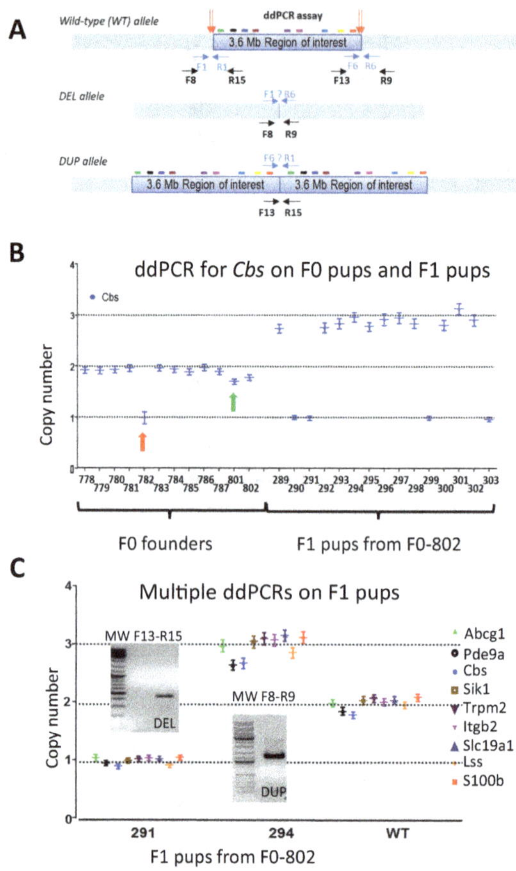

Fig. 4 Illustration of a "desperate" case, a case where the detection of SV rearrangements was particularly difficult. (**a**) Schematic drawing of both WT, DEL, and DUP alleles with the position of the initially designed primers that did not allow to detect a SV junction (in light blue) and new primers located further to the expected DSB sites (in black) selected after confirmation of the SV alleles by ddPCR assays. ddPCR assays performed over the 3.6 Mb region are in different colors. (**b**) Droplet digital PCR results. Left panel of the graph shows F0 results with one probe: a single copy of *Cbs* was detected in F0–782 (red arrow) whereas 1.8 copies of *Cbs* were detected in F0–801 (green arrow). On the right panel are F1 results for offspring of F0–802: one or three copies of *Cbs* were clearly detected showing that both deletion and duplication went germline. (**c**) ddPCR results with nine ddPCR probes ranging all over the region on two F1 pups (DEL: #291 and DUP: #294) that allowed confirming respectively the deletion and the duplication of the whole genomic region; a WT control (showing two ddPCR copies for each probe) is also shown. In integrated windows are the picture of the junction PCRs that were found after testing new primers (F13–R15; DEL junction and F8-R9; DUP junction) further from the hypothetic DEL and DUP junctions

1. The ideal case: the detection of the SV alleles is easy as junction PCRs and ddPCR mutually confirm each other. Four different scenarios can exist for a founder which can be classified as ideal. An example of an ideal case is shown in Fig. 6. In this case, the chance to obtain germline transmission of the SV allele detected in the F0 is high. The F0 founders showing these characteristics should preferably be crossed to generate the F1 segregated line(s). If the resulting F1 line is viable, it will very generally lead to generating the SV model of interest.
   (a) A DUP junction is confirmed by Sanger sequencing and >2 ddPCR copies counting are detected.
   (b) A DEL junction is confirmed by Sanger sequencing and <2 ddPCR copies counting are detected.
   (c) Both an INV5′ and an INV3′ junctions are confirmed by Sanger sequencing and ~2 ddPCR copies counting are detected.
   (d) Both DEL and DUP junctions are detected by junction PCR and ~2 ddPCR copies counting are detected (*see* Fig. 3a).
2. The "maybe" cases: SV alleles obtained are not certain in F0 founders. In some cases, PCR and ddPCR results do not provide easily interpretable results.
   (a) A DEL junction (but not a DUP) is observed by junction PCR and sequencing but ~2 ddPCR copies counting are detected.
   (b) A DUP junction (but not a DEL) is observed by junction PCR and sequencing but ~2 ddPCR copies counting are detected.
   (c) Only one INV junction is detected (5′ or 3′) and ~2 ddPCRDigital droplet PCR (ddPCR) copies counting are detected. If only one INV junction (5′ or 3′) is confirmed by sequencing, this must mean that the other junction also exists but the primersPrimers designed in a first instance do not allow to amplify the PCRPolymerase chain reaction (PCR) fragment of the new junction (deletionDeletion (DEL) of the sequence at the proximity of the new junction).

   These results can mean that a duplication or a deletion may have occurred on the other allele and balances the ddPCR copy counting but its junction is not detected with standard PCR. The forward or the reverse or both primers may be directed against a sequence that is deleted when the genomic rearrangement takes place. Alternatively, an unexpected insertion can prevent the PCR from working.

4. The “desperate” case: no detection of SV junctions in the F0. In some cases, the junction PCRs do not detect any rearrangement:
   (a) A clear modification of the ddPCR copy counting (different to 2) is detected but no junction PCR product is obtained. This likely means that a DUP and/or DEL occurred but their junctions remain undetected with standard PCR primers. The forward, the reverse, or both primers may be located inside a rearranged sequence: an indel has occurred near the DSB(s) when the genomic rearrangement took place. New PCR designs, moving the primers further from the expected junction, are thus performed to try to detect the SV junction. However, the determination of the real junctions remains not always possible: we have encountered cases for which we had a ddPCR confirmed DUP (many different ddPCR probes all confirming a ddPCR copy counting of 3) but we were unable to define a DUP junction (*see* an example in [13]).
   (b) On the opposite, it is also possible to observe founders that show a clear diminution of the ddPCR copy number but do not have a positive junction PCR. This means that the deletion is larger than expected (at least one PCR primer is inside the deletion) and/or that the DNA repair mechanism induces imperfect repairs. These events were first published by Allan Bradley’s team [14] and are regularly observed in our lab.
   (c) No PCR product for a new junction is obtained with the initially defined primers but the ddPCR copies counting is slightly different from 2 (*see* Fig. 4 as an example). In this case, SV junctions are not detected by PCR, and ddPCR does not allow one to clearly infer a type of SV rearrangement. Are these founders WT (or NHEJ) animals or do they contain DEL and DUP alleles with undefined junctions? Figure 4 illustrates such a case. After breeding a F0 founder (F0-802) that shows a slight decrease in ddPCR copy counting, F1 pups were obtained with 1 copy (DEL) or with 3 copies (DUP) (*see* Fig. 4b). The deletion or duplication of the whole allele was further confirmed by different ddPCR probes ranging all over the genomic fragment (*see* Fig. 4c). The sequence of the real DEL and DUP junctions was determined by moving the primers away from the theoretical junctions until a PCR product was obtained (Fig. 4c integrated pictures). When no other F0s are available, F0 founders showing these characteristics can be crossed and might generate

the F1 segregated line(s). If the resulting F1 line is viable, it will very generally lead to generating the SV model of interest (*see* **Note 5**).

3.9 Establishment of a Validated F1 Line

1. To establish a SV line (DEL, DUP, or INV), interesting candidate founders need to be bred with wild-type animals to obtain germline transmission. This breeding step allows the segregation of all the possible alleles and will introduce one parental WT allele (coming for the WT animal that is bred with the founder) that is a good internal control (i.e., one copy of the WT allele by ddPCR). The two main limitations for the obtention of any genomic rearrangement are the viability of the resulting allele and the ability of a founder to reproduce. A big genomic deletion can be lethal as it can contain essential genes. Alternatively, a founder with the expected SV can live but be sterile. The selection of the "best" founders to breed is thus really important. For example, a founder bearing both a DEL and a DUP (two copies detected by ddPCR) might be able to reproduce without noticeable problems as two copies of each gene are present ("balanced" [22]) whereas the DEL animal (bearing only one copy of the region of interest) or the DUP animal (bearing three copies of the region of interest) can be sterile or might die prematurely. It is thus important to keep the crossed founder alive until the viability and reproduction of F1 pups are confirmed.
2. A SV model may be considered having a validated SV allele when:
   (a) DUP allele: when a DUP junction is defined and a ddPCR copy counting of 3 is observed on F1 pups. All animals with the same DUP junction and copy counting correspond to the same line. Additional ddPCR probes can be performed at different locations all over the region of interest. A ddPCR copy counting of 3 should be obtained with all probes located on the genomic fragment of interest (*see* Fig. 4c for illustration).
   (b) DEL allele: when a DEL junction is defined and a ddPCR copy counting of 1 is obtained on F1 pups. All animals with the same DEL junction and copy number will give rise to a line. Like for the DUP, all animals with the same DEL junction and copy counting correspond to the same line. A panel of ddPCR probes is also advised to verify the proper deletion of the whole genomic fragment of interest (*see* Fig. 4c for illustration).
   (c) INV allele: both the 5′ and 3′ INV junctions are defined and a ddPCR copy counting of 2 is obtained on F1 pups.

Again, multiple ddPCR probes can be designed to verify the absence of unwanted internal deletion or duplication of the genomic fragment of interest.

3. When it was not possible to detect the junctions by PCR in F1 pups: In the F1 animals with a ddPCR copy count of 1 and 3, respectively, the DEL or DUP junction can be searched by selecting new PCR primers further to the sgRNAs expected DSB sites (*see* Fig. 4). Unfortunately, in some cases, a junction cannot still be found. Insertions (we have observed insertion of partial reverse transcribed mRNA) can occur at the junction site as well as very large deletions [7, 14] that render detection of the junction very difficult. Additional ddPCR can be performed in order to confirm that the whole genomic region is indeed deleted or duplicated.

4. Examples of other unexpected rearrangement alleles. Sometimes, aberrant/unexpected events occur and junction PCRs are not confirmed by ddPCR. Here, we describe three examples:

   (a) Case 1: a DUP junction PCR is obtained and confirmed by Sanger sequencing but the ddPCR copy number does not correspond to the junction identified, two copies are detected rather than the expected three copies. One hypothesis can be genomic instability, the duplication occurred but was unstable [23].

   (b) Case 2: a DUP junction PCR contains two closely related but distinct PCR products (observed after close analysis of the chromatogram) and four copies are detected by ddPCR. Two distinct DUP junctions exist and go in favor of a triplication (TRI) of the genomic region of interest (*see* Fig. 5). In the same litters, some F1 show a DUP junction PCR product than is not confirmed by ddPCR. This observation goes again in favor of instability of the rearrangement (*see* Fig. 5).

   (c) Case 3: The DEL junction is confirmed by Sanger sequencing but three copies are detected by ddPCR. These results are confirmed by re-sampling the animals and the use of three distinct ddPCR probes. We have not been able to find any explanation for this event that highlights the importance of a good validation of a line.

3.10 Establishment of the SV Line

Selected founders (*see* **Note 6**) are crossed with wild-type animals to establish the desired deletion, duplication, or inversion line. Using junction PCRs followed by Sanger sequencing as described above, the sequence of the extremities of each SV allele is precisely defined at the F1 stage as only one allele version is transmitted in one animal. All animals showing the same SV (same junction sequenced

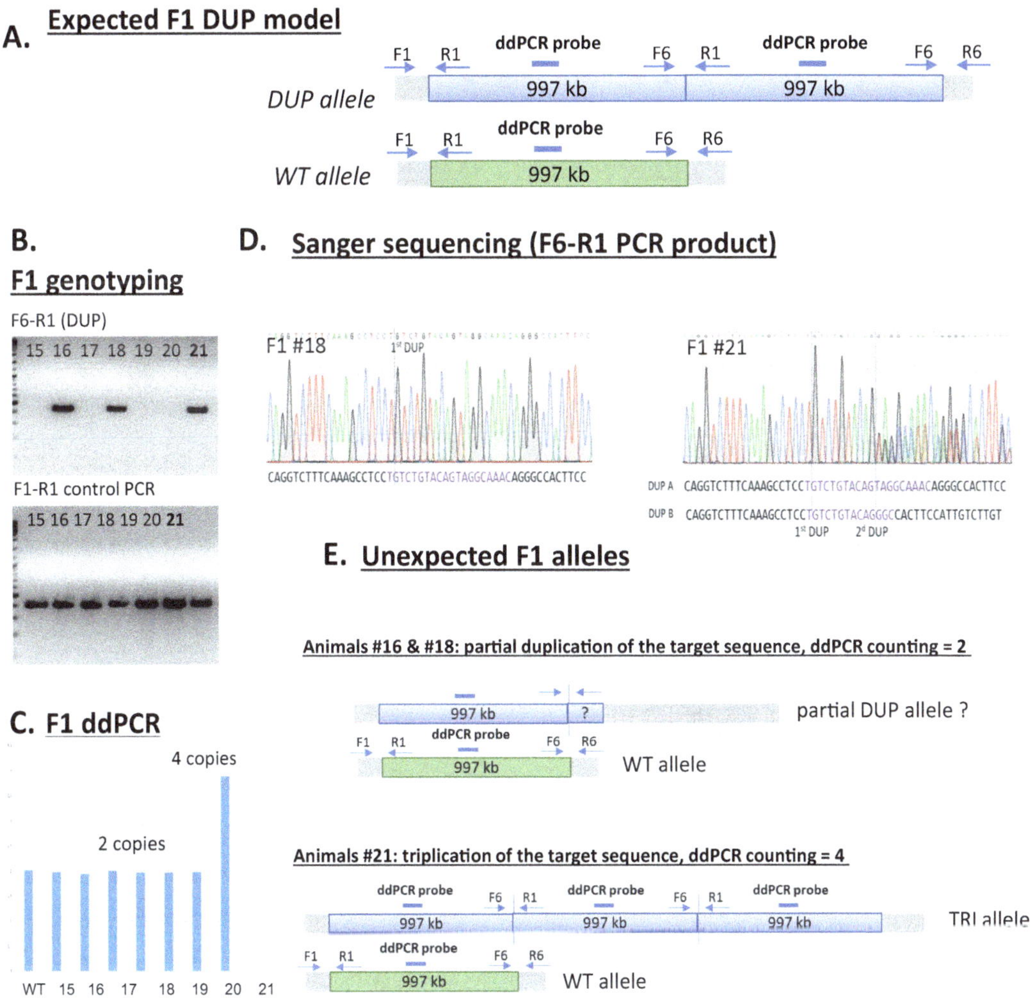

Fig. 5 An illustration of an atypical/aberrant case. The objective of this project was to duplicate a 997 kb region in the mouse genome. The F1 animal of interest is therefore a DUP heterozygote containing both a WT allele and a DUP allele. (**a**) The ddPCR probe is represented by a blue line and the positions of the primers are indicated with arrows. A founder was characterized and showed a DEL (F1–R6) and a DUP (F6–R1) junction (confirmed by Sanger sequencing) with 2.6 copies suggesting CRISPR/Cas9-mediated mosaic mutations. This founder was bred with a WT animal to establish the SV lines. (**b**) F1 genotyping: the junction PCR (F6–R1) gel picture is shown as well as the control PCR (F1–R1) that is specific to the WT allele and is thus always detected in all F1 pups. F1 pups #16, #18, and 21 show a strong F6–R1 PCR product indicating the presence of a DUP junction. (**c**) However, ddPCR counting did not result in detecting three copies (two copies for the DUP allele and one additional copy for the WT allele). Four copies were detected in the #21 F1 and two copies were detected in both #16 and #18 F1. (**d**) A careful examination of all chromatograms showed a clear difference between animals #16 or #18, and F1 #21. F1 #21 PCR product is a mix of two distinct but very similar PCR products that correspond to two slightly different DUP junctions. (**e**) No DEL junction was found by PCR and no animal with a ddPCR counting of one was found in any of the F1 pups. Extensive genomic rearrangements appear to have occurred, resulting respectively in a triplication of the region of interest (animal #21) or a partial duplication (animals #16 and #18)

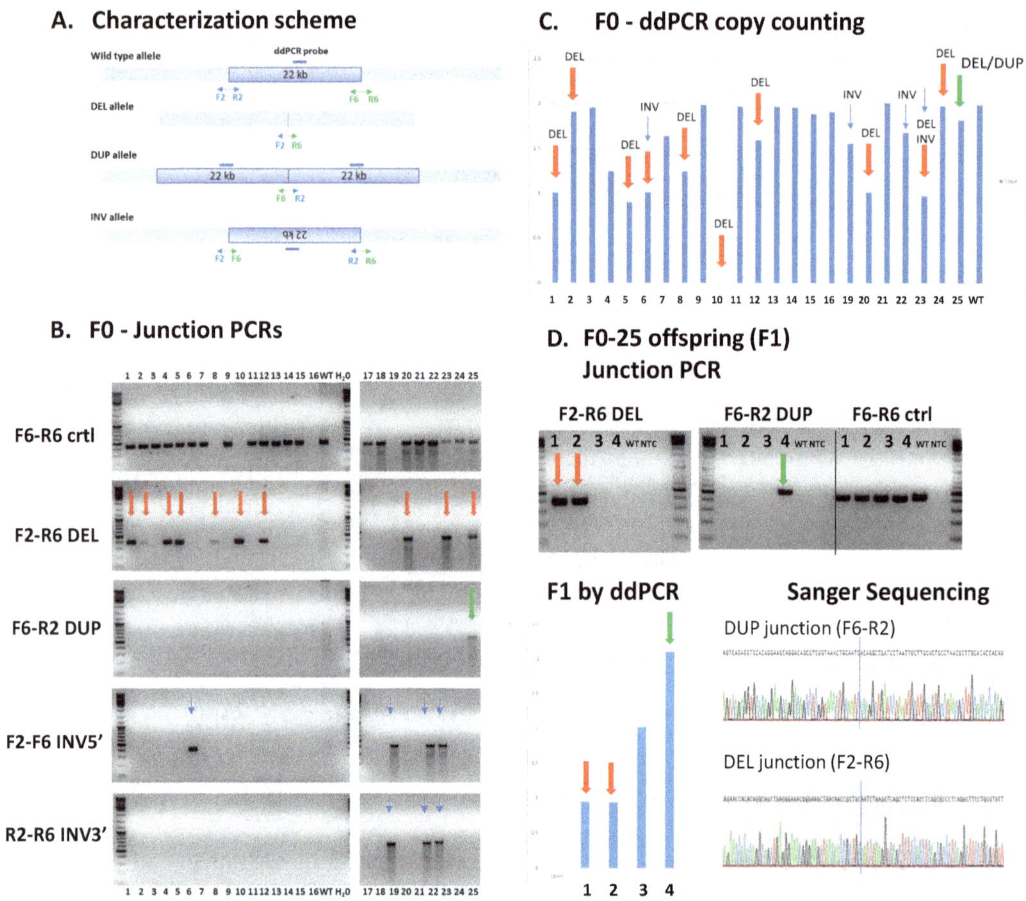

Fig. 6 Example of PCR, sequencing, and ddPCR screening results obtained with F0 founder and F1 pups on a region of interest of 22 kb region. (**a**) Scheme of the possible different alleles with the position of the primers and the ddPCR assay. (**b**) Picture of the junction PCR products obtained after analysis of tails biopsies of 25 F0 born after electroporation of CRISPR/Cas9 RNP in C57BL/6 N fertilized oocytes. PCR products were run on 2.5 % agarose gel. Ten F0 show a clear DEL junction (red arrow), one F0 (F0–25) shows a DEL and a DUP junction (green arrow), and its ddPCR copy number is close to 2. Four founders show an INV junction (light blue arrow); three of them have 5′ and 3′ junctions. (**c**) ddPCR results show a significative variation of the copy number for most of the F0. The majority of the F0 with a DEL junction have a clear diminution of their WT copy number (but not all). F0–10 has a copy number of 0 in correlation with the WT control PCR (F6–R6) which also does not show a PCR product. (**d**) Characterization of the first litter issued from F0–25. Two DEL (F1 #1 and #2) and one DUP (F1 #4) animals are fully characterized by junction PCR, Sanger sequencing, and ddPCR. The last animal (F1 #3) of the litter is WT

and same ddPCR quantification) will be at the origin of a new mouse or rat line. For large SV, multiple ddPCRs are designed on different parts of the region of interest to verify the absence of unwanted additional rearrangements in the SV allele (*see* Fig. 4c for an example). In addition, one can further verify the integrity of the chromosomal rearrangement with complementary techniques such as comparative genomic hybridization for copy number, molecular combing or fiber FISH, and optical genome mapping to look for enzymatic binding site along the region of interest.

3.11 Conclusions and Final Recommendations

We have shown that the outcome of CRISPR/Cas9-induced DSBs at each end of a genomic fragment leads to genomic rearrangements. These alleles are easily generated and allow the creation of rodent models for SV diseases. In this protocol, we showed that CRISMERE is an easy and efficient method to generate SV models. DEL, DUP, and INV can be easily generated and even triplications (TRI) alleles were achieved. Our protocol also highlights that a careful validation of each line is mandatory. The combination of junction PCR, its sequencing, and ddPCR allow to define precisely SV alleles but are not sufficient in some cases. Targeted locus amplification, CGH arrays, NGS, or Next Generation Cytogenomics (by Bionano Genomics) may be complementary approaches to validate the integrity of the SV allele gene by gene or even its full sequence.

4 Notes

1. The functionality of more than 100 targets was compared between the in vitro cleavage assay (SureGuide Cas9 Nuclease Kit, Agilent, 5190-7716) and in vivo results in founders. No correlation between effective in vitro cleavage and efficient cleavage in eggs was found, indicating that this control is not useful.
2. crRNA and tracrRNA can be ordered from different suppliers.
3. We also obtained good results with a concentration of 10 ng/μ L for all components. The advantage to work with RNA (rather than ribonucleoproteins, RNP) is that tubes with the appropriate mixes can be prepared in advance and stored for weeks at − 80 °C.
4. In some cases, we were able to detect more than one junction with the same PCR (mostly with DEL); in this case, double peaks are observed in the chromatograms. This means than more than one deleted allele exists in the founder [13].
5. When no good F0 candidates are available, it is worth crossing these animals as they may result in the awaited SV F1. It also means that a founder for which no junction PCR product is obtained and a ddPCR counting of 2 may not systematically be a wild-type animal.
6. Our data suggest that the timing of CRISPR/Cas9-mediated DSB during the egg division (G1 or G2 of the mitosis interphase) is changing the outcomes. DSB at the G2 stage will lead to an impressive potential variety of alleles (*see* Fig. 2). For

example, we have obtained up to six different alleles for one single founder. Of course, it is impossible to define if these alleles are all issued from the mitosis at the one-cell stage or are the results of the action of CRISPR later on. We did not find a limit to the size of the target region that can be rearranged using two distal DSBs. The only limitations for obtaining a specific SV allele seem to be the viability of the allele and the ability of a founder to reproduce and transfer the allele through the germline. For example, we have been able to duplicate a rat 24 Mb region that corresponds to the main human syntenic region for the human chromosome 21 [13]. However, we were not able to generate the rat deletion model. This deletion is known to be lethal in humans.

References

1. Ramirez-Solis R, Liu P, Bradley A (1995) Chromosome engineering in mice. Nature 378(6558):720–724
2. Vram B et al (2006) Modeling chromosomes in mouse to explore the function of genes genomic disorders and chromosomal organization. PLoS Genet 2(7):e86
3. Herault Y et al (1998) Engineering chromosomes in mice through targeted meiotic recombination (TAMERE). Nat Genet 20(4): 381–384
4. Adams DJ et al (2004) Mutagenic insertion and chromosome engineering resource (MICER). Nat Genet 36(8):867–871
5. Ruf S et al (2011) Large-scale analysis of the regulatory architecture of the mouse genome with a transposon-associated sensor. Nat Genet 43(4):379–386
6. Zheng B, Mills AA, Bradley A (1999) A system for rapid generation of coat color-tagged knockouts and defined chromosomal rearrangements in mice. Nucleic Acids Res 27(11): 2354–2360
7. Birling MC, Herault Y, Pavlovic G (2017) Modeling human disease in rodents by CRISPR/Cas9 genome editing. Mamm Genome 28(7–8):291–301
8. Kraft K et al (2015) Deletions inversions duplications: engineering of structural variants using CRISPR/Cas in mice. Cell Rep 10(5): 833–839
9. Li J et al (2015) Efficient inversions and duplications of mammalian regulatory DNA elements and gene clusters by CRISPR/Cas9. J Mol Cell Biol 7(4):284–298
10. Wang L et al (2015) Large genomic fragment deletion and functional gene cassette knock-in via Cas9 protein mediated genome editing in one-cell rodent embryos. Sci Rep 5:17517
11. Zhang L et al (2015) Large genomic fragment deletions and insertions in mouse using CRISPR/Cas9. PLoS One 10(3):e0120396
12. Korablev AN, Serova IA, Serov OL (2017) Generation of megabase-scale deletions inversions and duplications involving the Contactin-6 gene in mice by CRISPR/Cas9 technology. BMC Genet 18(Suppl 1):112
13. Birling MC et al (2017) Efficient and rapid generation of large genomic variants in rats and mice using CRISMERE. Sci Rep 7:43331
14. Boroviak K et al (2016) Chromosome engineering in zygotes with CRISPR/Cas9. Genesis 54(2):78–85
15. Lindner L et al (2021) Reliable and robust droplet digital PCR (ddPCR) and RT-ddPCR protocols for mouse studies. Methods 191:95–106
16. Concordet J, Haeussler M (2018) CRISPOR: intuitive guide selection for CRISPR/Cas9 genome editing experiments and screens. Nucleic Acids Res 46(W1):W242–W245
17. Cong L et al (2013) Multiplex genome engineering using CRISPR/Cas systems. Science 339(6121):819–823
18. Alghadban S et al (2020) Electroporation and genetic supply of Cas9 increase the generation efficiency of CRISPR/Cas9 knock-in alleles in

C57BL/6J mouse zygotes. Sci Rep 10(1): 17912
19. Teixeira M et al (2018) Electroporation of mice zygotes with dual guide RNA/Cas9 complexes for simple and efficient cloning-free genome editing. Sci Rep 8(1):474
20. Remy S et al (2017) Generation of gene-edited rats by delivery of CRISPR/Cas9 protein and donor DNA into intact zygotes using electroporation. Sci Rep 7(1):16554
21. Jacquot S et al (2019) Optimizing PCR for mouse genotyping: recommendations for reliable rapid cost effective robust and adaptable to high-throughput genotyping protocol for any type of mutation. Curr Protoc Mouse Biol 9(4):e65
22. Birchler JA, Veitia RA (2012) Gene balance hypothesis: connecting issues of dosage sensitivity across biological disciplines. Proc Natl Acad Sci U S A 109(37):14746–14753
23. Rodgers K, McVey M (2016) Error-prone repair of DNA double-strand breaks. J Cell Physiol 231(1):15–24

Chapter 13

Targeted Integration of Transgenes at the Mouse *Gt(ROSA)26Sor* Locus

Daniel Biggs, Chiann-mun Chen, and Benjamin Davies

Abstract

The targeting of transgenic constructs at single copy into neutral genomic loci avoids the unpredictable outcomes associated with conventional random integration approaches. The *Gt(ROSA)26Sor* locus on chromosome 6 has been used many times for the integration of transgenic constructs and is known to be permissive for transgene expression and disruption of the gene is not associated with a known phenotype. Furthermore, the transcript made from the *Gt(ROSA)26Sor* locus is ubiquitously expressed and subsequently the locus can be used to drive the ubiquitous expression of transgenes.

Here we report a protocol for the generation of targeted transgenic alleles at *Gt(ROSA)26Sor*, taking as an example a conditional overexpression allele, by PhiC31 integrase/recombinase-mediated cassette exchange of an engineered *Gt(ROSA)26Sor* locus in mouse embryonic stem cells. The overexpression allele is initially silenced by the presence of a loxP flanked stop sequence but can be strongly activated through the action of Cre recombinase.

Key words Gene overexpression, ROSA26, Embryonic Stem cell, PhiC31, RMCE, Cre recombinase

1 Introduction

New site-specific nuclease approaches [1] and the availability of archived mutant embryonic stem (ES) cell lines [2] have greatly facilitated our ability to explore gene function through loss-of-function mutagenesis in the mouse. Frequently, however, gene overexpression may be a more appropriate strategy for the analysis of gene function, in particular when compensatory effects are in operation to mask loss-of-function phenotypes [3]. Furthermore, changes in disease susceptibility and indeed certain genetic disease can result from overexpression or amplification of genes [4], and thus a gene overexpression strategy may provide an excellent means of modelling these diseases in the mouse.

Traditionally, gene overexpression has been achieved through the random integration of promoter-driven constructs or bacterial artificial chromosomes harboring the gene of interest, frequently by

Thomas L. Saunders (ed.), *Transgenesis: Methods and Protocols*,
Methods in Molecular Biology, vol. 2631, https://doi.org/10.1007/978-1-0716-2990-1_13,

pronuclear microinjection [5]. Despite the apparent simplicity of this approach, there are some major disadvantages associated with the methodology:

(i) Random insertion of a transgene can lead to mutagenesis of an endogenous sequence, complicating the phenotypic analysis of the resulting line [6]. More recent work using genomic sequencing to quantify the disruption at transgene insertion sites found significant levels of structural change and mutagenesis at the insertion sites associated with some commonly used Cre recombinase lines, generated by random insertion [7, 8].

(ii) The resulting transgene regulation can be influenced by the so-called position effects caused by the integration of the DNA construct within genomic loci under the influence of strong enhancer or silencer elements or within highly methylated loci which can lead to silenced, inconsistent, or ectopic expression patterns [9, 10].

(iii) Transgenic constructs frequently enter the genome as concatemer arrays. These repetitive arrays are associated with instable expression and variegated patterns of transgene expression frequently result [11].

These shortcomings necessitate the breeding and maintenance of multiple independent lines of mice, each bred from a specific founder animal, in order to ensure that the observed phenotype is causally linked to the inserted transgene. As a consequence, the cost and scale of such projects, both in terms of financial burden but also the number of animals used in a particular analysis, can be high.

To overcome these problems, a technique known as targeted transgenesis can be used to position constructs by gene targeting in precise locations at single copy within the genome [12, 13]. Selected insertion sites are chosen on the basis of their being permissive for transgene expression and their interruption not being associated with a phenotype. The *Gt(ROSA)26Sor* locus on chromosome 6 in the mouse has these particular characteristics and is frequently selected as a docking site for exogenous transgenes.

The locus encodes a gene of unknown function and was originally discovered by a reporter gene trap insertion, the expression of which was revealed to be ubiquitous throughout the developing embryo [14]. The presence of a ubiquitously expressed transcript which can be interrupted with no known consequences makes this locus an excellent choice, since transgenes inserted into the locus can be engineered to become driven by the endogenous ubiquitously expressed *Gt(ROSA)26Sor* promoter. The insertion of a floxed STOP cassette within the transcript has also enabled conditional transgene expression and is the basis of the widely used reporter lines used to signal Cre recombinase activity [15, 16]. Furthermore, the locus has been used to drive transcriptional

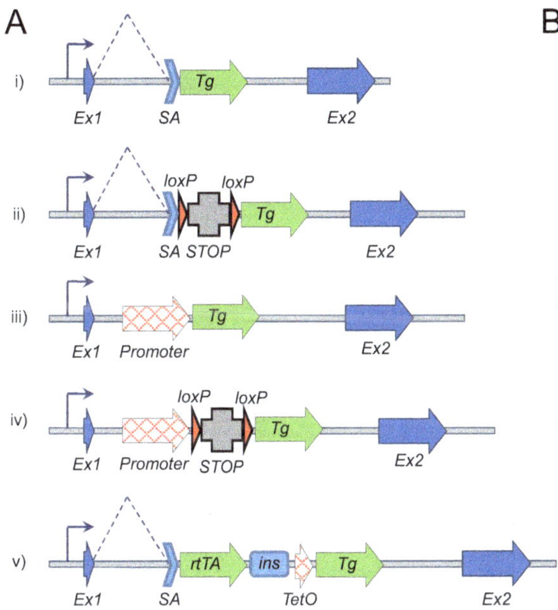

Fig. 1 Transgenic targeting at the *Gt(ROSA)26Sor* locus. (**a**) Common manipulations of *Gt(ROSA)26Sor*. (i) Integration of a transgene (Tg) whose expression becomes driven by the endogenous *Gt(ROSA)26Sor* promoter through an exogenous splice acceptor (SA) placed upstream of the transgene. (ii) Integration of a transgene (Tg) with a loxP flanked Stop sequence, allowing Cre-dependent transgene expression from the endogenous *Gt(ROSA)26Sor* promoter. (iii) Integration of a transgenic construct driven by an exogenous promoter. (iv) Integration of a transgenic construct, driven by an exogenous promoter, with a loxP flanked Stop sequence to allow Cre recombinase-dependent transgene expression. (v) Integration and expression of a transcriptional regulator (e.g., rtTA) which regulates expression of an inducible construct positioned downstream. Insulator sequences (ins) are frequently used to separate transcriptional units. (**b**) Recombinase Mediated Cassette Exchange. Recombination between compatible recombination sites within the genome and within a shuttle vector leads to the exchange of sequences. In general, a selection cassette (+/−) is replaced by the transgene (Tg)

regulators, for example, for the doxycycline inducible system [17, 18], which can allow inducible transgene expression from a co-integrated expression unit [19–22]. Exogenous promoters driving cDNA cassettes have also been inserted at *Gt(ROSA) 26Sor*, and their expression characteristics are frequently preserved [23, 24]. The locus has also been used to harbor RNA polymerase III promoters driving the expression of shRNAs for RNA interference applications [25, 26]. Figure 1a summarizes some of the more frequent targeted transgenic manipulations that have been performed at the *Gt(ROSA)26Sor* locus.

The manipulation of the *Gt(ROSA)26Sor* locus has been traditionally achieved by gene targeting in ES cells, which necessitates the generation of a targeting vector with homology arms upstream and downstream of the inserted transgene, together with a selection cassette [15]. Although the construction of the vectors has been simplified through the assistance of Gateway cloning [27, 28], nonetheless, each integration demands a separate construct and gene targeting event. To facilitate the positioning of transgenes within *Gt(ROSA)26Sor*, the locus has been adapted for recombinase-mediated cassette exchange (RMCE) methodologies [29]. These involve the positioning of recombinase sites, for example, loxP sites (for Cre recombinase) or FRT sites (for Flp recombinase) within the locus in an ES cell line which is then used as a docking-site cell line for further manipulation. These cell lines can be transfected with shuttle vectors encoding the transgene of interest together with corresponding recombinase sites which, when co-delivered with the cognate recombinase or integrase, results in the exchange of genomic sequence and subsequent integration of the transgene (*see* Fig. 1b). Frequently, a selection cassette is also exchanged, allowing selection of the desired integration event [30].

RMCE systems for *Gt(ROSA)26Sor* have been reported which rely on Cre recombinase [24, 31] and Flp recombinases [20, 24, 32], and frequently heterotypic recombinase sites are used to ensure the exchange/integration events are unidirectional. The use of Cre recombinase and Flp recombinase for the integration event, however, precludes the use of these enzymes for further manipulation of the locus, for example, for Cre-dependent transgene activation or FRT-dependent deletion of the selection cassette. To further increase the repertoire of manipulations achievable, PhiC31 integrase, a serine recombinase, has been used for RMCE at *Gt(ROSA)26Sor* [23, 26]. This enzyme catalyzes the recombination between attP and attB sites, generating attL or attR sites which are resistant to further processing; the recombination reaction is thus unidirectional. This class of enzyme is highly appropriate for achieving efficient transgene integration events and allows for subsequent manipulations of the locus by the more commonly used Cre and Flp recombinase systems, where many tissue-specific and ubiquitous deleter mouse lines are available.

Interestingly, direct integration of transgenes at the *Gt(ROSA) 26Sor* locus has been reported through the use of site-specific nucleases in combination with suitable targeting vectors. Double-strand breaks introduced into intron 1 of the *Gt(ROSA)26Sor* locus by site-specific zinc finger nucleases [33, 34], TALENs [35], and CRISPR/Cas9 nucleases [36–38] have been shown to allow repair by recombination with a targeting vector, achieving targeted integration of a transgene at this locus. This technique could circumvent the need for RMCE techniques. However, many laboratories are struggling with the targeted integration of large constructs using site-specific nucleases and are finding that the efficiency of achieving perfect integrations without additional rearrangements [39, 40], unexpected deletions [41], or even chromosome truncations [42] can be disappointingly low. Although we want to highlight the potential of these nuclease-assisted techniques, which can operate at frequencies which make the use of selection machinery unnecessary, we recommend a recombinase-driven approach because of the efficient, accurate, and robust nature of the integration reaction.

One great advantage of a nuclease-driven approach for transgene integration is that it can be applied directly in the mouse zygote, thus circumventing the need for ES cell manipulations [43]. Similarly, RMCE at *Gt(ROSA)26Sor* has also been shown to function directly within the zygote [44, 45]. Microinjection of zygotes, prepared from a line of genetically modified mice harboring the cassette exchange machinery at *Gt(ROSA)26Sor* with a shuttle vector and recombinase mRNA has been shown to result in efficient integration of transgenes. This technique could be of great utility for the production of targeted transgenic mouse model, although clearly a suitable line of mice must be housed and bred for this purpose.

Here we outline the protocol necessary to achieve the targeted integration of a transgene using the PhiC31 integrase system in ES cells, although the protocol is amenable to other recombinase systems. We have chosen to describe the generation of an ES cell line suitable for Cre recombinase-dependent transgene overexpression, although the method can easily be adapted for the insertion of any type of transgene. The protocol is split into two sections, one describing the cloning of the shuttle vector containing the transgene of interest and the second section describing the RMCE in ES cells.

ES cells generated using this system can be used for in vitro differentiation experiments to explore the consequence of transgene function in cell culture models. Alternatively, they can be used to generate mouse models via injection into preimplantation embryos. The single copy integration event in a fully characterized locus permits the generation of only a single line of genetically modified mice per construct. The technique is particularly appropriate when comparing variant transgenes as the functional differences can be attributed to transgene differences with greater confidence, since they are positioned at single copy within the exact same genomic context [31, 46].

2 Materials

2.1 Cloning of the Transgene of Interest

1. RS-PhiC ES cell line (*see* **Note 1**): This ES cell line, available from the authors [47], represents a docking-site cell line suitable for PhiC31 integrase-mediated RMCE for integration of transgenes at the *Gt(ROSA)26Sor* locus. The docking site engineered within this cell line includes the presence of an exogenous CAG promoter which is used to drive strong transgene expression. Upstream of this CAG promoter, an FRT site together with a splice acceptor sequence is positioned which

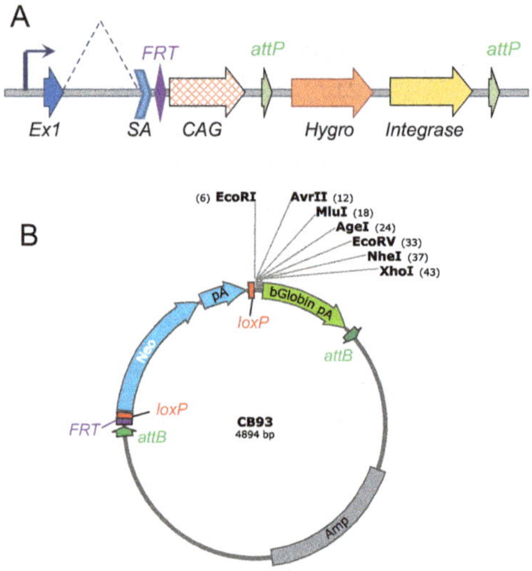

Fig. 2 The PhiC31 RMCE system. (**a**) Conformation of the locus within the RS-PhiC ES cell line. (**b**) A map of the CB93 shuttle vector with the position of the unique restriction sites used for cloning of the transgene

can be used to remove the CAG promoter by Flp recombinase action (*see* **Note 2**), allowing transgene expression to be driven by the weaker endogenous *Gt(ROSA)26Sor* promoter. Downstream of the CAG promoter, the open-reading frames for a hygromycin selection cassette, and PhiC31 integrase, linked together by a T2A self-cleaving peptide, are positioned, flanked by PhiC31 attP sites (Fig. 2a).

2. Plasmid CB93 (*see* **Note 1**): This plasmid, available from the authors [47], serves as the backbone shuttle vector for PhiC31 integrase-mediated cassette exchange using the above ES cell line. It contains a loxP flanked neomycin selection cassette together with a polylinker of unique restriction sites for the cloning of the transgene of interest, together with a polyadenylation sequence. This whole array is flanked by PhiC31 attB sites (*see* Fig. 2b).
3. Restriction enzymes.
4. Calf Intestine Phosphatase.
5. T4 DNA ligase.
6. High fidelity polymerase.
7. PCR cloning kit.
8. Gel extraction kit.
9. Plasmid mini- and midi-prep kits.
10. Chemically competent *E. coli*.

2.2 Recombinase-Mediated Cassette Exchange in ES Cells

1. Neomycin-resistant mitotically inactivated mouse embryonic fibroblasts (Mefs) are available from commercial suppliers (e.g., PMEF-NX, Merck-Milipore) but can be made in-house from mid-gestation Tg(DR4)1Jae/J (Jax stock number 003208) mouse embryos following standard protocols [48].
2. JM8F6 media: 500 mL KnockOut DMEM (Thermo Fisher Scientific), 50 mL ES cell qualified Foetal Bovine Serum, 6 mL 100× Penicillin–Streptomycin (10,000 units/mL penicillin and 10 mg/mL streptomycin), 6 mL 100× L-Glutamine (200 mM), 6 mL 100× MEM Non-Essential Amino Acids Solution, 550 μL 1000× β-mercaptoethanol, 60 μL LIF (10^7 units/mL).
3. 1000× β-mercaptoethanol: 50 μL of β-mercaptoethanol in 7.1 mL of Phosphate Buffered Saline (PBS) without Mg^{2+} and Ca^{2+}.
4. 2X + G Trypsin: 500 mL PBS without Mg^{2+} and Ca^{2+}, 0.1 *g* EDTA, 0.5 *g* D-Glucose, 5 mL Chicken Serum, 20 mL 2.5× Trypsin (146.6 mM NaCl/1.05 mM Trypsin).
5. 2× Freezing media: 2 mL JM8F6 Media, 2 mL ES cell qualified Foetal Bovine Serum, 1 mL hybrid-Max DMSO.
6. 2× lysis buffer: 400 ng/μL proteinase K, 40 ng/μL RNAse A in 0.1% SDS.
7. PBS without Mg^{2+} and Ca^{2+}.
8. 0.1 TE: 1 mM Tris HCl pH 7.5, 0.1 mM EDTA.
9. Neon transfection system (Thermo Fisher Scientific).
10. G418 sulfate, cell culture tested.
11. 0.1% gelatin.
12. Embryo tested light mineral oil.
13. 8 M LiCl.
14. Isopropanol (Molecular Biology Grade).
15. 70% ethanol.

3 Methods

An outline of the method used for the generation of targeted transgenic conditional overexpression mice is shown in Fig. 3.

3.1 Cloning of the Transgene of Interest

1. Design primers which will enable the cDNA of interest to be amplified by PCR from an appropriate cDNA source or from a commercial cDNA clone (*see* **Note 3**).
2. Consider the unique restriction sites available in the CB93 polylinker (*see* Fig. 2b)—an appropriate choice of two of these enzymes which do not cleave the cDNA of interest will greatly

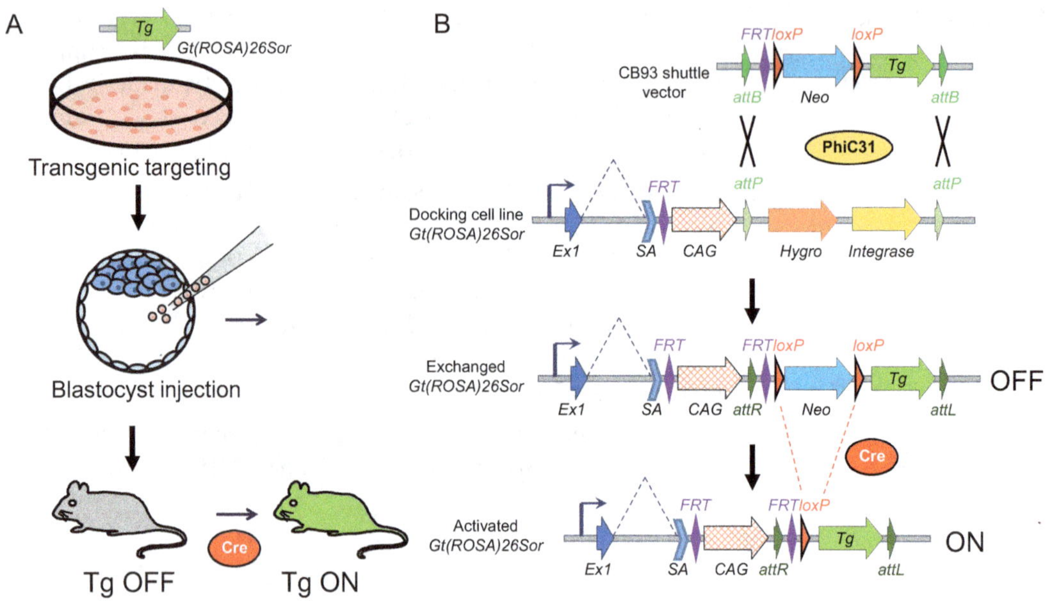

Fig. 3 Outline of the RMCE method. (**a**) Flow diagram showing the main steps in the protocol for the production of transgenic targeted mice, where strong transgene expression is activated by Cre recombinase. (**b**) Flow diagram showing the conformation of the alleles obtained at the various stages of the protocol

simplify the downstream cloning procedures—and build these as 5′ extensions into the PCR primers.

3. Ensure that a consensus Kozak sequence is present at the start methionine codon, to allow efficient translation of the cDNA. This can be designed as a 5′ extension into the forward primer used for cDNA amplification. An example of PCR primers suitable for the amplification of an example cDNA are shown in Fig. 4. Amplify the cDNA using a high-fidelity polymerase and clone the PCR product into a plasmid vector using either a blunt cloning kit or a TA cloning kit, depending on the characteristics of the polymerase enzyme use (*see* **Note 4**).
4. Sequence the amplified cDNA in both directions using sequencing primers present within the PCR cloning vector and validate the full sequence of the cDNA. Do not use the primers that were used to amplify the cDNA for sequencing as the whole sequence of the cDNA, including the extremities, must be verified. If mutations are found, select an independent clone or repeat the PCR amplification using fewer PCR cycles.
5. Digest 500 ng of CB93 with the selected unique restriction sites using the buffer systems and recommendations of the enzyme's supplier. After digestion, add 10 iu of Calf Intestinal Phosphatase to the vector and incubate at 37 °C for a further 30 minutes.

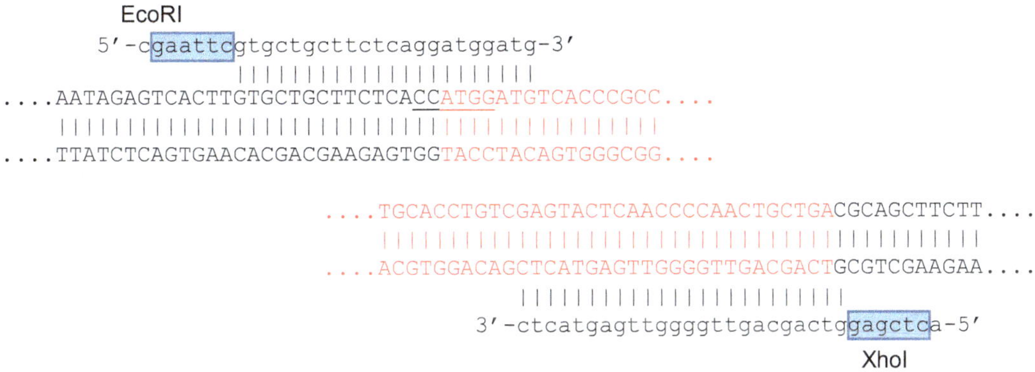

Fig. 4 Example of the primers used for amplification of the cDNA. The 5′ and 3′ sequence extremities of an example cDNA sequence cloned within a plasmid vector are shown. The coding region of the cDNA is shown in red. The position and sequence of the PCR primers used for the selective amplification of the cDNA is shown, along with the inclusion of unique restriction sites at the 5′ end of each of the primers for efficient cloning into the CB93 shuttle vector. A consensus Kozak sequence is underlined

6. Digest 1 μg of cloned PCR product with the selected unique restriction enzymes, using the buffer systems and recommendations of the enzymes' supplier.
7. Separate the digested fragments by electrophoresis on a 1% agarose gel.
8. The CB93 digest should effectively yield a single linearized fragment of 4.9 kb in size. Cut this fragment from the gel and use a gel extraction kit to purify the DNA. This fragment will serve as the vector for the ligation.
9. The digest of the cloned PCR product should yield two fragments, one corresponding to the size of the cloning vector and the other corresponding to the size of the cDNA. Cut this latter fragment from the gel and use a gel extraction kit to purify the DNA. This fragment will serve as the insert for the ligation.
10. Check the purity and concentration of the purified DNA, by running a sample (typically 1 μL of extracted fragment) on an agarose gel.
11. Ligate the vector and the insert together with T4 DNA ligase, according to the supplier's instructions. Typically, 100 ng of vector sequence can be ligated with the equivalent molar amount of insert DNA in final reaction volume of 10 μL.
12. Set up appropriate ligation controls—typically a reaction with the vector on its own and a reaction with the insert on its own, using the same quantities and conditions as in the experimental ligation.

Table 1
PCR and sequencing primers

Primer name	Sequence
CAG-F	5′-CAGCCATTGCCTTTTATGGT-3′
ExNeo2	5′-GTTGTGCCCAGTCATAGCCGAATAG-3′
attL-F1	5′-GCACTAGTTCTAGAGCGATCCCC-3′
Rosa-R1	5′-CGGGAGAAATGGATATGAAGTACTGGGC-3′
RMCE-seq-F	5′-CAATAGCAGGCATGCTGGGGATG-3′

13. Transform 5 μL of the completed ligation reaction, along with the control reactions, into competent *E. coli* bacteria, following the supplier's recommendation and plate the transformation on LB plates containing 100 μg/mL ampicillin.
14. If the ligation has been successful, there should be significantly more colonies on the ligation plate where both vector and insert were added to the reaction, compared to the controls where only vector or insert was present. If a clear difference exists, proceed with preparation of plasmid DNA from 4 to 6 colonies (*see* **Note 5**).
15. Inoculate 2 mL cultures of LB containing 100 μg/mL ampicillin with candidate bacterial clones and grow overnight.
16. Perform plasmid DNA preparation using a miniprep kit according to the supplier's instructions.
17. Digest 250 ng of each plasmid with an enzyme combination which would yield a clear pattern of digestion upon successful cloning of the cDNA insert into the CB93 vector. We typically use a restriction enzyme (SalI or KpnI) which cuts uniquely within CB93, upstream of the 5′ attB site in combination with an enzyme that cuts uniquely within the cDNA of choice.
18. Sequence candidate positive plasmids with the primer RMCE-seq-F (Table 1) which hybridizes to the polyadenylation sequence of the neomycin selection cassette and provides sequence through the cDNA sequence. Correctly cloned plasmid should thus reveal the sequence of the cDNA after the 5′ restriction site that was used for cloning.
19. Prepare a larger quantity of a correctly sequenced verified plasmid by using a commercial endotoxin free midiprep kit. Preserve the sterility of the plasmid preparation by performing the elution step with sterile elution buffer from the kit under the tissue culture hood.

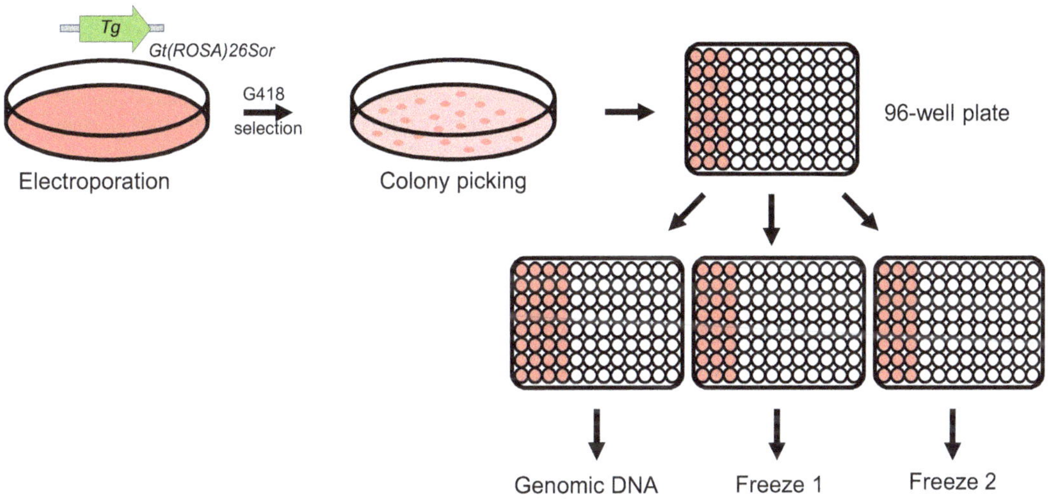

Fig. 5 Flow diagram showing the main steps in the ES cell targeting experiment

3.2 Recombinase-Mediated Cassette Exchange in ES Cells

An overview of the cell culture steps involved in this protocol is shown in Fig. 5.

3.2.1 ES Cell Electroporation

1. Thaw a frozen vial of 3 × 10^6 RS-PhiC ES cells by holding the cryovial in a 37 °C water bath until the ice crystal floats free of the side of the tube. Take the cryovial under the tissue culture hood and pipette the thawed cells into 5 mL of pre-warmed JM8F6 media. Centrifuge the thawed cell suspension at 180 *g* for 5 minutes. Resuspend the cell pellet carefully in 5 mL of JM8F6 media and plate onto a 25 cm^2 tissue culture flask, pre-coated with 1.25 × 10^6 mitotically inactivated mouse embryonic fibroblasts (Mefs) (*see* **Note 6**) and culture at 37 °C/5% CO_2.
2. Grow until 70% confluency, feeding every 24 hours—typically this takes 2–3 days.
3. Trypsinize the cells by washing twice in 3 mL PBS without Mg^{2+} and Ca^{2+}, removing the final PBS wash before adding 1 mL of 2X + G Trypsin and incubating for 10 minutes at 37 °C, agitating the cells half way through this incubation.
4. Neutralize the trypsinization reaction with 5 mL of JM8F6 media and resuspend firmly to obtain a single cell suspension (*see* **Note 7**).
5. Take 10 μL of the cell suspension and count the number of cells using either a hemocytometer or automated counting device. Remove 1.25 × 10^6 cells from the total cell count to compensate for the Mef cells on the flask and thus to obtain a cell count which corresponds to the number of ES cells present in the cell suspension (*see* **Note 8**).

6. While counting the cells, centrifuge the cell suspension at 180 *g* for 5 minutes.
7. Resuspend the cells in electroporation buffer R (Neon transfection system, Thermo Fisher Scientific), allowing 100 μL of buffer R per 1×10^6 ES cells.
8. Prepare 5 μg of shuttle vector in a sterile Eppendorf tube—assuming the midiprep concentration is greater than 500 ng/μ L, the volume will be less than 10 μL. Do not exceed a volume of 10 μL for the DNA, as this will compromise the efficiency of the electroporation.
9. Add 110 μL of cell suspension (1.1×10^6 cells) to the DNA and mix carefully by pipetting up and down, avoiding the introduction of air bubbles.
10. Electroporate the cells with 3 pulses of 1400v, 30 ms using a Neon 100 μL electroporation tip.
11. After electroporation, expel the cells into 10 mL of JM8F6 media in a 15 mL Corning tube and mix well.
12. Plate 3 mL of this cell suspension on a 6 cm dish pre-coated with 1.25×10^6 mitotically inactivated neomycin-resistant Mefs (*see* **Note 9**).
13. Feed the cells in JM8F6 media the following day.
14. Feed the cells daily in JM8F6 containing 210 μg/mL G418 for further 6 days.
15. Approximately 7 days after electroporation, clear G418-resistant colonies should be visible to the naked eye on the 6 cm dish. These are now ready for clonal isolation.
16. Per 6 cm dish (i.e., electroporation) to be picked, prepare 3 rows of a U bottom 96-well plate with 30 μL of 2X + G Trypsin.
17. Per 6 cm dish (i.e., electroporation) to be picked, prepare 3 flat-bottomed 96-well plate with 24 wells, with each well pre-coated with 1.7×10^4 mitotically inactivated neomycin-resistant Mefs (equivalent to 1.6×10^6 Mefs across a whole 96-well plate). These wells will be used for picking 24 colonies per 6 cm dish (i.e., electroporation), and the other two prepared 96-well plates will be used later in the protocol when the plates are replicated and frozen.
18. Wash the 6 cm dish 2× with 5 mL PBS without Mg^{2+} and Ca^{2+} and overlay the plate with 3 mL of PBS.
19. Using a dissection stereoscope, use a 20 μL tip to lift away each colony. This can be achieved easily by setting the pipette at 10ul, first using the point of the tip to cut around the colony, followed by suction to collect the excised colony into the

pipette tip. Each colony is then expelled into a single well on the U-bottomed 96-well plate containing 30 μL of 2X + G Trypsin (*see* **Note 10**).

20. Twenty-four colonies should be picked per construct. After picking this number, transfer the plate to the incubator for a 5 minutes incubation at 37 °C.
21. While incubating, replenish the media on the Mef-coated flat-bottomed 96-well plate with 100 μL fresh JM8F6 media per well.
22. Neutralize each well of the U-bottomed 96-well plate after trypsinization with 70 μL of JM8F6 media. Using an 8-well multichannel pipette, pipette each of the rows firmly up and down and transfer the entire contents to the freshly fed Mef-coated flat-bottomed 96-well plate and incubate at 37 °C/5% CO_2.
23. Feed daily in JM8F6 media containing 105 μg/mL G418 until the majority of the wells reach approximately 70% confluency (*see* **Note 11**).
24. Trypsinize the cells by washing twice with 150 μL PBS without Mg^{2+} and Ca^{2+}, removing the final PBS wash before adding 30 μL of 2X + G Trypsin to each well of the 96-well plate, using an 8-well multichannel pipette, and incubating for 10 minutes at 37 °C, agitating the cells half way through this incubation.
25. While trypsinizing, coat sufficient wells (24 per picked electroporation dish) on a flat-bottomed 96-well plate with gelatin by pipetting 50 μL of 0.1% gelatin solution, leaving for 5 minutes at room temperature and then aspirating the solution. Plate 140 μL of JM8F6 media onto each of the pre-gelatinized wells.
26. Also while trypsinizing, replenish the media on the two Mef-coated flat-bottomed 96-well plates (from **step 17**) with 140 μL of JM8F6 media per well.
27. Neutralize the trypsinization reaction with the addition of 150 μL of JM8F6 to each well, using an 8-well multichannel pipette. Pipette each of the 96-well plate columns of wells firmly up and down and transfer 3 × 60 μL of cell suspensions to the identical well positions on the pre-prepared gelatinized plate and the two Mef-coated 96-well plates. Incubate at 37 °C/5% CO_2.
28. Feed daily in JM8F6 media containing 105 μg/mL G418 until the majority of the wells reach approximately 70% confluency.
29. The two copies of the picked colonies growing on Mef-coated 96-well plates can now be frozen down, whereas the copy growing on the gelatin-coated 96-well plate should be left for a further 24 hours and used to prepare genomic DNA.

30. For the freezing copies (the Mef-coated 96-well plates), trypsinize the cells by washing twice with 150 μL PBS without Mg^{2+} and Ca^{2+}, removing the final PBS wash before adding 30 μL of 2X + G Trypsin to each well of the 96-well plate, using an 8-well multichannel pipette, and incubating for 10 minutes at 37 °C, agitating the cells half way through this incubation.
31. Neutralize the trypsinized cells by adding 70 μL of JM8F6 media. Add 100 μL of 2× Freezing media and resuspend each well firmly, again using an 8-well multichannel pipette.
32. Overlay the cell suspension with a drop of embryo tested light mineral oil and seal the plate using parafilm. Place the two plates in an ice-filled polystyrene box and freeze at −80 °C (*see* **Note 12**).

3.2.2 Preparation of Genomic DNA and PCR Screening for Correct RMCE Integration

1. After a further day of incubation, the majority of the gelatin-coated 96-well plate copy will be very confluent and the media should be orange to yellow in color, reflecting the large cell density. Remove the media and replenish with 100 μL of lysis buffer. Incubate overnight in the incubator at 37 °C.
2. The following day, remove the plate from the incubator and precipitate the DNA by adding 10 μL of 8 M LiCl and 100 μL of isopropanol to each well using a multichannel pipette. Shake the plate for 10 minutes at 350 rpm to mix the sample, the salt, and the alcohol to encourage an efficient precipitation. Small strands of genomic DNA may be seen by holding the plate up to the light, but these are not always visible.
3. Centrifuge the 96-well plate at 900 *g* for 20 minutes.
4. Pour off the supernatant by sharply inverting the plate over a sink and blotting the plate with 2–3 firm taps on paper towels. The DNA will have adhered tightly the plastic and there should be no danger of losing the sample.
5. Add 100 μL of 70% ethanol to each well and repeat **steps 3** and **4**.
6. Dry the pelleted DNA on the 96-well plate by placing the plate in a warm room or on a 37 °C heat block for 10–15 minutes. Be careful not to over-dry the DNA—visible signs of moisture should have just disappeared prior to processing the samples further.
7. Add 40 μL of 0.1 TE buffer per well and leave overnight at 4 °C to resuspend.
8. Prepare a suitable PCR master mix using the following primers in a multiplex reaction, CAG-F, ExNeo2, attL-F1, and ROSA26-R1 (sequences of the primers are shown in Table 1), and perform PCR using 1 uL of the prepared genomic DNA (no more than 100 ng of template genomic DNA

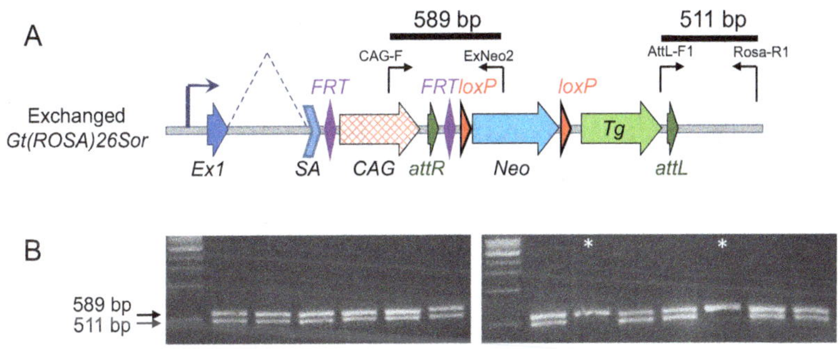

Fig. 6 PCR screening of transgenic targeted *Gt(ROSA)26Sor* ES cells. (**a**) Conformation of the correctly exchanged allele, showing the position of the screening primers used for the identification of the correct integration event. Correct 5′ attP x attB recombination results in an attR site and correct 3′ attP x attB recombination results in an attL site. (**b**) Example of a screen showing the PCR results from 13 clones. The majority of the clones reveal both PCR amplicons, suggesting correct cassette exchange. Two of the clones, marked by asterisks, are positive only at the 5′ end of the vector and thus do not represent correct exchange events

should be added to each well, dilute if necessary). The first two primers amplify a region between the CAG and the integrated neomycin cassette and yield a 589 bp amplicon upon successful integration. The latter two primers amplify a region corresponding to polyadenylation sequence at the 3′ end of the construct and the flanking sequence of *Gt(ROSA)Sor26* and yield a 511 bp amplicon (*see* **Note 13**).

9. PCR products should be analyzed by electrophoresis using a 2% agarose. Clones yielding an amplicon corresponding to the 5′ and the 3′ PCR screen are clones in which a true RMCE event has occurred and should be selected for further analysis. Details of the screen and an example of the results obtained is shown in Fig. 6.

3.2.3 Thawing of Correctly Recombined Clones and Further Validation by Southern Blot

An overview of the cell culture steps involved in this protocol is shown in Fig. 7.

1. Correctly recombined clones are thawed by warming a frozen copy of the plate, either by the body heat of the palm of the hand or by placing the frozen plate in a watertight autoclave bag and submerging in a 37 °C water bath.
2. Per clone to be thawed, prepare a well of a 24-well plate pre-coated with 1×10^5 mitotically inactivated neomycin-resistant Mefs (equivalent to 2.4×10^6 across a whole 24-well plate). Prior to thawing, replenish the media with 1 mL of fresh JM8F6 media.
3. When the ice crystals begin to float free from the edges of the thawed wells, place the plate under the tissue culture hood and encourage the candidate wells to thaw completely by warming the bottom of the candidate wells with your finger.

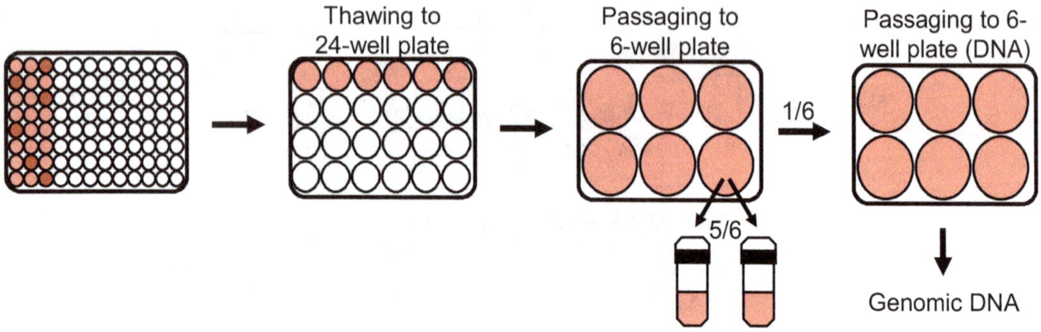

Fig. 7 Flow diagram showing the recommended steps for thawing and expansion of positive clones

4. The individual clones are in a final freezing volume of 200 μL, with a layer of mineral oil on top. Take a 200 μL pipette set on 190 μL and firmly pipette the thawed contents of the target clone up and down, trying to avoid disrupting the oil–media interface at the top of the media. The pipetting is recommended because the majority of the cells will have settled onto the bottom of the 96-well during the freezing process. During this pipetting action, some oil will inevitably transfer but this is harmless.
5. Transfer 190 μL of the thawed cell suspension and transfer this to a single well of a 24-well plate, prepared with fresh JM8F6 media. Distribute the 190 μL over the surface of the well and place the plate back in the incubator and culture overnight at 37 °C/5% CO_2.
6. Feed the cells daily in JM8F6 media, supplemented with 105 μg/mL G418. Typically, the wells will be approximately 70% confluent after approximately 3–4 days.
7. Trypsinize the cells by washing 2× in 0.5 mL of PBS without Mg^{2+} and Ca^{2+}, removing the final PBS wash before adding 150 μL of 2X + G Trypsin and incubating for 10 minutes at 37 °C, agitating the cells half way through this incubation.
8. Neutralize the trypsinization reaction by adding 3 mL of JM8F6 media to each well and resuspend firmly to obtain a single cell suspension.
9. Centrifuge the cell suspensions at 180 *g* for 5 minutes, resuspend the pellets in 3 mL of fresh JM8F6 media supplemented with 105 μg/mL G418, and plate each processed clone on a single well of a 6-well plate, pre-coated with 5×10^5 mitotically inactivated neomycin-resistant Mefs (equivalent to 3×10^6 across a whole 6-well plate). Place the plate back in the incubator and culture overnight at 37 °C/5% CO_2.
10. Feed the cells daily in JM8F6 media supplemented with 105 μg/mL G418 until the cells reach 70% confluency (usually 2–3 days).

11. Trypsinize the cells by washing 2× with 3 mL of PBS without Mg^{2+} and Ca^{2+}, removing the final PBS wash before adding 500 µL of 2X + G Trypsin and incubating for 10 minutes at 37 °C, agitating the cells halfway through this incubation.
12. Neutralize the trypsinization reactions by adding 5 mL of JM8F6 media to each well and resuspend firmly to obtain single cell suspensions.
13. Centrifuge the cell suspensions at 180 *g* for 5 minutes, resuspend each pellet in 600 µL of fresh JM8F6 media.
14. Transfer 500 µL of each cell suspension to a cryovial and add 500 µL of 2× Freezing Mix. Mix well and transfer 500 µL of this 1 mL mixture to a further tube. This results in two identical cryovials containing the equivalent cells of approximately half of a 6-well in 500 µL of volume, per cell clone. These should be labelled with the cell line name, the clone number, the passage number, and the freezing date. Transfer to a −80 °C freezer in a polystyrene box or equivalent freezing unit to achieve a controlled rate of slow freezing (approximate −1 °C per minute).
15. The remaining 100 µL should be resuspended in 3 mL of JM8F6 media and transferred to a freshly gelatinized well of a 6-well plate, pre-prepared by covering the surface of the well with 0.1% gelatin followed by 5 minutes incubation at room temperature, followed by aspiration of the gelatin solution. Place the plate back in the incubator and culture overnight at 37 °C/5% CO_2.
16. Grow the cells to 100% confluency, replacing the media daily.
17. Lyse the cells overnight in 1 mL of lysis buffer.
18. Collect the cell lysates into individual sterile microfuge tubes and prepare genomic DNA by adding 100 µL of 8 M LiCl, 1 mL of isopropanol and mixing by inversion until a strand of genomic DNA becomes visible and can be fished out of the alcohol mix using a plastic pipette.
19. Rinse the DNA strand which should adhere to the plastic pipette in 500 µL of 70% ethanol; remove the pipette, checking that the DNA strand is still adhered; and place into 50 µL of 0.1 TE buffer. Mix the pipette in the 0.1 TE buffer until the DNA strand detaches from the pipette.
20. Incubate overnight at 4 °C while shaking to fully resuspend the genomic DNA.
21. This genomic DNA can be used to verify the genotype of the cell clones—we recommend first repeating the PCR screen using a dilution of this DNA as a template (concentrations of this preparation will be much higher than the original 96-well plate preparation) to ensure that the correct clones have been thawed, expanded, and archived.

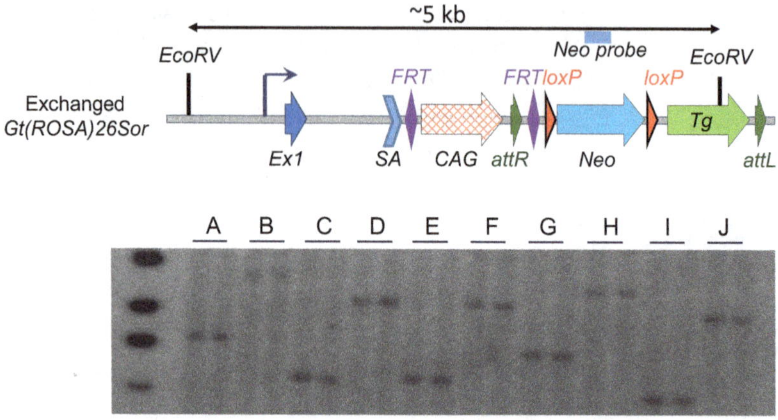

Fig. 8 Southern blot confirmation of positive clones. Top panel shows the conformation of the correctly exchanged allele with the position of example restriction sites (EcoRV) used for the digestion of genomic DNA. The single fragment to be detected by a probe against the neomycin gene is shown. Note the enzyme used and the size of the unique fragment will vary depending upon the sequence integrated. Bottom panel shows a Southern blot obtained when analyzing two independent clones for ten different transgenes. Each different transgene reveals a different size of detected fragment, due to the different transgene sequences used

22. We then recommend using the remainder of the genomic DNA for Southern blot analysis using a probe against the neomycin resistance gene, following standard procedures (*see* **Note 14**). An example of the Southern analysis is shown in Fig. 8. As an alternative to Southern blotting, quantitative PCR analysis can be used to assess the copy number of the neomycin resistance gene [49].

3.2.4 Downstream Steps

1. Following RMCE, the ES cell clones harbor the transgene of interest but no expression will result due to the presence of the loxP flanked neomycin selection cassette, acting as a transcriptional or translational stop. These ES cells can be used to generate a line of genetically modified mice by microinjection into or aggregation with pre-implantation embryos (*see* **Note 15**). Alternatively, the cells can be used for in vitro differentiation experiments.
2. Transgene overexpression can be activated by introducing the Cre recombinase into either cells by transfection or electroporation or mice by breeding with Cre recombinase expressing transgenic strains. The resulting recombination between the two loxP sites flanking the neomycin cassette will excise this intervening sequence linking the CAG promoter to the cDNA of interest and turning on strong overexpression in cells and tissues where Cre recombinase has been active. Figure 9 shows an example of Cre recombinase-dependent activation of transgene expression, using the nonagouti gene as a transgene.

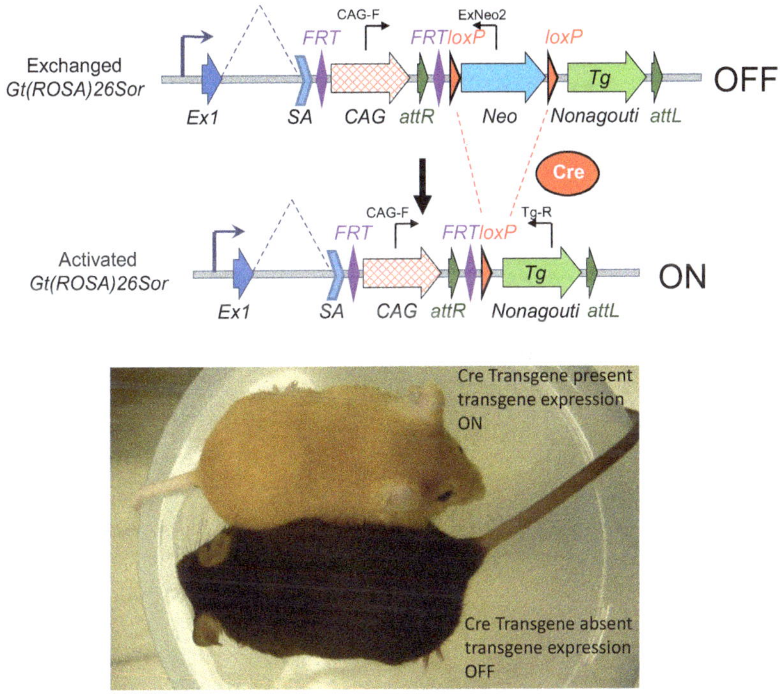

Fig. 9 Cre recombinase activation of transgene expression. Top panel shows the conformation of the exchanged allele, before and after the action of Cre recombinase. Cre recombinase deletes the neomycin selection cassette which acts as a transcriptional STOP. Genotyping primers are shown, the Tg-R primer is designed to be specific for a sequence at the 5′ end of the integrated transgene. The lower panel shows an example of the system used for Cre recombinase-dependent activation of a nonagouti transgene, which, when activated, leads to a strong yellow pigmentation phenotype in mouse

3. To detect this genomic deletion event and activation of the transgene, we recommend a PCR screen which uses the CAG-F primer, binding within the CAG promoter, in combination with a specific reverse primer which binds within the 5′ of the cDNA of interest. A suitably sized amplicon (depending on the position of the primer) will indicate successful deletion (*see* **Note 16**).

4 Notes

1. The RS-PhiC ES cell line and accompanying CB93 shuttle vector represent one of many suitable systems that have been published for RMCE at *Gt(ROSA)26Sor*. The RS-PhiC ES cell line is particularly easy to use as the cells express the enzyme performing the exchange event (PhiC31) constitutively from within the locus. When an appropriate shuttle vector is introduced into the cells, as outlined in the protocol, the exchange that occurs also excises this integrase cassette. Other ES cell

lines and accompanying shuttle vectors are available which use PhiC31 integrase [26], Flp recombinase [20, 24, 32], and Cre recombinase [24, 31] to achieve the RMCE event. Manipulation of these cell lines would necessitate a different transfection regime (e.g., most probably the shuttle vector for these systems would need to be co-introduced with an expression vector for the integrase or recombinase) and a different screening regime, but the broad steps of cell culture outlined in this protocol would still be appropriate for these alternative RMCE systems.

2. The RS-PhiC ES cell line includes the provision of a splice-acceptor and an FRT site upstream of the introduced CAG promoter. If an additional FRT site is included within the shuttle vector (this is present within the CB93 shuttle vector backbone (*see* Fig. 2b)), this can be used to delete the CAG promoter from the engineering *Gt(ROSA)26Sor* locus. The splice acceptor positioned in intron 1 remains and can be used to drive expression of transgenes from the endogenous *Gt(ROSA)26Sor* promoter. This additional capability is not described in this protocol, to avoid complication. However, this Flp recombinase feature has been fully described elsewhere [47].

3. For short cDNA sequences, gene synthesis by a commercial supplier can be chosen. In this case, the 5′ and 3′ extremities of the cDNA should incorporate the unique restriction sites which will allow cloning into the enzyme sites present at the polylinker within CB93. Ensure that a Kozak sequence is present around the start methionine codon.

4. Where gene synthesis has been chosen, we recommend ordering the cDNA cloned into a vector. Alternatively, linear fragments can be synthesized and cloned into a blunt cloning vector. For PCR, we recommend the use of polymerases with a proof-reading activity for the amplification of cDNA to avoid PCR generated mutations. Frequently these generate DNA products which are blunt. Cloning of these PCR products must thus proceed using a blunt cloning kit. Alternatively, some PCR polymerases add an adenosine triphosphate residue to the 3′ ends of the double stranded PCR product—in such cases, a TA cloning kit should be used.

5. If there is no difference in colony number between the experimental ligation and the control ligations, consider repeating the isolation of the vector and insert fragments. The most likely cause is that the individual vector and insert fragments contain appreciable (yet invisible on a gel) amounts of undigested plasmid, which contribute to a high background of colonies. Ensuring a thorough and complete digest is essential to avoid this kind of contaminating plasmid.

6. RS-PhiC ES cells are derived from JM8F6 ES cells which typically require Mef feeder layers for their optimal culture. For selection purposes, the Mefs used must harbor an appropriate selection cassette (in this protocol, neomycin resistance is used). Mefs can be prepared from embryos harboring the required selection cassette and mitotically inactivated using either mitomycin C or by irradiation and either used fresh or frozen for later use. Alternatively, inactivated Mefs can be purchased from commercial suppliers. The Tg(DR4)1Jae/J (Jax stock number 003208) mouse strain, which harbors resistance cassettes for neomycin, puromycin, and hygromycin and has an inactive *Hprt* gene is particularly useful for Mef preparation as the derived Mef cultures can be used for diverse selection regimes.
7. If large clumps of cells are visible after 10 minutes of incubation in 2X + G Trypsin, consider leaving the cells for a further 2 minutes of incubation at 37 °C. Alternatively, pipette the cells up and down firmly while in 2X + G Trypsin, prior to neutralizing with the JM8F6 media.
8. When counting the cells, it will become clear whether a single cell suspension has been generated or whether there are still large clumps of cells. Small clumps of cells are occasionally seen and are no cause for concern, providing the majority of cells are in single suspension. If large amounts of clumps are visible, consider re-treating the pelleted cells with 2X + G Trypsin and repeating the neutralization and vigorous resuspension.
9. The exact efficiency of colony recovery will depend on many factors, but efficiencies vary according to the size of the DNA construct to be inserted. Generally, we find plating approximately 1/3 of the electroporation reaction is sufficient to yield colonies which are of suitable density for picking. Potentially, this fraction can be adjusted based on the expected number of clones—increasing the proportion of the electroporation to be plated where the insert size is large.
10. Colonies can be visualized in the 2X + G Trypsin drop on the U-bottomed 96-well plate under the stereoscope to ensure that they have been transferred.
11. If all the colonies are picked at the same size and the efficiency of the trypsinization is equal across all the colonies, then an approximately equal number of cells should be plated. These should grow at an equal rate and the wells of the 96-well plate should all approach confluency synchronously. Clearly, this represents an ideal situation, and frequently some wells have fewer or slow-growing cells and other wells have more or fast-growing cells. Overall, a decision must be made to passage the

plate when the average confluency of the plate is 70%—some wells will be passaged prematurely and other wells may be overgrown, but the majority of the plate should be at the correct confluency for passaging.

12. Numerous products exist for the freezing and archiving of cells on a 96-well plate format and any of these products would be suitable for this step. Some systems have individual capped tubes within the 96-well plate format, enabling a seal against the environment and the ability to thaw individual clones. For the freezing, it is important to obtain a steady and slow rate of chilling—approximately −1 °C per minute is ideal. Suitable freezing vessels or containers which help achieve this controlled rate can be used.
13. We frequently perform these two screening PCRs in a single multiplex reaction. Optimization may be needed, depending on exact PCR conditions and reagents used, and subsequently it may be a good idea to perform the two PCRs separately in the first instance. Invariably, all colonies are positive for the 5′ PCR, reflecting the reconstitution of a functional CAG driven neomycin cassette (which has been selected for), and approximately 70% of the colonies are positive for the 3′ PCR. 5′ and 3′ positive colonies correspond to true RMCE events, and it is these clones that should be selected. Clones which are 5′ positive but negative for the 3′ screening PCRs represent clones in which the whole of the shuttle vector sequence, including the backbone, have integrated into the *Gt(ROSA) 26Sor* locus—although these represent site-specific integration events, the integration of the vector plasmid sequence is not considered useful, and thus these clones should be discounted.
14. The exact details of the Southern blot screen will depend on the sequence integrated, but hybridization of a suitable restriction digest of the genomic DNA should yield a single band of predictable size. The presence of multiple bands is indicative of additional random integration and clones revealing this pattern of hybridization should be discounted.
15. When the targeted ES cells are to be used for the generation of lines of genetically modified mice, additional quality control steps can be advisable. We recommend following the requirements and advice of your local transgenic facility who can advise about what measures are appropriate prior to commencing in vivo work. Examples of required quality control are karyotype analysis and pathogen screening of the cultured ES cells.
16. When Cre recombinase is introduced into the system, the floxed neomycin selection cassette is deleted and the expression of the transgene is strongly activated. As well as screening for the removal of this cassette, as outlined, we also recommend

screening for the original un-recombined allele, i.e., using the CAG-F and ExNeo2 PCR. We find that Cre recombinase delivery, no matter how efficient, can lead to mosaic outcomes, and subsequently it is important to genotype mice or ES cell clones carefully. Non-mosaic Cre-deleted clones should be positive for the PCR using the CAG-F primer and a reverse primer in the gene of interest and should be negative for the 5′ RMCE screening PCR (CAG-F/ExNeo2).

References

1. Wang H, Yang H, Shivalila CS et al (2013) One-step generation of mice carrying mutations in multiple genes by CRISPR/Cas-mediated genome engineering. Cell 153(4): 910–918. https://doi.org/10.1016/j.cell.2013.04.025
2. Bradley A, Anastassiadis K, Ayadi A et al (2012) The mammalian gene function resource: the International Knockout Mouse Consortium. Mamm Genome 23(9–10):580–586. https://doi.org/10.1007/s00335-012-9422-2
3. Barbaric I, Miller G, Dear TN (2007) Appearances can be deceiving: phenotypes of knockout mice. Brief Funct Genomic Proteomic 6(2):91–103. https://doi.org/10.1093/bfgp/elm008
4. Knight JC (2014) Approaches for establishing the function of regulatory genetic variants involved in disease. Genome Med 6(10):92. https://doi.org/10.1186/s13073-014-0092-4
5. Palmiter RD, Brinster RL (1986) Germ-line transformation of mice. Annu Rev Genet 20: 465–499. https://doi.org/10.1146/annurev.ge.20.120186.002341
6. Beier DR, Morton CC, Leder A et al (1989) Perinatal lethality (ple): a mutation caused by integration of a transgene into distal mouse chromosome 15. Genomics 4(4):498–504
7. Cain-Hom C, Splinter E, van Min M et al (2017) Efficient mapping of transgene integration sites and local structural changes in Cre transgenic mice using targeted locus amplification. Nucleic Acids Res 45(8):e62. https://doi.org/10.1093/nar/gkw1329
8. Goodwin LO, Splinter E, Davis TL et al (2019) Large-scale discovery of mouse transgenic integration sites reveals frequent structural variation and insertional mutagenesis. Genome Res 29(3):494–505. https://doi.org/10.1101/gr.233866.117
9. Cranston A, Dong C, Howcroft J et al (2001) Chromosomal sequences flanking an efficiently expressed transgene dramatically enhance its expression. Gene 269(1–2):217–225
10. Dobie KW, Lee M, Fantes JA et al (1996) Variegated transgene expression in mouse mammary gland is determined by the transgene integration locus. Proc Natl Acad Sci U S A 93(13):6659–6664
11. Garrick D, Fiering S, Martin DI et al (1998) Repeat-induced gene silencing in mammals. Nat Genet 18(1):56–59. https://doi.org/10.1038/ng0198-56
12. Bronson SK, Plaehn EG, Kluckman KD et al (1996) Single-copy transgenic mice with chosen-site integration. Proc Natl Acad Sci U S A 93(17):9067–9072
13. Wallace H, Ansell R, Clark J et al (2000) Pre-selection of integration sites imparts repeatable transgene expression. Nucleic Acids Res 28(6):1455–1464
14. Zambrowicz BP, Imamoto A, Fiering S et al (1997) Disruption of overlapping transcripts in the ROSA beta geo 26 gene trap strain leads to widespread expression of beta-galactosidase in mouse embryos and hematopoietic cells. Proc Natl Acad Sci U S A 94(8):3789–3794
15. Soriano P (1999) Generalized lacZ expression with the ROSA26 Cre reporter strain. Nat Genet 21(1):70–71. https://doi.org/10.1038/5007
16. Mao X, Fujiwara Y, Orkin SH (1999) Improved reporter strain for monitoring Cre recombinase-mediated DNA excisions in mice. Proc Natl Acad Sci U S A 96(9):5037–5042
17. Belteki G, Haigh J, Kabacs N et al (2005) Conditional and inducible transgene expression in mice through the combinatorial use of Cre-mediated recombination and tetracycline induction. Nucleic Acids Res 33(5):e51. https://doi.org/10.1093/nar/gni051
18. Yu HM, Liu B, Chiu SY et al (2005) Development of a unique system for spatiotemporal and lineage-specific gene expression in mice. Proc Natl Acad Sci U S A 102(24): 8615–8620. https://doi.org/10.1073/pnas.0500124102
19. Miyazaki S, Yamato E, Miyazaki J (2004) Regulated expression of pdx-1 promotes

in vitro differentiation of insulin-producing cells from embryonic stem cells. Diabetes 53(4):1030–1037

20. Haenebalcke L, Goossens S, Dierickx P et al (2013) The ROSA26-iPSC mouse: a conditional, inducible, and exchangeable resource for studying cellular (De)differentiation. Cell Rep 3(2):335–341. https://doi.org/10.1016/j.celrep.2013.01.016
21. Mao J, Barrow J, McMahon J et al (2005) An ES cell system for rapid, spatial and temporal analysis of gene function in vitro and in vivo. Nucleic Acids Res 33(18):e155. https://doi.org/10.1093/nar/gni146
22. Vidigal JA, Morkel M, Wittler L et al (2010) An inducible RNA interference system for the functional dissection of mouse embryogenesis. Nucleic Acids Res 38(11):e122. https://doi.org/10.1093/nar/gkq199
23. Chen CM, Krohn J, Bhattacharya S et al (2011) A comparison of exogenous promoter activity at the ROSA26 locus using a PhiiC31 integrase mediated cassette exchange approach in mouse ES cells. PLoS One 6(8):e23376. https://doi.org/10.1371/journal.pone.0023376
24. Tchorz JS, Suply T, Ksiazek I et al (2012) A modified RMCE-compatible Rosa26 locus for the expression of transgenes from exogenous promoters. PLoS One 7(1):e30011. https://doi.org/10.1371/journal.pone.0030011
25. Seibler J, Kuter-Luks B, Kern H et al (2005) Single copy shRNA configuration for ubiquitous gene knockdown in mice. Nucleic Acids Res 33(7):e67. https://doi.org/10.1093/nar/gni065
26. Hitz C, Wurst W, Kuhn R (2007) Conditional brain-specific knockdown of MAPK using Cre/loxP regulated RNA interference. Nucleic Acids Res 35(12):e90. https://doi.org/10.1093/nar/gkm475
27. Hohenstein P, Slight J, Ozdemir DD et al (2008) High-efficiency Rosa26 knock-in vector construction for Cre-regulated overexpression and RNAi. PathoGenetics 1(1):3. https://doi.org/10.1186/1755-8417-1-3
28. Nyabi O, Naessens M, Haigh K et al (2009) Efficient mouse transgenesis using Gateway-compatible ROSA26 locus targeting vectors and F1 hybrid ES cells. Nucleic Acids Res 37(7):e55. https://doi.org/10.1093/nar/gkp112
29. Seibler J, Schubeler D, Fiering S et al (1998) DNA cassette exchange in ES cells mediated by Flp recombinase: an efficient strategy for repeated modification of tagged loci by marker-free constructs. Biochemistry 37(18):6229–6234. https://doi.org/10.1021/bi980288t
30. Turan S, Zehe C, Kuehle J et al (2013) Recombinase-mediated cassette exchange (RMCE) – a rapidly-expanding toolbox for targeted genomic modifications. Gene 515(1):1–27. https://doi.org/10.1016/j.gene.2012.11.016
31. Chen SX, Osipovich AB, Ustione A et al (2011) Quantification of factors influencing fluorescent protein expression using RMCE to generate an allelic series in the ROSA26 locus in mice. Dis Model Mech 4(4):537–547. https://doi.org/10.1242/dmm.006569
32. Sandhu U, Cebula M, Behme S et al (2011) Strict control of transgene expression in a mouse model for sensitive biological applications based on RMCE compatible ES cells. Nucleic Acids Res 39(1):e1. https://doi.org/10.1093/nar/gkq868
33. Hermann M, Maeder ML, Rector K et al (2012) Evaluation of OPEN zinc finger nucleases for direct gene targeting of the ROSA26 locus in mouse embryos. PLoS One 7(9):e41796. https://doi.org/10.1371/journal.pone.0041796
34. Perez-Pinera P, Ousterout DG, Brown MT et al (2012) Gene targeting to the ROSA26 locus directed by engineered zinc finger nucleases. Nucleic Acids Res 40(8):3741–3752. https://doi.org/10.1093/nar/gkr1214
35. Kasparek P, Krausova M, Haneckova R et al (2014) Efficient gene targeting of the Rosa26 locus in mouse zygotes using TALE nucleases. FEBS Lett 588(21):3982–3988. https://doi.org/10.1016/j.febslet.2014.09.014
36. Quadros RM, Harms DW, Ohtsuka M et al (2015) Insertion of sequences at the original provirus integration site of mouse ROSA26 locus using the CRISPR/Cas9 system. FEBS Open Bio 5:191–197. https://doi.org/10.1016/j.fob.2015.03.003
37. Wefers B, Bashir S, Rossius J et al (2017) Gene editing in mouse zygotes using the CRISPR/Cas9 system. Methods (San Diego, CA) 121–122:55–67. https://doi.org/10.1016/j.ymeth.2017.02.008
38. Chu VT, Weber T, Graf R et al (2016) Efficient generation of Rosa26 knock-in mice using CRISPR/Cas9 in C57BL/6 zygotes. BMC Biotechnol 16:4. https://doi.org/10.1186/s12896-016-0234-4
39. Shin HY, Wang C, Lee HK et al (2017) CRISPR/Cas9 targeting events cause complex deletions and insertions at 17 sites in the mouse genome. Nat Commun 8:15464. https://doi.org/10.1038/ncomms15464
40. Boroviak K, Fu B, Yang F et al (2017) Revealing hidden complexities of genomic rearrangements generated with Cas9. Sci Rep 7(1):

12867. https://doi.org/10.1038/s41598-017-12740-6

41. Kosicki M, Tomberg K, Bradley A (2018) Repair of double-strand breaks induced by CRISPR-Cas9 leads to large deletions and complex rearrangements. Nat Biotechnol 36(8):765–771. https://doi.org/10.1038/nbt.4192
42. Cullot G, Boutin J, Toutain J et al (2019) CRISPR-Cas9 genome editing induces megabase-scale chromosomal truncations. Nat Commun 10(1):1136. https://doi.org/10.1038/s41467-019-09006-2
43. Skarnes WC (2015) Is mouse embryonic stem cell technology obsolete? Genome Biol 16:109. https://doi.org/10.1186/s13059-015-0673-6
44. Ohtsuka M, Ogiwara S, Miura H et al (2010) Pronuclear injection-based mouse targeted transgenesis for reproducible and highly efficient transgene expression. Nucleic Acids Res 38(22):e198. https://doi.org/10.1093/nar/gkq860
45. Ohtsuka M, Miura H, Mochida K et al (2015) One-step generation of multiple transgenic mouse lines using an improved Pronuclear Injection-based Targeted Transgenesis (i-PITT). BMC Genomics 16(1):274. https://doi.org/10.1186/s12864-015-1432-5
46. Adam J, Yang M, Bauerschmidt C et al (2013) A role for cytosolic fumarate hydratase in urea cycle metabolism and renal neoplasia. Cell Rep 3(5):1440–1448. https://doi.org/10.1016/j.celrep.2013.04.006
47. Dolatshad H, Biggs D, Diaz R et al (2015) A versatile transgenic allele for mouse overexpression studies. Mamm Genome 26(11–12):598–608. https://doi.org/10.1007/s00335-015-9602-y
48. McElroy SL, Reijo Pera RA (2008) Preparation of mouse embryonic fibroblast feeder cells for human embryonic stem cell culture. CSH Protoc 2008:pdb.prot5041. https://doi.org/10.1101/pdb.prot5041
49. Frendewey D, Chernomorsky R, Esau L et al (2010) The loss-of-allele assay for ES cell screening and mouse genotyping. Methods Enzymol 476:295–307. https://doi.org/10.1016/s0076-6879(10)76017-1

Chapter 14

Improved Genome Editing via Oviductal Nucleic Acids Delivery (*i*-GONAD): Protocol Steps and Additional Notes

Masahiro Sato, Ayaka Nakamura, Marie Sekiguchi, Takashi Matsuwaki, Hiromi Miura, Channabasavaiah B. Gurumurthy, Shigeru Kakuta, and Masato Ohtsuka

Abstract

The clustered regularly interspaced short palindromic repeats (CRISPR) technology has made it possible to produce genome-edited (GE) animals more easily and rapidly than before. In most cases, GE mice are produced by microinjection (MI) or by in vitro electroporation (EP) of CRISPR reagents into fertilized eggs (zygotes). Both of these approaches require ex vivo handling of isolated embryos and their subsequent transfer into another set of mice (called recipient or pseudopregnant mice). Such experiments are performed by highly skilled technicians (especially for MI). We recently developed a novel genome editing method, called "GONAD (*G*enome-editing via *O*viductal *N*ucleic *A*cids *D*elivery)," which can completely eliminate the ex vivo handling of embryos. We also made improvements to the GONAD method, termed "improved-GONAD (*i*-GONAD)." The *i*-GONAD method involves injection of CRISPR reagents into the oviduct of an anesthetized pregnant female using a mouthpiece-controlled glass micropipette under a dissecting microscope, followed by EP of the entire oviduct allowing the CRISPR reagents to enter into the zygotes present inside the oviduct, in situ. After the *i*-GONAD procedure, the mouse recovered from anesthesia is allowed to continue the pregnancy to full term to deliver its pups. The *i*-GONAD method does not require pseudopregnant female animals for embryo transfer, unlike the methods relying on ex vivo handling of zygotes. Therefore, the *i*-GONAD method can reduce the number of animals used, compared to the traditional methods. In this chapter, we describe some newer technical tips about the *i*-GONAD method. Additionally, even though the detailed protocols of GONAD and *i*-GONAD have been published elsewhere (Gurumurthy et al., Curr Protoc Hum Genet 88:15.8.1–15.8.12, 2016 Nat Protoc 14:2452–2482, 2019), we provide all the protocol steps of *i*-GONAD in this chapter so that the reader can find most of the information, needed for performing *i*-GONAD experiments, in one place.

Key words *i*-GONAD, CRISPR, Genome editing, Mouse, Electroporation, Oviduct, Knockout, Knock-in, Fertilized egg, Zygotes, Preimplantation embryos, Ribonucleoprotein, gRNA

Thomas L. Saunders (ed.), *Transgenesis: Methods and Protocols*,
Methods in Molecular Biology, vol. 2631, https://doi.org/10.1007/978-1-0716-2990-1_14,

1 Introduction

Genetically engineered mouse (GEM) models are widely used not only for analysis of in vivo gene function but also as a model animal for human diseases. Recent development in genome editing technology, especially the clustered regularly interspaced short palindromic repeats (CRISPR)/CRISPR-associated endonuclease (Cas) (CRISPR-Cas) tool, have made it possible to create GEM models more rapidly and easily than before [1]. The CRISPR system is composed of two components: Cas protein (a nuclease that cleaves DNA) and a guide RNA (gRNA) that directs the nuclease to a specific site in the genome. The gRNA for *Streptococcus pyogenes* Cas9 (SpCas9) protein (one of the most commonly used CRISPR nuclease) consists of crRNA and tracrRNA complexes [2]. The 20 bases on the 5′ end of crRNA sequence is complementary to the target sequence. The genomic sequence must have a protospacer adjacent motif (PAM) sequence (NGG for SpCas9) on the 3′ side of the gRNA target sequence, which allows the Cas9/gRNA complex to recognize the target sequence and introduce a double-stranded DNA break (DSB). Such DSB can be repaired using either non-homologous end joining (NHEJ) or homology directed repair (HDR) pathway. NHEJ pathway results in introduction of small insertion or deletion (indel) mutation in the absence of repair template, and the indels often cause frameshift mutations or lead to premature stop codon, which can be used as gene knockout (KO). With the HDR pathway, which is less error-prone than NHEJ, it is possible to knock-in (KI) foreign DNA sequence at the cleavage site by supplying donor DNA that contains homology arms at both ends of the sequence for KI.

The microinjection (MI) method, by which CRISPR components are physically injected into fertilized eggs (zygotes), is commonly used to produce KO and KI mice without the need for the embryonic stem (ES) cell-based step [1, 3]. Although the MI method is a simpler procedure than the ES cell-based gene targeting method, it still requires ex vivo embryo manipulation steps: namely, isolation of zygotes, MI, in vitro culture of zygotes, and transfer of the MI-treated eggs to pseudopregnant animals. It generally takes large amount of time (at least several months to years) to master the techniques of ex vivo embryo handling and MI. The MI method, in particular, is an extremely difficult technique that can only be performed by highly skilled personnel and requires micromanipulator which is also a very expensive equipment. Therefore, most researchers, who want to produce genome-edited (GE) animals, rely on transgenic core facilities in their institutes or commercial organizations that provide GE animal production services. Even though the recently developed in vitro electroporation (EP) method has gained popularity (that can exclude the need for micromanipulator and MI methods) [4, 5], the ex vivo embryo manipulation steps are still unavoidable.

We recently developed a method called "*G*enome editing via *O*viductal *N*ucleic *A*cids *D*elivery (GONAD)" and improved it further (improved-GONAD (*i*-GONAD)), which lead to easy and convenient production of GE animals without the need for ex vivo embryo manipulation [6, 7]. In this method, the genome editing reagents (Cas9/gRNA complex and donor DNA) are injected into the oviducts of E0.7 day pregnant female mice, which contain fertilized eggs, followed by in vivo EP of the entire oviducts [8]. The EP-treated females are then allowed to continue their pregnancies to deliver the GE founder mice. Notably, unlike the conventional methods for GE animal production that rely on MI and in vitro EP, there is no need to sacrifice female mice to collect their zygotes and to maintain vasectomized males (used for preparation of pseudopregnant mice). Therefore, the number of mice required for the *i*-GONAD experiment is much less compared to the methods that rely on ex vivo zygote handling. Furthermore, *i*-GONAD method is extremely simple compared to the conventional method, and therefore, it is more accessible to most researchers including students who are interested in creating GE mice themselves [9].

The *i*-GONAD method enables production of KO mice by introducing indel mutation or of KI mice using short single-stranded DNA (ssDNA) (~200 bases), with relative ease. Although the efficiency for inserting long inserts still remains to be improved [10], we have succeeded in knocking in of a reporter gene cassette (~900 bases) using longer ssDNA as donor DNA [7]. In addition, there are reports on the successful large deletion and chromosomal inversion models [11, 12]. Furthermore, *i*-GONAD can be used to generate floxed alleles for conditional KO mice [13]. On the other hand, it is still difficult to introduce double-stranded DNA such as plasmid DNA into fertilized eggs by EP-based gene delivery methods (whether ex vivo or in vivo), which makes it difficult to create KI or transgenic mice by using DNA fragments of larger than 1 kb in size. For inserting longer than 0.9 kb cassettes, we do not recommend using EP-based approaches. The most reliable approach for longer cassettes would be via MI; using long ssDNA as donors, up to 2 kb inserts can be efficiently inserted using *Easi*-CRISPR approach [14, 15].

Here, we present step-by-step protocol of the *i*-GONAD method for CRISPR-based creation of KO mice with indel mutations or of KI mice using single-stranded oligodeoxynucleotide (ssODN). Since the basic procedure of *i*-GONAD described here is very similar to our previous protocol paper [16, 17], we will provide additional technical tips in the Subheading 4 of this chapter.

2 Materials

2.1 Mice

1. MCH(ICR) (CLEA Japan) (10–20 weeks old female and 10–50 weeks old male) (*see* **Note 1**).
2. C57BL/6 J (CLEA Japan) (7–10 weeks old female and 10–50 weeks old male).

   Male mice are kept in individual cages, and females can be group-housed at five to six per cage. Housing density of animals should follow institute-approved animal protocol.

2.2 Equipment

Equipment required for *i*-GONAD procedure is shown in Fig. 1.

1. Stereomicroscope.
2. LED light device.
3. Electroporator (CUY21EditII, BEX).
4. Electrodes (LF650P3, BEX).
5. Puller (P-97/IVF, Sutter Instruments, or PN-31, Narishige) (*see* **Note 2**).
6. Multi-panel heater.
7. Shaking heat block.
8. Heat block.
9. Thermal cycler.
10. Gel imaging system.

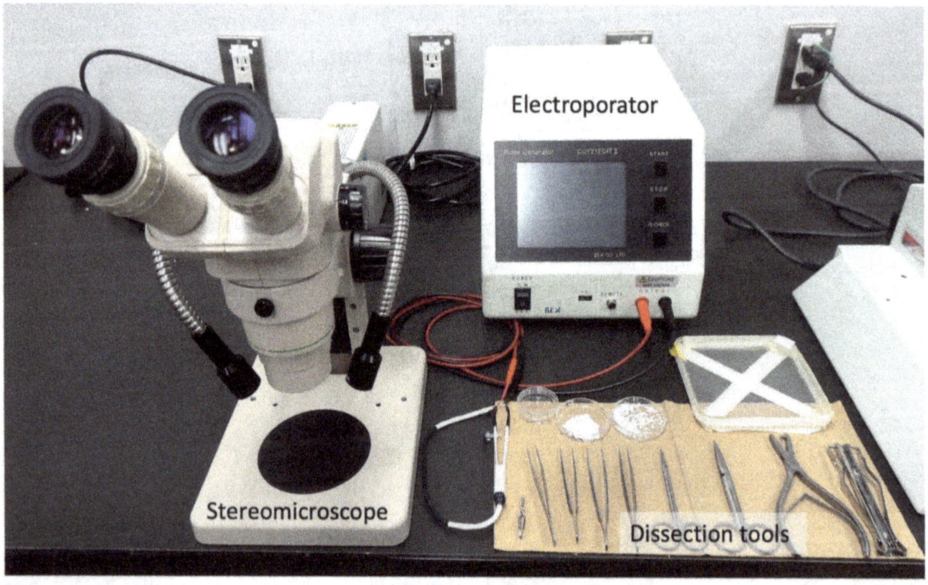

Fig. 1 Tools used for *i*-GONAD procedure. A stereomicroscope, an electroporator, and dissection tools are shown

11. Forceps.
12. Glass capillaries.
13. Aorta clamp.
14. Scissors.
15. Vinyl tape.
16. Square Petri dish.
17. Paper towel.
18. Kimwipe.
19. Suture wound clips.
20. Cotton ball.
21. Mouth pipette.
22. Mouth pipette: Make your own pipette using mouthpieces, silicon tubes, glass tubes, micro caps, etc.

2.3 Reagents

1. Anesthetic solution: Mix 0.75 mL of medetomidine hydrochloride (1.0 mg/mL), 0.8 mL of midazolam (5.0 mg/mL), 1.0 mL of butorphanol tartrate (5.0 mg/mL), and 7.45 mL of normal saline to make 10 mL of anesthetic solution.
2. Atipamezole hydrochloride: Mix 0.15 mL of atipamezole hydrochloride (5.0 mg/mL) and 9.85 mL of normal saline to make 10 mL.
3. Ethanol.
4. 1× PBS (phosphate-buffered saline) (pH 7.2).
5. Alt-R S.p. Cas9 Nuclease V3 (Integrated DNA Technologies (IDT)).
6. Nuclease-Free Duplex Buffer (IDT).
7. gRNA: Dissolve crRNA (IDT) for Cas9 in Nuclease-Free Duplex Buffer to 200 μM (*see* **Note 3**). Dissolve tracrRNA (IDT) for Cas9 in Nuclease-Free Duplex Buffer to 200 μM. Mix equal amounts of crRNA and tracrRNA, incubate at 94 °C for 2 min, and leave at room temperature for 10 min before use as gRNA. Store at −80 °C until use.
8. Nuclease-free water.
9. ssODN: Dissolve ssODN in nuclease-free water to a concentration of 10 μg/μL. Store at −80 °C until use.
10. Opti-MEM medium.
11. Fast Green FCF (Nacalai Tesque): Dissolve in Opti-MEM medium to a concentration of 0.2%. Store at −20 °C or −80 °C until use.
12. Allele-In-One Mouse Tail Direct Lysis Buffer (Allele Biotechnology).
13. Primer set (for genotyping).

14. 50× TAE.
15. SeaKem ME agarose.
16. 6× gel loading buffer.
17. DNA Ladder.
18. Ethidium bromide.

3 Methods

3.1 Preparing Pregnant Female Mice

1. House female mice, of the appropriate estrus cycle, with male mice after 3:00 pm on the previous day of the *i*-GONAD procedure (Fig. 2a) (*see* **Note 4**).
2. Next day morning (around 9:00 am), examine the females and segregate the ones with vaginal plugs (indicative of mating) into separate cages (Figs. 2b and 3) until they are used for the *i*-GONAD procedure. Begin the *i*-GONAD procedure at 4:00 pm, which corresponds to late zygote stage (*see* **Notes 5** and **6**).

3.2 Preparation of Genome Editing Solution

1. Mix each reagent in a 1.5-mL microcentrifuge tube (Table 1) (*see* **Notes 7–9**). The tube should be kept on ice.
2. Just before performing the *i*-GONAD procedure, incubate the tube containing the genome editing solution at 37 °C for 10 min. Then, maintain it at room temperature throughout the *i*-GONAD procedure (~3 h).

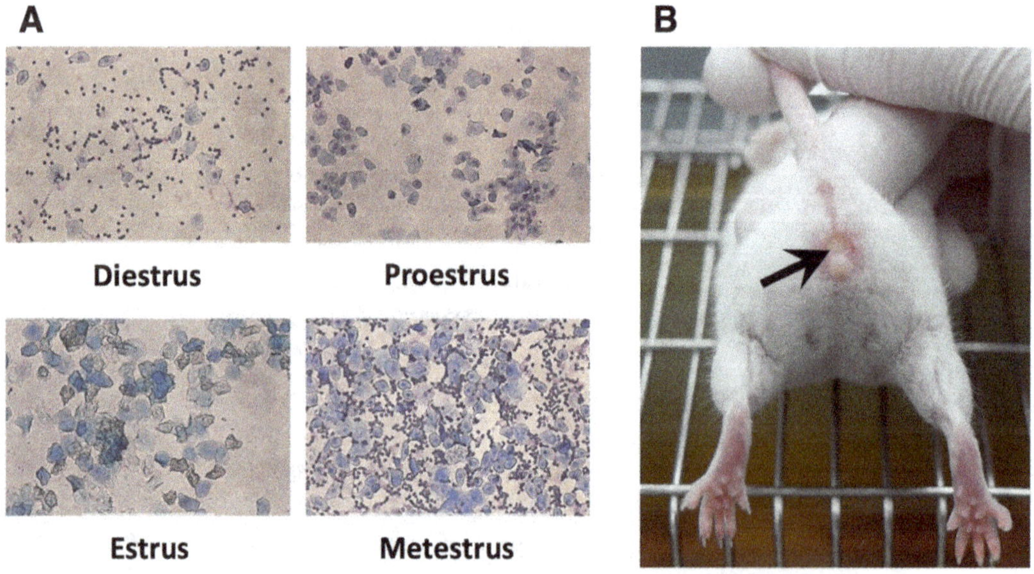

Fig. 2 Confirmation of the estrus cycle by the smear method. Photos after checking vaginal smears are shown in (**a**). The presence of vaginal plug (arrow) generated after mating is discernible (**b**)

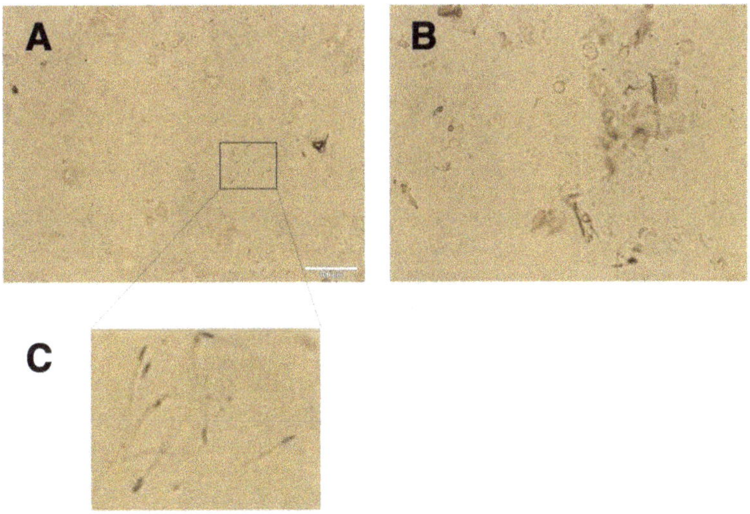

Fig. 3 Confirmation of successful mating by checking the presence of sperm in the vaginal smears. Presence of sperm is apparent in one specimen (**a**), but not in the other specimen (**b**). (**c**) Magnification of a part shown in (**a**)

Table 1
Preparation of genome editing solution (for 3 females)

Reagent (original concentration)	Volume (μL)	Final concentration
crRNA/tracrRNA (100 μM)	3	30 μM
ssODN (10 μg/μL)[a] (for KI)	(2)	2 μg/μL
Cas9 protein (10 mg/mL)	1	1 mg/mL
Fast Green FCF (0.2%)	1	0.02
Opti-MEM	5 (3 for KI using ssODN)	
Total	10	

[a]If you do not use donor DNA (ssODN for KI), add Opti-MEM to compensate for the volume

3.3 Surgical Procedure to Expose the Oviduct

1. Measure the weight of the plugged females and calculate the amount of anesthetic to inject. Inject the anesthetic through intraperitoneal or subcutaneous route using a 1-mL syringe with a needle (0.1 mL/10 g). Return the mice to the cage.
2. Assess anesthesia by pinching the toes of the mice with forceps. Once deep anesthesia is achieved, which is indicated by no pinch response, place the mouse on a square Petri dish and anchor the feet with vinyl tape pieces (Fig. 4a).
3. Spray the back of the anesthetized mouse with 70% ethanol (Fig. 4b). Make incision in the middle of the dorsal skin with scissors (10–15 mm in length) (Fig. 4c,d).

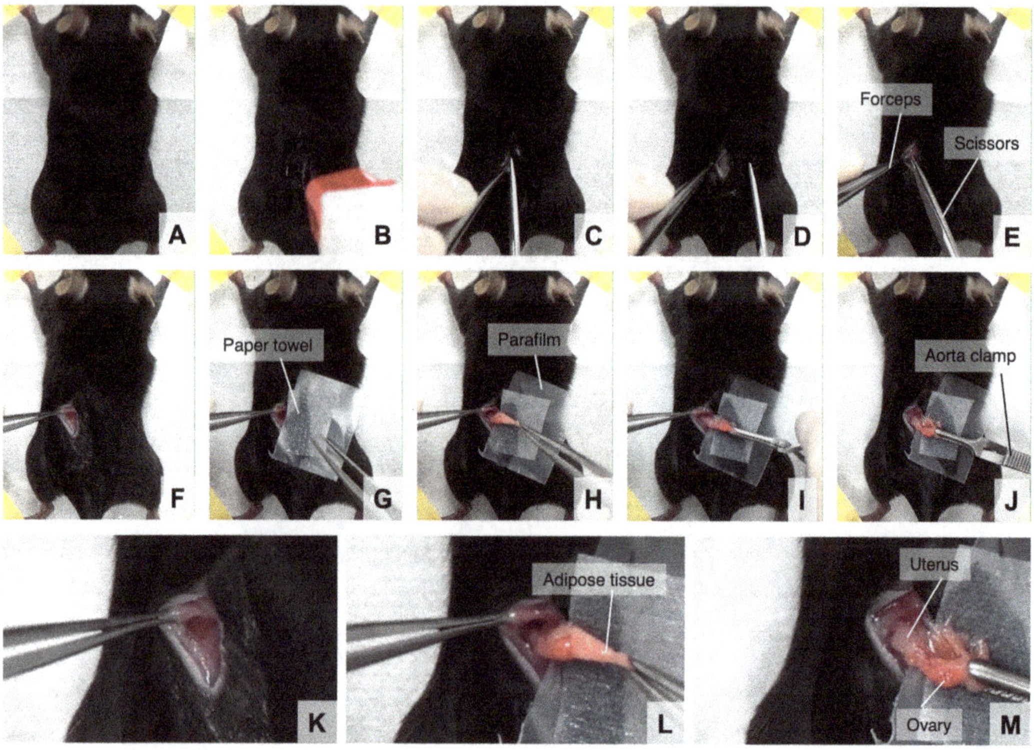

Fig. 4 Steps to expose the oviduct from an anesthetized mouse. The procedure for exposing the left oviduct is shown. (**k**–**m**) are the magnifications of dissection parts shown in (**f**, **h**, **j**), respectively

4. Grasp the left muscle layer on the inner side of the skin with forceps and lift it up (Fig. 4e). Then, make a 5–7 mm incision in the muscle layer with scissors (Fig. 4f).
5. Place a piece of parafilm to the dorsal incision and subsequently a paper towel next onto the parafilm (Fig. 4g) (*see* **Note 10**).
6. Grasp the adipose tissue (connected to ovary/oviduct/uterus) with forceps and pull it outward from the incision (Fig. 4h).
7. The ovary, oviduct, and part of the uterus are exposed along with the adipose tissue. They are placed on a paper towel, and the adipose tissue is pinched and fixed in place with an aorta clamp to prevent these tissues from returning to inside of the body (Fig. 4i,j).
8. Under the stereomicroscope, rotate the aorta clamp and adjust it so that the ampulla of the oviduct (where the fertilized egg resides) is located in the center of the microscope view (Fig. 4m) (*see* **Note 11**).

3.4 Injection of Genome Editing Solution into the Oviduct

1. Collect 1.5 μL of the genome editing solution (prepared in Subheading 3.2) from the bottom of the tube using a pipette tip (attached to the Pipetman) and transfer it to the upper inner wall of the tube.
2. Cut the tip of the capillary with scissors while observing under the stereomicroscope (*see* **Note 12**).
3. Attach the capillary to the tip of a mouth pipette and aspirate 1.5 μL of the solution prepared in Subheading 3.2, **step 1**.
4. While observing under the stereomicroscope, hold forceps in one hand and gently grasp the oviduct. Hold a mouth pipette with a capillary in the other hand and insert the tip of the capillary into the oviductal lumen (Fig. 5a,b) (*see* **Note 13**).

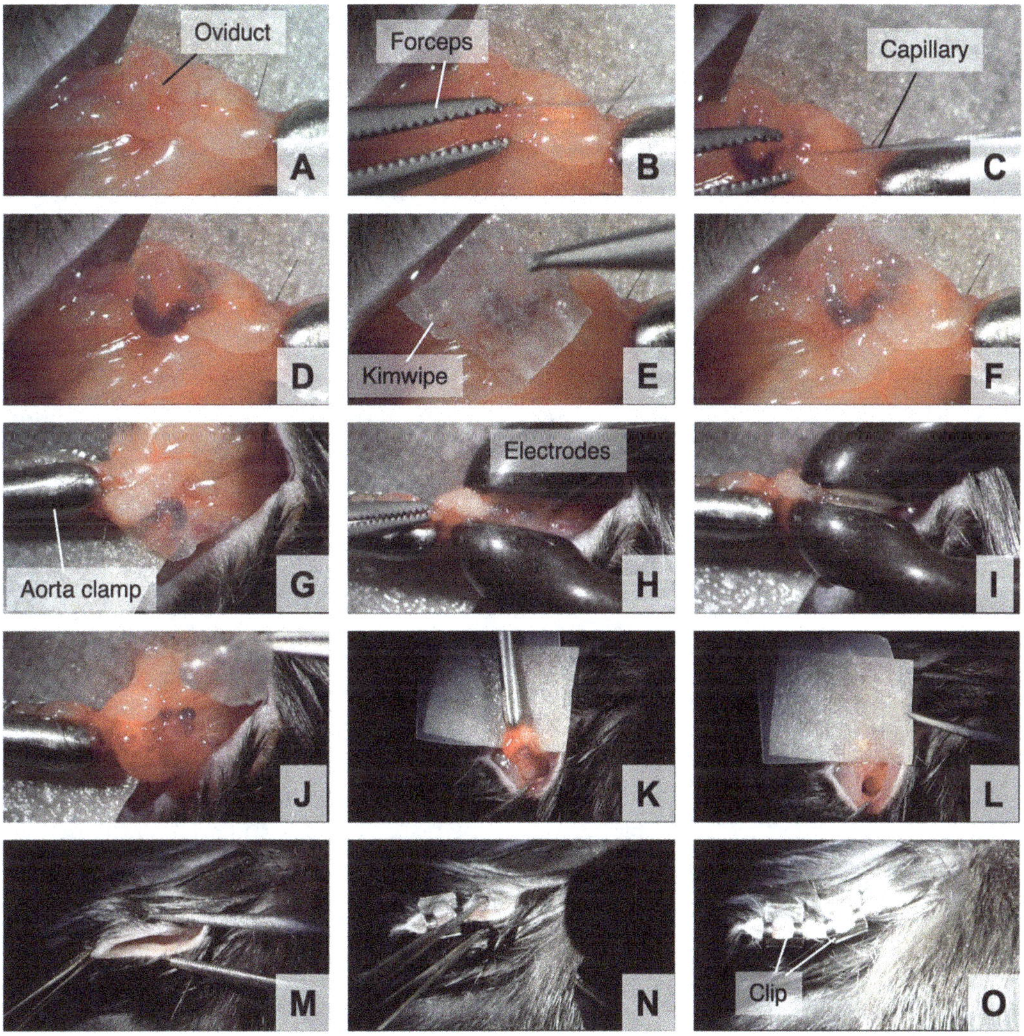

Fig. 5 Intraoviductal injection of a solution and subsequent electroporation toward the injected oviduct

5. Fix the capillary tip in the oviductal lumen and inject the solution into there by blowing it strongly with a mouth pipette (Fig. 5c) (*see* **Notes 14–19**).
6. After the solution is injected, pull the capillary out of the oviduct and proceed to the next step (in vivo EP) immediately.

3.5 In Vivo EP

1. After injecting the solution into the oviductal lumen (Fig. 5d), cover the entire oviduct with a Kimwipe piece soaked in 1× PBS (Fig. 5e–g).
2. Soak the tips of the tweezer-type electrodes in 1× PBS.
3. Place the oviduct covered with a Kimwipe piece between the electrodes (Fig. 5h) (*see* **Note 20**).
4. EP is performed using an electroporator generating square wave pulses (Fig. 5i) (*see* **Note 21**).
5. The output of the electric pulse can be confirmed by the generation of minute bubbles between the electrodes (which is observable under a stereomicroscope: Fig. 5i). The electroporation can also be confirmed by noting the actual electrical conditions (current, voltage, and resistance) displayed on the electroporator (*see* **Note 22**).
6. Remove the electrodes from the oviduct and gently wipe the electrode with a cotton ball soaked in 70% ethanol to remove any adhering blood.
7. Remove the Kimwipe pieces, aorta clamp, and paper towels in that order (Fig. 5j–l).
8. Using forceps, return the ovary, oviduct, uterus, and adipose tissue to their original position in the body (Fig. 5l).

3.6 I-GONAD Procedure on the Right Oviduct

1. Perform *i*-GONAD procedure on the other (right) oviduct (as in Subheadings 3.3, **step 4**, 3.4, and 3.5, **step 8**) (*see* **Note 23**).

3.7 Postoperative Procedures

1. After completion of the *i*-GONAD procedure on both oviducts, the dorsal incision on the skin is closed with a clip (Fig. 5m–o).
2. Inject atipamezole hydrochloride by intraperitoneal or subcutaneous route (0.1 mL/10 g). Transfer the treated mice to the cage placed on a hot plate (37 °C) and wait until they are awake.
3. The *i*-GONAD-treated mice can be analyzed either at mid-embryonic stage (Fig. 6) or after birth, depending on the purpose (*see* **Notes 24–27**).

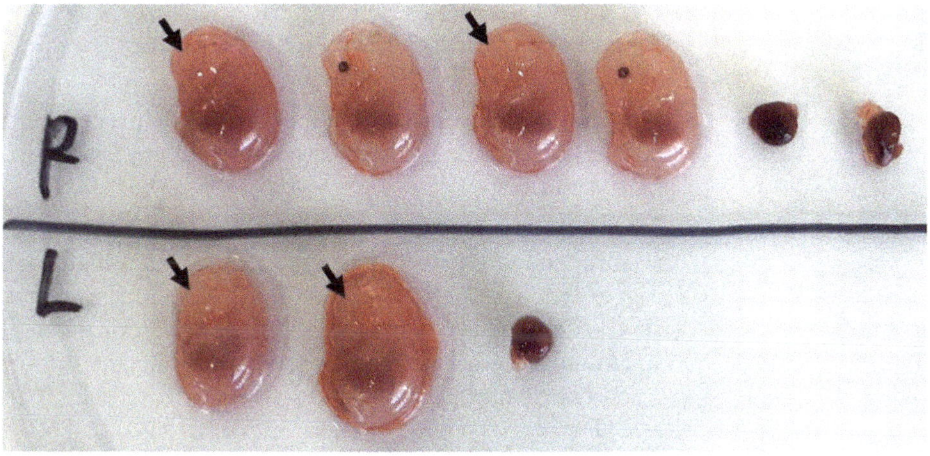

Fig. 6 *Tyrosinase* KO experiment using C57BL/6 J strain. Fetuses and placental remnants were recovered at day 14.5 of pregnancy from the right (R) and left (L) uteri after *i*-GONAD. (Note that the successfully GE fetuses lack pigmented eyes (arrows))

3.8 Genotyping of Founder (F0) Individuals

1. Collect samples of a fetal tail or limb tip (2–3 mm long) or a 2-mm ear piece from a newborn mouse in a 1.5-mL microtube.
2. Add 40 μL of Allele-In-One Mouse Tail Direct Lysis Buffer and incubate at 55 °C for 3 h or overnight using a heat block.
3. Incubate in a heat block at 85 °C for 45 min.
4. Centrifuge at 21,000 g for 2 min at room temperature, and use them for PCR reaction as templates.
5. Prepare the PCR reaction mixture according to the manufacturer's instructions.
6. Perform the PCR reaction using a thermal cycler to amplify the target region.
7. Electrophoresis of PCR products together with DNA markers (DNA Ladder) on a 1% agarose gel at 135 V for 30–60 min. Stain the gel with ethidium bromide (0.5 μg/mL) for 20 min. Take photographs and analyze using a gel imaging system.
8. If necessary, perform T7 endonuclease I (T7EI) assay, restriction enzyme cleavage analysis, or sequencing of PCR products (*see* **Notes 28–31**).

4 Notes

1. For beginners, it is recommended to use the MCH (ICR) hybrid strain to master the technique.
2. Pull capillary to the left or right using a puller, and use it for injecting the solution into the oviduct. If a needle puller is not available, they can be generated by heating capillary with a Bunsen burner flame.

3. Design crRNA using web tools. Most tools serve the purpose. Two examples are listed here: CRISPOR (http://crispor.tefor.net/) or CHOPCHOP (http://chopchop.cbu.uib.no/).
4. We do not recommend superovulation procedure using eCG-hCG because it often causes reduced implantation, pregnancy, and delivery rates.
5. In some cases, it may be hard to detect vaginal plugs. It is recommended to use experienced females (that had prior mating experience) or those recognized as being at appropriate stage of estrus cycle after checking vaginal smears (Fig. 2a). Examining vaginal smears can serve as a confirmatory method to ensure mating, if vaginal plug detection method is inconclusive (Fig. 3). In our experience, 52.5% (21/40) of C57BL/6 J females, which were recognized as being in the pre-estrus cycle after vaginal smear examination, had vaginal plugs when mated with C57BL/6 J males. This efficiency may be further improved by using experienced females. In the case of the MCH (ICR) hybrid strain, it is easy to identify pre-estrus individuals through inspection of vaginal coloration, and thus examining vaginal smears may not be necessary. Our success rate in identifying mated females by vaginal plug examination of MCH (ICR) hybrid strain is >90%.
6. We typically use about 3 to 5 plugged females per one session of *i*-GONAD experiments. We suggest to start with mating of about 6–7 females, which may yield about 3 to 5 successfully mated females. Note that inbred strains like C57BL/6 J may require more mating pairs. For a beginner, two pregnant mice per session may be preferable, whereas experienced researchers can handle as many as eight females per session.
7. When three plugged females are used, 10 μL of CRISPR EP solution is generally sufficient (since 1.5 μL is used for each oviduct). We recommend preparing higher amount of solution for beginners.
8. Use of a filtered tip is recommended since the CRISPR EP solution contains RNA.
9. The *i*-GONAD method uses higher concentration of Cas9 protein (1 μg/μL) compared to the MI or in vitro EP methods. However, we have also confirmed that the use of lower concentrations of Cas9 protein (e.g., 100 to 500 ng/μL) is enough to obtain GE individuals [17].
10. In order to prevent the leakage of electric current from the aorta clamp to the mouse due to the wet fur by alcohol disinfection during skin incision, place folded paper towels (four or more folds) or parafilm sheets on the incision area (Fig. 4g). This might improve the pregnancy rate.

11. Take precaution not to cut any blood vessels during exposure of the oviduct. If bleeding occurs, use a piece of paper towel to remove blood as much as possible before injecting the genome editing solution.
12. The optimal inner diameter of the capillary tip is 50–70 μm.
13. Be sure to insert the capillary in the direction from upstream (ovary side) to downstream (uterus side) of the ampulla of the oviduct. If the solution is injected from downstream (uterus side) to upstream (ovary side) of the ampulla, the embryos may be hampered to move into the uterus for their implantation.
14. The oviduct can be secured in place by holding with forceps during the injection procedure (which may prevent the back-flow of the injected solution to some extent), but it is not always necessary. It is suggested to spread the injected solution throughout the ampulla by gentle squeezing, and the spread of solution is easily recognized by green coloration inside the oviduct (Fig. 5c).
15. This is one of the very important steps that determine the success of the *i*-GONAD. As with any surgical technique, mastering *i*-GONAD can take a few attempts. Some researcher may learn the technique by performing the procedure on just a couple mice, whereas some may require many more attempts. Note that injection may not be successful in a single trial even after mastering the technique. The beginners may try their first injection at positions one or two turns upstream (ovary side) from the ampulla. If this is not successful, a second injection attempt on the same mouse can be done at a position closer to the ampulla.
16. During injection, body fluid or blood may flow into the capillary and clog the needle tip, but this can often be overcome by cutting the tip of the capillary a bit more. Alternatively, it is possible to use a new capillary. For this reason, it is advised to prepare extra genome editing solution, as a back-up.
17. After the capillary is inserted into the oviduct, sometimes the solution might not be easily injected when you blow on it, suggestive of needle clogging. Another possibility is that the tip of the capillary is tightly attached to the inner wall of the oviduct. To overcome this, pull the capillary slightly backward and try blowing; sometimes it may need stronger blowing to ensure successful injection.
18. If you find that the solution is leaked out of the oviduct, you might have pierced the opposite side of the oviductal wall with the capillary. In this case, move the capillary back and forth slightly to make sure that the capillary tip is inside the oviduct. In case that the capillary pierces the oviductal wall, give up injection from that site and try to inject from another region of the oviduct (which is closer to the ampulla).

19. During the injection process, you may encounter that air bubbles are introduced along with the solution into the oviductal lumen, which is often caused by over-blowing. However, this may not be a serious matter, because we have successfully obtained GE fetuses from the *i*-GONAD procedures performed in the presence of such air bubbles.
20. The green color of the Fast Green FCF can be seen through the Kimwipe piece, allowing to confirm that the oviduct is correctly held between the electrodes.
21. The electroporation settings for the CUY21EditIII (BEX) electroporator are as follows: square (mA), (+/−); PdV, 80 V; PdA, 100–150 mA; Pd ON, 5.00 ms; Pd OFF, 50 ms; Pd N, 3; decay 10%; decay type, Log. As for the PdA value, GE pups can be obtained relatively stably under any current value between 100 and 150 mA for the MCH (ICR) strain, but when using the C57BL/6 J strain, it is better to apply current value of 100 mA. It is advisable to explore the optimal EP conditions when researchers use other mouse strains that are not described in our previous experimental protocols for *i*-GONAD.
22. The actual voltages can be around 15–35 V when EP is performed at 100 mA current setting.
23. There is no need to make a new dorsal skin incision. The second side oviduct can be accessed through the same (dorsal) incision of the skin. However, a new incision on the right side is necessary for the muscle layer.
24. When *i*-GONAD-treated mice are under some stress or have small number of fetuses, they may not give birth naturally. To address such a situation, we recommend preparing foster mother mice. If a female does not give birth on the scheduled delivery date (usually on day 20 of pregnancy; day 0 is designated as the day when vaginal plug is recognized), Caesarean section should be performed to retrieve the pups in the morning of day 21 of pregnancy. The rescued pups are then nursed by the foster mother mice. Our experience is that ICR strain mothers sometimes do not take care of small inbred newborns, while BALB/c strain mice are good mothers. Therefore, we suggest preparing BALB/c surrogate mothers than ICR surrogates.
25. Environmental enrichment (e.g., Shepherd Shack) may be appropriate to prevent the pups from being eaten by mothers.
26. If the EP conditions are not appropriate, the fertilized eggs are easily injured, leading to loss of developing fetuses. Since the optimal conditions vary depending on the mouse strain and the equipment used, it is wise to perform a few *i*-GONAD procedures, using your EP equipment, to determine the optimal conditions.

27. If a gene of interest (GOI) to be targeted is essential for embryonic development, it is highly likely that only a few pups are obtained because homozygous offspring die during gestation. Knowledge about the embryonic lethality of GOI before starting *i*-GONAD experiment may be helpful in such situations.
28. In general, genotyping of the resulting GE mice, especially F0 mice, is complex because of mosaicism [18]. There are several methods to identify KO alleles due to indel mutations, such as the T7EI assay and restriction enzyme cleavage analysis, but the most reliable method would be to sequence the target region by Sanger sequencing or next-generation sequencing. In addition, the sequencing of the next generation (F1 individuals) should be confirmed when establishing GE strains.
29. The genome editing efficiencies obtained by the *i*-GONAD method in our laboratory range between 35–100% and 15–60%, for creating KO (indel mutations) or KI using ssODN, respectively.
30. Based on our experience during the past 3–4 years, in offering *i*-GONAD workshop, and from the participants' learning experience, we recommend that *Tyrosinase* gene (*Tyr*) targeting experiment serves as a very good example for any new researcher to try and establish the *i*-GONAD method in their laboratories (Fig. 6).
31. The genome editing efficiency by the *i*-GONAD method is comparable to that obtained by the MI method. For example, the efficiency of *Tyr* mutation repair using Cas9 protein (KI using ssODN) was 49% by the *i*-GONAD method and 52% by the MI method [7].

Acknowledgements

This work was supported by the 2014 Tokai University School of Medicine Research Aid, Research and Study Project of Tokai University General Research Organization, 2016–2017 Tokai University School of Medicine Project Research, Grant-in-Aid for challenging Exploratory Research (15K14371) from JSPS, and Grant-in-Aid for Scientific Research (B) (21H02393) from JSPS to MO.

References

1. Harms DW et al (2014) Mouse genome editing using the CRISPR/Cas system. Curr Protoc Hum Genet 83:1571–15727
2. Cong L et al (2013) Multiplex genome engineering using CRISPR/Cas systems. Science 339:819–823
3. Gurumurthy CB et al (2016) CRISPR/Cas9 and the paradigm shift in mouse genome manipulation technologies. In: Turksen K (ed) Genome editing. Springer Nature, Cham, pp 65–77
4. Wang W et al (2016) Delivery of Cas9 protein into mouse zygotes through a series of electroporation dramatically increases the efficiency of model creation. J Genet Genomics 43:319–327
5. Hashimoto M, Yamashita Y, Takemoto T (2016) Electroporation of Cas9 protein/sgRNA into early pronuclear zygotes generates non-mosaic mutants in the mouse. Dev Biol 418:1–9
6. Takahashi G et al (2015) GONAD: genome-editing via oviductal nucleic acids delivery system: a novel microinjection independent genome engineering method in mice. Sci Rep 5:11406
7. Ohtsuka M et al (2018) I-GONAD: a robust method for in situ germline genome engineering using CRISPR nucleases. Genome Biol 19: 25
8. Sato M et al (2018) In vivo genome editing targeted towards the female reproductive system. Arch Pharm Res 41:898–910
9. Ohtsuka M, Sato M (2019) i-GONAD: a method for generating genome-edited animals without ex vivo handling of embryos. Develop Growth Differ 61:306–315
10. Gurumurthy CB, Lloyd KCK (2019) Generating mouse models for biomedical research: technological advances. Dis Model Mech 12: dmm029462
11. Iwata S et al (2019) Simple and large-scale chromosomal engineering of mouse zygotes via in vitro and in vivo electroporation. Sci Rep 9:14713
12. Iwata S et al (2021) An efficient i-GONAD method for creating and maintaining lethal mutant mice using an inversion balancer identified from the C3H/HeJJcl strain. G3 Genes Genomes Genet 11:jkab194
13. Shang R, Zhang H, Bi P (2021) Generation of mouse conditional knockout alleles in one step using the i-GONAD method. Genome Res 31: 121–130
14. Quadros RM et al (2017) Easi-CRISPR: a robust method for one-step generation of mice carrying conditional and insertion alleles using long ssDNA donors and CRISPR ribonucleoproteins. Genome Biol 18:92
15. Miura H, Quadros RM et al (2018) Easi-CRISPR for creating knock-in and conditional knockout mouse models using long ssDNA donors. Nat Protoc 13:195–215
16. Gurumurthy CB et al (2016) GONAD: a novel CRISPR/Cas9 genome editing method that does not require ex vivo handling of embryos. Curr Protoc Hum Genet 88:15.8.1–15.8.12
17. Gurumurthy CB et al (2019) Creation of CRISPR-based germline-genome-engineered mice without ex vivo handling of zygotes by i-GONAD. Nat Protoc 14:2452–2482
18. Gurumurthy CB, Saunders TL, Ohtsuka M (2021) Designing and generating a mouse model: frequently asked questions. J Biomed Res 35:76–90

Chapter 15

Gene Targeting in Rat Embryonic Stem Cells

Hongsheng Men, Daniel J. Davis, and Elizabeth C. Bryda

Abstract

Rat germline-competent embryonic stem (ES) cell lines have been available since 2008, and rat models with targeted mutations have been successfully generated using ES cell-based genome targeting technology. This chapter will focus on the procedures of gene targeting in rat ES cells.

Key words Rat ES cells, Nucleofection, Gene targeting, ES cell selection

1 Introduction

Germline-competent ES cells, which are capable of generating functional gametes in chimeric animals, were established in mice in the early 1980s [1]. Mouse models created through targeted gene modifications in mouse ES cells have revolutionized biomedical research [2, 3] and made mice the preferred animal model species over rats for biomedical research for over three decades. Germline-competent ES cells in rats were finally established in 2008 [4, 5]. This laid the foundation for creation of rat models with targeted mutations using rat ES cells similar to ES cell-based mouse genome modifications [6, 7]. The basic procedures for rat genome editing using ES cell targeting involves targeted modification(s) of rat ES cells through homologous recombination, selection of correctly targeted ES cell clones, creation of chimeric animals using targeted ES cells, and breeding chimeric animals to establish the mutant line.

2 Materials

Prepare all stock solutions using cell culture grade water (Lonza 17-724Q) or high grade (ACS or higher) ethanol and cell culture tested reagents. All stock solutions are prepared at room

Thomas L. Saunders (ed.), *Transgenesis: Methods and Protocols*,
Methods in Molecular Biology, vol. 2631, https://doi.org/10.1007/978-1-0716-2990-1_15,

temperature unless indicated otherwise and with procedures described previously with slight modifications [8].

2.1 Primary Mouse Embryonic Fibroblast (PMEF) Medium

1. Fetal Bovine Serum (FBS) (HyClone SH30070).
2. GlutaMAX™-I solution (Invitrogen 35050-061).
3. Penicillin/streptomycin solution (Invitrogen 15140-122).
4. Glasgow Minimum Essential Medium with L-glutamine (GMEM) (Sigma G5154).
5. Sterile filtration units 0.2 uM pore size, surfactant-free cellulose acetate membrane (Fisher Scientific 09-740-39A).

2.2 Rat ES Cell Culture Medium Components

1. Apo-Transferrin (Sigma T1147).
2. 7.5% bovine serum albumin (BSA) solution (Invitrogen 15260-037).
3. Insulin (Sigma I1882).
4. Progesterone (Sigma P8783).
5. Putrescine (Sigma P5780).
6. Sodium selenite (Sigma S5261).
7. Dulbecco's Modified Eagle Medium: Nutrient Mixture F-12 (DMEM/F12, Sigma D6421).
8. B27™ supplement (Invitrogen 17504-044, 50×), serum free.
9. CHIR99021 (Stemgent 04-0004).
10. PD0325901 (Stemgent 04-0006).
11. Neurobasal medium (Invitrogen 21103-049).
12. GlutaMAX™-I (Gibco 35050061).
13. β-mercaptoethanol (Sigma ES-007-E).
14. Accutase solution (Sigma A6964).
15. Multichannel pipetter (20–200 uL).
16. Pipetters: 2–20 uL, 20–200 uL, 100–1000 uL.
17. Sterile tips for pipetters.
18. Hemocytometer.
19. Phase contrast inverted microscope with 40×, 100×, and 200× magnification.
20. Dissecting stereo microscope that will fit in a laminar flow cell culture hood.
21. Biological safety cabinet
22. Screw cap cryovial (2 mL).
23. Mr. Frosty® freezing container (Fisher Scientific 5100-0036).
24. Isopropyl alcohol.

25. −20 °C freezer.
26. −80 °C freezer.
27. Liquid nitrogen storage system for cryovials.

2.3 Rat ES Cell Targeting and Cryobanking

1. 50 mg/mL G418 stock (Sigma G8168).
2. Tris–HCl, pH 8.9.
3. 0.9% Triton X-100 (Sigma T8787).
4. 0.9% Nonidet P40 (US Biological N3500).
5. Cell culture grade water.
6. Proteinase K stock solution (Qiagen 19131).
7. DMSO (Sigma 2438).
8. Rat ES cells: F344-Tg.EC4011/Rrrc (RRRC 654).
9. PMEF feeder cells (MilliporeSigma PMEF-CF), strain CF-1, mitomycin C treated, passage 3.
10. PMEF feeder cells (MilliporeSigma PMEF-N), strain FVB, neomycin resistant, mitomycin C treated, passage 3.
11. P3 Primary Cell 4D-Nucleofector® X Kit L (Lonza V4XP-3012).
12. 4D-Nucleofector® (Lonza).
13. Amber 1.5 mL microtubes (Eppendorf®).
14. 37 °C water bath.
15. 50 mL sterile centrifuge tubes.
16. Cell culture centrifuge.
17. Sterile tissue culture pipettes, 1 mL, 5 mL, 10 mL.
18. Vacuum source for media aspiration.
19. Sterile disposable cell culture dishes, 60 mm, 100 mm.
20. Sterile disposable flat bottom 96 well plates for cell culture.
21. 37 °C, 5% CO_2, 95% relative humidity cell culture incubator.
22. 0.2 mL PCR tube strips.
23. Centrifuge for 0.2 mL PCR tube strips.
24. Sterile Pasteur pipettes.

3 Methods

3.1 Prepare Rat ES Cell Medium Stock Solutions

1. Prepare Apo-Transferrin stock solution: Dissolve 500 mg apo-Transferrin in 5 mL cell culture grade water overnight at 4 °C. Prepare 1 mL aliquots in 1.5 mL Eppendorf tubes and store at −20 °C for 1 year.
2. 7.5% BSA stock solution: Prepare 1 mL aliquots in 1.5 mL Eppendorf tubes and store at −20 °C for 1 year.

Table 1
Preparation of 10 mL N2 stock solution (100×)

Components	Volume	Final conc
apo-Transferrin stock (100 mg/mL)	1 mL	10 mg/mL
BSA (7.5%)	0.67 mL	5 mg/mL
Progesterone (0.6 mg/mL)	33 μL	2 μg/mL
Putrescine (160 mg/mL)	100 μL	1.6 mg/mL
Sodium selenite (3 mM)	10 μL	3 μM
DMEM/F12	8.187 mL	N/A

3. Insulin stock solution (25 mg/mL): Dissolve 100 mg insulin in 4 mL 0.01 M HCl overnight at 4 °C. Prepare 100 μL aliquots in 0.5 mL amber Eppendorf tubes. Store at −20 °C for 1 year.
4. Progesterone stock solution (0.6 mg/mL): Dissolve 6 mg progesterone in 10 mL high-grade ethanol. Prepare 0.5 mL aliquots in 1.5 mL Eppendorf tubes and store at −20 °C for 1 year.
5. Putrescine stock solution (160 mg/mL): Dissolve 1.6 g putrescine in 10 mL cell culture grade water. Prepare 0.5 mL or 1 mL aliquots in 1.5 mL Eppendorf tubes and store at −20 °C for 1 year.
6. Sodium selenite stock solution (3 mM): Prepare this solution in a fume hood. (1) 30 mM stock: dissolve 25.9 mg sodium selenite in 5 mL cell culture grade water. (2) 3 mM stock: add 0.5 mL of 30 mM stock into 4.5 mL cell culture grade water. Prepare 0.5 mL aliquots in 1.5 mL Eppendorf tubes and store at −20 °C for 1 year.
7. N2 stock (100×, insulin-free): Mix all the components as shown in Table 1 to make 10 mL N2 stock. Prepare 1 mL aliquots in 1.5 mL Eppendorf tubes and store at −20 °C for 1 year.
8. B27™ supplement: prepare 1 mL aliquots in 1.5 mL Eppendorf tubes. Store at −20 °C until the expiration date.
9. CHIR99021 stock (1000×): dissolve 2 mg CHIR99021 in 1.43 mL DMSO. Make 200 μL aliquots in 0.5 mL amber Eppendorf tubes, label and store at −20 °C for 1 year.
10. PD0325901 (stock (1000×): dissolve 2 mg PD0325901 in 8.2 mL DMSO. Make 200 μL aliquots in 0.5 mL amber Eppendorf tubes, label and store at −20 °C for 1 year.

3.2 Rat ES Cell Medium with Two Inhibitors (N2B27 + 2i)

1. To a 250 mL beaker, add 100 mL DMEM/F12, 100 mL Neurobasal medium, 1 mL N2 stock, 2 mL B27, 1 mL GlutaMAX™-I, 200 μL CHIR99201, 200 μL PD0325901, 2 mL β-mercaptoethanol.
2. With constant stirring, add 100 μL insulin stock drop-by-drop to avoid precipitation (*see* **Note 1**).
3. Filter with a filtration unit and keep at −4 °C (*see* **Note 2**) and use within 1 month.

3.3 Prepare PMEF Feeder Cell Media

For 250 mL PMEF medium, add 25 mL (10%) FBS, 2.5 mL GlutaMAX™-I solution, and 2.5 mL penicillin/streptomycin solution to 220 mL GMEM and filter with a 0.2 uM pore sized filtration unit. Store at 4 °C and use within 1 month.

3.4 ES Cell Targeting and Cryobanking Solutions

1. Lysing buffer stock (without proteinase K): 40 mM Tris–HCl, pH 8.9, 0.9% Triton X-100, 0.9% Nonidet P40 in cell culture grade water.
2. Proteinase K stock: add lyophilized proteinase K to cell culture grade water and dissolve to prepare a 20 mg/mL stock solution.
3. Working lysing buffer: add 200 μL 20 mg/mL proteinase K stock solution to 10 mL lysing buffer stock (400 μg/mL proteinase K final) and mix by inversion.
4. Freezing medium: (1) 10% DMSO (Sigma 2438) in PMEF medium for routine ES cell freezing: To prepare 50 mL freezing medium, add 5 mL DMSO into 45 mL PMEF medium and filter to sterilize.
5. Freezing medium (2) 12% DMSO in PMEF medium for positive ES cell freezing after selection: to prepare 50 mL freezing medium, add 6 mL DMSO into 44 mL PMEF medium and filter to sterilize.

3.5 PMEF Feeder Cell Culture

One day prior to ES cell culture, thaw one vial of CF-1 PMEF feeder cells in a 37 °C water bath. Wash the PMEF cells once by mixing with 3 mL PMEF medium in a 50 mL tube and centrifuge the cell suspension at 200 × g for 3 min. Aspirate the supernatant and resuspend cells in an appropriate volume of PMEF medium. Plate the cells in a volume of 4 mL cell suspension in each 60 mm dish at the density suggested by the manufacturer (*see* **Note 3**). Transfer the dishes to an incubator set at 37 °C with 5% CO_2 and maximal humidity (90–100%). Change PMEF medium the following day and then every 48 h thereafter.

3.6 Rat ES Cell Culture

Thaw a vial of rat ES cells in a 37 °C water bath. Wash the ES cells once by mixing with 3 mL N2B27-2i medium and centrifuge at 200 × g for 3 min. Aspirate the supernatant and resuspend the cells in 4 mL N2B27-2i. Aspirate the PMEF medium from a 60 mm feeder cell dish and transfer the ES cell suspension to the dish. Return the dish to the incubator and refresh the medium daily with fresh N2B27-2i (*see* **Note 4**).

3.7 ES Cell Passage

When ES cell colonies reach ~70% confluency, remove the medium from the ES cell culture plates (*see* **Note 4**) and add 2 mL Accutase. Return the plate to an incubator and incubate for 5–8 min at 37 °C. Following incubation, gently pipette the solution to break up any remaining cell clumps (minimize air bubbles). After the culture becomes a single cell suspension, collect the suspension into a 50 mL centrifugation tube and centrifuge at 200 × g for 3 min. Aspirate the supernatant and re-suspend the cells in the appropriate volume of N2B27-2i. Aspirate the PMEF medium from the appropriate number of dishes with PMEF monolayers and transfer the ES cell suspension to each dish. For rat ES cell colonies in a 60 mm dish with ~70% confluence, cells can be passaged to other 60 mm dishes at a ratio of 1:5 or 1:6.

3.8 ES Cell Freezing

Process ES cell colonies into single cell suspension as described in Subheading 3.7. After Accutase treatment, quickly count the concentration of ES cells using a hemocytometer to determine the total cell number in the suspension. Pellet the ES cells by centrifugation at 200 × g for 3 min and resuspend the ES cells in an appropriate volume of freezing medium to reach a concentration of 1 × 106 cells/mL. ES cells are usually frozen at 1 mL aliquot (1 × 106 cells) in a NALGENE 1 mL cryovial. Place these vials into a "Mr. Frosty" freezing container containing 100% isopropyl alcohol. Place the freezing container(s) into a −80 °C freezer overnight and transfer vials to liquid nitrogen the next day for long-term storage.

3.9 G418 Killing Curve

The G418 killing curve is usually conducted in a 6-well plate (*see* **Note 5**). Estimate the number of ES cells needed for this procedure and grow enough ES cells as described in Subheading 3.7 before proceeding (*see* **Note 6**).

1. One day prior to culturing ES cells, grow neomycin-resistant PMEF (*see* **Note** 7) monolayers in a 6-well plate with 2 mL PMEF medium per well.
2. Culture ES cells on PMEF monolayers with 2 mL N2B27-2i in each well.
3. After 48 h of ES cell culture, check the ES cells to make sure the colonies reach 60–70% confluence.

Table 2
G418 stock (50 mg/mL) in various working concentrations in 6-well plate

Well	Working concentration (μg/mL)	G418 stock to 2 mL N2B27-2i (μL)
1	50	2
2	100	4
3	150	6
4	200	8
5	250	10
6	0	0

4. Start the selection by replacing the old medium in each well with 2 mL fresh N2B27-2i and then add the corresponding volume of G418 into each well to reach the desired concentrations (50, 100, 150, 200, 250 μg/mL of G418) using 50 mg/mL G418 stock solution (Table 2). One well without G418 serves as a control.
5. Observe each wells every day for 1 week.
6. Refresh N2B27-2i and add G418 with desired concentrations to each well as initial treatment (Table 2) every 3 days.
7. The minimal G418 concentration to use is the lowest concentration that kills 100% of the cells in 1 week from the start of antibiotic selection.

3.10 Nucleofection

The following protocol is based on the use of Lonza 4D-Nuclefector® system to insert a floxed gene carried on a targeting vector (Fig. 1) into the genome of rat ES cells.

1. Estimate ES cells needed for nucleofection (*see* **Note 8**) and grow enough ES cells as described in Subheading 3.7 before nucleofection.
2. One day prior to nucleofection, estimate the number of 100 mm dishes (selection dishes) needed and plate neomycin-resistant PMEF feeder cells into a corresponding number of 100 mm dishes according to procedures described in Subheading 3.1 (*see* **Note 9**) and replace PMEF medium with 10 mL N2B27-2i medium on the day of nucleofection.
3. Nucleofection solution: Add Supplement 1 (supplied with Lonza's 4D-Nucleofector® X Kit L) to Nucleofector® Solution according to the user's manual, and mix gently.
4. Disassociate ES cell colonies into single cells by treating the ES cells with appropriate amount of Accutase solution as described in Subheading 3.7. Collect the ES cell solution into 50 mL conical tubes and centrifuge at 200 × g for 3 min to pellet the ES cells.

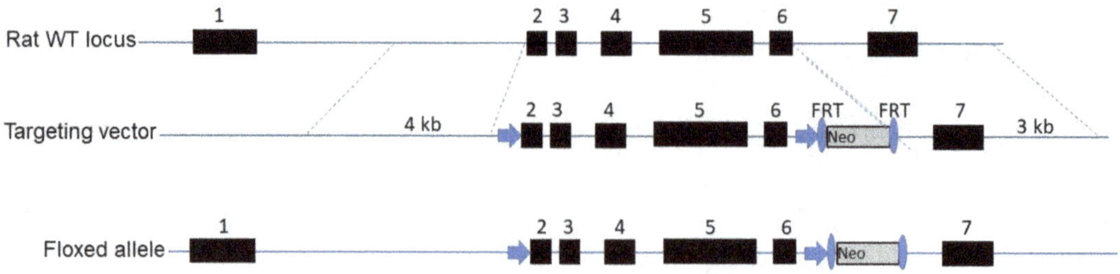

Fig. 1 Schematic of basic strategy to generate a floxed rat allele. LoxP sites are indicated by arrows. FRT sites are indicated by ovals, black boxes are exons. Neo = neomycin resistance gene (*see* **Note 10**) and dotted lines indicate homology arm regions

5. Resuspend the ES cells with appropriate amount of supplemented P3 Primary Cell 4D-Nuceofector Solution (ES cell solution) based on the treatment groups (see below) so that each 90 μL aliquot contains ~3 × 106 cells (*see* **Note 8**). Prepare 90 μL aliquots in individual 1.5 mL Eppendorf tubes.
6. Experimental and control groups in 100 μL single nucleocuvette:
   (a) Experimental groups: Prepare tubes containing 1–5 μg linearized plasmids (*see* **Note 11**) in 10 μL supplemented P3 Primary Cell 4D-Nuceofector Solution (DNA solution). Mix the DNA solution with 90 μL ES cell solution and transfer the 100 μL solution into an Amaxa certified cuvette. Use program CA-210 on 4D Nucleofector device to nucleofect the ES cells (*see* **Note 12**).
   (b) Positive control: Prepare a tube containing 2 μg pmaxGFP® vector (*see* **Note 13**) in 10 μL supplemented P3 Primary Cell 4D-Nuceofector Solution. Mix this solution with 90 μL ES cell solution and transfer the 100 μL solution into an Amaxa certified cuvette. Use program A-13/A-23/A-24/A-30 on 4D Nucleofector device to nucleofect the rat ES cells.
   (c) Negative control 1: No electroporation control. Prepare a tube containing 2 μg linearized plasmid in 10 μL supplemented P3 Primary Cell 4D-Nuceofector Solution. Mix this solution with 90 μL ES cell solution and transfer the 100 μL solution into an Amaxa certified cuvette. Do not nucleofect the ES cells.
   (d) Negative control 2: No DNA control. Transfer 100 μL ES cell Solution into an Amaxa certified cuvette. Use program A-13/A-23/A-24/A-30 on 4D Nucleofector device to nucleofect the rat ES cells.

7. Nucleofection: Follow the procedures described in Lonza 4D-Nuclefector™ system user's manual. After nucleofection, take the cuvette out of the holder. Add 500 μL of prewarmed N2B27-2i medium and transfer the cells into a pre-equilibrated 100 mm selection dishes containing 10 mL N2B27-2i and neomycin-resistant PMEF monolayers.

3.11 ES Cell Selection

1. One day after nucleofection, refresh the selection dishes with 10 mL fresh N2B27-2i medium.
2. Add optimal concentration of G418 to the culture as determined by the kill curve to kill any ES cells that do not have the DNA targeting construct stably integrated into their genome. A "pulse" selection protocol is used: 48 h in N2B27-2i with G418 followed by 24 h of G418-free N2B27-2i medium [6, 7].
3. Repeat an additional two cycles for a total 9 days of selection.
4. Colonies are ready for pickup at the end of selection. Colonies that survive the selection process are presumed to have stable integration of the DNA from the targeting vector.

3.12 Colony Pick-Up

Every selected colony is split and transferred to a well on each of two 96-well plates: one plate for freezing and one plate for genotyping.

1. Estimate the number of duplicate 96-well plates needed based on the colonies ready for pick-up. Seed each well of the 96-well plates with CF-1 PMEF cells in 200 μL PMEF medium the day before colony pick-up (*see* **Note 14**).
2. On the day of colony pick-up, replace PMEF medium with 100 μL N2B27-2i in each well of the duplicate 96-well plates and clearly mark the two plates (one as freezing plate and the other as genotyping plate) to avoid confusion. Place the plates into a 37 °C incubator until colony pick-up (Fig. 2).

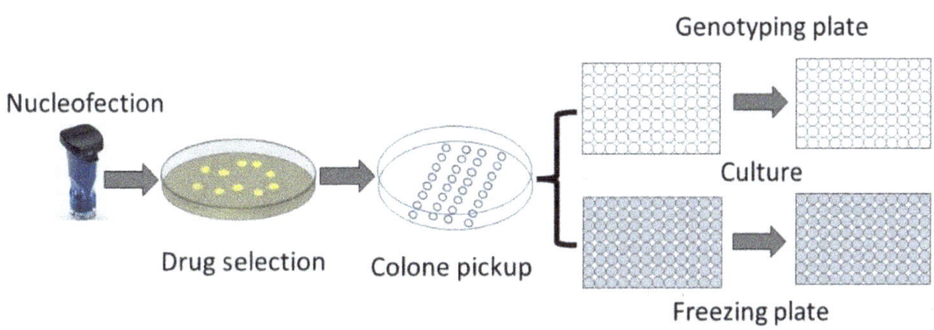

Fig. 2 Nucleofection, drug selection, and positive colony pick-up and preparation for genotyping and freezing

3. Place a dissecting stereomicroscope, 100 mm sterile dishes, two 96-well plates (freezing and genotyping plates), warmed Accutase, and the 100 mL selection dish with positive colonies into a laminar flow hood.
4. Using a multichannel pipette, make a row of eight 20 μL Accutase drops on a 100 mm petri dish (lid or the dish).
5. Use a P10 or P20 pipette with a 10 μL tip (set the volume at 5 μL) to pick up eight colonies a time under a dissection microscope.
6. Place ES cell colonies one per drop. Incubate 3–5 min at room temperature.
7. At the end of Accutase incubation, use a multichannel pipette to pipette up and down the drops three times to dissociate ES cell colonies and then transfer these droplets into a 96-well plate designated for freezing and gently pipette up and down 1–2 times to mix the ES cells with medium.
8. Take 50 μL from each well and transfer ES cells into the genotyping plate.
9. Continue the same process for the remaining colonies that are ready for pick-up.
10. Return the 100 mL selection dish into a 37 °C incubator to grow the remaining small sized colonies for later pick-up.
11. Add additional 100 μL of fresh N2B27-2i medium to each well in the freezing plate and the genotyping plate to make 150 μL total in each well.
12. Place the two plates back in the incubator to propagate the ES cells and refresh the culture with 200 μL fresh N2B27-2i every 48 h.

3.13 Preparation for Freezing and Genotyping

1. Prepare the working lysing buffer as described in Subheading 3.4.
2. Use two (duplicate) PCR tube holders and clearly label 0.2 mL PCR tube strips based on the layout of 96-well plates and place the labelled tubes in the tube holders. Label the tube holders as either freezing or genotyping.
3. Start with the 96-well plate designated for freezing.
4. Use a multichannel pipette (set volume to 200 μL) to detach the colonies in each well in a row by pipetting and then transfer the colonies into respective PCR tube strips. Make sure to change tips between rows.
5. Continue this process until all the colonies in the freezing plate are transferred into 0.2 mL PCR tube strips.
6. Centrifuge the tubes at 200 × g for 5–6 min at room temperature in a benchtop centrifuge with plate holders.
7. Aspirate the supernatant from each tube as much as possible by vacuum using a Pasteur pipette connected to a 200 μL tip.

8. After removing the supernatant from the tubes, add 30 μL/well Accutase and gently mix by pipetting one to two times using a multichannel pipette.
9. Make sure to change tips between strips of tubes.
10. Place the freezing tubes in an incubator and incubate for 5–6 min to disassociate the cell colonies.
11. After Accutase treatment, place 150 μL freezing medium with 12% DMSO directly into each tube (10% DMSO final) in the freezing tube strips using a multichannel pipette.
12. Place the tubes into a Styrofoam box and then into a −80 °C freezer overnight.
13. The tubes are transferred into the vapor phase in a LN2 Dewar for long-term storage. This provides an archived source for all positive clones.
14. To prepare the genotyping plate for processing, remove supernatant from the wells of the genotyping plate.
15. Place 30 μL lysing buffer into each well of the genotyping plate and mix well by pipetting.
16. Transfer the content in each well into respective 0.2 mL strip PCR tubes.
17. Place the tubes in a programmable thermocycler and incubate at 65 °C for 20 min, 95 °C for 10 min, and then stored 4 °C → ∞.
18. After lysing, the contents in the strip PCR tubes are ready for genotyping. The tubes can also be frozen in a −20 °C freezer for genotyping at a later time.
19. Positive clones are initially screened by PCR using an appropriate assay specific to the integrated DNA. Positive clones confirmed by this method should be further analyzed by additional methods including comprehensive nucleotide sequencing of the integrated DNA sequence.

4 Notes

1. Insulin has low solubility at physiological pH. Adding insulin stock drop-by-drop into the medium with constant stirring will avoid insulin recrystallization.
2. Some components in the N2B2-2i ES cell medium are light sensitive; keep the medium away from light if possible.
3. MilliporeSigma EmbryoMax® PMEFs (Cat #: PMEF-CF) have 5–6 × 106 cells per vial. Each 60 mm dish has 21.5 cm^2 growth area and the suggested cell seeding density is ~0.8 × 106 cells. Therefore, one vial of cells can plate 6 to eight 60 mm dishes.

4. Rat ES cells are low in integrin expression. Therefore, ES cell colonies are loosely attached to the PMEF feeder monolayer, and some colonies are even floating. If there are floating colonies present in the culture, collect them into a 15 mL conical tube and centrifuge at 200 × g for 3 min to collect them. Aspirate the supernatant and resuspend in 1–1.5 mL Accutase solution. Transfer this solution back to the culture dish.
5. A "kill curve" is used to determine the concentration of antibiotics required to kill cells within 3–5 days.
6. Each well in a 6-well plate has 9.6 cm^2 growth area and the recommended cell seeding density is 0.3 × 106 cells/well.
7. MilliporeSigma EmbryoMax® neomycin-resistant PMEFs (Cat #: PMEF-N) have 5–6 × 106 cells per vial. Adjust concentrations according to the growth areas of the respective culture wares to be plated.
8. The suggested minimal number of cells per 100 μL nucleofection sample is 3 × 106 cells in Lonza's P3 Primary Cell 4D-NucleofectorTM X Kit L.
9. Each 100 mm dish has 56.7 cm^2 growth area and the recommended cell seeding density is 2.2 × 106 cells.
10. Neo: bacterial gene for neomycin phosphotransferase, used for positive knock-in selection.
11. This is the range of concentrations suggested by Lonza. Test the optimized concentration for your specific plasmid.
12. Lonza continues to optimize their nucleofector protocols. Please refer to Lonza's protocols for the appropriate program for nucleofection of specific cell types or contact their customer service.
13. This is the plasmid supplied with Lonza's P3 Primary Cell 4D-Nucleofector™ X Kit L and is used to easily visualize successful nucleofection. Other comparable plasmids can be used as positive controls as well.
14. Each well of a 96-well plate has an area of 0.32 cm^2 and the recommended cell seeding density is 1 × 104 cells/well.

Acknowledgements

This work was supported by a grant from National Institutes of Health: P40 OD011062 (ECB).

References

1. Evans MJ, Kaufman MH (1981) Establishment in culture of pluripotential cells from mouse embryos. Nature 292(5819):154–156
2. Capecchi MR (2005) Gene targeting in mice: functional analysis of the mammalian genome for the twenty-first century. Nat Rev Genet 6(6):507–512
3. Wu S et al (2007) Toward simpler and faster genome-wide mutagenesis in mice. Nat Genet 39(7):922–930
4. Buehr M et al (2008) Capture of authentic embryonic stem cells from rat blastocysts. Cell 135(7):1287–1298
5. Li P et al (2008) Germline competent embryonic stem cells derived from rat blastocysts. Cell 135(7):1299–1310
6. Tong C (2010) Production of p53 gene knockout rats by homologous recombination in embryonic stem cells. Nature 467(7312): 211–213
7. Tong C et al (2011) Generating gene knockout rats by homologous recombination in embryonic stem cells. Nat Protoc 6(6):827–844
8. Nichols J, Ying QL (2006) Derivation and propagation of embryonic stem cells in serum- and feeder-free culture. Methods Mol Biol 329:91–98

Chapter 16

Rat Embryonic Stem Cell Transgenesis

Elizabeth C. Bryda, Hongsheng Men, and Barbara J. Stone

Abstract

The availability of reliable germline competent rat embryonic stem cell (ESC) lines that can be genetically manipulated provides an important tool for generating new rat models. Here we describe the process for culturing rat ESCs, microinjecting the ESCs into rat blastocysts, and transferring the embryos to surrogate dams by either surgical or non-surgical embryo transfer techniques to produce chimeric animals with the potential to pass on the genetic modification to their offspring.

Key words Embryonic stem cells, Rat, Blastocyst, Transgenesis, Embryo transfer, Chimera, NSET

1 Introduction

While the rat is the preferred animal model species for a number of disciplines [1, 2], it lost traction to the mouse because of the inability to perform sophisticated genetic manipulations in rats similar to those that were possible in the mouse. This was primarily due to the lack of availability of rat embryonic stem cells (ESCs). The discovery of methods to isolate and successfully culture rat embryonic stem cells changed the playing field and allowed genetic manipulations in the rat that included the ability to create targeted knock-ins and knock-outs, as well as generate conditional and inducible rat models [3–6]. While the advent of genome editing technologies such as CRISPR/Cas9 allow knock-outs and knock-ins to be efficiently generated directly in zygotes [7, 8], there are still some types of genetic alterations such as targeted knock-in of larger constructs/transgenes that are best accomplished through manipulation of ESCs. Here we describe the pipeline for using ESCs to generate chimeric animals with the potential to transmit genetic modifications for the establishment of new rat models.

Thomas L. Saunders (ed.), *Transgenesis: Methods and Protocols*,
Methods in Molecular Biology, vol. 2631, https://doi.org/10.1007/978-1-0716-2990-1_16,

2 Materials

In general, care should be taken to use endotoxin-free reagents and disposables whenever possible. Individual components of medium are prepared using embryo transfer water (Lonza, 17-724Q). Most reagents used in the various media should not be freeze-thawed.

2.1 Embryonic Stem Cell (ESC) Culture

1. Primary Mouse Embryonic Fibroblasts (PMEFs): MEF feeder cells, strain CF1 (Sigma-Aldrich, PMEF-CF1), mitomycin C-treated, 5 vials, 5–6 × 10^6 cells/vial.
2. 70% ethanol.
3. Water bath, 37 °C.
4. Incubator, 37 °C with 5% CO_2 and maximal humidity (90–100%).
5. 5 mL serological pipettes.
6. 0.2 μm Nalgene filter bottle units (Thermo Scientific 568-0020).
7. Benchtop centrifuge.
8. 15 and 50 mL conical tubes.
9. 60 mm tissue culture plates.
10. 1.5 and 0.5 mL regular and amber Eppendorf tubes.
11. PMEF medium: 10% FBS (HyClone SH30070), 1% GlutaMAX™-I (Invitrogen 35050-061), 1% penicillin/streptomycin in GMEM (Sigma G5154). Filter sterilize using a 0.2 μm filter. Store at 4 °C and use within 1 month.
12. Rat ESCs: for best results, use low passage (P3-15), well-characterized cell lines that have been shown to be germline competent. Several ES cell lines are available through the Rat Resource and Research Center (www.rrrc.us).
13. Accutase® (Sigma A6964).
14. Apotransferrin (Sigma T1147) 100 mg/mL stock solution: dissolve 500 mg in 5 mL sterile water overnight at 4 °C. Prepare 1 mL aliquots in 1.5 mL Eppendorf tubes and store at −20 °C for up to 1 year.
15. BSA 7.5% solution (Invitrogen 15260-037): prepare 1 mL aliquots in 1.5 mL Eppendorf tubes and store at −20 °C for up to 1 year.
16. Progesterone (Sigma P8783) 0.6 mg/mL stock solution: dissolve 6 mg in 10 mL high-grade (ACS or higher) ethanol. Prepare 50 μL aliquots in 0.5 mL amber Eppendorf tubes and store at −20 °C for up to 1 year.

17. Putrescine (Sigma P5780) 160 mg/mL stock solution: dissolve 1.6 g in 10 mL sterile water. Prepare 0.5 mL or 1 mL aliquots in 1.5 mL amber Eppendorf tubes and store at −20 °C up to 1 year.
18. Sodium selenite (Sigma S5261) 3 mM stock solution: prepare this solution in a fume hood. Dissolve 25.9 mg in 5 mL sterile water to make a 30 mM stock and then add 0.5 mL of this stock into 4.5 mL sterile water to obtain a 3 mM stock solution. Prepare 0.5 mL aliquots of 3 mM stock in 1.5 mL Eppendorf tubes and store at −20 °C for up to 1 year.
19. Insulin (Sigma I1882) 25 mg/mL stock solution: dissolve 100 mg insulin in 4 mL sterile 0.01 M HCl overnight at 4 °C. Prepare 100 μL aliquots in 0.5 mL amber Eppendorf tubes. Store at −20 °C for up to 1 year.
20. CHIR99021 (Stemgent 04-0004) 1396 ug/mL in DMSO (Sigma D2438). Inhibitor for N2B27 + 2i medium. Prepare 200 μL aliquots in 0.5 mL amber Eppendorf tubes. Store at −20 °C for up to 1 year.
21. PD0325901 (Stemgent 04-0006) 241.1 ug/mL in DMSO. Inhibitor for N2B27 + 2i medium. Prepare 200 μL aliquots in 0.5 mL amber Eppendorf tubes. Store at −20 °C for up to 1 year.
22. N2 Stock solution: make insulin-free 100× N2 stock solution using final concentrations of 10 mg/mL apotransferrin, 0.5% BSA, 3 μM sodium selenite, 1.6 mg/mL putrescine, 1.98 ug/mL progesterone, and 1× DMEM/F12 (Sigma D6421). The N2 stock solution can be stored at −20 °C in 1 mL aliquots for up to 1 year.
23. N2B27 + 2i medium: to prepare 200 mL of medium, add 100 mL DMEM/F12 and 1 mL of 100× N2 stock solution to 1 beaker and 100 mL Neurobasal™ medium (Invitrogen 21103-049), 2 mL B27 supplement (Invitrogen 17504-044), and 1 mL of GlutaMax™ -I to a second beaker. Combine the contents of both beakers, mix, and while stirring constantly, add 100 μL of insulin drop-by-drop to the solution to prevent precipitation. Add 200 μL each of the inhibitors, CHIR99021 and PD0325901. Add 2 mL of β-mercaptoethanol (Millipore ES-007-E). Mix well. Sterilize with a 0.2 μm Nalgene filter unit. Prepare 40 mL aliquots in conical tubes and store at 4 °C. Should be used within 1 month.
24. N2B27 + 2i + 20 mM HEPES (Gibco, 15630-080). The addition of HEPES provides better buffering capacity and is used for steps that involve manipulation of cells/embryos outside the CO_2 incubator. HEPES should be added fresh to N2B27 + 2i at the time of use.

2.2 Blastocyst Collection

1. LHRHa: [des-Gly^{10}, D-Ala^{6}]-Leuteinizing Hormone Releasing Hormone (LHRHa) ethylamide acetate salt hydrate (Sigma L-4513). This an LHRH agonist. Make 200 μg/mL stock by diluting 5 mg of LHRHa in 25 mL of 1× modified Dulbecco's Phosphate-Buffered Saline (without calcium chloride and magnesium chloride) (Sigma 8537). Alternatively, an injectable PBS (Covetrus C1880725) can be used. Store at −80 °C as 1 mL aliquots for up to 3 months.
2. Rats: Sexually mature (8–10 weeks of age) female rats and sexually mature, intact, preferably proven breeder males (at least 10 weeks of age).
3. Insulin syringes with needles for hormone injections.
4. Otoscope with light source for visualization of copulatory plug.
5. Rat Stock solution: This Rat Stock solution is used in mR1ECM medium. To make 500 mL: use embryo-grade water (Lonza, 17-724Q) and add 23.376 g NaCl (Sigma S5886), 1.1928 g KCl (Sigma P5405), 6.7576 g D-Glucose (Sigma G6152), 0.3750 g Penicillin G Potassium Salt (Sigma P7794), 0.2500 g Streptomycin Sulfate (Sigma S1277), 12.637 g Sodium Lactate (60% syrup) (Sigma L7900), 1.4702 g $CaCl_2$-$2H_2O$ (Sigma C7902), 0.5083 g $MgCl_2$-$6H_2O$ (Sigma M2393).
6. mR1ECM (modified Rat 1-cell Embryo Culture Medium). To make 100 mL: use embryo-grade water (Lonza) and add 0.01 g PVA (Sigma, P8136), 0.21 g $NaHCO_3$ (Sigma S5761), 0.0055 g Sodium Pyruvate (Sigma P4562), 1 mL 100× MEM NEAA (Invitrogen 11140-050), 2 mL 50× MEM EAA (Invitrogen 11130-051), 0.05 mL GlutaMAX™-I, 10 mL Rat Stock solution. Gas the solution with 5% CO_2. Check the pH: it should be approximately 7.4. If the pH is <7.0, do not use. Measure the osmolality: it should be between 235 and 255 mOsm. If the osmolality is not in this range, do not use. Sterile filter the medium using a 0.2 μm Nalgene filter unit. This medium can be stored up to 4 weeks at 4 °C.
7. mR1ECM + 20 mM HEPES +10% FBS. To make 100 mL: use embryo-grade water (Lonza) and add 0.01 g PVA, 0.042 g $NaHCO_3$, 0.0055 g Sodium Pyruvate, 1 mL 100× MEM NEAA, 2 mL 50× MEM EAA, 0.05 mL GlutaMAX™-I, and 10 mL Rat Stock solution. Adjust the pH to approximately 7.4 using 10 N NaOH. Measure the osmolality: it should be between 240 and 255 mOsm. If the osmolality is not in this range, do not use. Sterile filter the medium using a 0.2 μm Nalgene filter unit. This medium can be stored up to 4 weeks at 4 °C. Note that the concentration of $NaHCO_3$ is different in this formulation than in the mR1ECM without HEPES. The HEPES and FBS should be added fresh at the time of use.

8. mR1ECM + 10% FBS. FBS should be added fresh to mR1ECM (*see* Subheading 2.2, **item 6**) at the time of use.
9. 70% ethanol.
10. 35 mm petri dishes.
11. 1 mL syringes.
12. 30G blunt needles.
13. Incubator, 37 °C with 5% CO_2 and maximal humidity (90–100%).
14. Curved serrated forceps for dissection.
15. Dissection scissors.
16. Pure mineral oil (Thermo Scientific AC415080010). Test each bottle before use. Culture C57B6/NJ (Jackson Laboratory) mouse embryos from zygote to blastocysts in KSOM medium (Sigma, MR-106-D) under oil. If 80% or more of the mouse embryos develop to blastocysts, the mineral oil is considered acceptable for use.
17. Aspirator tube (Sigma, A5177). Consists of a flexible silicone rubber nosepiece, hard plastic mouthpiece, and 15 inches of latex tubing.
18. 125 μL Stripper tips (CooperSurgical, MXL3-125).

2.3 Microinjection of ESCs into Blastocysts

1. mR1ECM + 20 mM HEPES +10% FBS.
2. mR1ECM +10% FBS.
3. 35 and 60 mm petri dishes/lids.
4. Pure mineral oil.
5. Incubator, 37 °C with 5% CO_2 and maximal humidity (90–100%).
6. Eppendorf TransferMan NK2 micromanipulators.
7. Nikon Eclipse T*i* inverted microscope.
8. Hoffman Modulation Contrast (HMC) system.
9. Eppendorf Transfertip (Millipore Sigma, EP5195000079).
10. Aspirator tube assembly.
11. 125 μL Stripper tips.
12. P-10 pipette.

2.4 Embryo Transfer

1. Rats: female Sprague Dawley rats 8–24 weeks old are recommended. Sterile, vasectomized males at least 10 weeks of age or older.
2. LHRHa: See Subheading 2.2 for details about preparation. Thaw frozen aliquots at room temperature immediately before use.
3. Insulin syringe with needle.
4. Otoscope with light source.

2.5 Surgical Embryo Transfer

1. Ketamine/Xylazine working stock: add 2 mL of 100 mg/mL XylaMed (VetOne) to a 10 mL vial of 100 mg/mL Ketathesia (Covetrus) and mix thoroughly. There are many sources of these drugs.
2. Banamine working stock: add 0.4 mL of 50 mg/mL (Merck, 065477) to 9.6 mL of sterile saline and mix thoroughly.
3. Lubricating ointment: for eyes.
4. Fur clippers.
5. 70% ethanol.
6. Betadine.
7. Cotton swabs.
8. 150 mm petri dishes.
9. Fine dissection scissors.
10. Curved, serrated forceps.
11. Bulldog clamps.
12. Gauze.
13. Leica (M165 C) dissecting microscope.
14. Fiber-optic lamp for illumination while using a dissecting microscope.
15. mR1ECM + 20 mM HEPES +10% FBS.
16. 125 μL Stripper tip.
17. Aspirator tube assembly.
18. Saline, sterile.
19. Autoclips for wound closure.
20. Atipamezole (Zoetis): used as a ketamine reversal. Concentration is 5 mg/mL. There are many sources of this drug.

2.6 Non-surgical Embryo Transfer

1. rNSET device with specula (ParaTechs #70010).
2. P-2 pipette.
3. Incubator, 37 °C.
4. Stereomicroscope with transmitted and reflected illumination source with 20× and 40× magnification.
5. 60 mm tissue culture dishes.
6. Embryo handling pipette assembly.
7. 1 mL syringes.
8. 26G needles.
9. M2 medium (Sigma-Aldrich M7167) or mR1ECM + 20 mM HEPES +10% FBS.
10. Oxytocin for cervical dilation, Oxoject (Henry Schein #034974).

3 Methods

3.1 Embryonic Stem Cell (ESC) Culture

1. Preparation of ESCs for injection into blastocysts should begin at least 7 days prior to the injection day, beginning with ensuring that feeder cell plates are prepared at least 1 day prior to culturing ESCs.
2. Plate PMEF feeder cells. Multiple feeder plates are prepared since the ESCs will need to be passaged. One day prior to beginning ESC culture, quickly thaw a vial of PMEF-CF-1 mouse feeder cells in a 37 °C water bath. Once thawed, wipe down the outside of the vial with 70% ethanol to sterilize. Working aseptically in a hood and using a 5 mL serological pipette, transfer the PMEFs from the vial into a 50 mL conical tube. Rinse the vial 3× with 1 mL of warm (37 °C) PMEF medium and transfer to the 50 mL conical to ensure all cells have been transferred. Centrifuge the cell suspension at 200 × g for 3 min at room temperature. Aspirate the supernatant and resuspend the cells in the appropriate volume of warm (37 °C) PMEF medium so that 5 mL of medium with a cell density of 1×10^6 PMEF cells can be added to each 60 mm plate (*see* **Note 1**). Transfer the plates to an incubator at 37 °C with 5% CO_2 and maximal humidity (90–100%). Change the PMEF medium the following day and every 48 h thereafter if necessary (for any extra plates not immediately being used). The plates are ready for use when the feeder cells form a confluent uniform monolayer (*see* **Note 2**).
3. Plate rat ESCs. Quickly thaw a vial of rat ESCs (1×10^6 cells/vial) in a 37 °C water bath. Once thawed, wipe down the outside of the vial with 70% ethanol to sterilize. Working aseptically in a hood and using a 5 mL serological pipette, transfer the cells into a 15 mL conical tube. Rinse the vial 3× by adding 1 mL of warm (37 °C) N2B27 + 2i medium to the vial and then transferring to the 15 mL conical tube. Centrifuge the cell suspension at 200 × g for 4 min at room temperature. Aspirate the supernatant and resuspend the cells in 4 mL warm (37 °C) N2B27 + 2i medium. Aspirate the medium from a 60 mm feeder cell plate and add the ESC suspension to the plate. Replace the medium daily (*see* **Note 3**). Undifferentiated rat ESCs grow in compact colonies with smooth boundaries (Fig. 1). Over time, it is not unusual for colonies to become detached from the feeder layer.
4. Passage rat ESCs. Rat ESCs will typically need to be passaged every 48 h. Detach ESC colonies from the feeder layer by gently pipetting the ESC culture medium up and down several times. Collect the ESC colonies into a 15 mL conical tube and centrifuge at 200 × g for 4 min at room temperature. Aspirate the supernatant, resuspend the ESC colonies in 3 mL of Accutase® solution and incubate 5–8 min at 37 °C. Following the

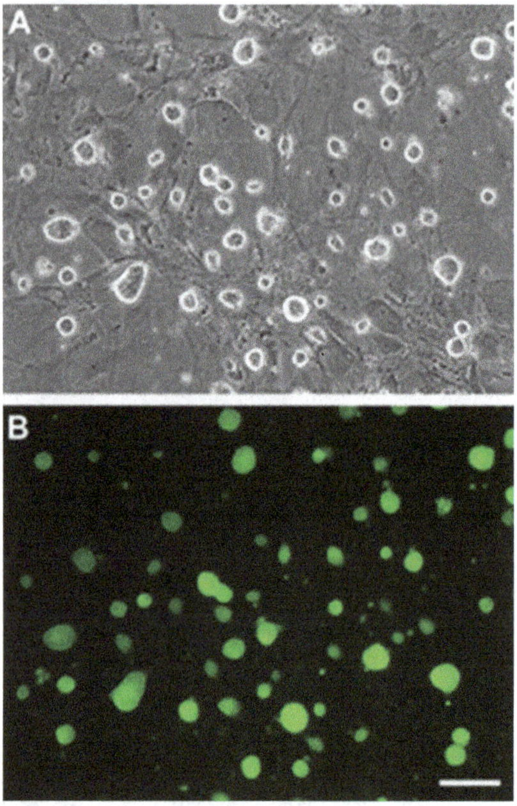

Fig. 1 Rat embryonic stem cells. The morphology of RRRC# 561: SD-Tg (GFP) 2BalRrrc-ES1/Rrrc ESCs is shown and is representative of other rat ES cell lines. (**a**) Phase contrast image shows cultured ESCs forming compact colonies with smooth edges. (**b**) Fluorescence microscopy image of same field of view as (**a**). Cultured ESCs express the EGFP transgene. Scale bar represents 100 mm

incubation, gently pipette the solution 3–5 times to break up any remaining cell clumps (minimize air bubbles). After the ESC colonies become a single cell suspension, centrifuge at 200 × g for 4 min at room temperature. Aspirate the supernatant and re-suspend the cells in an appropriate volume of N2B27 + 2i (*see* **Note 4**). Aspirate the PMEF medium from the feeder cell plates and transfer 5 mL of the ESC suspension to each 60 mm tissue culture plate.

5. On the day of injection, detach rat ESC colonies from the feeders by gently pipetting the medium up and down followed by collection into a 15 mL conical tube. Centrifugation at 200 × g for 3 min at room temperature and remove the supernatant. Add 3 mL of Accutase® solution to the pelleted ESC colonies to disassociate them into a single cell suspension. Centrifugation at 200 × g for 3 min at room temperature. Resuspend the cell pellet in 0.5 or 1 mL N2B27 + 2i + 20 mM HEPES. Transfer the ESC solution into a 1.5 mL Eppendorf tube and incubate on ice until needed.

3.2 Blastocyst Collection

All applicable international, national, and institutional guidelines for the care and use of animals should be followed.

1. The timing of the following procedures is based on a 12:12 h light:dark cycle (lights come on at 6:30 am and go off at 6: 30 pm.).
2. The choice of strain/stock for the rats used to generate blastocysts should be guided by the genetic background of the ESCs (*see* **Note 5**).
3. Four days prior to mating (9 days prior to blastocyst collection), sexually mature female rats (typically 8–10 weeks of age) are estrus cycle synchronized hormonally with an intraperitoneal (IP) injection of 40 μg LHRHa in a 200 μL injection volume. The injections are typically given between 8:00 and 9:00 am.
4. By 1:00–3:00 pm 4 days after females have received LHRHa, they are expected to be in proestrus. Each female is paired overnight 1:1 with a sexually mature male to allow mating.
5. The following morning at the beginning of the light cycle (6: 30 am), females are removed from the mating cage, examined for the presence of copulatory plug, and housed for an additional 4.5 days (*see* **Note 6**).
6. Prepare one or two 35 mm petri dishes for embryo collection. In each dish, place five to six 30 μL drops of mR1ECM + 10% FBS and overlay with mineral oil. Place in a 37 °C, 5% CO_2 incubator to allow equilibration. These plates can be prepared the day before blastocyst collection or at least 2 h prior to collection.
7. Euthanize blastocyst donor female rats one at a time (*see* **Note 7**).
8. Collect blastocysts. To collect the blastocysts, place a euthanized female donor rat on the benchtop and spray the abdominal area with 70% ethanol. Grasp the abdominal skin with serrated forceps and make a lateral incision at the midline using scissors. Push the intestines away to reach the uterine horns (they are positioned laterally) to collect the oviduct-uterine tract including the cervix. Place the tracts into a 35 mm petri dish with 2 mL mR1ECM + 20 mM HEPES +10% FBS and wash twice. Transfer the tract into a small volume of flushing medium (mR1ECM + 20 mM HEPES +10% FBS) in a fresh dish. Make a lengthwise cut at the utero-tubal junction. Recover blastocysts from each uterine horn by flushing from the cervix to the utero-tubal junction with 0.2–0.5 mL flushing medium using a 30-gauge blunt needle attached to a 1 mL syringe. Using a 125 μL Stripper

tip attached to an aspirator tube assembly, pick up the embryos by mouth pipetting through the aspirator tube and transfer them to one drop of 30 μL mR1ECM + 10% FBS in a 35 mm petri dish. Wash the embryos by moving them through two additional medium drops in the same petri dish. Transfer to a fresh 30 μL mR1ECM + 10% FBS drop under mineral oil and culture in a 37 °C, 5% CO_2 incubator until needed.

3.3 Microinjection of ESCs into Blastocysts

1. The procedures described here are based on the use of Eppendorf TransferMan NK2 (Eppendorf, Hamburg, Germany) micromanipulators on a Nikon Eclipse T*i* inverted microscope (Nikon Cooperation, Tokyo, Japan) equipped with a Hoffman Modulation Contrast (HMC) system (*see* **Note 8**).
2. Use a 60 mm petri dish lid and make an injection plate by placing several injection drops (10 μL mR1ECM + 20 mM HEPES +10% FBS) in the center of the lid. Overlay each injection drop with mineral oil.
3. Place the injection plate on the stage of a Nikon inverted microscope. Under 40× magnification, lower the holding pipette (Eppendorf) into one injection drop until it touches the bottom of the lid and slowly bring down the beveled ESC Eppendorf Transfertip into the same focal plane with the holding pipette. Adjust the Transfertip (injection needle) so that it is also in parallel with the lid, in the same focal plane with the holding pipette and aligned in parallel with the holding pipette (Fig. 2). Check the position of the setup under 400× magnification.
4. After initial setup, switch back to 40× magnification. Using a 125 μL Stripper tip attached to an aspirator tube assembly, transfer 10 blastocysts into an injection drop. Using a P-10 pipette, add 1–3 μL ESC solution (depending on the concentration of ESC solution) from Subheading 3.1, **step 5** to the injection drop.
5. Under 400× magnification, hold a blastocyst with the holding pipette under negative pressure. Then select 10–12 ESCs with round, smooth shapes and use negative pressure to draw them into the Transfertip. Inject the ESCs into the blastocoel of the blastocyst (Fig. 2). Repeat the injection procedure for each blastocyst in the drop.
6. After injection of each group of 10 blastocysts, transfer the blastocysts to a 30 μL drop of mR1ECM +10% FBS under oil in a 35 mm petri dish and culture for approximately 1 h at 37 ° C with 5% CO_2 and maximal humidity (90–100%).
7. Continue the procedure until all blastocysts have been injected.

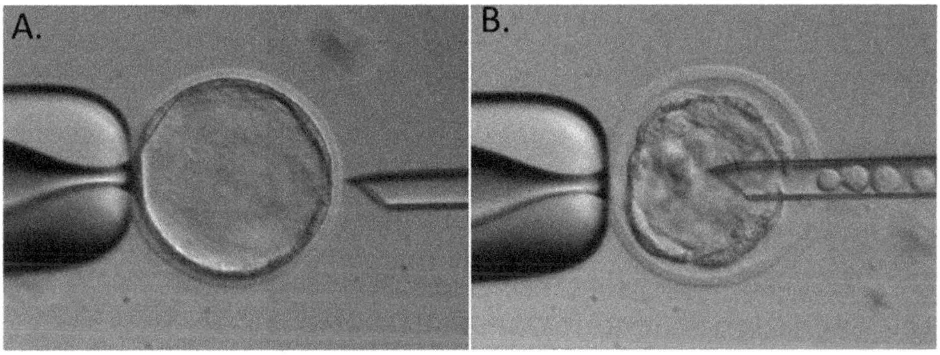

Fig. 2 Blastocyst injection: (**a**) Stabilize rat blastocyst with the holding pipette (left) and align the Transfertip (right) containing rat ESCs with the blastocyst. (**b**) Inject 10–12 ESCs into the blastocoel of each blastocyst

3.4 Embryo Transfer

1. Prepare pseudopregnant recipients. Four days prior to mating, Sprague Dawley (SD) female rats (8–10 weeks of age) are estrus cycle synchronized by intraperitoneal (IP) injection with 40 μg of LHRHa at 9 am (*see* **Note 9**).
2. Pair estrus synchronized females 1:1 with vasectomized male rats in the late afternoon 4 days after injection with LHRHa.
3. Remove the female from the mating cage the following morning and check for the presence of a vaginal copulation plug. An otoscope can be used for easier visualization. Females with a plug are considered 0.5 days post coitum (dpc) pseudopregnant and can be used as surrogates to receive manipulated blastocysts via either surgical or non-surgical embryo transfer 3 days later (*see* **Note 10**).

3.5 Surgical Embryo Transfer

1. For this procedure, approximately 20–30 blastocysts are transferred into the uterine horns of 3.5 dpc pseudopregnant SD females (10–15 blastocysts per uterine horn).
2. Recipient females are anesthetized by intramuscular administration of 5.5 mg ketamine and 1.0 mg xylazine per 100 g body weight. Lubricating ointment should be placed on both eyes to prevent corneal ulceration.
3. To manage surgical pain and inflammation, the recipients are then given Banamine (flunixin meglumine) subcutaneously: 0.2 mg for rats under 250 g and 0.3 mg for rats over 251 g.
4. Shave the back of the rat using fur clippers and wipe the shaved area with 70% ethanol to clean and remove loose hairs. Scrub the surgical area using betadine-soaked cotton swabs (3 times, alternating between 70% ethanol and betadine). Gently pick up the rat and place it on the lid of a 150 mm petri dish.
5. Sterile equipment and techniques should be used for all surgical procedures. Make a 1 cm long transverse incision in the skin using fine dissection scissors about 1 cm to the left of the spinal

cord, at the level of the last rib. Slide the skin to one side and using sterile scissors, bluntly dissect the dermal layer away from the underlying connective tissue until the incision is over the ovary or fat pad. They should both be visible through the body wall.

6. Using serrated forceps, pick up the body wall (dorsolateral) over the fat pad and make a small cut to create a hole (para-lumbar fossa). Gently grab the fat pad using serrated/curved blunt forceps to exteriorize the oviduct, ovary, and part of the uterine horn. Grip the fat pad with a bulldog clamp being careful to avoid the reproductive organs. Lay the fat pad on top of gauze to avoid retraction of the reproductive tract back into the abdominal cavity; the bulldog clamp will act as an additional weight.
7. Pick up the petri dish with the rat and place it on a Leica dissecting microscope stage and illuminate the tissues with a fiber-optic lamp.
8. Retrieve the embryos from the incubator and wash them in 2 mL of mR1ECM+ 20 mM HEPES +10% FBS in a 35 mm petri dish to remove the mineral oil. Load the embryos into a 125 μL Stripper tip attached to an aspirator tube assembly with as little medium as possible.
9. Use an insulin needle to puncture a hole in the uterine horn near the utero-tubal junction, being careful to avoid large blood vessels.
10. Insert the Stripper tip into the pre-punctured hole and expel approximately 10–15 blastocysts into the uterine horn. As soon as air bubbles are seen entering the uterine horn, carefully pull the tip out. Apply sterile saline on the tissues using a dropper and gently replace them into the abdominal cavity.
11. Repeat the same procedure on the opposite side of the animal. Close the skin with an autoclip. Inject the rat with atipamezole (5 mg/mL) at 1 mg/kg subcutaneously post-op to reverse the anesthesia. Allow the rat to recover with thermal maintenance as required by institutional guidelines. Rats are typically awake and ambulatory within approximately 5–10 min.

3.6 Non-surgical Embryo Transfer

1. Inject 2 IU oxytocin IP to 3.5 dpc pseudopregnant rats approximately 2 h prior to scheduled embryo transfer. Oxytocin is used to dilate the cervix so that the rNSET can pass through and reach the uterine horn (*see* **Note 11**).
2. Just prior to the rNSET procedure, place a 50–70 μL drop of medium for each rat transfer scheduled in a 60 mm tissue culture dish. Equilibrate the medium to 37 °C. Under the microscope while using reflective lighting, load up to 20 blastocysts into the drop using a standard embryo handling pipette (*see* **Note 12**).

3. Place the rNSET device onto a P-2 pipette that has been set to 1.8 μL (*see* **Note 13**).
4. Using the embryo handling pipette, group the embryos in the center of the drop. Load the embryos into the rNSET device by pressing the P-2 pipette plunger to first stop, lowering the rNSET tip into the drop and slowly pulling the embryos into the device. Carefully set the P-2 pipette to 2.0 μL to create a small air bubble at the tip of the rNSET (this helps keep embryos in place). Gently place the P-2 pipette with embryos on its side on the benchtop near the rat, avoiding surface contact with the tip of the catheter (*see* **Note 14**).
5. Hold the rat still for the procedure. This is best done by holding the base of the tail using the thumb and forefinger of your non-dominant hand while gently positioning the animal on the benchtop between your arm and your side. Angle the tail upward. Place the speculum into the vagina (*see* **Note 15**).
6. Carefully pick up the P-2 pipette and insert the embryo-loaded rNSET tip through the speculum, past the cervix, and into the uterine horn. Once the rNSET hub contacts the speculum, expel embryos by pressing the plunger to the first stop only (*see* **Note 16**).
7. After embryo delivery, wait for 5 s and slowly remove the rNSET and speculum (*see* **Note 17**). No recovery or post-operative monitoring is required and the rat may be returned to the vivarium.

4 Notes

1. While we typically prepare 60 mm tissue culture plates, other size plates can be used. Guidelines for volumes and cell densities based on plate size are provided by the manufacturer in their protocols.
2. Plates of feeders can usually be maintained and used up to 10 days post initial plating.
3. Care must be taken when changing the medium in ESC plates due to the poor attachment of rat ESCs to the feeder cells. While rat ESC colonies usually attach to the feeders well when they are small, this changes over time. Aspirate slowly when changing the media to prevent the unwanted removal of ESC colonies. Alternatively, change only half the medium volume (~2 mL) or only add a half volume (2 mL) of fresh medium to each culture dish daily.
4. For culture using 60 mm plates, rat ESCs are routinely passaged at a ratio of 1:5 or 1:6 so add 25–30 mL of N2B27 + 2i to the cells recovered from each 60 mm plate and replate on 5–6 new 60 mm plates.

5. The ability of the ESCs to successfully integrate within the inner cell mass of the recipient blastocyst has been shown to be strain/stock dependent, and therefore the choice of blastocyst strain/stock must be considered carefully [3, 4, 9–11].
6. A 70–80% mating success rate is usually seen.
7. We euthanize with CO_2 using a flow rate of 30% chamber volume per minute. In our experience, the length of time exposed to CO_2 may affect embryo quality.
8. The HMC system allows the use of a plastic dish to set up the injection plate. If using an inverted microscope equipped with a differential interference contrast system, a glass lid or chamber is needed.
9. Depending on institutional protocols, the use of anesthesia for injections is optional and will not have deleterious effects on the embryo transfer procedure. Most recipients will be in estrus on day 4 or 5 post injection. If possible, extra female recipients should be estrus synchronized.
10. Plug checks should be performed as early as possible in the morning as vaginal plugs will fall out over time. If no plug is found, the male and female may be left together for a second night, as needed, for embryo transfer scheduling. Any recipient female who is not used for embryo transfer may be reused as a recipient after waiting at least 2 weeks from the last date of mating.
11. Depending on institutional protocols, the use of anesthesia for injections is optional and will not have deleterious effects on the embryo transfer procedure. It is possible that not all recipients will dilate appropriately so it is recommended to prepare extra recipients.
12. The use of an oil overlay to reduce medium evaporation is not recommended as oil contamination in the uterine horn may affect embryo transfer. Thus, embryos are prepared in the drop of medium just prior to embryo transfer to avoid complications with medium evaporation.
13. The Rat NSET™ Device consists of a tapered catheter which can be inserted past the cervix to deposit embryos into the uterine horn. The cervix must be dilated for passage of the catheter to occur. The procedure is minimally invasive and does not require the use of anesthesia or analgesia. The embryo transfer procedure can be performed in under 30 s. No post-procedure monitoring is required. Compared to surgical embryo transfers, this procedure dramatically reduces the time and expertise required for the production of genetically modified rats. Studies at ParaTechs resulted in pregnancy rates of 80% for Sprague Dawley and 90% for F344 rNSET recipients [12]. Birth rates of pups from transfers of 11–20 embryos were 39% and 36%, respectively [12].

14. Visualization of the embryos under a low magnification may aid in loading of embryos into the device.
15. For ease of use, the speculum can be pre-moistened with sterile saline. The excess liquid is removed by gently shaking the speculum prior to insertion. The speculum must be correctly positioned for the rNSET catheter to pass through the cervix. If the speculum is pushed out during the procedure, simply reinsert. Practicing the holding technique and passage of the rNSET device catheter through the cervix without embryos is highly recommended. No anesthesia or analgesia is required for this procedure. The use of anesthesia is optional and will not have deleterious effects on the embryo transfer procedure.
16. When the catheter passes through the cervix, the rNSET catheter will easily slide into place. Do not introduce excess air. Do not release the plunger prior to rNSET removal from the uterine horn. This procedure may require multiple attempts to find the cervix. Each attempt should be performed gently, taking care not to bend the tip. This can be done repeatedly without embryo loss.
17. The speculum can be carefully removed by hand, with forceps, or will fall out on its own. The animal can be returned to the vivarium immediately as no recovery or post-procedure monitoring is required.

Acknowledgments

This work was supported by the Office of Research Infrastructure Programs, Office of the Director, of the National Institutes of Health under award numbers P40 OD011062 (ECB) and R43/44OD018231 (BJS). In addition, the NSET technology was supported in part by an award from the Kentucky Cabinet for Economic Development, Office of Commercialization and Innovation under the grant agreement KSTC-184-512-10-096 with the Kentucky Science and Technology Corporation. The content is solely the responsibility of the authors and does not necessarily represent the official views of the National Institutes of Health or other funding agencies.

References

1. Abbott A (2004) Laboratory animals: the renaissance rat. Nature 428(6982):464–466
2. Gill TJ 3rd et al (1989) The rat as an experimental animal. Science 245(4915):269–276
3. Buehr M et al (2008) Capture of authentic embryonic stem cells from rat blastocysts. Cell 135(7):1287–1298
4. Li P et al (2008) Germline competent embryonic stem cells derived from rat blastocysts. Cell 135(7):1299–1310
5. Tong C et al (2011) Generating gene knockout rats by homologous recombination in embryonic stem cells. Nat Protoc 6(6):827–484

6. Tong C et al (2010) Production of p53 gene knockout rats by homologous recombination in embryonic stem cells. Nature 467(7312): 211–213
7. Chenouard V et al (2021) Advances in genome editing and application to the generation of genetically modified rat models. Front Genet 12:615491
8. Remy S et al (2017) Generation of gene-edited rats by delivery of CRISPR/Cas9 protein and donor DNA into intact zygotes using electroporation. Sci Rep 7(1):16554
9. Kawamata M, Ochiya T (2010) Generation of genetically modified rats from embryonic stem cells. Proc Natl Acad Sci U S A 107(32): 14223–14228
10. Men H, Bauer BA, Bryda EC (2012) Germline transmission of a novel rat embryonic stem cell line derived from transgenic rats. Stem Cells Dev 21(14):2606–2612
11. Men H, Bryda EC (2013) Derivation of a germline competent transgenic Fischer 344 embryonic stem cell line. PLoS One 8(2): e56518
12. Stone BJ, Steele KH et al (2020) A nonsurgical embryo transfer technique for fresh and cultured blastocysts in rats. J Am Assoc Lab Anim Sci 59(5):488–495

Chapter 17

CRISPR/Cas9-Mediated Genome Editing in Zebrafish

Jeffrey G. Daniel, Xinge Yu, Allison C. Ferguson, and Jordan A. Shavit

Abstract

The CRISPR/Cas9 system is a powerful tool for genome editing in zebrafish. This workflow takes advantage of the genetic tractability of zebrafish and will allow users to edit genomic sites and produce mutant lines using selective breeding. Established lines may then be employed by researchers for downstream genetic and phenotypic analyses.

Key words CRISPR/Cas9, Zebrafish, Genome editing, Mutagenesis

1 Introduction

The zebrafish (*Danio rerio*) is an established vertebrate model organism for modeling genetic diseases [1, 2]. Colonies of zebrafish are inexpensive to maintain and can quickly produce large numbers of transparent embryos. Zebrafish development has been well-characterized [3], allowing researchers to easily observe and describe the developmental and pathophysiological effects of genetic lesions. Furthermore, the entire zebrafish genome is in its 11th iteration [4], facilitating the easy design and usage of targeted genetic tools. Forward and reverse genetic methods are readily available for use in zebrafish embryos. The first large-scale screen (ENU mutagenesis) for deleterious mutations was described in the December 1996 issue of *Development* [5], which identified several mutant classes, but was phenotype-driven. Since then, genetic methods for zebrafish have become widely varied, with the most recent and robust development being CRISPR (clustered regularly interspaced short palindromic repeats)/Cas9-mediated genome editing.

Prior to the advent of genome editing, morpholino oligonucleotides (MOs), designed to bind to mRNA and inhibit splicing or translation of target genes, were the only method for targeted inhibition of a gene [6–8]. MOs are especially useful for

Thomas L. Saunders (ed.), *Transgenesis: Methods and Protocols*,
Methods in Molecular Biology, vol. 2631, https://doi.org/10.1007/978-1-0716-2990-1_17,

knockdown (generating the so-called morphants) but are not ideal for large-scale genetic screens. Additionally, MOs do not directly modify the genome and thus may not be appropriate for genes active later in development. Zinc finger nucleases (ZFNs) and transcription activator-like effector nucleases (TALENs) are two classes of engineered proteins capable of targeted genomic modification [9–13]. ZFNs, TALENs, and CRISPR generate double-stranded DNA breaks, which are then repaired primarily through non-homologous end-joining (NHEJ). NHEJ does not always restore the original sequence, thus introducing insertions and deletions (indels) that may lead to nonsense mutations. Despite the mutational capability of ZFNs and TALENs, they can be cumbersome to produce. The disadvantages of these systems make CRISPR the most attractive method for targeted gene disruption.

Although the first papers detailing CRISPR usage were published in the last decade [14–17], researchers began dissecting the CRISPR/Cas9 system in the 1980s in bacteria and archaea [18–23]. CRISPR-associated proteins (e.g., Cas9) work in tandem with a dual guide RNA (gRNA) complex consisting of a gene-specific CRISPR RNA (crRNA) and a trans-activating CRISPR RNA (tracrRNA). The CRISPR/Cas9 ribonucleoprotein (RNP) complex scans the genomic sequence for a protospacer-adjacent motif (PAM) site and then generates a double-strand break (DSB) if the gRNA matches the target sequence. The cell then repairs the DSB via NHEJ as described above, potentially generating indels. CRISPR/Cas9 can be employed in zebrafish by injecting multiple gene-specific gRNAs along with either Cas9 mRNA or protein. Despite the strengths of the CRISPR/Cas9 system, several drawbacks still exist. Cas9 can have off-target effects [24], which may be mitigated by gRNA prediction algorithms (e.g., CHOPCHOP or CRISPRscan) [25, 26]. The resulting sequence following NHEJ repair is unpredictable, so CRISPR-generated mutants may have a non-deleterious sequence [27]. CRISPR can also be used to generate "crispants" after injection at the one-cell stage, which can produce a knockdown phenotype similar to morphants [28–30]. Crispants will likely have allelic mosaicism resulting from asymmetric edits of the zygotic alleles.

2 Materials

2.1 Equipment

1. Glass capillary with filament (Narishige, model GD-1).
2. Micropipette puller (Sutter Instruments, model P-97).
3. Plunger-based or pneumatic microinjector setup (e.g., Nanoliter 2000, World Precision Instruments; PLI-100, Harvard Apparatus).
4. Micromanipulator with magnetic base.

5. Low power stereomicroscope.
6. Microloader tips.
7. Fine forceps (DuMont).
8. Stage micrometer slide to calibrate injection volume.
9. Injection mold (World Precision Instruments, Z-MOLDS).
10. Electrophoresis equipment.
11. UV lightbox.
12. Mini quick spin RNA columns (11814427001; Sigma-Aldrich).

2.2 Kits

1. MinElute PCR purification kit (Qiagen).
2. In vitro transcription kit, T7 (Ambion).

2.3 Buffers and Chemicals

1. Mineral oil.
2. Embryo medium (E3; 0.33 mM magnesium sulfate, 5 mM sodium chloride, 0.17 mM potassium chloride, and 0.33 mM calcium chloride).
3. Injection medium (200 mM potassium chloride, 8.3 mM HEPES).
4. Agarose.
5. MS-222 (tricaine methanesulfonate).
6. Taq polymerase.
7. Lysis buffer (0.5 μM EDTA, 1 μM tris pH 8.0, 0.1% Triton).
8. Proteinase K (proK; 10 mg/mL stock, 10 μg/mL final in lysis buffer).
9. 100% ethanol.
10. 70% ethanol in nuclease-free water.
11. Phenol/chloroform/isoamyl alcohol (P/C/IA; 25:24:1).
12. Chloroform.
13. Cas9 enzyme with nuclear localization sequence (NEB M0386).

3 Methods

3.1 CRISPR/Cas9 sgRNA Design for Genomic Edits

Cas9 single guide RNAs (sgRNA) are composed of two main components (reviewed in Knott and Doudna [31]): a backbone sequence (trans-activating CRISPR RNA; tracrRNA) that binds to the Cas9 enzyme and a gene-specific, 17–20 base pair CRISPR RNA (crRNA). Cas9 generates a double-strand break (DSB) between the third and fourth nucleotides upstream of the PAM sequence (NGG). The protocols described here will enable the induction of small or large genomic indels using individual or multiple sgRNAs.

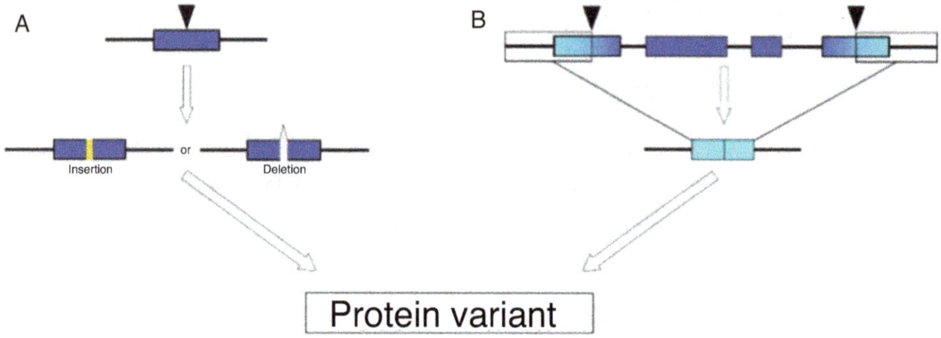

Fig. 1 Genome editing using a one or two sgRNA scheme. (**a**) Indels are generated through the NHEJ cellular repair process that follows a Cas9-mediated DSB (black arrowhead) in a single exon. (**b**) sgRNA target sites (black arrowheads) are designed to generate a DSB at the 5′ and 3′ ends of the locus (within or beyond proximal and distal exons). These mutations result in frameshift mutations (**a**) or large deletions (**b**) and typically produce null alleles. F_0 and F_1 fish are screened for indels at the target site(s) as described in Fig. 3

1. Individual sgRNAs will typically produce indels ranging from less than ten to hundreds of base pairs. Out-of-frame mutations often produce stop codons, which may result in nonsense-mediated decay, although there is always the possibility of truncated protein production with some residual function. To prevent this, whole gene deletion can be performed using sgRNAs targeted to the 5′ and 3′ ends of a locus (Fig. 1).
2. CHOPCHOP and CRISPRscan are two popular sgRNA design tools [25, 26, 32], but many others are available depending on user preference (reviewed in Torres-Perez et al. [33]). Sequencing of the target site (and PCR primer binding sites) in samples from the laboratory's fish colony is advisable as common variation in zebrafish can reduce the efficiency if there are variants or indels in the selected regions.
3. sgRNAs can be produced via in vitro transcription (IVT) or ordered from various manufacturers, such as IDT (Integrated DNA Technologies) or Synthego.

3.2 sgRNA Production by In Vitro Transcription (IVT)

PCR is the easiest and fastest method to generate a DNA template for sgRNA production. This consists of two primers (Table 1): a gene-specific primer containing template for the crRNA sequence with a leading T7 promoter and a common primer used to produce the sgRNA scaffolding section (modified protocol [34]).

1. Set up the following 100 μL reaction to produce the DNA template for synthesis of each sgRNA: 5 μL common primer (10 μM stock), 5 μL gene-specific primer (10 μM stock), 40 μL nuclease-free water, and 50 μL Taq 2× master mix.
2. Thermocycler protocol: 95 ° C, 3 min; (95 °C, 30 s; 45 °C, 30 s; 72 °C, 30 s) ×30 cycles; 72 °C, 5 min.

Table 1
Sequences of common and target-specific primers

Gene-specific primer (crRNA segment)	TAATACGACTCACTATAGGNNNNNNNNNNNNNNNNNN gttttagagctagaa
Common primer (tracrRNA segment)	AAAAGCACCGACTCGGTGCCACTTTTTCAAGTTGATAACGG ACTAGCCTTATTTTAACTTGCTATttctagctctaaaac

The region of overlap is shown in lowercase letters. Gene-specific sequences obtained from CRISPRscan replace the Ns in the crRNA sequence

3. Run 1 μL of the PCR product on an agarose gel to ensure proper template synthesis (117 bp amplicon).
4. If the correct amplicon was produced, purify the sgRNA template by phenol–chloroform extraction as follows.
5. Add 100 μL nuclease-free water to the PCR reaction to bring the volume up to 200 μL.
6. Add an equal volume of P/C/IA and vortex for 1 min.
7. Centrifuge at maximum speed for 5 min at room temperature (RT).
8. Remove as much of the top (aqueous) layer as possible without disturbing the bottom (organic) layer and place it into a new tube.
9. Add one volume chloroform and vortex for 1 min.
10. Centrifuge at maximum speed for 5 min (RT).
11. Transfer as much of the top layer as possible to a new tube.
12. Add 0.1 volume 3 M sodium acetate and 2.5 volumes 100% ethanol.
13. Chill at −80 °C for at least 60 min.
14. Centrifuge at maximum speed for 30 min (4 °C).
15. Remove as much of the supernatant as possible without disturbing the pellet.
16. Add 70% ethanol and centrifuge at maximum speed for 10 min (4 °C).
17. Remove the supernatant and allow the pellet to air dry.
18. Resuspend in 20 μL nuclease-free water and check concentration on a spectrophotometer.
19. Use T7 IVT kit according to manufacturer's instructions.
20. Column-purify the entire sgRNA IVT reaction using RNA spin columns according to the manufacturer's instructions.
21. Check concentration on a spectrophotometer.
22. Store at −80 °C in 2 μL aliquots.

3.3 Generation of Knockout Lines

In order to produce germline mutations, one cell embryos are injected with either sgRNAs and Cas9 mRNA or assembled sgRNA/Cas9 ribonucleoprotein complexes (RNPs). Embryos should be collected as soon as possible as edits generated earlier in embryonic development may reduce larval genetic chimerism and increase the chance of germline transmission of mutations (*see* **Note 1**). The microinjection protocols described here have been used with both a pneumatic- and plunger-based pump system (*see* **Note 2**).

1. Set up crosses using dividers to separate males and females on the afternoon prior to the day of injection.
2. Using a micropipette puller, prepare an adequate number of injection needles from glass capillary tubes, and break off each tip using fine tweezers.
3. For best results, prepare the injection mix fresh on the day of injection (*see* **Note 3**).
4. If using Cas9 protein, incubate with sgRNA at 37 °C for 5 min. to assemble the RNP complexes.
5. Harvest single-cell embryos and inject with 1–3 nL of the injection mix using a microinjection system.
6. Crispants (F_0 generation) with confirmed edits are grown to sexual maturity and outcrossed to non-injected fish to test for germline transmission in a subset of F_1 embryos (*see* Subheading 3.4; Fig. 2 depicts a basic breeding plan overview).
7. The remaining F_1 embryos from F_0 adults with confirmed germline transmission are raised to produce heterozygous mutants.
8. Stable lines heterozygous for embryonic lethal mutations may be generated and expanded by selecting heterozygotes from either F_1 mutant carrier outcrosses or incrosses.
9. Subsequent generations of carriers are then selectively bred to maintain the mutation of interest and stock genetic fitness.

3.4 Crispant Genotyping

Genomic edits made by RNPs may be genotyped using primers flanking the cut site(s). Here, we describe genotyping schemes for indels at a single site as well as large genomic deletions.

1. Individual sgRNA targets in F_0 crispants (injected embryos) are initially screened for editing efficiency using target-specific primer pairs at 2 days post-fertilization (dpf) or later. Embryos are collected and placed into PCR tubes, then the embryo medium is replaced with 20 μL lysis buffer + proK. Use the following thermocycler protocol for embryo lysis: 55 °C, 120 min; 95 °C, 5 min; hold @ 4–10 °C (*see* **Note 4**).
2. Indels produced from only one cut site enable amplification and genotyping with a single set of primers. However, large deletions require multiplexed systems to distinguish between wild-type and mutant alleles (Fig. 3).

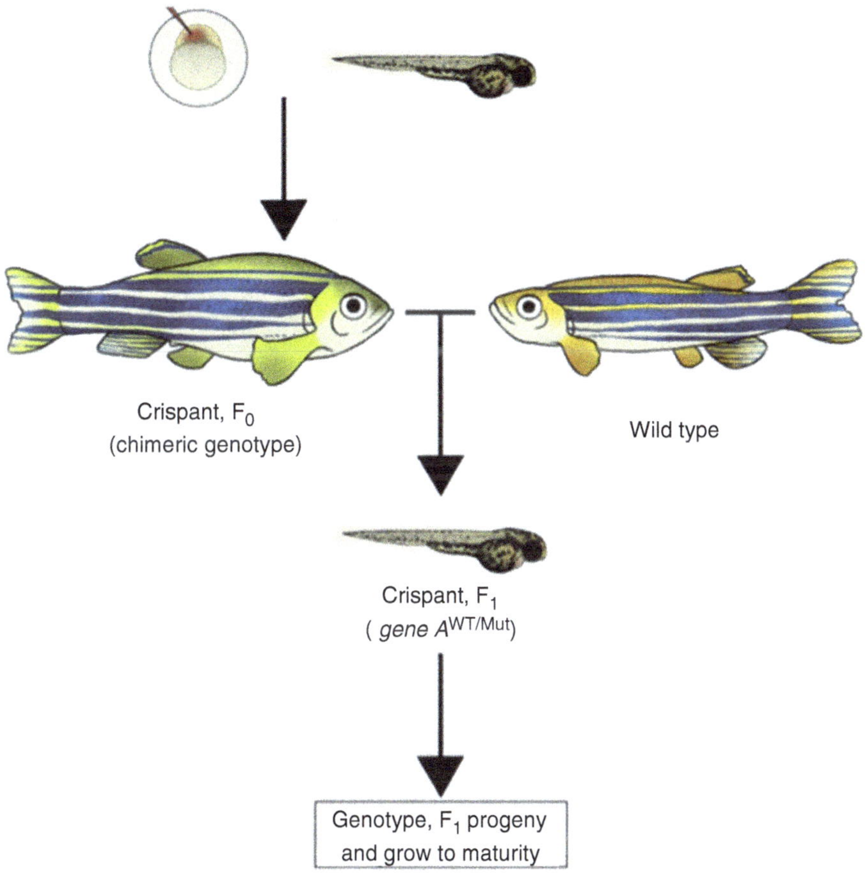

Fig. 2 General scheme for knockout line production. One-cell embryos are injected with Cas9 (mRNA or protein) and sgRNA(s), checked for edits by PCR, and grown to sexual maturity. F_0 adults may then be outcrossed with wild-type individuals, and F_1 offspring screened for germline transmission of mutations. Adults with confirmed genotypes may then be incrossed to generate homozygous mutants for downstream genetic analysis

3. To genotype large deletions, two multiplex genotyping systems are described here. The first uses three primers; a primer pair flanking the two CRISPR target sites will amplify the mutant band. The third uses a primer specific to the wild-type allele and is located within the predicted deletion on either the 3′ or 5′ end of the locus (Fig. 3). Alternatively, one can design primers to amplify a wild-type-specific amplicon within the deleted region for a 4-primer system (Fig. 3). Primers should preferentially be designed to bind within exons, as variation in introns is common and can lead to false-positive results.
4. BLAT search [35] primer sequences against the zebrafish genome (UCSC [36] or Ensembl genome browser) to ensure specificity to the target gene.

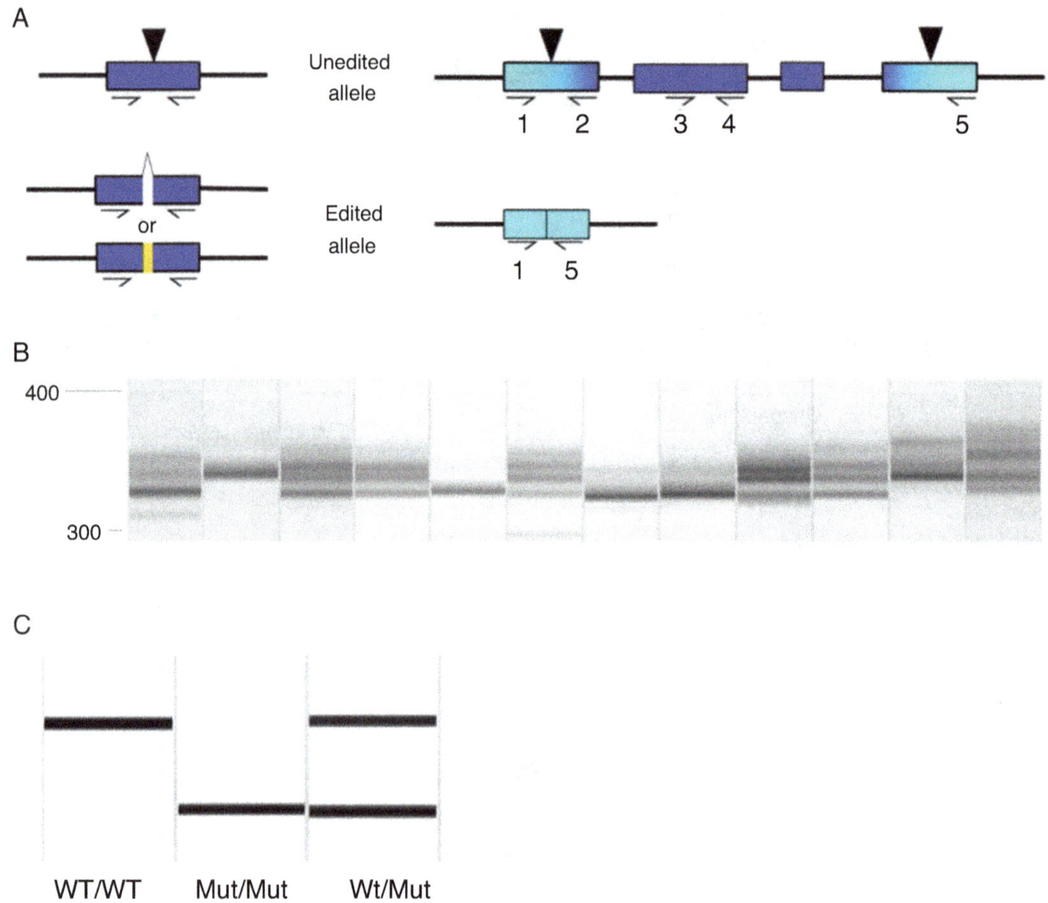

Fig. 3 Two primer systems for genotyping Cas9-mediated indels. (**a**) Indels produced by genome editing of a single target exon. Cas9 generates a double-strand break, which stimulates NHEJ and results in either an insertion or deletion. A single pair of genotyping primers are indicated by horizontal arrows. In the case of a large deletion, genotyping cannot be done with a single set of primers, as for most loci the wild-type allele will be too large for PCR amplification. Primers 1, 2, and 5 constitute the 3-primer system, and primers 1 and 3–5 make up the 4-primer system. (**b**) Digital electrophoresis shows examples of band patterns from edited F_0 alleles, demonstrating mosaic deletions and insertions (WT amplicon = 327 bp). Progeny of injected crispants (F_1) may be screened for specific lesions using the 3- or 4-primer systems. (**c**) Representative results of a multiplexed reaction with the 3- or 4-primer system using F_1 progeny of injected crispants, assuming a larger wild-type band. Reactions producing an amplicon should be sequenced to confirm the presence of the large deletion

5. Amplify the genomic region of interest using 1 μL of the embryo lysate in a 20 μL PCR reaction and analyze by electrophoresis (Fig. 3) to look for band patterns that differ from uninjected controls, indicating successful genome editing (*see* **Note 5**).
6. Each primer pair should be individually optimized before combining into 3- or 4-primer systems.

4 Notes

1. Gene targets that are embryonic lethal may complicate the establishment of a knockout line upon injection of Cas9 protein or ubiquitously expressed Cas9 mRNA constructs. An alternative option is to inject Cas9 mRNA under the control of the *nanos* 3′ UTR, which restricts Cas9 expression to germ cells [34].
2. The plunger-based system does not require needle calibration; however, one may confirm that the correct volume is being dispensed in either system using a stage micrometer. Place a small amount of mineral oil onto the micrometer, then float a single injection volume on top of the oil such that the liquid forms a sphere. Measure the diameter of the sphere formed by the injection mix and calculate the volume using the formula $V = (\pi/6)d^3$ (V = volume, divide by 10^6 to obtain nanoliters; d = diameter in microns).
3. Our lab has had success with an equimolar sgRNA:Cas9 ratio. We recommend using 50 ng/μL of sgRNA as a starting point. OligoCalc [37] may be used to estimate the molarity of an sgRNA. Commercial preparations of Cas9 are often sold with defined molarities. If using Cas9 mRNA, we recommend 250 ng/μL as a starting point, then titrating based on toxicity and cutting efficiency.
4. If a region of interest proves to be difficult to amplify with Taq, an alternative polymerase (i.e., one with proofreading activity) may yield better results.
5. Electrophoresis band patterns of edited alleles from injected fish may appear as smears rather than discrete bands. Qualitative differences in band patterns suggest that genomic edits were successful.

Acknowledgments

We would like to thank Dr. Raghunath Azhwar for helpful discussions. This work was supported by NIH grants R35HL150784 and R01ES032255 to J.A.S.

References

1. Santoriello C, Zon LI (2012) Hooked! Modeling human disease in zebrafish. J Clin Invest 122:2337–2343
2. Kretz CA, Weyand AC, Shavit JA (2015) Modeling disorders of blood coagulation in the zebrafish. Curr Pathobiol Rep 3:155–161
3. Kimmel CB et al (1995) Stages of embryonic development of the zebrafish. Dev Dyn 203: 253–310
4. Howe K et al (2013) The zebrafish reference genome sequence and its relationship to the human genome. Nature 496:498–503

5. Haffter P et al (1996) The identification of genes with unique and essential functions in the development of the zebrafish Danio rerio. Development 123:1–36
6. Flynt AS, Rao M, Patton JG (2017) Blocking zebrafish MicroRNAs with Morpholinos. Methods Mol Biol 1565:59–78
7. Huang P et al (2012) Reverse genetic approaches in zebrafish. J Genet Genomics 39:421–433
8. Nasevicius A, Ekker SC (2000) Effective targeted gene 'knockdown' in zebrafish. Nat Genet 26:216–220
9. Doyon Y et al (2008) Heritable targeted gene disruption in zebrafish using designed zinc-finger nucleases. Nat Biotechnol 26:702–708
10. Sander JD et al (2011) Targeted gene disruption in somatic zebrafish cells using engineered TALENs. Nat Biotechnol 29:697–698
11. Meng X et al (2008) Targeted gene inactivation in zebrafish using engineered zinc-finger nucleases. Nat Biotechnol 26:695–701
12. Gaj T, Gersbach CA, Barbas CF 3rd (2013) ZFN TALEN and CRISPR/Cas-based methods for genome engineering. Trends Biotechnol 31:397–405
13. Ma D, Liu F (2015) Genome editing and its applications in model organisms. Genomics Proteomics Bioinformatics 13:336–344
14. Cong L et al (2013) Multiplex genome engineering using CRISPR/Cas systems. Science 339:819–823
15. Hwang WY et al (2013) Efficient genome editing in zebrafish using a CRISPR-Cas system. Nat Biotechnol 31:227–229
16. Jinek M et al (2012) A programmable dual-RNA-guided DNA endonuclease in adaptive bacterial immunity. Science 337:816–821
17. Mali P et al (2013) RNA-guided human genome engineering via Cas9. Science 339: 823–826
18. Barrangou R et al (2007) CRISPR provides acquired resistance against viruses in prokaryotes. Science 315:1709–1712
19. Bolotin A et al (2005) Clustered regularly interspaced short palindrome repeats (CRISPRs) have spacers of extrachromosomal origin. Microbiology (Reading) 151:2551–2561
20. Ishino Y et al (1987) A nucleotide sequence of the iap gene responsible for alkaline phosphatase isozyme conversion in Escherichia coli and identification of the gene product. J Bacteriol 169:5429–5433
21. Mojica FJ et al (1995) Long stretches of short tandem repeats are present in the largest replicons of the Archaea Haloferax mediterranei and Haloferax volcanii and could be involved in replicon partitioning. Mol Microbiol 17:85–93
22. Mojica FJ et al (2005) Intervening sequences of regularly spaced prokaryotic repeats derive from foreign genetic elements. J Mol Evol 60: 174–182
23. Pourcel C, Salvignol G, Vergnaud G (2005) CRISPR elements in Yersinia pestis acquire new repeats by preferential uptake of bacteriophage DNA and provide additional tools for evolutionary studies. Microbiology (Reading) 151:653–663
24. Tsai SQ et al (2015) GUIDE-seq enables genome-wide profiling of off-target cleavage by CRISPR-Cas nucleases. Nat Biotechnol 33:187–197
25. Moreno-Mateos MA et al (2015) CRISPRscan: designing highly efficient sgRNAs for CRISPR-Cas9 targeting in vivo. Nat Methods 12:982–988
26. Labun K et al (2019) CHOPCHOP v3: expanding the CRISPR web toolbox beyond genome editing. Nucleic Acids Res 47:W171–W174
27. Burger A et al (2016) Maximizing mutagenesis with solubilized CRISPR-Cas9 ribonucleoprotein complexes. Development 143:2025–2037
28. Cox DBT et al (2017) RNA editing with CRISPR-Cas13. Science 358:1019–1027
29. Abudayyeh OO et al (2017) RNA targeting with CRISPR-Cas13. Nature 550:280–284
30. Kushawah G et al (2020) CRISPR-Cas13d induces efficient mRNA knockdown in animal embryos. Dev Cell 54:805–817
31. Knott GJ, Doudna JA (2018) CRISPR-Cas guides the future of genetic engineering. Science 361:866–869
32. Montague TG et al (2014) CHOPCHOP: a CRISPR/Cas9 and TALEN web tool for genome editing. Nucleic Acids Res 42:W401–W407
33. Torres-Perez R et al (2019) WeReview: CRISPR tools-live repository of computational tools for assisting CRISPR/Cas experiments. Bioengineering (Basel) 6:63
34. Vejnar CE et al (2016) Optimized CRISPR-Cas9 system for genome editing in zebrafish. Cold Spring Harb Protoc. https://doi.org/10.1101/pdb.prot086850
35. Kent WJ (2002) BLAT--the BLAST-like alignment tool. Genome Res 12:656–664
36. Kent WJ et al (2002) The human genome browser at UCSC. Genome Res 12:996–1006
37. Kibbe WA (2007) OligoCalc: an online oligonucleotide properties calculator. Nucleic Acids Res 35:W43–W46

Chapter 18

Generation of Rabbit Chimeras by Eight-Cell Stage Embryo Injection

Dongshan Yang, Jun Song, Jie Xu, Jifeng Zhang, and Y. Eugene Chen

Abstract

Pluripotent stem cell (PSC) injection to the blastocyst stage embryos is a widely used method to evaluate the pluripotency through chimeric contribution. It is routinely used to produce transgenic mice. However, PSC injection to the blastocyst stage embryos in rabbits is challenging. At this stage, the in vivo developed rabbit blastocysts possess a thick mucin layer that is inhibitory for microinjection, whereas in vitro developed rabbit blastocysts that lack such mucin layer often fail to implant after embryo transfer. In this chapter, we describe a detailed protocol of rabbit chimera production through mucin-free eight-cell stage embryo injection procedure.

Key words Rabbit, Pluripotent stem cells, Eight-cell stage embryo injection, Chimera

1 Introduction

Pluripotent stem cells (PSCs)-mediated gene targeting in mice has revolutionized the study of developmental biology and human medicine. The PSCs can be derived from preimplantation embryos (i.e., embryonic stem cells, ESCs) or reprogrammed from differentiated cells (i.e., induced pluripotent stem cells, iPSCs). The PSCs possess robust proliferative ability and pluripotency. They have the capability to differentiate into various cell types, including germ cells. Although the recent development of designer nuclease-based gene editing methods, especially the CRISPR/Cas9 technology, have substantially decreased the needs for PSC-based gene targeting approach, this method remained as a valuable technique for generating sophisticated genetic models as well as for stem cell research.

Rabbit (*Oryctolagus cuniculus*) is a classic research animal model and is increasingly becoming a translational model of choice, serving to bridge the gap between rodent models and larger animal models [1, 2]. Unfortunately, PSC technology in rabbits has lagged

Thomas L. Saunders (ed.), *Transgenesis: Methods and Protocols*,
Methods in Molecular Biology, vol. 2631, https://doi.org/10.1007/978-1-0716-2990-1_18,

considerably behind compared to mice. Although a number of groups, including us [3–8], have reported the successful establishment of rabbit ES cells and iPS cells, which exhibit the characteristic features of primed pluripotency similar to those of human pluripotent stem cells, and some cell lines can be passaged for at least 50 times or more than 1 year, almost all of the established cell lines have low or no chimeric competency, and no germline transmission has been reported yet. Such a low rate of chimera generation may result either from the failure of ESCs to colonize the preimplantation embryo, or from the failure of incorporated ESCs to truly participate in embryo development because of competition with host cells, or from both [9].

A unique challenge for rabbit chimera production is that rabbit embryos develop mucin layer, a layer of glycoprotein accumulated during oviductal transport. An in vivo developed rabbit blastocyst has a mucin coat of 50–60 μm thick [10], making microinjection formidable. In vitro cultured embryos do not develop mucin layer. However, mucin-free rabbit blastocysts often fail to implant, leading to poor embryo transfer results. Alternatively, in mice, PSCs can be injected into eight-cell stage embryos to produce chimeras. It has been reported that when injected at eight-cell stage, ESCs decrease the number of differentiative divisions and the presence of ESCs reduces the number of ICM cells of the host embryo origin [11]. Indeed, the eight-cell stage injection method has been used to generate fully ESC-derived mice [12, 13] and resulted in higher germline contribution compared with the blastocysts injection method [14]. We previously reported that in vitro cultured mucin-free eight-cell stage rabbit embryos give rise to a high pregnancy rate after being transfered into oviduct, a rate very similar to those achieved by transferring pronuclear stage embryos [15]. Based on this knowledge and the success of eight-cell stage embryo injection in mice, we developed an eight-cell stage embryo injection procedure in rabbits. Briefly, rabbit pronuclear stage embryos were cultured in vitro until eight-cell stage for PSC injection, and then the injected mucin-free embryos were transferred into oviducts of synchronized recipients to generate chimera.

2 Materials

2.1 Equipment

1. Inverted microscope with a fluorescent light source and digital camera.
2. Vibration isolation platform.
3. Thermo Plate for microscope stage.
4. Micromanipulator.
5. Microinjector.
6. Stereomicroscope.

7. Micropipette puller.
8. Microforge.
9. Ceramic tile for scoring glass.
10. CO_2 incubator.
11. Alcohol lamp.
12. Embryo moving pipette (pulled Pasteur pipette) assembled into a hand- or mouth-operated system.
13. Autoclave-steam sterilizer.
14. Anesthesia machine.
15. TPR temperature/pulse oximeter/respiration system.
16. Water-heated mat.
17. Vacuum clipper system: Clipper connected to vacuum.
18. Skin staple handle.
19. Skin staple remover.
20. DNA thermal cyclers.
21. Refrigerated microcentrifuge.
22. Spectrophotometer.
23. Horizontal gel electrophoresis system.
24. Ultra low freezer.
25. Surgery pack: Thumb Tissue Forceps 5.5 "1 × 2 teeth, Operating Scissors 6.5" Sharp/Sharp Straight, Operating Scissors 4.5 "Blunt/Blunt Straight, Forceps Thumb Dressing 5.5", Adson Dressing Forceps 4.75, "Iris Scissors 3 1/2" Straight, Scalpel Handle Stainless Steel #3—For #10–15 Blades, Towel Clamps 5.5, "Needle Holder 5.25", Crile Forceps 5.5 "Straight, Thumb Dressing Forceps 8".

2.2 Consumables

1. Glass capillary: thin wall borosilicate tubing with filament, 0.78 mm inner diameter, 1 mm out diameter.
2. Injection pipette: TransferTip® (ES) or equivalent (20° tip angle, 15 μm inner diameter, 1 mm flange, sterile, Eppendorf Catalog No. 5195000079).
3. 4-Well Dish.
4. Petri dish, 35 mm, 60 mm.
5. Centrifuge tubes, sterile, 15 mL, 50 mL.
6. Microcentrifuge tubes, 0.5 mL, 1.5 mL, certified RNase, DNase, DNA & Pyrogen free.
7. Syringes, 1 cc, 5 cc Luer Slip.
8. Embryo transfer tube: Tom Catheter 3.5Fr × 4.5″, Sterile.
9. Scalpel Blades #10 Stainless Steel.

10. Powder Free Surgeon Glove, Sterile.
11. Surgery Drape: Poly-lined Sterile 18″ × 26″.
12. Suture #4-0 3/8 Circle Rev. Cut 19 mm / 36″.
13. Surgery Gown: XLarge 47″ Non-Reinforced Sterile with Towel.
14. Staple Cartridges Surgi-Close 6 Cartridges × 20 Staples.
15. Soft-E Collar XSmall (Cat/Dog 5–9#).

2.3 Media and Solutions

1. HM: 25 mM HEPES, 10% fetal bovine serum in Medium 199 with Hanks' salts.
2. EBSS-complete medium: 1× MEM non-essential amino acid, 1× BME amino acid, 1 mM L-glutamine, 0.4 mM sodium pyruvate, 10% FBS in Earle's Balanced Salt Solution with calcium, magnesium, and phenol red.
3. Mineral oil: BioReagent, suitable for embryo cell culture.
4. FSH (follicle-stimulating hormone): 10 mg/mL in saline.
5. HCG (human chorionic gonadotropin): 200 IU/mL in saline.
6. Euthasol (euthanize solution): Sodium pentobarbital 390 mg/mL, sodium phenytoin 50 mg/mL.
7. Anesthetic: Isoflurane inhalant anesthetic.
8. Analgesic: carprofen injectable, 50 mg/mL.
9. Digestion buffer for DNA extraction: 5 mM EDTA, pH 8.0, 200 mM NaCl, 100 mM Tris–HCl, pH 8.0, 0.2% sodium dodecyl sulfate (SDS).
10. Molecular biology Kits: PCR purification kit, RNA purification kit.

2.4 Animals

1. Embryo donors: 4 months to 2 years old New Zealand White (NZW) does (*see* **Note 1**).
2. Fertile males: 6 months to 2 years old NZW buck rabbits. They should be replaced every 1–2 years.
3. Recipients: 5 months to 2-year-old NZW does (*see* **Note 1**).

3 Methods

3.1 Superovulation

1. Subcutaneously inject sexually matured NZW female rabbits with FSH twice/day with a dosage of 3 mg for the first two injections, 5 mg for the next two injections, and 6 mg for the last two injections.
2. Alternatively, PMSG can be used instead of FSH. Intramuscularly inject 150 IU of PMSG in a single dosage (*see* **Note 2**).
3. At 72 h after the first FSH injection, or PMSG injection, inject 100–200 IU HCG intravenously to induce ovulation.

4. Put a superovulated doe into a male's cage immediately after hCG injection.
5. Observe and record the successful mating. It will happen within minutes (*see* **Note 3**).
6. Put the female back to her cage.

3.2 Zygote Collection

1. Prepare a 30 mm petri dish with 2 mL HM medium covered by mineral oil.
2. Keep the dishes on a heated stage at 38.5 °C.
3. Prepare a 4-well dish containing 500 μL EBSS-complete medium in each well, cover with 400 μL mineral oil (for embryo culture).
4. Prepare a 30 mm dish with four 80 μL drops of EBSS-complete medium covered with mineral oil (for embryo wash).
5. Place the dishes in 5% CO_2 incubator at 38.5 °C overnight.
6. Prepare two 50 mL centrifuge tubes with 25 mL HM medium and prewarm in 37 °C water bath.
7. At 17–19 h after hCG injection, euthanize the superovulated rabbits with an overdose intravenous injection of Euthasol (150 mg/kg). Recover the oviducts with a part of uterus in 25 mL HM medium.
8. Wash the oviducts in a new 25 mL HM medium. Use an Iris scissor to remove the fat tissue around the oviduct to make it straight, hold the uterus side with one pair of tissue forceps, insert a blunt needle connected to a syringe filled with 5 mL of HM medium, and flush the oviduct from uterus to infundibulum.
9. Collect the flushed media in a 60 mm dish. Swirl the dish gently to gather eggs in the center.
10. Collect eggs under a stereomicroscope, wash the eggs in the 2 mL pre-warmed HM dish prepared in **step 1**, then wash three times in EBSS-complete medium, and incubate at 38.5 °C, 5% CO_2 in air overnight (*see* **Note 4**).

3.3 PSCs Injection

1. Prepare PSC single-cell suspension according to the PSC cell culture condition (depends on the sources of the PSC). The prepared PSC single-cell suspension in the PSC medium is usable for several hours if kept on ice or stored at 4 °C fridge (*see* **Note 5**).
2. Prepare holding pipette by hand pulling a piece of the borosilicate glass capillary tubes in the flame of an alcohol lamp. The outside diameter of the drawn-out region of the pipette should be 120–200 μm (i.e., close to the diameter of the zona pellucida).

3. Break the pipette at ~1–2 cm from the shoulder of the pipette by scoring the glass lightly with a ceramic tile at the desired position.
4. Fire-polish the pipette tip with the microforge until the pipette tip shrinks to an inner diameter of ~40 μm. Bend the pipette ~20-degree angle by approaching the side of the pipette to the heating glass bead on the microforge.
5. Before microinjection, prepare the microinjection platform by placing a 50 μL drop of HM medium and several 50 μL drops of PSC culture medium with PSC single-cell suspension prepared in Subheading 3.3, **step 1** on a reversed lid of a 60 mm dish.
6. Cover the drops with mineral oil and place them on the microscope stage of the inverted microscope.
7. Place holding pipette into the left arm of the micromanipulator connected to CellTram 4 m Air syringe or equivalent and position in the drop of HM medium.
8. Adjust the holding pipette angle to make the tip parallel to the stage.
9. Place injection pipette into the right arm of the micromanipulator connected to a CellTram 4r Oil syringe or equivalent and position it into the drop of HM medium with the tip parallel to the stage and in line with the holding pipette.
10. Transfer 10–30 embryos to the HM drop on the platform.
11. Move the holding pipette out of the platform and move the stage to locate the injection tip to the PSC medium drop.
12. With the injection micrometer syringe, uptake a small amount of injection media into the injection pipette. This will allow a buffer area between the cells and the oil and help prevent injection of oil inside the embryo.
13. Lower the injection pipette to the bottom of the dish to load 15–75 PSCs inside the injection pipette with as little injection media as possible.
14. Back in the HM medium drop, move the holding pipette back and use the holding pipette to capture an embryo and align the injection needle, embryo and holding pipette along the x-axis.
15. Under 200× magnification, advance the injection pipette through the zona and inject 10–15 PSCs per embryo. After injection, withdraw the needle. Release the embryo and repeat till all the embryos in the drop are injected (*see* Fig. 1a).
16. Wash the injected embryos three times in EBSS-complete medium, and incubate at 38.5 °C, 5% CO_2 in air for 2 h and transfer to oviducts of recipients, or the injected embryos can be cultured for additional 2 days in the incubator until embryos reach the blastocyst stage for in vitro analysis.

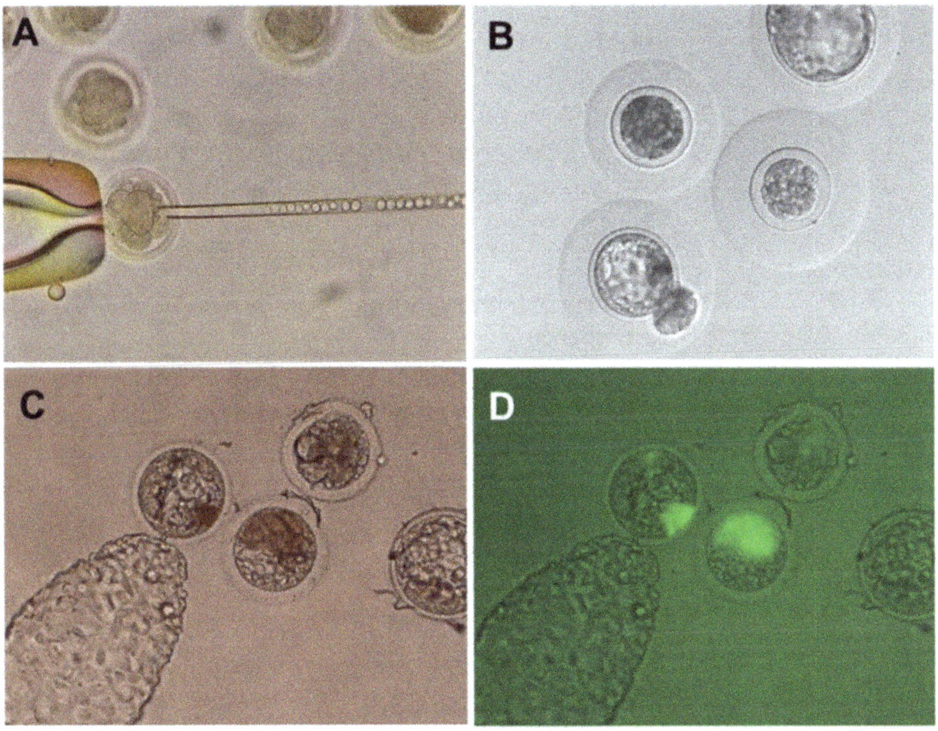

Fig. 1 Rabbit eight-cell stage embryo injection. (**a**) Injection of rabbit ES cells into eight-cell stage rabbit embryos. (**b**) Injected rabbit embryos regained mucin layer 1 day after transferred into oviduct of a recipient. (**c**) Rabbit embryos injected with GFP-positive ES cells developed to blastocyst stage at 2 days after cultured in vitro. (**d**) Fluorescent microscopy showed the GFP-positive ES cells were incorporated into the chimera embryos. Magnificent: 200×

3.4 Embryo Transfer

Surgery is performed under sterile conditions, with all surgical tools, media, supplies, etc. sterile. For each animal, a fresh, sterile, surgical pack, containing sterile drapes and tools, is used. Personnel involved in the surgical procedures are scrubbed, gloved, gowned, capped, and masked. The rabbit is placed on the surgical table on a water-heated mat maintained at 37 °C and covered with a sterile drape.

1. Select a female NZW rabbit in good health, weighing approx. 3–4 kg and at least 5 months of age. Check vagina to select does with red or pink vulva color (*see* **Note 6**).
2. To prepare recipients, inject 100 IU of hCG intravenously 24 h after donors were mated and injected with hCG (*see* **Note 7**).
3. Before surgery, weigh the rabbit and inject 4 mg/kg carprofen subcutaneously as an analgesic (*see* **Note 8**).
4. Connect anesthesia machine to oxygen supply and scavenging system.
5. Restrain the rabbit on its neck and back firmly, put on an anesthesia mask connected with the anesthesia machine to the rabbit's mouth (*see* **Notes 9** and **10**).

6. On the anesthesia machine, adjust the flow meter at 100–200 mL/min and vaporizer at 5% to induce anesthesia. Squeeze the footpad to check for lack of pedal reflex which is an indication of adequate anesthesia.
7. Adjust the evaporator to 2–5% to maintain anesthesia. Monitor temperature, pulse, and respiration during procedure and adjust vaporizer as needed.
8. Clip the rabbit's flanks with a vacuum clipper system.
9. Put the rabbit under anesthesia on a water-heated mat in lateral recumbence.
10. Wash the shaved flank with antiseptic scrub (Povidone Scrub) three times, followed by antiseptic solution (70% alcohol), and wipe with a sterile gauze sponge.
11. The surgical incision site can be approximated by drawing an imaginary equilateral triangle over the flank with vertices at the wing of the ilium, the greater trochanter, and the ovary.
12. Make a 2–3 cm horizontal (anterior posterior) skin incision with a scalpel blade over the anticipated location of the ovary. Incise the external and then internal abdominal oblique muscles.
13. Gently elevate the peritoneum with Crile forceps and sharply incised.
14. Grasp the fat pad attached to the oviduct with Thumb Dressing forceps to pull the ovary and oviduct out of the peritoneal cavity.
15. Under a stereomicroscope, 10–15 embryos are loaded into an embryo transfer tube (Tom Catheter) attached to a 1 mL syringe with HM medium.
16. Insert the embryo transfer tube through the infundibulum into the oviduct of recipients. Make sure the tube passed the big turn of the oviduct and unload the embryos inside. Withdraw the transfer tube gently and carefully.
17. Put the ovary and oviduct back to the peritoneal cavity, close the peritoneum and muscle layer, and then subcutaneous fascia layer with absorbable suture (e.g., Vicryl). Close the skin with skin staples.
18. Apply the same procedure to the other side to transfer another batch of 10–15 embryos (*see* **Note 11**).
19. At the end of the surgery, turn the vaporizer and gas supply off, place the animal in the recovery area with thermal support until fully recovered.
20. Put Elizabethan Soft-E Collar on the neck to prevent the rabbit from biting the wound.

21. After surgery, inject a dose of carprofen subcutaneously at 4 mg/kg/day as an analgesic for at least 5 days.
22. At 10–14 days post-operation, remove the Soft-E Collar and remove the skin staples with a skin staple remover.
23. 1–2 weeks before the expected delivery date, move the recipient rabbit into a breeding cage with nesting materials (*see* **Note 12**).

3.5 Evaluation of the Chimera

The chimera can be identified based on the marker in the PSC donor (coat color, green fluorescence protein (GFP), enzymes, and DNA markers, etc.). DNA makers including the GFP marker can be detected by PCR genotyping:

1. Hold 1- to 3-week-old kit firmly with one hand, with the other cut off a tiny piece of ear skin (around 2–5 mm) with a scissor.
2. Digest the ear skin tissue with 300 μL digestion buffer with 1 mg/mL proteinase K in a 1.5 mL Eppendorf tube at 55 °C overnight.
3. Add 1 mL of 100% ethanol into each tube, mix completely. Centrifuge at 13,000 g for 12 min, and then pour out the ethanol.
4. Wash the DNA pellets by adding 0.8 mL of 70% ethanol into the tubes, then centrifuge at 13,000 × g for 8 min.
5. Pour out the ethanol from the tube, and then suck the remaining ethanol with a pipette.
6. Dissolve the DNA pellets with 300 μL TE buffer or distilled water.
7. Store the DNA at 4 °C or −20 °C and use it as a template for PCR.
8. Germline transmission can be evaluated by natural breeding of the chimera animal with a wild-type counterpart.

4 Notes

1. We recommend using donor rabbits older than 16 weeks. However, NZW rabbits as early as 12 weeks of age are old enough to be stimulated for superovulation though the embryo quality may be compromised. The female recipients should be good mothers as they must carry transferred embryos to term and nurse newborns to wean at 6 weeks of age. They should also allow the inspection of newborn kits and accept kits from another mother. We recommend using experienced mothers as recipients if available or using recipients older than 5 months of age.

2. In rabbits, PMSG and FSH have been commonly used for superovulation. Though the PMSG program is simpler and less laborious, the FSH program is more reliable [16].
3. To ensure successful fertilization, put the female in more male's cages to record at least 2–3 successful mating.
4. It's not necessary to remove cumulus cells using hyaluronidase, since most of the cumulus cells will be detached from the rabbit oocytes at the time zygotes are recovered. For the same reason, the rabbit zygotes are collected by oviduct flush but can't be released by ampulla tearing as in mice. If there are residual cumulus cells attached to the zona, they will be detached after overnight culture.
5. An excellent and quite pure PSC is necessary for microinjection to ensure that cells introduced to the subzonal space of embryos is PSC cells mainly. A special procedure that eliminates feeder cells and enriches PSCs may be required if the PSC is cultured on feeder cells (refer to Lee [17]); alternatively, a maker like EGFP in the donor PSC may help distinguish the PSC versus feeder cells.
6. Female rabbits don't have an estrus cycle and are classified as induced ovulators. A doe is receptive when she accepts service. In recipient preparation, however, we can't test its receptivity by mating. The color of the vulva is instead used as a strong indication of receptivity. It is recommended to select females with red and pink vulva as recipients.
7. Embryos cultured in vitro may develop slower than that in vivo. To allow the in vitro cultured mucin-free embryos to regain the mucin layer, the injected eight-cell stage embryos were transferred back to the oviducts which may delay their movement to uterus, so the ovulation of the recipients were induced 24 h after the donors were mated to synchronize the development of the transferred embryos with the uterus status at implantation. Rabbit ovulation is normally induced by the stimuli associated with coitus and occurs 10–12 h after mating. By mimicking the natural mating, ovulation can be induced artificially by various techniques, including mechanical stimulation of the vagina and injection of hCG. In our experience, hCG injection is more reliable.
8. It is not recommended to fast the rabbit before surgery, as the risk of aspiration due to vomiting in rabbits is negligible because rabbits lack the vomiting reflex and are physically almost incapable of regurgitation. Feeding rabbit before surgery helps the gastrointestinal (GI) tract remain active, which will speed recovery. Rabbits may suffer liver damage when the GI tract is empty.
9. Because rabbits lack the vomiting reflex, intubation is not necessary for anesthesia.

10. Rabbit can tolerate restraint. It is not necessary to use injectable anesthetics to induce anesthesia in rabbits.
11. In rabbits, the embryos never move from one uterine horn to the other. Bilateral embryo transfer is necessary to fully utilize the recipients. The average rabbit litter size is 7–8 in natural mating. A minimum of two implanted embryos seems to be necessary to carry a pregnancy to term. In our experience, when the litter size is larger than 10, the survival rate of the kits will be compromised.
12. The average gestation period in rabbits is 30–32 days. Newborn kits weigh 30–100 g at birth, are hairless, and have closed eyelids. The doe normally nurses her kits once a day, usually in the morning. Kits open their eyes at about 10–12 days and begin to eat solid food at about 3 weeks. Rabbit kits are weaned at 4–6 weeks.

References

1. Fan J, Watanabe T (2003) Transgenic rabbits as therapeutic protein bioreactors and human disease models. Pharmacol Ther 99(3): 261–282
2. Esteves PJ et al (2018) The wide utility of rabbits as models of human diseases. Exp Mol Med 50(5):1–10
3. Fang ZF et al (2006) Rabbit embryonic stem cell lines derived from fertilized, parthenogenetic or somatic cell nuclear transfer embryos. Exp Cell Res 312(18):3669–3682
4. Wang S et al (2007) Generation and characterization of rabbit embryonic stem cells. Stem Cells 25(2):481–489
5. Honda A et al (2008) Stable embryonic stem cell lines in rabbits: potential small animal models for human research. Reprod Biomed Online 17(5):706–715
6. Osteil P et al (2013) Induced pluripotent stem cells derived from rabbits exhibit some characteristics of naive pluripotency. Biol Open 2(6): 613–628
7. Xue F et al (2012) Recombinant rabbit leukemia inhibitory factor and rabbit embryonic fibroblasts support the derivation and maintenance of rabbit embryonic stem cells. Cell Reprogram 14(4):364–376
8. Du F (2015) Derivation of rabbit embryonic stem cells from vitrified-thawed embryos. Cell Reprogram 17(6):453–462
9. Tancos Z et al (2012) Generation of rabbit pluripotent stem cell lines. Theriogenology 78(8):1774–1786
10. Chen CH et al (2012) Dynamic profiles of Oct-4, Cdx-2 and acetylated H4K5 in in-vivo-derived rabbit embryos. Reprod Biomed Online 25(4):358–370
11. Humięcka M et al (2016) ESCs injected into the 8-cell stage mouse embryo modify pattern of cleavage and cell lineage specification. Mech Dev 141:40–50
12. Huang J et al (2008) Efficient production of mice from embryonic stem cells injected into four- or eight-cell embryos by piezo micromanipulation. Stem Cells 26(7):1883–1890
13. Poueymirou WT et al (2007) F0 generation mice fully derived from gene-targeted embryonic stem cells allowing immediate phenotypic analyses. Nat Biotechnol 25(1):91–99
14. Guo J et al (2014) Contribution of mouse embryonic stem cells and induced pluripotent stem cells to chimeras through injection and coculture of embryos. Stem Cells Int 2014: 409021
15. Lin TA et al (2011) Open-pulled straw vitrification differentiates cryotolerance of in vitro cultured rabbit embryos at the eight-cell stage. Theriogenology 75(4):760–768
16. Kitajima S, Liu E, Fan J (2009) Rabbit transgenesis. In: Houdebine L-M, Fan J (eds) Rabbit biotechnology: rabbit genomics, transgenesis, cloning and models. Springer, Heidelberg
17. Lee KH (2014) Generating chimeric mice from embryonic stem cells via vial coculturing or hypertonic microinjection. Methods Mol Biol 1194:77–111

Chapter 19

Genome Editing in Pigs

David Preisinger, Thomas Winogrodzki, Bernhard Klinger, Angelika Schnieke, and Beate Rieblinger

Abstract

The generation of genetically engineered (GE) pigs for disease modeling and xenotransplantation has been massively facilitated by the discovery of the CRISPR/Cas9 system. For livestock, genome editing is a powerful tool when used in combination with either somatic cell nuclear transfer (SCNT) or microinjection (MI) into fertilized oocytes. To generate either knockout or knock-in animals using SCNT, genome editing is carried out in vitro. This has the advantage that fully characterized cells are being employed to generate cloned pigs, predetermining their genetic makeups. However, this technique is labor-intensive and, hence, SCNT is better suited for more challenging projects such as the generation of multi-knockout- and knock-in pigs. Alternatively, CRISPR/Cas9 is introduced directly into fertilized zygotes via microinjection to produce knockout pigs more rapidly. Finally, the embryos are each transferred into recipient sows to deliver GE piglets.

Both techniques, SCNT and MI, are technically challenging and therefore require skilled expertise, especially when applied for porcine embryos. Here, we present a detailed laboratory protocol for the generation of knockout and knock-in porcine somatic donor cells for SCNT and knockout pigs via microinjection. We describe the state-of-the-art method for isolation, cultivation, and manipulation of porcine somatic cells, which can then be used for SCNT. Moreover, we describe the isolation and maturation of porcine oocytes, their manipulation by microinjection, and the embryo transfer into surrogate sows.

Key words CRISPR/Cas9, Genome editing, Knockout, Knock-in, Porcine somatic cells, Zygote, Pig, Microinjection, Somatic cell nuclear transfer, Embryo transfer

1 Introduction

Since mouse models resemble human disease phenotypes often inaccurately, the need for a more suitable species to be used for preclinical studies has become increasingly apparent. Pigs share many similarities with humans in terms of size, physiology, organ

The authors David Preisinger and Thomas Winogrodzki have contributed equally to this work and share first authorship and the authors Angelika Schnieke and Beate Rieblinger have contributed equally to this work and share last authorship.

Thomas L. Saunders (ed.), *Transgenesis: Methods and Protocols*,
Methods in Molecular Biology, vol. 2631, https://doi.org/10.1007/978-1-0716-2990-1_19,

anatomy, and lifespan. Furthermore, it enables long-term diagnostic studies and multiple collections of liquid or tissue biopsies from a single animal, which is consistent with the 3Rs rules: replacement, reduction, and refinement. Therefore, it has emerged as the animal species of choice for translational research. To model human diseases accurately and design organ donors for the emerging field of xenotransplantation, genetically engineered (GE) pigs play a crucial role. Thus, the demand for GE pigs has increased considerably over the last decade [1, 2].

In the absence of fully functional ES cells, gene targeting had to be carried out in somatic cells which is discouragingly inefficient [3]. The advent of CRISPR/Cas9 has revolutionized the production of genetically modified pigs. Sequence-specific induction of DNA double-strand breaks enables efficient generation of GE pigs and saves breeding time, since modifications often are homozygous. Multiplexed genome editing allows even simultaneous inactivation of several porcine genes and genome-wide disruption of multiple functional porcine endogenous retrovirus (PERV) copies [4, 5]. Moreover, the generation of Cas9-expressing pigs provides the most recent advance in the toolbox of porcine genome engineering [6].

While the possibilities for manipulating the porcine genome are becoming more diverse, the production of genetically modified pigs itself is hampered by slow progress in porcine embryology. The two major techniques to generate genetically modified animals, microinjection and somatic cell nuclear transfer (SCNT), are shown in Fig. 1. Both are inefficient and underwent only minor changes since their original description in 1985 and 2000, respectively [7, 8]. Nevertheless, slight advances in porcine embryology were made recently, such as optimized media compositions leading to higher in vitro maturation efficiencies of porcine oocytes and the generation of pluripotent stem cells with enhanced differentiation potential [9, 10].

For SCNT, somatic donor cells are first modified in culture. This allows cells to be analyzed in advance for correct genome editing or insertion of transgenes at a predetermined position. Then, the nuclei of modified donor cells are transferred into enucleated, in vitro maturated (IVM) oocytes prior to embryo transfer [11]. Hence, this approach reduces the proportion of non-correctly targeted animals considerably and is superior to microinjection for challenging attempts, such as multiplexed genome editing and knock-in strategies. However, genome engineering of somatic cells and subsequent SCNT are labor-intensive and cloned offspring often suffers from developmental defects caused by abnormal imprinting [12].

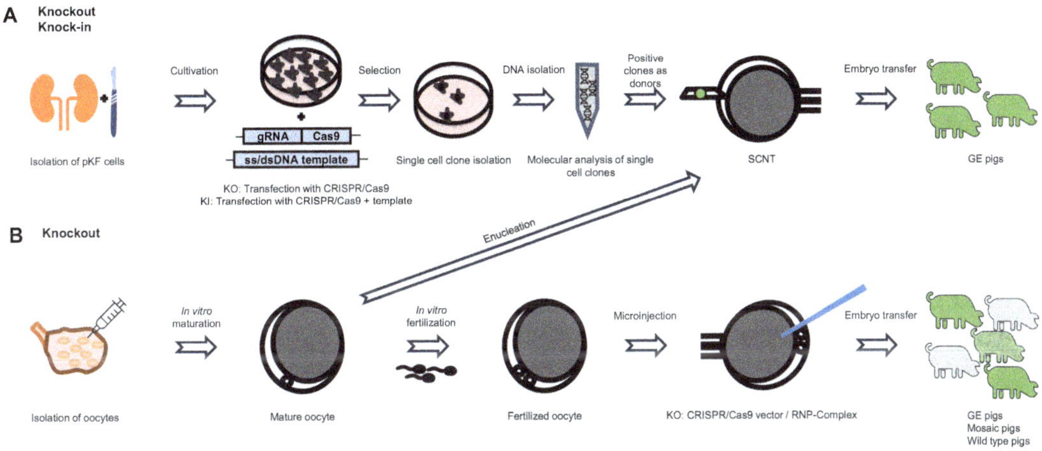

Fig. 1 Genome editing workflow in pigs. (**a**) Generation of genetically engineered pigs via genome editing in porcine primary cells and subsequent use of correctly edited cells for SCNT. First, kidney fibroblasts (pKF) are isolated and cultured. Cells are subsequently transfected with CRISPR/Cas9 components (Cas9 and gRNA encoded on a single plasmid). If a knock-in is desired, the respective DNA template (ssDNA or dsDNA containing a single base substitution or a transgene of interest) is additionally delivered to the cells. Antibiotic selection could then be carried out if the transfected constructs carried a selectable marker cassette. Afterward, single-cell clones are isolated and analyzed on molecular level and positive cell clones are used for SCNT. This results in 100% genome-edited offspring. (**b**) Generation of genome-edited pigs via microinjection. If a gene knockout is desired, genome editing could be carried out by microinjection, providing a faster method compared to genome editing in cells and subsequent SCNT. In vitro maturated and fertilized oocytes are microinjected with CRISPR/Cas9 components (Cas9 and gRNA encoded on a single plasmid or assembled as RNP complex). The constructs/complexes are microinjected into the cytoplasm of a zygote near the polar bodies. Similar to SCNT, injected embryos are transferred into a surrogate sow which gives birth to certain percentage of GE pigs

For a more rapid generation of GE pigs, especially for knockouts, microinjection is the method of choice. It is of superior speed because no previous genome editing of somatic cells is required. Instead, CRISPR/Cas9 constructs are physically introduced into or near the zygotes' pronuclei prior to embryo transfer. The strong pigmentation of porcine oocytes can make it diffucult to visualize the pronuclear membrane in the zygote. However, the disadvantage is a high probability of mosaicism in the progeny. Therefore, for each project, an evaluation of the most appropriate method for the production of genetically modified pigs should be carried out.

Here, we describe an effective state-of-the-art protocol for the isolation, production, and preparation of genetically modified somatic cells as donors for SCNT in the first part. In the second part, we give information from first to last about the rapid generation of transgenic pigs already at the embryonic stage. This reaches from the isolation of slaughterhouse-derived oocytes, to their maturation to pronuclear microinjection, and finally to embryo transfer.

2 Materials

2.1 Generation of Genetically Engineered Pigs Via SCNT

1. Favorable cells for SCNT: Porcine kidney fibroblast cells (pKF) and porcine adipose-derived mesenchymal stem cells (pADMSC).
2. pKF growth medium: Dulbecco's Modified Eagle Medium (DMEM) supplemented with the following:
   (a) 20% (v/v) fetal calf serum (FCS).
   (b) 1% (v/v) non-essential amino acids (NEAA).
   (c) 1% (v/v) sodium pyruvate.
   (d) 1% (v/v) L-alanine-L-glutamine.
   (e) 1% (v/v) penicillin-streptomycin (pen-strep) (only for isolation and first 3 days in culture).

   Stored at 4 °C and pre-warmed to 37 °C before use.
3. pADMSC growth medium: Advanced DMEM supplemented with the following:
   (a) 20% (v/v) FCS.
   (b) 1% (v/v) NEAA.
   (c) 1% (v/v) L-alanine-L-glutamine.
   (d) 1% (v/v) pen-strep (only for isolation and first 3 days in culture).

   Stored at 4 °C and pre-warmed to 37 °C before use.
4. Serum starvation medium: pKF growth medium or pADMSC growth medium containing only 0.5% (v/v) FCS instead of 20% (v/v).
5. Reduced serum medium (Opti-MEM).
6. Phosphate buffered saline (PBS): 137 mM NaCl (8.0 g/l), 2.7 mM KCl (0.2 g/l), 10 mM Na_2HPO_4 (1.42 g/l), and 1.8 mM KH_2PO_4 (0.24 g/l). Autoclaved and stored at room temperature for immediate use or at 4 °C for long-term storage.
7. Freezing medium: 70% (v/v) FCS, 20% (v/v) growth medium, and 10% (v/v) dimethyl sulfoxide (DMSO). Stored at −20 °C.
8. Accutase cell dissociation reagent containing 0.5 mM EDTA. Store at 4 °C after thawing.
9. Selection medium: Growth medium mixed with respective antibiotic selection chemical (e.g., puromycin, hygromycin, blasticidin S, G418). Prepared shortly before use.
10. Lipofection transfection reagent.
11. DNA vectors containing CRISPR/Cas9 components for genome editing (such as pSpCas9(BB)-2A-Puro (PX459) (Addgene plasmid #62988) containing your gRNA(s) of

interest); additionally, for CRISPR/Cas9-mediated knock-in, ss/ds DNA templates harboring the transgene or mutation of interest and, if the experimental design allows, a selectable marker cassette (*see* **Notes 1** and **2**).

12. Standard cell culture dishes (T-150 flasks, 10 cm dishes, 24-well plates, 6-well plates).
13. Sterile bacterial petri dishes, scalpels, and Erlenmeyer flasks containing magnetic stirrer for cell isolation.
14. Collagenase type 1A diluted in PBS to 1 mg/mL.
15. Controlled CO_2 incubator at 5% CO_2 and 37 °C with humidified air.
16. Cell culture microscope.
17. Laminar flow hood.
18. Sterile plastic cloning cylinders, silicon grease, and tweezers for single-cell cloning.

2.2 Generation of Genetically Engineered Pigs Via Microinjection

All solutions are produced using embryo-grade water unless otherwise noted.

2.2.1 Collection of Ovaries and In Vitro Maturation of Oocytes

1. Water bath (38.5 °C).
2. Thermal container filled with transport solution for ovary transport.
3. Transport solution: PBS, supplemented with each 1% pen-strep and amphotericin B, pre-warmed to 38.5 °C.
4. Sieve for rinsing ovaries.
5. Hexadecyltrimethylammonium bromide (CETAB) stock: 25 g CETAB dissolved in 125 mL ethanol (95%) and filled up to 500 mL with ddH_2O. Stored at room temperature.
6. CETAB working solution: PBS, pre-warmed to 38.5 °C mixed with 1% CETAB. Prepared shortly before use.
7. PBS for rinsing ovaries, stored at 38.5 °C.
8. Glass beaker, 500 mL.
9. Heating plate, set to 38.5 °C.
10. 15 mL falcons for follicular fluid collection.
11. 10 mL syringe and 18G needle.
12. Controlled CO_2 incubator at 5% CO_2 and 38.5 °C with humidified air—for storage of follicular fluid during collection.
13. Working medium: HEPES-buffered TCM199 supplemented with the following:
    (a) 10% FCS.
    (b) 1% Pen-strep.

    Stored at 4 °C and pre-warmed to 38.5 °C before use.

14. "4+"-medium: 3.05 mM glucose, 0.1% (w/v) polyvinyl alcohol, 0.91 mM sodium pyruvate, and 1% pen-strep. Prepared freshly every week and stored at 4 °C. This solution is used to prepare the following stocks:
15. Cysteine stock (×1000): 57 mM Cysteine. Stored at −20 °C in 10 μL aliquots.
16. Epidermal growth factor (EGF) stock (×1000): 1 μg/mL. Supplemented with 0.1% (w/v) BSA. Stored at −20 °C in 10 μL aliquots.
17. Fibroblast growth factor 2 (FGF2) stock (×1000): 4 μg/mL. Supplemented with 0.1% (w/v) BSA. Stored at −20 °C in 10 μL aliquots.
18. Insulin-like growth factor 1 (IGF1) stock (×1000): 2 μg/mL. Supplemented with 0.1% (w/v) BSA. Stored at −20 °C in 10 μL aliquots.
19. Leukemia-inhibitory factor (LIF) stock (×1000): 2 μg/mL. Supplemented with 0.1% (w/v) BSA. Stored at −20 °C in 10 μL aliquots.
20. Equine chorion gonadotropin (eCG) and human chorion gonadotropin (hCG)-mixed stock (×100): 1000 IU/mL each (e.g., Suigonan and Ovogest). Stored at −20 °C in 100 μL aliquots.
21. In vitro maturation medium (termed "FLI" medium [9]): "4+"-medium supplemented with 1× cysteine, EGF, FGF2, IGF1, LIF, and eCG–hCG mix each. Prepared freshly before in vitro maturation of oocytes.
22. 10 cm bacteriological petri dish divided into equal sections by indentations.
23. Stereo microscopes equipped with a 38.5 °C heating plate (e.g., Stemi 508, Carl Zeiss GmbH, with HT-50, Minitube).
24. Mouth pipette: Rubber tube connected to a plastic mouthpiece and to a 0.22 μm syringe filter respectively. The syringe filter itself is further linked to a purchased or homemade glass capillary of about 300 μm diameter (*see* **Note 3** and Fig. 2).
25. Pasteur pipettes.
26. Class II laminar flow hood.
27. Nunc 4-well IVF plate (Thermo Fisher Scientific).
28. Pipettes.
29. Scissors.
30. Cell culture plates, 60 mm × 10 mm and 35 mm × 10 mm (Corning).
31. Controlled Triple gas incubator at 5% CO_2, 90% N_2, and 38.5 °C with humidified air—for in vitro maturation and embryo cultivation (e.g., Heracell VioS 160i, Thermo Fisher Scientific) (*see* **Note 4**).

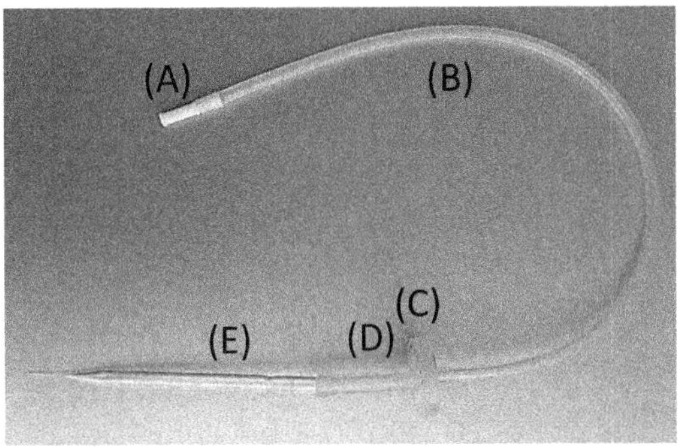

Fig. 2 Fully assembled mouth pipette. A plastic mouthpiece (**a**) is connected to a rubber tube (**b**). The other side of the tube is connected to a filter to prevent ingestion of specimen (**c**). A second rubber tube (**d**) is connecting the filter to a glass capillary drawn from glass pipettes (**e**)

2.2.2 In Vitro Fertilization

1. Porcine zygote medium 5 (PZM-5, Fujihira Industry) (*see* **Notes 5** and **6**).
2. Porcine fertilization medium (PFM, Fujihira Industry) (*see* **Notes 5** and **6**).
3. Mineral oil (*see* **Note 7**).
4. Working medium, pre-warmed to 38.5 °C (*see* Subheading 2.2.1).
5. Androstar plus sperm dilution medium (Minitube) freshly prepared and pre-warmed to 38.5 °C.
6. Centrifuge with insert for 15 mL falcon tubes.
7. Microscope slides and cover slips.
8. Neubauer improved cell counting chamber.
9. Counting device.
10. Thermometer.
11. Sperm of a breeding boar, deep-frozen at sperm concentration optimized for in vitro fertilization (IVF) and stored in liquid nitrogen (*see* **Note 8**).
12. Long Tweezers for safe handling of deep-frozen sperm.
13. Inverted microscope for the assessment of sperm concentration and quality.
14. Box filled with water (38.5 °C) to thaw sperm.
15. Hyaluronidase stock (×50): 50 mg/mL. Stored at −20 °C in 40 μL aliquots.
16. 5 mL Polystyrene Round-Bottom Tube with lid enabling gas exchange.

2.2.3 Microinjection

1. Micropipette puller (e.g., P 1000 Flaming Brown, Sutter Instruments) (*see* **Note 9**).
2. Borosilicate glass with filaments (e.g., from Sutter Instrument).
3. Vacuum pump.
4. Micromanipulator (e.g., Transfer man NK2, Eppendorf).
5. Stereomicroscope, equipped with a heating plate at 38.5 °C (e.g., Axiovert 200 M, Carl Zeiss, with HT-50, Minitube).
6. Microinjection unit (e.g., CellTram vario pro and FemtoJet Express, Eppendorf).
7. Holding pipette (e.g., Vacutip, Eppendorf).
8. Low Tris EDTA buffer: 10 mmol/L Tris–HCl, pH 7.6, and 0.25 mmol/L EDTA, pH 8.0.
9. Injection solution containing genome engineering components diluted in low Tris EDTA buffer (concentrations might have to be optimized for specific purposes):
   (a) 5 ng/μL CRISPR/Cas9 vector optionally with 5 ng/μL linearized repair template.
   (b) RNA-protein complexes (RnPs): 50 ng/μL Cas9 protein, 100 ng/μL sgRNA.

2.2.4 Embryo Transfer

1. Twelve-hour fasted and synchronized surrogate sow (*see* **Notes 10** and **11**).
2. HEPES-free TCM-199 medium incubated overnight in triple gas incubator at 38.5 °C in a Polystyrene Round Bottom Tube.
3. HEPES-free TCM-199 incubated overnight in triple gas incubator at 38.5 °C in a 15 mL falcon tube without lid.
4. Sterilized 15 cm glass dish.
5. 80% Ethanol.
6. PBS.
7. Transportable incubator, pre-heated to 38.5 °C.
8. Sterile, flexible central vein catheter.
9. Intravenous catheters.
10. 10% Ketamine hydrochloride.
11. 2% Xylazine.
12. Skin disinfectant.
13. Operation table, tiltable to 30°.
14. Stereo-microscopes equipped with a 38.5 °C heating plate (e.g., HT-50, Minitube).
15. Anesthetics: 5 mg/kg bodyweight (BW) azaperone, 25 mg/kg BW ketamine.
16. Peri-operative: 0.4 mg/kg BW meloxicam, 15 mg/kg BW amoxicillin.

17. Heating plate, set to 38. 5 °C.
18. Bucket filled with warm water and soap.
19. Disposable razors.
20. Iodine solution.
21. Scalpels.
22. Suction compress.
23. Self-adhesive surgery drape.
24. Cauterizer.
25. Suture material (Surgicryl 910 DS 12, 1 (5/0), 45 cm).
26. Silver spray.

3 Methods

3.1 Generation of Genetically Engineered Pigs Via SCNT

All solutions are pre-warmed to 37 °C unless otherwise noted.

3.1.1 Isolation of Primary Porcine Kidney Fibroblasts (pKF)

1. Directly process freshly harvested kidney (preferably from a young animal) or store at 4 °C in PBS until usage (*see* **Notes 12** and **13**).
2. Pre-warm pKF growth medium containing pen-strep, 10 mL collagenase type 1A in PBS (1 mg/mL), and a magnetic stirring plate to 37 °C.
3. Place kidney into a petri dish, remove the outer conjunctive tissue with a scalpel, and isolate a tissue cube of approximately 0.5 cm edge length from the renal cortex.
4. Cut tissue into 1 mm^3 pieces using a scalpel.
5. Add 10 mL of collagenase type 1A, resuspend several times, and transfer the tissue pieces in collagenase into a 100 mL Erlenmeyer flask containing a magnetic stirrer bar. Stir for 30 min at 37 °C (*see* **Note 14**).
6. Stop reaction by adding 10 mL of pKF growth medium.
7. Transfer the cell suspension into a 50 mL centrifuge tube and centrifuge for 5 min at 300 rcf.
8. Aspirate the supernatant, resuspend in 20–25 mL pKF growth medium containing pen-strep, and transfer the suspension into a T150 culture flask.
9. Incubate at 37 °C, 5% CO_2.
10. Change medium daily for at least 2 days.
11. Change to regular pKF growth medium lacking pen-strep at the third day after isolation.

12. Exclude the presence of mycoplasma by PCR after cell cultivation in antibiotic-free pKF growth medium for at least 2 days.
13. Once almost confluent, detach cells by adding 4 mL Accutase followed by incubation for 10–15 min at 37 °C. Stop reaction by adding an equal volume of medium, centrifuge for 5 min at 300 rcf, aspirate the supernatant and resuspend in 10 mL cryomedium. Distribute quickly to 10 cryovials and freeze to −80 °C (*see* **Notes 15** and **16**).

3.1.2 Isolation of Adipose-Derived Mesenchymal Stem Cells (pADMSC)

1. Prepare two 10 cm dishes. Pre-warm pADMSC growth medium containing pen-strep, PBS, 10 mL collagenase type 1A in PBS (1 mg/mL), and a magnetic stirring plate to 37 °C.
2. Place 1 cm^3 of freshly isolated fat tissue in a 10 cm dish with 10 mL PBS and remove remaining skin and blood vessels.
3. Transfer the fat tissue in a new 10 cm dish, add 10 mL collagenase type 1A, and cut the tissue into 1 mm^3 pieces.
4. Transfer the tissue pieces in collagenase into a 100 mL Erlenmeyer flask containing a magnetic stirrer bar. Place onto a stirring plate and stir for 20 min at 37 °C.
5. Filter the cell suspension through a 100 μm strainer to remove remaining fat clumps. Mix the suspension with 10 mL pADMSC growth medium and centrifuge for 10 min at 1000 rcf.
6. Upon removal of the supernatant, resuspend the cell pellet in pADMSC growth medium and plate cells on a T150 flask.
7. Incubate at 37 °C and 5% CO_2 overnight.
8. Wash gently with PBS. Culture the cells for two more days in antibiotic-containing pADMSC growth medium and change medium daily.
9. Exclude the presence of mycoplasma by PCR after cultivation in antibiotic-free pADMSC growth medium for at least 2 days.
10. Once almost confluent, detach cells by adding 4 mL Accutase, incubate for 10–15 min at 37 °C, and stop the reaction by adding an equal volume of growth medium. Centrifuge for 5 min at 300 rcf, aspirate the supernatant, and resuspend the cell pellet in 10 mL cryomedium. Distribute quickly to 10 cryovials and freeze to −80 °C (*see* **Note 17**).

3.1.3 Transfection of Primary Cells Via Lipofection

1. Thaw a low passage (1–3) of pKF cells or pADMSC and seed on a T150 flask.
2. When cells are almost confluent (70–95%), detach and seed at a density reaching approximately 80% confluency at the next day (*see* **Note 18**).

   (a) If single-cell clones are intended, seed cells on 10 cm dishes.
   (b) If knockout pools are intended, seed on six-well plates.
3. Check cells for appropriate confluency at the next day. Aspirate medium, wash twice, and add only enough reduced serum medium to cover the cells:
   (a) 4 mL for a 10 cm dish.
   (b) 1 mL for a six well.
4. Prepare transfection complexes according to manufacturer's instructions and transfect cells (*see* **Note 19**):
   (a) Amount of DNA depends on the vessel size to be transfected:
      (i) 10 cm dish: 2–5 μg DNA.
      (ii) 6-well plates: 0.5–2 μg DNA.
   (b) For knock-in: Optimal ratio of ss/ds DNA template to CRISPR/Cas9 vector has to be evaluated empirically but usually ranges from 2:1 to 5:1.
   (c) Prepare an additional reaction containing 1 μg of GFP vector as transfection control.
5. Cultivate cells with transfection complexes in the incubator at 37 °C for 4–6 h and then add growth medium.
6. Change medium and check for successful transfection the next day.

3.1.4 Determination of Adequate Antibiotic Concentrations for Selection

1. An adequate antibiotic concentration for selection has to be determined individually for each cell isolate and antibiotic. Seed the isolated cells on a six-well plate at the same passage that will later be used for transfection reaching a density of 90–100% at the next day.
2. Select antibiotics according to the experimental design:
   (a) To generate knockout cells: Selection for transient uptake of CRISPR/Cas9 vectors which contain a puromycin resistance cassette (such as pSpCas9(BB)-2A-Puro (PX459), (Addgene plasmid # 62988)). This is a short 48 h puromycin selection aiming at killing all non-transfected cells, but minimizing integration of Cas9.
   (b) To generate knock-in cells: Knock-in cells are generated by providing CRISPR/Cas9 components together with a repair template containing the mutation or the transgene to be introduced and, if the experimental design allows, a selection cassette. If no insertion of a selection cassette is intended, transiently Cas9-expressing cells can be selected as described above, which enriche also knock-in cells,

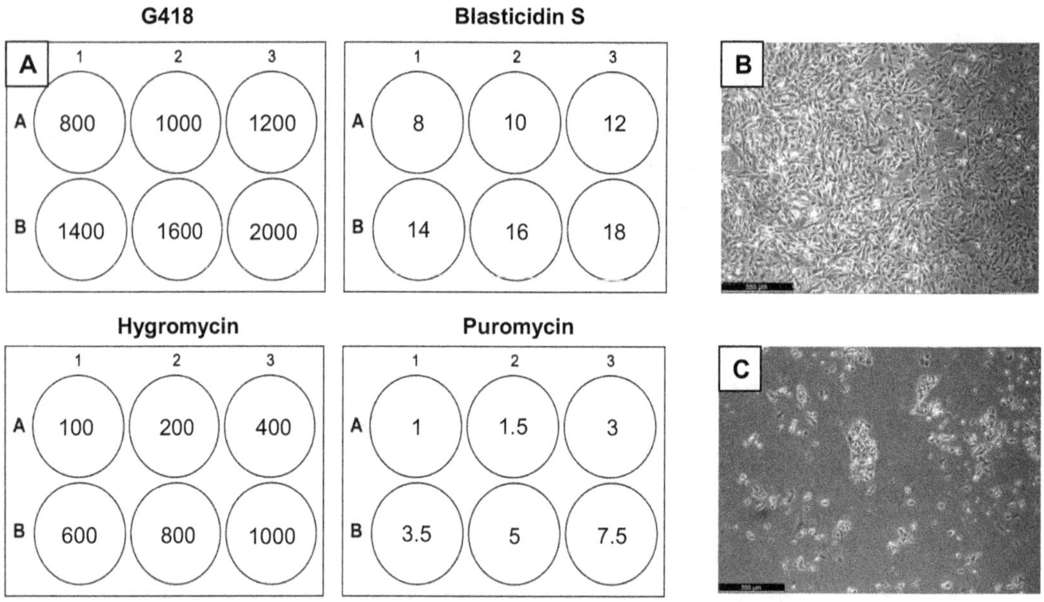

Fig. 3 Killing curve. (**a**) Pipetting schemes of classical killing curve experiments. Each well of a 6-well plate containing medium with different selection antibiotic concentrations. Common concentration in μg/mL are indicated. (**b**) pKF cells at the start of the experiment. (**c**) pKF cells at day 7 of selection with hygromycin showing remaining colonies and hence indicating a too low antibiotic concentration for sufficient selection

albeit with lower efficiencies compared to knockout cells. Selection for stable genomic integration of a selection cassette uses blasticidin S, hygromycin, or G418 for approximately 7 days.

3. Add culture medium containing the appropriate antibiotic concentration and select cells for a suitable time period.
4. Change selection medium every second day and check cell morphology and viability at the end of selection. Lowest concentration where all of the non-transfected cells die is the most appropriate (*see* Fig. 3).

3.1.5 Selection, Isolation, and Screening of CRISPR/Cas9-Mediated knockout Cells

1. To select for cells transiently expressing CRISPR/Cas9 components and puromycin, start puromycin selection 24 h post-transfection (*see* **Note 20**).
2. Add culture medium containing the determined concentration of antibiotics and select for 48 h. Then, remove antibiotics and change medium frequently to remove dead cells.
3. Detach the cells, and count and seed 1500 cells onto a 15 cm dish to grow single-cell colonies.
4. Culture colonies until they reach a size suitable for transfer (*see* **Note 21** and Fig. 4).

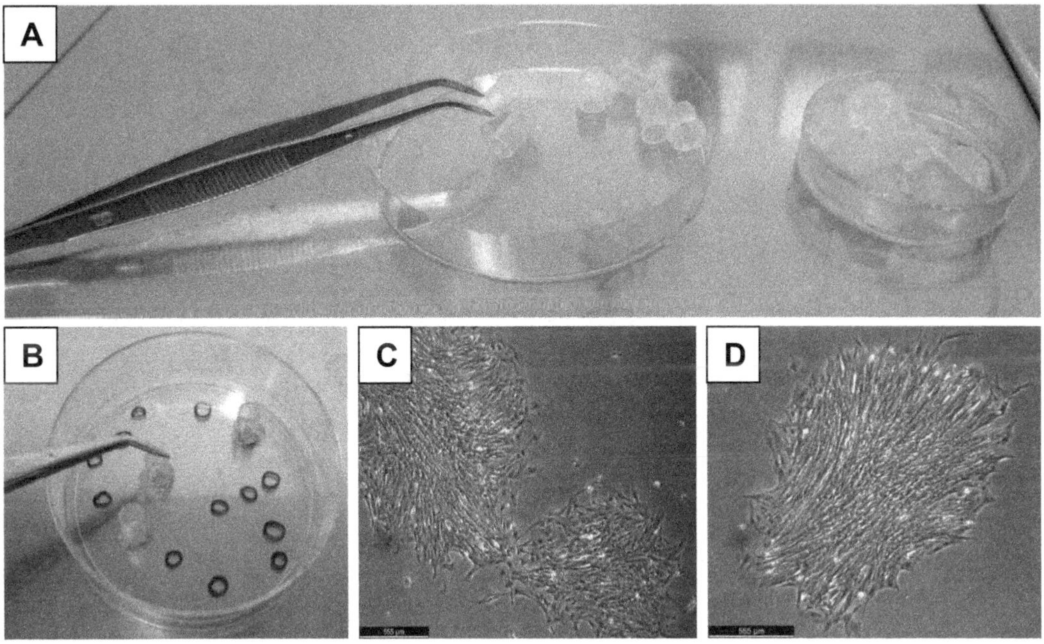

Fig. 4 Single-cell clone isolation. (**a**) Tools for single-cell clone isolation: tweezers, silicone grease, and cloning cylinders. (**b**) Suitable colonies marked on bottom of a 10 cm dish. (**c**) Unsuitable colonies. Two cell colonies located in close proximity inhibit isolation of individual single-cell clones. (**d**) Perfect single clone. A few hundred fibroblast cells, spindle shaped, orientation in one direction and no adjacent colonies

5. At the day of single-cell clone isolation, prepare sterile silicon grease, cloning cylinders, tweezers, and 24-well plates for colony transfer (*see* **Note 22**). Make sure to use only individual colonies. Mark colonies on the bottom of the tissue culture plate with a lab marker.
6. Aspirate medium and wash cells with PBS. Cover the bottom of a cloning cylinder with grease and press it onto the culture dish so it surrounds a cell colony using tweezers. Repeat until all suitable cell colonies are surrounded with cloning cylinders. Subsequently, add a few drops of Accutase into each cylinder until the bottom is covered and incubate for approximately 10 min at 37 °C. Cells should appear roundish. Add culture medium into each cylinder, resuspend, and transfer the cells into 24-well plates using a 1000 μL pipette (*see* **Note 23** and Fig. 4).
7. Culture cells until they are 50–90% confluent, then detach, use half of the cells for DNA isolation, and transfer the residual ones into 6-well plates (*see* **Note 24**).
8. Perform PCR across the sgRNA target site and check for INDELS at the target site by subsequent sequencing of the amplicon.

9. Cell clones carrying INDELS that lead to an early stop codon/ functional gene knockout can be either frozen and stored at − 80 °C or used directly for SCNT.

3.1.6 Selection, Isolation, and Screening of CRISPR/Cas9-Assisted Knock-in Clones

1. Start selection 24 h post-transfection. Change media every second day and wash cells with PBS if there are substantial numbers of dead cells in the supernatant. Maintain puromycin selection for 48 h to select for transient CRISPR/Cas9 expression. For selection of a stably integrated marker cassette, maintain selection until single-cell colonies without any background are clearly visible under the microscope (usually 7–9 days). Exchange selection medium to regular culture medium after selection.
2. Culture colonies until they reach a size suitable for transfer and isolate them as described in Subheading 3.1.5.
3. Screen for correct integration of the DNA template via PCR across the 5′ and 3′ junctions.
4. Once cells are 80–90% confluent again, freeze three vials each and store at −80 °C.

   Use correct clones for SCNT.

3.1.7 SCNT

1. Prepare donor cells for SCNT: Thaw donor cells 5–7 days prior SCNT, seed on 96-well plates at a density reaching 90–100% confluency 48 h prior to SCNT, and change to serum starvation medium 48 h before SCNT (*see* **Note 25**).
2. Collect ovaries and maturate oocytes as described in Subheading 3.2.1.
3. Enucleate oocytes, inject modified donor cells into the perivitelline space of the enucleated oocytes, and fuse both cells. After oocyte activation via an electric pulse, transfer embryos into a synchronized recipient sow. See Kurome M. et al. [11] for a detailed protocol of somatic cell nuclear transfer and also Subheading 3.2.4 for embryo transfer.

3.2 Generation of Genetically Engineered Pigs Via Microinjection

All solutions are pre-warmed to 38.5 °C unless otherwise noted.

3.2.1 Collection of Ovaries and In Vitro Maturation of Oocytes

1. Prepare FLI medium (*see* **Note 26**).
2. Fill 500 μL FLI medium each into a well of a NUNC 4-well plate. Store in triple gas incubator for a minimum of 2 h to allow the medium to equilibrate.
3. Collect ovaries from a nearby slaughterhouse and transport in a pre-warmed thermal container filled with transport solution and antibiotics (*see* **Notes 27** and **28**).
4. Rinse ovaries several times with pre-warmed PBS containing 1% CETAB, followed by rinsing with pre-warmed PBS only.

Fig. 5 Porcine ovary in follicular stage of the estrous cycle. The asterisk indicates an ovarian follicle with a size of approximately 3–5 mm suitable for COC aspiration

5. Transfer ovaries to a glass beaker filled with pre-warmed PBS, placed on a heating plate at 38.5 °C.
6. Carefully aspirate follicles of at least 3.5 mm size using an 18G needle attached to a 10 mL disposable syringe (*see* Fig. 5).
7. Collect aspirated follicular fluid in a pre-warmed 15 mL tube and store it at 38.5 °C until all follicular fluid is aspirated. Cumulus-oocyte-complexes (COCs) will sediment at the bottom of the tube.
8. Discard supernatant and mix sedimented COCs with 5–10 mL pre-warmed working medium. Pipet suspension onto a 10 cm-bacteriological petri dish that has been subdivided into equal segments by indentations to facilitate screening for high-quality COCs.
9. Select COCs with dark, evenly granulated cytoplasm and several compact layers of cumulus cells using a mouth pipette and a stereo microscope (*see* **Notes 29** and **30** and Fig. 6). Discard low-quality COCs.
10. Wash selected COCs twice in working medium and once in equilibrated FLI medium. To minimize medium transfer, 50 COCs are first pipetted dropwise onto a pre-warmed lid of a petri dish, centered, and then transferred with as little media as possible into a well of a 4-well plate for rinsing. Next, groups of 50 COCs are placed in each well of the NUNC 4-well plate, filled with 500 μL of equilibrated FLI maturation medium.
11. Incubate in a triple gas incubator for 45 h at 38.5 °C (*see* **Note 31**).

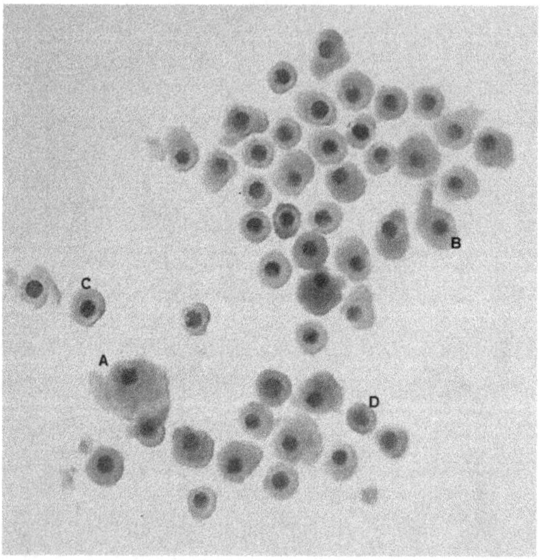

Fig. 6 Representative picture of oocytes isolated from liquor folliculi for in vitro maturation. COCs are classified using a four-grade scale and a representative of each grade is marked (A-D). Grade A includes COCs with heterogeneous cytoplasm and complete cumulus oophorus, grade B comprises COCs with incomplete but compact cumulus oophorus with more than five layers and homogenous cytoplasm, grade C contains COCs with heterogeneous cytoplasm and more than three cumulus cell layers or more than five layers in some parts, and grade D includes COCs with cumulus oophorus partly or totally absent and cytoplasm strongly heterogeneous

3.2.2 In Vitro Fertilization

1. The day prior to in vitro fertilization (IVF), prepare two NUNC 4-well plates per 100 COCs to be fertilized. Each well of the first plate is filled with 500 μL PFM for in vitro fertilization, each well of the second plate is filled with 500 μL PZM5 covered by a thin layer of mineral oil for storing zygotes after microinjection. Fill 1 mL PFM into a polystyrene round-bottom tube with a lid enabling gas exchange and equilibrate all media in a triple gas incubator.
2. On the day of IVF, assess successful maturation of COCs (*see* **Notes 32** and **33** and Fig. 7).
3. Rinse COCs twice in working medium.
4. Rinse 50 COCs in one well of the PFM plate and transfer them to the adjacent clean well. Repeat the process with the next 50 COCs in the remaining wells of the plate. Store oocytes in a triple gas incubator at 38.5 °C until **step 12**.
5. Thaw a deep-frozen sperm straw in a water bath at 38.5 °C for 30 s.
6. Dry sperm straw, open it, and collect the content in a 15 mL falcon.

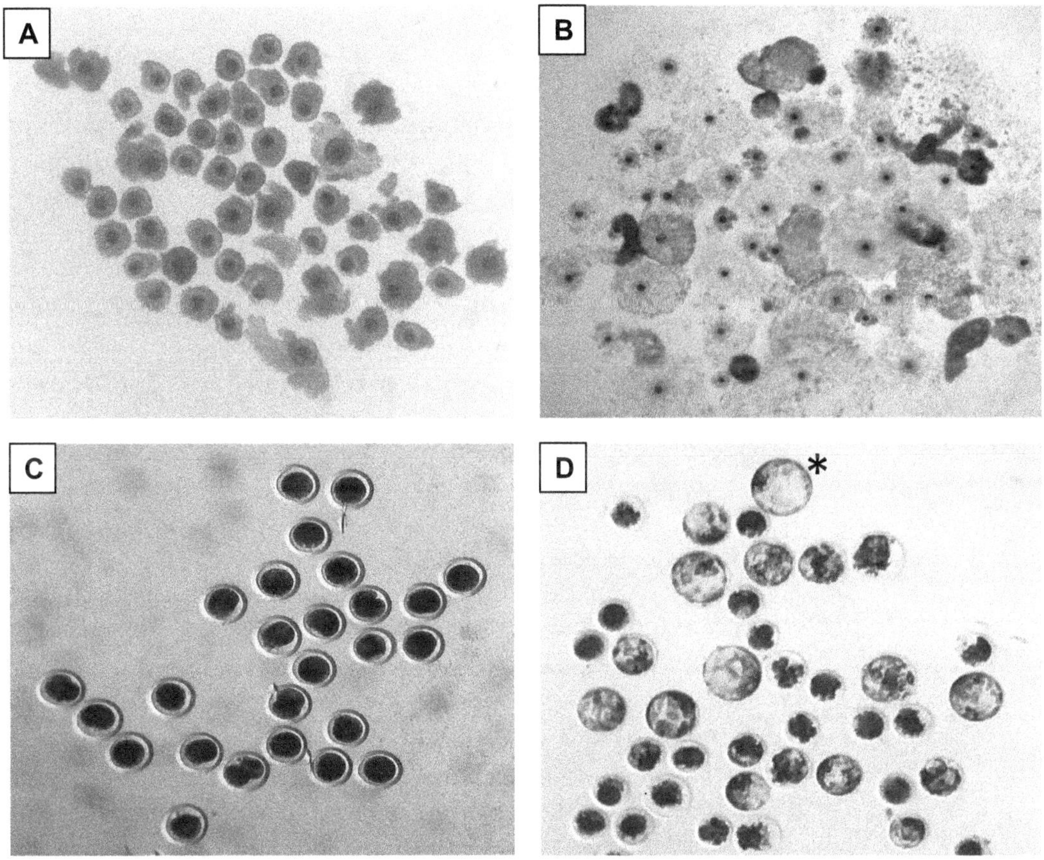

Fig. 7 In vitro development from oocytes to blastocysts. (**a**) Immature COCs freshly isolated from liquor folliculi. (**b**) Maturated oocytes. The state of maturity is characterized by an increased cumulus cell number surrounding the oocyte. (**c**) Zygotes after in vitro fertilization and removal of cumulus cells. (**d**) Embryos after in vitro cultivation in PZM5 medium. The asterisk marks a blastocyst

7. Mix sperm with 2 mL of pre-warmed sperm diluent and centrifuge at 1000 rcf for 5 min. Discard supernatant.
8. Repeat **step** 7.
9. Resuspend the sperm pellet in 500 μL of PFM from the polystyrene round-bottom tube. Remove appropriate volumes for **steps 10** and **11**. Store the residual sperm suspension in a triple gas incubator at 38.5 °C. Remove the tube lid to enable gas exchange.
10. Pipet a drop of sperm suspension directly onto a pre-warmed microscope slide and determine sperm quality regarding motility and morphology (*see* **Note 34**).

11. Determine sperm concentration using a Neubauer improved counting chamber and a sperm suspension diluted 1:50 in H_2O.
12. Transfer an optimized sperm count into a well containing 50 COCs in PFM and co-incubate for 7 h in the triple gas incubator at 38.5 °C (*see* **Note 35**).
13. Remove cumulus cells and sperm from IVF zygotes by gently pipetting up and down in working medium, supplemented with 1 mg/mL hyaluronidase.
14. Rinse all zygotes twice in working medium.
15. Select zygotes with dark, evenly distributed cytoplasm showing extrusion of two polar bodies (*see* **Note 36**).
16. Store zygotes in a drop of working medium covered with mineral oil on a heating plate until microinjection.

3.2.3 Microinjection

1. Prepare injection solution and store on ice (*see* **Note 37**). For RNP-complexes, mix components and allow complex formation for 15 min at room temperature, then store on ice.
2. Place an empty injection needle into the injection solution and wait until filled by capillary forces.
3. Assemble micromanipulator holding pipette and the filled injection needle with their respective micromanipulation units. Open the tip of the injection needle by gently tapping it against the holding pipette (*see* **Note 38**).
4. Transfer groups of 20–30 zygotes into a 4 μL drop of working medium, covered with mineral oil to avoid medium evaporation.
5. Bring individual zygotes, holding pipette, and injection needle into the same horizontal plane. Fix zygotes by carefully applying negative pressure with the holding pipette and position the polar body extrusion at twelve or six o' clock orientation. Gently insert the injection needle into the cytoplasm of each zygote and deliver approximately 10 pl of injection solution (*see* **Notes 39** and **40**).
6. Rinse injected zygotes in PZM5 and place groups of 50 zygotes in each well of the NUNC 4-well plate previously filled with PZM5. Cultivate in triple gas incubator for 12–36 h until embryo transfer or for 7 days (to assess blastocyst development and targeting efficiency), respectively.

3.2.4 Preparations for Embryo Transfer

1. The day prior to embryo transfer: Plug in a transportable incubator, add 15 mL HEPES-free TCM-199 medium into a tube without lid, and fill 2 mL HEPES-free TCM-199 medium

into a Polystyrene Round-Bottom Tube with lid enabling gas exchange. Equilibrate in the triple gas incubator at 38.5 °C.

2. On the day of embryo transfer: Rinse embryos in working medium and transfer embryos into the Polystyrene Round-Bottom Tube (*see* **Note 41**).
3. Transport embryos and the additional HEPES-free TCM-199 medium in the transportable incubator to the surgery facility.
4. Prepare three 35 mm corning dishes on a heating plate set to 38.5 °C and fill with HEPES-free TCM-199 medium accordingly:
   (a) Dish A contains 2 mL medium.
   (b) Dish B contains 1.5 mL medium.
   (c) Dish C contains no medium.
5. Transfer embryos to dish C using an unmodified glass pipette connected to a suction bulb and rinse the transport tube with medium from dish B to avoid losing embryos.
6. Wash embryos by collecting the individual embryos and transferring them into dish B using a mouth pipette.
7. Transfer embryos into dish A and release them close to each other.
8. Suck 1 mL HEPES-free TCM-199 medium into a sterile catheter, followed by 0.2 mL air. Next, collect all embryos within the smallest volume of medium possible, followed by 0.2 mL air. Finally, suck 0.2 mL of HEPES-free TCM-199 medium. This segmentation facilitates the assessment of whether all embryos have been transferred during the subsequent embryo transfer.
9. Place filled catheter onto a sterile, pre-warmed 15 cm glass plate. Cover and transport to surgery room. Place onto heating plate at 38.5 °C until embryo transfer.

3.2.5 Embryo Transfer

1. Fast synchronized recipient 12 h before surgery.
2. Anesthetize sow by intravenous (i.v.) application of ketamine hydrochloride (1.2 mg/10 kg) and xylazine (0.5 mL/10 kg).
3. Apply 0.4 mg/kg BW meloxicam (i.m.) and 15 mg/kg BW amoxillin (i.m.) perioperative.
4. Clean abdominal area with warm water and soap. Shave this area and disinfect with iodine solution.
5. Elevate recipient on a surgery table and fix in a 30 °C head down position.
6. Place self-adhesive surgery drape and carry out skin incision at height of the second to last pair of teats at the Linea Alba.

7. Inspect correct ovulation state by visual assessment of antral follicles or fresh ovulation sites.
8. Insert embryo-containing sterile catheter into the oviduct. Release 70–200 embryos into the oviduct while gently withdrawing the catheter.
9. Close abdomen in three layers using single stitches for muscle and subcutaneous tissue and running stitches for skin.

Alternatively, endoscopic embryo transfer can be carried out. A detailed protocol of endoscopic embryo transfer has been described elsewhere [11].

4 Notes

1. If it is desired to knockout several genes at once, a multiplexed gRNA vector can be used. Since not all gRNAs show equal efficiency, thorough analysis of the manipulated somatic cells prior to subsequent SCNT ensures that the offspring carries all desired knockouts.
2. Instead of DNA vectors, also in vitro assembled RNP complexes can be employed.
3. A homemade glass capillary is made by heating the middle part of the taper of a soda lime glass Pasteur pipette over a Bunsen burner flame. Once the Pasteur pipette becomes malleable, it is removed from the flame and quickly pulled 5–10 cm depending on the application. The final position is held for approximately 3 s to allow the glass to harden. A breaking point is notched with a grinding stone and the glass capillary is broken there. A capillary with a wider opening is used to pick up cumulus-oocyte complexes, while a capillary with a narrower opening is used to pick up cumulus-free oocytes. In general, a longer capillary with a consistent diameter after tapering allows for more controlled mouth pipetting. The glass capillary must be cooled down before use.
4. It is recommended to use the incubator exclusively for embryological work to prevent contamination. Frequent opening of the incubator door should be avoided to maintain a constant environment.
5. The required volumes of PZM5 and PFM should be equilibrated in the triple gas incubator at 38.5 °C 24 h before usage to allow media equilibration.
6. It must be ensured that PZM5 and PFM were maintained cool during the entire delivery. Experience shows that the cold chain fails especially at the customs office. Prior contact with the authorities is therefore strongly recommended.

7. Store embryo-grade mineral oil in the dark. At best, aliquot and use freshly.
8. Sperm quality varies between ejaculates from different boars and between ejaculates from the same boar. Therefore, a careful selection of the highest quality sperm must be made in advance. Good sperm quality is characterized by high movement of single sperms, together with optimal sperm morphology [13].
9. Program 21 on a P 1000 Flaming Brown micropipette puller (Sutter Instruments) was used to ensure a holding pipette diameter of 150 μm.
10. The highest efficiency of embryo transfer is achieved in young adult sows, but prepubertal gilts aged 6 to 7 months weighing 110–130 kg are a possible alternative.
11. Synchronization of recipients is achieved by supplementing feed with Altrenogest (Regumate) for 15 days. 24 h after last Regumate administration, 750 IU eCG is injected intramuscularly, and 750 IU of hCG is injected intramuscularly 80 h after ECG injection. One to two days after hCG administration, embryo transfer is performed.
12. Also, ear or embryonic tissues can be used for isolation of ear fibroblast or fetal fibroblasts, respectively. Use pKF growth medium for cultivation.
13. A fresh kidney can be stored in PBS at 4 °C for up to 2 days. Cell isolation should be carried out at the beginning of the week to guarantee daily medium exchange during the first 3 days of isolation.
14. Placing the Erlenmeyer flask off-center on the stirring plate allows additional shear forces to mince the tissue more thoroughly.
15. Accutase is a gentler cell detachment solution than trypsin/EDTA and is well suited for detachment of pKF and pADMSC cells.
16. Always freeze cells at a slow rate of −1 °C per minute using a freezing container. Store cells at −80 °C for short term or in the vapor phase above liquid nitrogen below −135 °C for long term.
17. pADMSCs grow slowly in the beginning but more rapidly after some days of cultivation.
18. Confluent cells leave cell cycle, thus editing efficiency is reduced.
19. We recommend Lipofectamine™ 2000 transfection reagent (Invitrogen) or similar for lipofection. Use 1–2 μL of transfection reagent per 6-well plates and 6 μL per 10 cm dish. Adding the lipofection-DNA mixture dropwise directly onto the cell

surface may increase transfection efficiency. If you use synthetic guides or RNP complexes, follow the manufacturer's instructions for the appropriate transfection method.

20. Alternatively, if a fluorescence activated cell sorting (FACS) machine is available, a CRISPR/Cas9 plasmid containing a fluorescence gene as selection marker (such as pSpCas9(BB)-2A-GFP (PX458); Addgene Plasmid # 48138)) can be used for transfection and fluorescent cells can then be selected via FACS. Moreover, if cell surface proteins are being inactivated, magnetic cell separation (MACS) or FACS can be applied to select for cells with a functional homozygous gene knockout.
21. At the last day of selection, single-cell clones of around 50–100 cells should be clearly visible and outside the colonies the plate should be empty.
22. Cloning cylinders can be obtained inexpensively by cutting 0.5 mL centrifuge tubes into plastic rings followed by autoclaving.
23. The number of colonies depends on the transfection and selection efficiency. If too many colonies grow, transfect cells at lower density or reduce DNA amount to enable picking of individual single-cell clones. Some clones do not survive transfer. Use only normally shaped, well growing, and detaching cells.
24. If cells grow in dense colonies after single clone isolation, detach and resuspend a few times to distribute evenly in the well. Cells that are too long in the center of a colony stop proliferating.
25. Confluency and serum starvation ensure that cells enter G_0 cell cycle phase which is required for SCNT.
26. 2 mL medium will be needed for in vitro maturation of 100 COCs. Between 300 and 500 oocytes are typically isolated from 100 ovaries.
27. Although in vivo–derived oocytes or zygotes are an option, most GE animals are generated using IVM oocytes (SCNT) or IVM/IVF embryos (MI) from slaughterhouse material due to animal welfare reasons.
28. The most important factors determining oocyte quality are speed and temperature. An optimal workflow should take a maximum of 3 h from collection of the ovaries to the start of in vitro maturation. Ambient temperature of the oocytes should stay close to 38.5 °C. Unlike other animal species, porcine embryology is very temperature sensitive. Maintaining a temperature of 38.5 °C throughout the process is key to successful embryo production in pigs. Therefore, it is essential to measure the temperature of the transport solution

immediately after returning to the laboratory and to optimize the workflow with regard to time if necessary.

29. It is important to transfer as little debris as possible to achieve high ratios of matured oocytes and minimize contamination risk.
30. High-quality COCs are characterized by four- to five-layers of surrounding cumulus cells, as well as a dark, evenly granulated cytoplasm. COCs can be classified using a four-grade scale: A being the most suitable and D being the least suitable oocyte for IVM (*see* Fig. 6).
31. It is essential that oocytes, zygotes, and embryos are removed from the incubator only as short as necessary.
32. Successful maturation is confirmed by visual assessment of cumulus expansion, evenly granulated cytoplasm, and polar body extrusion from a sample group of oocytes. Cumulus cell expansion is often accompanied by an increase in viscosity and is easily misinterpreted as a contamination. However, this is a maturing feature of COCs (*see* Fig. 7b).
33. Careful resuspension four to five times with a pipette facilitates the subsequent transfer of the COCs with a mouth pipette.
34. Count spermatozoa covering a distance of at least 5 μm/s in a straight line or in wide circle. For subsequent calculation of the live sperm count, the ratio of moving sperm is determined.
35. A high sperm to oocyte ratio increases not only the fertilization rate in IVM but also the degree of polyspermy. Therefore, the optimal live sperm concentration must be determined for each ejaculate individually. Empirically, the optimal sperm-to-oocyte ratio is about 7500 live sperm per oocyte.
36. Fertilized oocytes are characterized by the extrusion of two polar bodies. Unfortunately, the second extrusion can often be poorly visible. Therefore, oocytes with at least one visible polar body are selected for microinjection.
37. Minimal volume of injection solution should be 20 μL, at best 100 μL, to allow optimal filling of injection needles.
38. Successful opening of the needle can be tested by moving the injection needle from the working medium into the surrounding oil and ejecting the injection solution there. If bubbles form, the needle is open. It is recommended to test the optimal opening size on a single oocyte. Too narrow will lead to frequent blockage, while too large an opening will reduce zygote development due to excessive membrane damage.
39. Successful injection into the cytoplasm can be visually confirmed by observing movement of the intracellular lipid droplets caused by influx of injection solution. During injection,

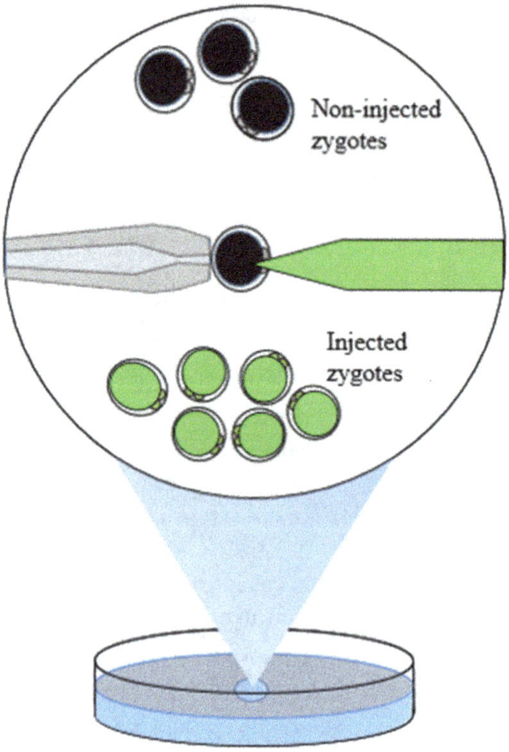

Fig. 8 Micromanipulation drop. Non-injected zygotes are positioned at the top of the drop, whereas injected ones are moved to the lower part of the drop to prevent mingling

the expansion of the ooplasm should be kept below 10% of the total volume to minimize cellular damage.

40. After injection, position injected zygotes in the lower part of the droplet to separate them from un-injected zygotes (*see* Fig. 8).
41. Transfer of oil during this wash step should be avoided to minimize the chance of transferring oil during embryo transfer.

References

1. Flisikowska T, Kind A, Schnieke A (2014) Genetically modified pigs to model human diseases. J Appl Genet 55:53–64. https://doi.org/10.1007/s13353-013-0182-9
2. Lu T, Yang B, Wang R et al (2019) Xenotransplantation: current status in preclinical research. Front Immunol 10:3060. https://doi.org/10.3389/fimmu.2019.03060
3. Ryu J, Prather RS, Lee K (2018) Use of gene-editing technology to introduce targeted modifications in pigs. J Anim Sci Biotechnol 9:5. https://doi.org/10.1186/s40104-017-0228-7
4. Niu D, Wei H-J, Lin L et al (2017) Inactivation of porcine endogenous retrovirus in pigs using CRISPR-Cas9. Science 357:1303–1307. https://doi.org/10.1126/science.aan4187
5. Fischer K, Rieblinger B, Hein R et al (2020) Viable pigs after simultaneous inactivation of porcine MHC class I and three xenoreactive antigen genes GGTA1, CMAH and B4GALNT2. Xenotransplantation 27:e12560. https://doi.org/10.1111/xen.12560
6. Rieblinger B, Sid H, Duda D et al (2021) Cas9-expressing chickens and pigs as resources for genome editing in livestock. Proc Natl Acad

Sci U S A 118:e2022562118. https://doi.org/10.1073/pnas.2022562118

7. Hammer RE, Pursel VG, Rexroad CE et al (1985) Production of transgenic rabbits, sheep and pigs by microinjection. Nature 315: 680–683. https://doi.org/10.1038/315680a0
8. Onishi A, Iwamoto M, Akita T et al (2000) Pig cloning by microinjection of fetal fibroblast nuclei. Science 289:1188–1190. https://doi.org/10.1126/science.289.5482.1188
9. Yuan Y, Spate LD, Redel BK et al (2017) Quadrupling efficiency in production of genetically modified pigs through improved oocyte maturation. Proc Natl Acad Sci U S A 114: E5796–E5804. https://doi.org/10.1073/pnas.1703998114
10. Gao X, Nowak-Imialek M, Chen X et al (2019) Establishment of porcine and human expanded potential stem cells. Nat Cell Biol 21:687–699. https://doi.org/10.1038/s41556-019-0333-2
11. Kurome M, Kessler B, Wuensch A et al (2015) Nuclear transfer and transgenesis in the pig. Methods Mol Biol 1222:37–59. https://doi.org/10.1007/978-1-4939-1594-1_4
12. Niemann H, Tian XC, King WA et al (2008) Epigenetic reprogramming in embryonic and foetal development upon somatic cell nuclear transfer cloning. Reproduction 135:151–163. https://doi.org/10.1530/REP-07-0397
13. Sexual, Reproductive Health and Research (2021) WHO laboratory manual for the examination and processing of human semen. World Health Organization

Chapter 20

Generation of Genome-Edited Chicken Through Targeting of Primordial Germ Cells

Alewo Idoko-Akoh and Michael J. McGrew

Abstract

Genome editing technology facilitates the creation of specific and precise genetic changes to unravel gene function and rapidly transfer unique alleles between chicken breeds in contrast to lengthy traditional crossbreeding methods for the study of poultry genetics. Innovations in genome sequencing technology have made it possible to map polymorphisms associated with both monogenic and multigenic traits in livestock species. We, and many others, have demonstrated the use of genome editing to introduce specific monogenic traits in chicken through targeting of cultured primordial germ cells. In this chapter, we describe materials and protocols for performing heritable genome editing in the chicken through targeting of in vitro propagated chicken primordial germ cells.

Key words Genome editing, Poultry, Chicken, Primordial germ cells

1 Introduction

In 2015, the Avian Phylogenetics Consortium initiated a project to sequence the genomes of 10,000 bird species [1]. Since then, ongoing technological developments in genome sequencing has led to the generation of over 500 avian reference genomes as at the time of writing this manuscript [2]. This development in avian genomics presents the opportunity to exploit the application of genome editing technology to investigate gene regulatory function, perform rapid allele transfer between breeds, and introduce novel alleles into bird species. The methods for performing precise genome editing in zygotes are well described for many mammalian species but are not readily applicable in birds due to evolutionary differences in reproductive biology. The development of methods for in vitro propagation of chicken primordial germ cells (hereafter referred to as PGC or PGCs) was a significant step in actualizing defined and heritable genetic modification in avian species [3]. PGCs present a cell lineage that can be targeted to introduce

Thomas L. Saunders (ed.), *Transgenesis: Methods and Protocols*,
Methods in Molecular Biology, vol. 2631, https://doi.org/10.1007/978-1-0716-2990-1_20,

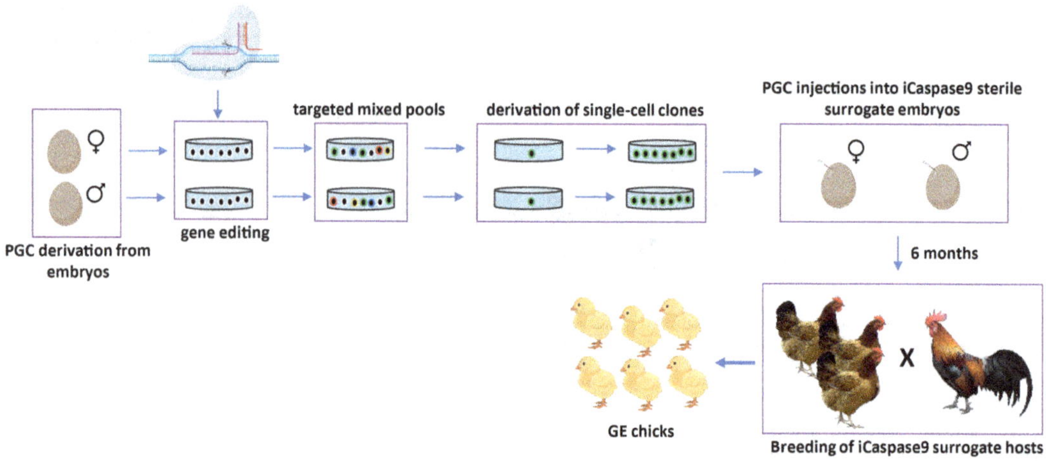

Fig. 1 Workflow for the generation of genome-edited (GE) chickens through targeting of cultured PGCs

genetic changes that will be heritable in offspring derived from the modified germ cells. We and others have demonstrated the generation of genome-edited chickens from targeted cultured chicken PGCs [4–6]. CRISPR/Cas9-aided gene editing has been used to perform seamless a footprint-less transfer of monogenic traits between chicken breeds [7] and to study genes that control avian germ cell and gonadal development [8, 9]. We previously described a serum-free and feeder-free optimized medium suitable for chicken PGC derivation, long-term PGC culture, and single-cell culture for clonal derivations [10]. The use of this optimized medium eliminates the risk of pathogen contamination associated with animal-derived products. We have also generated sterile i-Caspase9 sterile surrogate hosts through genome editing of PGCs to facilitate the generation of offspring homozygous for the desired genotype genetic change in a single breeding generation [7].

In this chapter, we will describe methods for the derivation and genome editing of chicken PGCs in serum-free, feeder-free medium to generate clonal targeted PGCs and subsequent PGC injection into surrogate host embryos. The methods described in this chapter may be applied in other avian species if culture methods for the PGCs from these species are developed. However, the application of genome editing using cultured PGCs for the majority of other avian species is still not possible. The overall pathway to proceed from embryo to cultured PGC, to genome-edited PGC, to surrogate host chicken carrying the engineered chicken germ cells is shown in Fig. 1.

2 Materials

2.1 Avian Knockout DMEM (KO-DMEM): A No-Calcium, Low Osmolarity DMEM for Culturing Embryonic Cells

1. Avian KO-DMEM (Life Technologies 041-96570M) or.
2. Alternatively, prepare ~50 ml of avian KO-DMEM replacement as follows:
   (a) 37.5 ml of calcium-free DMEM (Life Technologies: 21068028).
   (b) 11.6 ml of distilled water (Life Technologies: 15230089).
   (c) 0.5 ml of 50× MEM amino acids solution (Life Technologies: 11130051).
   (d) 0.5 ml of 100 mM sodium pyruvate (Life Technologies: 11360070).
   (e) 0.5 ml of 100× MEM vitamin solution (Life technologies: 11120052).

2.2 PGC Basal Medium

Prepare 50 ml of PGC basal medium as follows:

1. 47 ml of avian KO-DMEM.
2. 1 ml of 50× B27 supplement (Life Technologies: 17504044).
3. 0.5 ml of 100× MEM non-essential amino acids (Life Technologies: 11140050).
4. 0.5 ml of 100× GlutaMAX™-I (Life Technologies: 35050061).
5. 0.5 ml of 100× EmbryoMax® nucleosides (Merck Millipore: ES-008-D) (*see* **Note 1**).
6. 0.2 ml of 100 mM sodium pyruvate (Life Technologies: 11360070).
7. 0.1 ml of 50 mM 2-mercaptoethanol (Life Technologies: 31350010).
8. 0.075 ml of 100 mM CaCl2 dissolved in tissue culture grade distilled water. Use solution filtered through a 0.22 μm syringe filter (Merck Millipore; SLGPO33RS).
9. 0.5 ml of 20% Ovalbumin (Sigma-Aldrich: A5503). Prepare 5 ml of the solution as follows:
   (a) Add 1000 mg of ovalbumin to 5 ml of KO-DMEM.
   (b) Put on a roller for up to 20 min or until the solution is clear for complete dissolution.
   (c) Filter the clear solution through a 0.22 μm syringe filter.
   (d) Aliquot and store at 4 °C.

10. 0.1 ml of 50 mg/ml heparin sodium (Sigma-Aldrich: H3149). Prepare 5 ml of the solution as follows:
    (a) Add 250 mg of heparin sodium to 5 ml of KO-DMEM.
    (b) Put on a roller for up to 20 min or until the solution is clear for complete dissolution.
    (c) Filter the clear solution through a 0.22 μm syringe filter.
    (d) Aliquot and store at 4 °C.
11. 0.1 ml of penicillin/streptomycin (10,000 U/ml) (Life Technologies: 15140122) (optional).

2.3 5000× Vitamin B12 (Sigma-Aldrich: V6629) Prepared as Follows (See Note 2)

1. Prepare 500,000× stock (3.4 mg/ml): dissolve 100 mg of vitamin B12 in 29.41 ml tissue culture grade distilled water.
2. Filter the stock solution through a 0.22 μm syringe filter and store at −20 °C.
3. Prepare 5000× stock (0.034 mg/ml): add 100 μl of the 500,000× stock solution to 10 ml of tissue culture grade distilled water.
4. Filter the solution through a 0.22 μm syringe filter and store at −20 °C.

2.4 PGC Growth Factors

1. Recombinant human FGF2 (R & D systems: 234-FSE-025) is shipped on dry ice as a lyophilized powder containing Tris–NaCl and bovine serum albumin (BSA) as a carrier. Prepare in a sterile tissue culture hood as follows:
    (a) Spin 25 μg FGF2 vial for 15 s in a centrifuge.
    (b) Filter tissue culture grade phosphate buffered saline (PBS) through a 0.22 μm syringe filter into a sterile 15 ml conical tube.
    (c) Add 5 μl of sterile 20% ovalbumin to 1 ml of sterile PBS to prepare a 0.1% ovalbumin/PBS solution.
    (d) Add 1 ml of 0.1% ovalbumin/PBS solution to the 25 μg FGF2 vial to achieve a stock concentration of 25 μg/ml (*see* **Note 3**).
    (e) Allow to stand for 10 min, and then pipette up and down six times.
    (f) Make aliquots and store at −80 °C. Upon thawing, do not refreeze and use within 4 weeks.
2. Recombinant human activin A (Peprotech: 120–14). Prepare in a sterile tissue culture hoods as follows:
    (a) Spin 5 μg activin A vial for 15 s in a centrifuge.
    (b) Filter tissue culture grade PBS through a 0.22 μm syringe filter into a sterile 15 ml conical tube.
    (c) Add 50 μl of sterile 20% ovalbumin to 10 ml of sterile PBS to prepare a 0.1% ovalbumin/PBS solution.

(d) Add 200 μl of 0.1% ovalbumin/PBS solution to the 25 μg FGF2 vial to achieve a stock concentration of 25 μg/ml.

(e) Allow to stand for 10 min, and then pipette up and down six times.

(f) Make aliquots and store at −80 °C. Upon thawing, do not refreeze and use within 4 weeks.

3. Ovotransferrin, also referred to and sold as conalbumin (Sigma-Aldrich: C7786). Prepare 5 ml of 10 mg/ml ovotransferrin solution as follows:

   (a) Add 50 mg of ovotransferrin to 5 ml of KO-DMEM.

   (b) Put on a roller for up to 20 min or until the solution is clear for complete dissolution.

   (c) Filter the clear solution through a 0.22 μm syringe filter.

   (d) Aliquot and store at 4 °C.

4. Chicken serum (Biosera: CH-515 or Sigma-Aldrich: C5405).

2.5 Complete PGC Culture Medium

Prepare any of the following complete PGC culture medium by adding the following growth factors [10].

1. Prepare 10 ml of FAOT complete PGC medium by adding the following growth factors (*see* **Note 4**):

   (a) 10 ml of PGC basal medium.

   (b) 1.6 μl of human FGF2 (25 μg/ml).

   (c) 10 μl of human activin A (25 μg/ml).

   (d) 10 μl of ovotransferrin (10 mg/ml).

   (e) 2.0 μl of 5000× vitamin B12.

2. Prepare 10 ml of FACSOT complete PGC medium by adding the following growth factors (*see* **Note 5**):

   (a) 10 ml of PGC basal medium.

   (b) 1.6 μl of human FGF2 (25 μg/ml).

   (c) 10 μl of human activin A (25 μg/ml).

   (d) 20 μl of chicken serum.

   (e) 5 μl of ovotransferrin (10 mg/ml).

   (f) 2.0 μl of 5000× vitamin B12.

3. Prepare 10 ml of FABOT complete PGC medium by adding the following growth factors (*see* **Note 6**):

   (a) 10 ml of PGC basal medium.

   (b) 1.6 μl of human FGF2 (25 μg/ml).

   (c) 10 μl of human activin A (25 μg/ml).

   (d) 10 μl of human BMP4 (25 μg/ml).

   (e) 10 μl of ovotransferrin (10 mg/ml).

   (f) 2.0 μl of 5000× vitamin B12.

2.6 PGC Derivation, Culture, and Cryopreservation

1. For blood-derived PGCs, incubated chicken eggs containing embryos at Hamburger and Hamilton (HH) stages 15–16+ [11].
2. Scissors and tweezers sterilized in 70% ethanol.
3. Borosilicate microcapillary tubes (Harvard apparatus).
4. Aspirator apparatus (Sigma-Aldrich: A5177) consisting of plastic mouthpiece, latex tubing (15 inches) and silicone rubber nosepiece.
5. REDExtract-N-Amp™ tissue polymerase chain reaction (PCR) Kit (Sigma-Aldrich; XNAT).
6. 48-well tissue culture plates (Thermo Scientific: 150687) and 24-well tissue culture plates (Thermo Scientific: 142475).
7. FAOT complete PGC medium (*see* **Note 7**).
8. STEM-CELLBANKER® (Amsbio: 11897). Alternatively, use locally prepared cryopreservation medium prepared as follows:
   (a) 500 μl of DMSO (Sigma-Aldrich).
   (b) 1 ml of chicken serum.
   (c) 7.5 μl of 100 mM $CaCl_2$.
   (d) 8.5 ml of KO-DMEM.
   (e) Filter through a 0.22 μm syringe filter and store at −20 °C.
9. 1.8 ml polypropylene cryogenic tubes (Thermo Scientific: 377267).
10. Mr. Frosty™ freezing container (Thermo Scientific: 51000-0001).

2.7 PGC Transfection, Selection, and Clonal Expansion

1. Lipofectamine 2000 (Invitrogen: 11668019).
2. Opti-MEM I reduced serum medium (Invitrogen: 31985-062).
3. 0.1 mg/ml puromycin dihydrochloride (Sigma-Aldrich: P7255) dissolved in double-distilled water.
4. PX459 and PX458 CRISPR/Cas9 vectors [12, 13].
5. 24-well tissue culture plates (Thermo Scientific: 142475), 48-well tissue culture plates (Thermo Scientific: 150687), and 96-well tissue culture plates (Thermo Scientific: 167008).

2.8 PGC Injection into Surrogate Embryos

1. Incubated chicken eggs containing embryos at HH stages 15–16+.
2. Scissors and tweezers sterilized in 70% ethanol.
3. Glass microcapillary tubes (Harvard apparatus, borosilicate, 1.5 mm).
4. Aspirator apparatus (Sigma-Aldrich: A5177) consisting of plastic mouthpiece, latex tubing (15 inches), and silicone rubber nosepiece.

5. 25 mM B/B compound in DMSO (Takara Bio).
6. 0.5 mM B/B compound in ethanol (Takara Bio).
7. Penicillin–streptomycin (10,000 U/ml) (Life Technologies: 15140122).
8. Penicillin/streptomycin-B/B-compound mixture: Prepare by adding 30 μl of 0.5 mM B/B compound to 270 μl of Penicillin–streptomycin (10,000 U/ml).
9. Fast Green dye (Sigma-Aldrich F7252).
10. Leukosilk tape (BSN Medical).

2.9 Equipment and General Reagents

1. Humidified incubator with 5% CO_2.
2. Category 2 biological safety cabinet.
3. Horizontal laminar flow hood.
4. Stereomicroscope set up in a horizontal laminar flow hood.
5. Egg incubator set at 37.8 °C and 50% humidity.
6. 70% ethanol.
7. Finely textured tissue paper (KIMTECH Science: 7558).
8. Sharp-tip stainless steel forceps/tweezers (~140 mm length with straight, fine pointed ends).
9. Blunt-end stainless steel forceps/tweezers (~140 mm length with straight, blunt round ends with serrated jaws).
10. Stainless steel scissors (~140 mm length with straight, fine pointed ends).
11. 1.5 ml snap cap microcentrifuge tubes (Pyrogen-free, nuclease-free).
12. 1.5 ml screw cap microcentrifuge tubes (Pyrogen-free, nuclease-free).
13. Chicken egg holder.
14. Double-distilled water.
15. Autoclaved PBS.
16. Moving-coil microelectrode puller (Campden Instruments Ltd – model 753).
17. UV sterilizer cabinet.
18. Benchtop centrifuge (Sigma Laborzentrifugen GmbH: Model 1–14).
19. Benchtop microcentrifuge (Eppendorf 5452 MiniSpin centrifuge).
20. BD FACSAria III cell sorter (BD Biosciences).

3 Methods

3.1 Collection of Blood from Embryos (See Note 8)

1. In a sterile biological safety cabinet, add 500 μl of double-distilled water into each peripheral well of a 48-well tissue culture plate to reduce evaporation from the culture medium. Add 300 μl of FAOT medium into a well in the plate and repeat this for the number of desired wells.
2. Pull microcapillary tubes using a moving-coil microelectrode puller to create needles. Sterilize the pulled microcapillary tubes under UV in a UV sterilizer cabinet (*see* Fig. 2).
3. Fertile chicken eggs are incubated for 2.5 days to obtain stage 16 HH embryos. Embryos can be between stages 15 and 16+ HH but not older than stage 17 HH.
4. Sterilize the stereomicroscope and other tools in the horizontal laminar flow hood using 70% ethanol.
5. Using the stereomicroscope, break the pulled end of the micro-capillary needless using sterilized sharp-tip forceps to create a narrow opening (~100 μm). Insert the unpulled end of the microcapillary tube into the aspirator tube. Insert a sterile 1.0 ml filter pipette tip into the other end of the aspirator tube. Ensure that the exposed needle end of the microcapillary tube and the pipette tip do not touch laboratory surfaces (*see* Fig. 3).
6. Take an egg from the incubator and sterilize by lightly wiping using 70% ethanol and gloved hands. Using the blunt-end forceps, create a window on the blunt end of the egg.

Fig. 2 Pulled microcapillary tube. Green arrow points to the wide end of the microcapillary tube that is inserted into the aspirator tube. To aspirate 1–2 μl of blood or PGC suspension, do not exceed the point indicated by the red arrow on the needle end

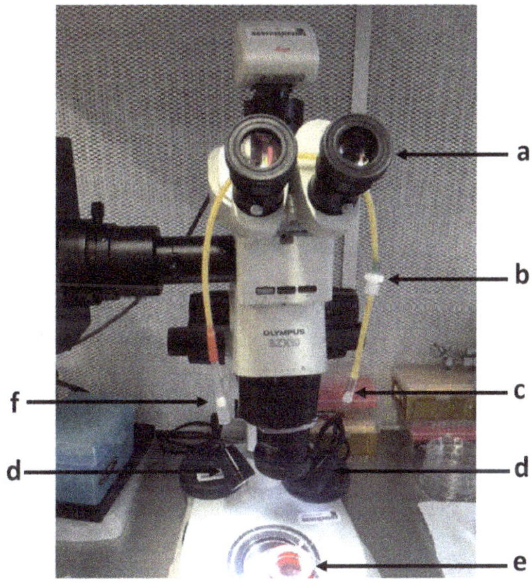

Fig. 3 Aspirator apparatus draped over the eyepiece of a stereomicroscope (**a**) positioned in a sterile horizontal laminar flow hood. The aspirator tube may be fitted with a 0.2 μm filter unit (**b**). Insert the wide end of the pulled microcapillary tube into the aspirator tube on one end (**c**). Use a sterile pair of forceps to break part of the needle end of the capillary to make it patent. Dissecting lights (**d**) provide illumination for good visibility. Place the egg in an egg-stand (**e**) to stabilize it. Drape the assembled aspirator apparatus over the microscope to prevent it from coming in contact with laboratory surfaces. Insert a 1 ml filter pipette tip (**f**) into the other end of the aspirator tube which will be used to aspirate blood or inject PGCs. Use this positioning during blood collection and PGC injection experiments

7. Using a sterile blunt-end forceps, gently remove the shell membrane to expose the embryo. The heart should be visibly beating for efficient sampling of embryonic blood.
8. Blow a small bubble in the egg albumin to demonstrate that the microcapillary needle is patent.
9. Insert the needle into the dorsal aorta at a shallow angle (10–30°) and aspirate 1–2 μl of embryonic blood into the microcapillary tube.
10. Gently express the aspirated blood into the FAOT medium in a well of the 48-well plate. Repeat this procedure for the desired number of embryos. Do not mix blood samples from different embryos in a well if the goal is to obtain pure PGC lines.
11. Collect embryonic tissue from sampled embryos for extraction of genomic DNA to perform W chromosome PCR sexing: Stabilize the embryo with the blunt-end forceps and use the fine forceps to tear a small piece of the visible vitelline membrane and underlying tissues. Place the excised tissue in a sterile

1.5 ml snap cap microcentrifuge tube and store at −20 °C prior to genomic DNA extraction and subsequent PCR-aided sex determination. Wash down the forceps and scissors with sterile PBS and then 70% ethanol between embryos.

12. At the end of the experiment, wash the aspirator tube in tap water and spray with 70% ethanol. Allow to air-dry before storage.

3.2 PCR-Aided Determination of the Sex of Chicken Embryos

The sex of a PGC culture can be determined by sexing the embryo that was sampled. Use the REDExtract-N-Amp™ Tissue PCR Kit (Sigma-Aldrich; XNAT) for sex screening of embryos as follows:

1. Add 100 μl of extraction solution to the embryonic tissue collected in sterile 1.5 ml snap cap microcentrifuge tube.
2. Add 25 μl of tissue preparation solution.
3. Vortex and then incubate at room temperature for 10 min.
4. Pierce the lid of the closed microcentrifuge tube and heat at 95 °C for 3 min.
5. Add 100 μl of neutralization solution.
6. Spin the sample for 2 min at 13,000 revolutions per minute (rpm) (equivalent to 12,470 g) in a benchtop centrifuge and gently collect the lysate.
7. Make a 1:10 dilution of the lysate in nuclease-free water for PCR.
8. Prepare PCR reaction using REDExtract PCR mix as shown in Table 1 (consult the manufacturer's manual if needed):
9. PCR primers are designed to amplify a repeated section of the female W sex chromosome which is present in many copies. Amplification of a section of the W chromosome may be performed using the following primer pair:

   Left primer: 5′ – CCCAAATATAACACGCTTCACT – 3′.

   Right primer: 5′ – GAAATGAATTATTTTCTGGCGAC – 3'.

Table 1
Reagents for REDExtract PCR reaction mix preparation

Reagent	Volume (μl)
REDExtract PCR mix	10.0
Left primer (50 pmol/μl)	0.4
Right primer (50 pmol/μl)	0.4
Nuclease-free water	5.2
Sample lysate (1:10 dilution)	4.0
Total volume	*20.0*

10. Use the following thermal cycling profile for the W chromosome primer pair above:
    (a) 94 °C for 5 min.
    (b) 35 cycles of 94 °C for 30 s, 60 °C for 30 s, and 72 °C for 30 s.
    (c) 60 °C for 10 min.
11. Include a no-template negative control, a positive male DNA control, and positive female DNA control.
12. Run the PCR reaction on 1% agarose gel in 1× TAE buffer using an electrophoresis condition of 100 V for 1 h. A PCR product of approximately 200 base-pairs will only be present in female samples.
13. Also perform PCR amplification of a housekeeping gene to confirm the integrity of the DNA samples. Amplification of the GAPDH gene may be performed using the following primer pair:

    Left primer: 5′ – TGTGACTTCAATGGTGACA – 3′.

    Right primer: 5′ – CAGATCAGTTTCTATCAGC – 3′.
14. Include the controls used for the W PCR and use the following thermal cycling profile for the GAPDH primer pair above:
    (d) 94 °C for 5 min.
    (e) 35 cycles of 94 °C for 30 s, 60 °C for 30 s, and 72 °C for 30 s.
    (f) 60 °C for 10 min.
15. Perform gel electrophoresis as described above. The PCR product for the GAPDH primer pair is 700 base-pairs and should be present in all samples except the negative no-template control. If the GAPDH PCR product is absent in a tissue sample, then the DNA may be degraded. The sex PCR assays can also be performed on genomic DNA collected from PGCs, but the rate of success of culture derivations for male and female embryos sampled cannot be determined if the embryos are not sexed.

3.3 *In Vitro* Propagation of PGCs

1. The 1–2 μl embryonic blood is added to the 300 μl of FAOT culture medium in a single well of the 48-well plates. Transfer the plate to a humidified incubator set at 37 °C and 5% CO_2.
2. Refresh the culture medium every 48 h. Without disturbing the cells aggregated in the bottom center of the well, remove 90 μl of the culture medium by placing the pipette tip against the wall of the well. Replenish the medium by gently adding 100 μl of fresh FAOT medium.
3. PGCs are visible after a week and several hundred cells will be apparent by week two (*see* Fig. 4). Blood cells are mostly lysed by week three.

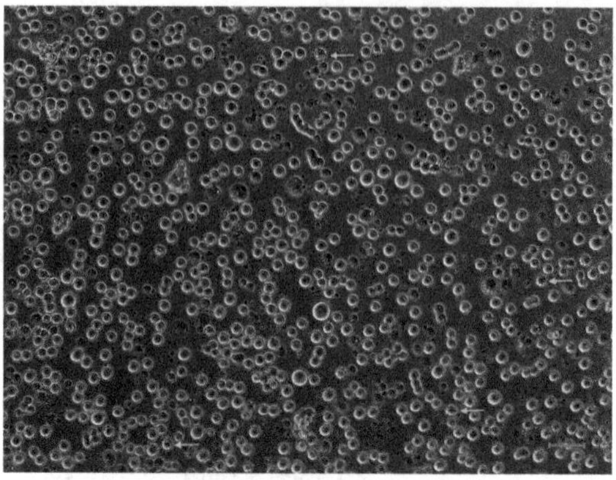

Fig. 4 Micrograph of PGCs in culture. Green arrowheads show PGCs with blebs bulging from the cell membrane. 10x magnification. Scale bar—50 μm

4. After about a week, the culture medium can be pipetted up and down 10 times to break up PGC clumps but only after refreshing the medium. Repeat if necessary during subsequent feedings.
5. Count cells in cultures at 3–4 weeks and determine the cell number. Cultures containing more than 50,000 cells are successful derivations (*see* **Note 9**).
6. After 4 weeks, discard cultures in 48-well plates that do not contain up to 50,000 PGCs. These are failed derivations.
7. The color of the culture medium changes to yellow (becomes acidic) as PGCs reach confluency. PGCs in confluent wells (containing >50,000 cells) should be transferred to a 24-well plate.
8. The PGC cultures are propagated in 500–600 μl FAOT medium/well in 24-well plates. Add 1 ml of double-distilled water into each peripheral well of a 24-well plate to reduce evaporation from the culture medium. Top up the transferred PGC culture to 500 μl with fresh FAOT medium.
9. Refresh the culture medium every 2 days:
   (a) Gently pipette up and down five times without forming bubbles.
   (b) Transfer medium to a sterile 1.5 ml screw cap microcentrifuge tube.
   (c) Centrifuge in a benchtop microcentrifuge at 1600 rpm (200 g) for 4 min. This is the standard centrifugation condition for PGCs.
   (d) Carefully remove the supernatant and resuspend the visible cell pellet in 500 μl of fresh FAOT medium.

10. Propagate PGCs to a maximum density of 300,000 cells/well in a 24-well plate. Either split into two or three wells at a seeding density of 50,000 cells/well. Alternatively, discard or cryopreserve excess PGCs.

3.4 Cryopreservation of PGCs

1. Only the PGC cultures that are free of contamination, healthy (>90% cell viability), and of optimal confluency should be cryopreserved. For optimum results, cells should be in log phase of growth with 50–80% confluency representing approximately 120,000–200,000 cells in 500 μl of FAOT medium in a well of a 24-well plate.
2. Count the number of cells in the culture. Gently pellet the cells by standard centrifugation, remove supernatant, and gently resuspend pellet in STEM-CELLBANKER® cryopreservation medium at a concentration of 100,000 cells/100 μl. If using locally prepared 5%DMSO-10%FBS cryopreservation medium, add an equal volume of cryopreservation medium at room temperature dropwise to the cells (100,000 cells/100 μl) to prevent cell lysis.
3. Dispense at least 200 μl of PGC cryopreservation mixture into a 1.8 ml polypropylene cryogenic tube.
4. Place the cryogenic tubes in a Mr. Frosty™ freezing container for controlled cooling (at the rate of −1 °C/min) to −80 °C.
5. After 6 h of storage at −80 °C, transfer the frozen cryogenic tubes to an ultra-low temperature freezer for long-term storage at <−150 °C. Avoid storing PGCs at −80 °C for more than 1 week as this may reduce viability upon thawing.

3.5 Thawing of PGCs

1. Thaw the cryogenic tube containing the frozen PGCs at room temperature or by holding in your gloved hand.
2. Add room-temperature PGC basal medium (4 times the original volume of the cryopreservation mixture) to the frozen cells in a dropwise manner. For instance, if the volume of cryopreservation mixture in the cryogenic tube is 200 μl, add 800 μl of PGC basal medium.
3. Allow to stand at room temperature for up to 5 min and gently mix the cell suspension.
4. Transfer the thawed cell suspension into a 1.5 ml microcentrifuge tube and pellet the cells by standard centrifugation.
5. Remove the supernatant and gently resuspend the cell pellet in 500 μl of FAOT medium. Transfer the cell suspension to a well in a 24-well plate. Add 1 ml of double-distilled water to each peripheral well of the 24-well plate to reduce evaporation from the culture medium and place in a humidified incubator set at 37 °C and 5% CO2 to culture PGCs as described previously.

3.6 Transfection of PGCs

1. 100,000 cells (but do not exceed 150,000 cells) are used per transfection (*see* **Note 10**).
2. Add 2 μg (up to 4 μg can be used) of plasmid DNA to be transfected to approximately 150 μl Opti-MEM I medium. If transfecting short single-stranded DNA oligonucleotides (ssODN), do not use more than a total of 400 ng or 10 μM per transfection to prevent toxicity (*see* **Note 11** for use of ssODN templates). The total volume of the DNA/Opti-MEM mixture will be 150 μl in a 1.5 ml screw cap microcentrifuge tube.
3. In a separate 1.5 ml screw cap microcentrifuge tube, add 2 μl (3 μl is also suitable) of Lipofectamine 2000 to 148 μl (or 147 μl if using 3 μl of Lipofectamine 2000) of Opti-MEM I medium to have a total volume of 150 μl. Let the Lipofectamine/Opti-MEM mixture incubate at room temperature for 15–20 min.
4. Gently combine the DNA/Opti-MEM mixture with the Lipofectamine/Opti-MEM mixture to have a total volume of 300 μl. Gently pipette the DNA/Lipofectamine mixture up and down five times and incubate for 25 min at room temperature. Do not vortex or mix violently.
5. While the Opti-MEM mixture is incubating, pellet the cells to be transfected by standard centrifugation at 1600 rpm (200 g) for 4 min. Remove the supernatant.
6. Gently resuspend the cell pellet in 750 μl of Opti-MEM I medium and centrifuge at 1600 rpm (200 g) for 4 min to wash the cells. Remove the supernatant.
7. Gently resuspend cell pellet at a concentration of 100,000 cells per 50 μl of Opti-MEM I medium (do not exceed 150,000 cells).
8. Add 50 μl of the Opti-MEM I cell suspension to the 300 μl DNA/Lipofectamine Opti-MEM mixture and gently pipette up and down five times. The transfection mixture has a final volume of 350 μl in a 1.5 ml screw cap microcentrifuge tube. Do not vortex or mix violently.
9. Incubate the transfection mixture with the microcentrifuge tube standing upright in an incubator at 37 °C and 5% CO_2 for 6 h to overnight. The screw cap should be set on loosely to allow CO_2 and oxygen to enter the microcentrifuge tube through the threads.
10. Tighten the screw cap on the microcentrifuge tube and centrifuge the transfection mixture at 2200 rpm (300 g) for 10 min to pellet the cells. A cell pellet may not be visible.
11. Remove as much as possible the supernatant from the microcentrifuge tube because the transfection solution is toxic.

12. Gently resuspend the cell pellet in 500 μl of room-temperature FAOT medium and transfer to a single well in a 24-well plate. Add 1 ml of double-distilled water to each peripheral well of the 24-well plate to reduce evaporation from the culture medium and place in a humidified incubator set at 37 °C and 5% CO_2.

3.7 Selection for CRISPR-Transfected Cells (See Note 12)

1. For CRISPR/Cas9 plasmids expressing a fluorescent reporter protein, a high level of expression should be achieved between 48 and 72 h after transfection. Fluorescence-activated cell sorting may be performed within this period to select for transfected cells. Single cell sorting may be directly performed at this point but the efficiency and cell survival may be low. We recommend sorting and expanding the mixed pool of transfected cells to allow recovery and optimal cell health before performing single-cell cultures.
2. If the CRISPR/Cas9 plasmid bears an antibiotic selection gene, the selection reagent may be added 24 h after transfection (*see* **Note 13**).
3. Before adding the selection reagent, the culture medium may be refreshed if the cells have not been perturbed by slowly and gently withdrawing 250 μl of the culture medium by directing the pipette tip at the wall of the well. Centrifuge this solution at 1600 rpm (200 g) for 4 min and discard the supernatant. Resuspend any pellet with 260 μl of FAOT medium and add this solution back to the side of the well.
4. For CRISPR/Cas9 plasmids bearing a puromycin resistance gene, add 2.0 μl of 0.1 mg/ml puromycin solution to the transfected cells and incubate for 48 h (*see* **Note 14**).
5. Centrifuge the transfected cells at 1600 rpm (200 g) for 4 min. Discard the supernatant and resuspend the cell pellet in 500 μl of PGC basal medium.
6. Centrifuge the cells again for 1600 rpm (200 g) for 4 min. Discard the supernatant, resuspend the cell pellet in 500 μl of fresh FAOT medium, and transfer to a new well in a 24-well plate. If there is massive cell death and very few live cells, resuspend the puromycin-treated cells in 300 μl of fresh FAOT medium and culture in a 48-well plate. Add 500 μl of double-distilled water into each peripheral well of the 48-well tissue culture plate to reduce evaporation from the culture medium. Place the plate in a humidified incubator set at 37 °C and 5% CO_2 and continue to culture.
7. Refresh the culture medium every 48 h:
   (a) 24-well plate: without disturbing the cells aggregated in the center of the well, remove 250 μl of the culture medium by placing the pipette tip at wall of the well. Centrifuge at 1600 rpm (200 g) for 4 min. Discard the

supernatant and resuspend any pellet in 260 μl of FAOT medium. Gently adding this solution back to the side of the well.

(b) 48-well plate: without disturbing the cells aggregated in the center of the well, remove 90 μl of the culture medium by placing the pipette tip at wall of the well. Replenish the medium by gently adding 100 μl of fresh FAOT medium. Transfer the culture to a 24-well plate when the cell count is >50,000 cells.

8. Untransfected PGCs that are not expressing the puromycin resistance gene will die out within 5 days from the time of the addition of puromycin. The culture medium will contain a lot of visible cellular debris and dead cells. Clumps of living cells will gradually become apparent and may be clonal populations.
9. Propagate PGCs to a maximum density of 300,000 cells/well in a 24-well plate. Split into two or three wells at a seeding density of 50,000 cells/well to continue expansion. Collect cell pellets for genomic DNA extraction to analyze gene editing. Cryopreserve the remaining PGCs as described previously.

3.8 Genome Analysis of Transfected PGCs

1. Use >100,000 cells for extraction of genomic DNA.
2. Propagate PGCs to a maximum density of 300,000 cells/well in a 24-well plate.
3. Centrifuge the PGCs at 1600 rpm (200 g) for 4 min. Discard the supernatant. The cell pellet can be stored at −20 °C until ready for use.
4. Use the QIAMP DNA Micro kit (Qiagen; 56304) to extract genomic DNA according to the manufacturer's instruction. PCR amplification of the target site is performed using the purified DNA.
5. Gene deletions using two CRISPR/Cas9 gRNAs may be immediately assessed by running the PCR products in 1–2% agarose gels and checking for the estimated difference in product size comparing with PCR products from wild-type cells.
6. Single-site targeting may be assessed by performing Sanger sequencing of the PCR products. The online TIDE analysis suite (https://tide.nki.nl/) can be used to estimate INDEL frequency through analysis of the Sanger sequencing chromatogram files (.AB1 file format).
7. Proceed to single-cell clonal culture.

3.9 Single-Cell Clonal Culture

1. Use expanded transfected PGC cultures that are free of contamination, healthy (>90% cell viability), and of optimal confluency. For optimum results, cells should be in the log phase of growth with 50–80% confluency representing approximately

120,000–200,000 cells in 500 μl of FAOT medium in a well of a 24-well plate (*see* **Note 15**).

2. Add 200 μl of double-distilled water into each peripheral well of a 96-well tissue culture plate to reduce evaporation from the culture medium. Add 50 μl of FAOT medium into each of the inner 60 wells in the plate. Conditioned FAOT medium or 50% conditioned FAOT medium (prepared by mixing conditioned FAOT medium with fresh FAOT medium) may be used throughout for single-cell culture maintenance when cell growth appears to be retarded (*see* **Note 16**).
3. Single-cell plating may be performed manually by hand through serial dilution until a single cell is seeded in a well of a 96-well plate. Allow the plate to sit for 10 min and then observe under the microscope to ensure that each selected well contains a single PGC (*see* **Note 17**).
4. Alternatively, single-cell plating may be easily and quickly performed using the BD FACSAria III cell sorter. Seed a single PGC into each well containing 50 μl of FAOT medium in a 96-well plate. Transfer the plate into a humidified incubator set at 37 °C and 5% CO_2 and incubate for 48 h.
5. Add 50 μl of FAOT medium to each well for a total culture volume of 100 μl and incubate for another 48 h.
6. Again, add 50 μl of FAOT medium to each well to achieve a total culture volume of 150 μl and incubate for another 48 h.
7. Subsequently, the culture medium may be refreshed every 48 h by slowly and gently withdrawing 45 μl of the culture medium by directing the pipette tip at the wall of the well. Gently replace with 50 μl of fresh FAOT medium. Successful cultures in 96-well plates take 2–3 weeks to reach approximately 30–50% confluency.
8. Once the cell confluency reaches approximately 50%, transfer the PGC culture to a well in a 48 well-plate and increase the culture volume to 300 μl with fresh FAOT medium. Add 500 μl of double-distilled water into each peripheral well of the 48-well tissue culture plate to reduce evaporation from the culture medium. Transfer the plate into a humidified incubator set at 37 °C and 5% CO_2.
9. Refresh the culture medium every 48 h. Without disturbing the cells aggregated in the center of the well, remove 90 μl of the culture medium by placing the pipette tip at wall of the well. Replenish the medium by gently adding 100 μl of fresh FAOT medium.
10. After about a week, the culture medium can be pipetted up and down five times to break up PGC clumps but only after refreshing the medium.

11. Count the cells in the cultures. Cultures with more than 50,000 cells are successful clonal derivations. The color of the culture medium changes to yellow (becomes acidic) as PGCs reach confluency. Successful clonal derivations take 1 week to reach 50,000 cells in a 48-well plate upon transfer from a 96-well plate.
12. Transfer PGCs in confluent wells (containing >50,000 cells) of the 48-well plate to a 24-well plate. Increase the volume of each PGC culture to 500 μl/well with fresh FAOT medium. Add 1 ml of double-distilled water into each peripheral well of the 24-well plate to reduce evaporation from the culture medium.
13. Propagate the clonal PGCs to a maximum density of 300,000 cells/well in the 24-well plate. Split into two or three wells at a seeding density of 50,000 cells/well. Collect cell pellets for genomic DNA extraction to confirm gene editing. Cryopreserve the remaining PGCs as described above.
14. Overall, it takes 3–5 weeks to establish a clonal line from a single PGC. Higher efficiency is obtained for cloning male PGCs compared to female PGCs.

3.10 Injection of PGCs into Surrogate Host Embryos

1. Thaw the cryopreserved clonal PGCs 5–7 days before the intended injection date and propagate to a maximum density of 300,000 cells/well in a 24-well tissue culture plate.
2. Pull microcapillary tubes using a moving coil microelectrode puller to create pointed needles. Sterilize the pulled microcapillary tubes under UV in a UV sterilizer cabinet.
3. Fertile chicken eggs are incubated upside down (pointy end up) for 2.5 days to obtain stage 16 HH embryos. Embryos can be between stages 15 and 16+ HH but not older than stage 17 HH.
4. Surface-sterilize the stereomicroscope and other tools in the horizontal laminar flow hood using 70% ethanol.
5. Under the stereomicroscope, break off a small portion of the pulled end of the microcapillary tube using sterilized sharp-tip forceps to create a needle. Insert the unpulled end of the microcapillary tube into the aspirator tube. Insert a sterile 1 ml filter pipette tip into the other end of the aspirator tube. Ensure that the exposed needle end of the microcapillary tube and the pipette tip do not touch the laboratory surfaces. Alternatively, a needle beveller can be used to create a bevelled opening.
6. The cultured PGCs are pelleted by standard centrifugation and resuspended in KO-DMEM at a concentration of 5000 cells/μl.
7. (Optional) If using iCaspase9 sterile embryos, add 1.0 μl of B/B compound to 50 μl of the PGC suspension and maintain at room temperature.

8. A neutral dye solution such as Fast Green dye (Sigma) is added to the PGC suspension to aid visualization of the injection. Add 0.5 μl of 0.1% dye to 50 μl of PGC suspension.
9. Take an egg from the incubator and sterilize lightly using 70% ethanol egg. Using the blunt-end forceps, create a small window on the pointy end of the egg.
10. Using a sterile blunt-end forceps, gently remove the shell membrane to expose the embryo. The heart should be visibly beating.
11. (Optional) Prior to injection, aspirate 1 μl of endogenous blood to make space for the injection in the circulatory system of the embryo.
12. Flick PGC suspension to resuspend PGCs as the cells settle rapidly.
13. Aspirate 1–2 μl of the PGC suspension into the microcapillary tube. If PGC solution will not enter the needle, break off a small portion of the tip and repeat aspiration.
14. Insert the needle into the dorsal aorta at a shallow angle (10°–30°) and inject the PGC suspension into the vascular system. The dye should enable visualization of the vascular system filling with the PGC solution.
15. If using iCaspase9 sterile hosts, gently inject 50 μl of penicillin/streptomycin-B/B-compound mixture on top of the embryo. For other embryos, use 50 μl penicillin/streptomycin solution.
16. Seal the egg with Leukosilk tape and incubate the manipulated egg blunt end up with rocking until hatching.
17. Carefully remove the microcapillary tube and dispose in a sharps bin.
18. After successful hatching, collect chorio-allantoic membrane (CAM) samples from each egg for DNA extraction for sex determination of the surrogate host as described in Subheading 3.2.
19. Raise hatched chicks to sexual maturity and breed to generate the G1 generation of genome-edited chickens.

4 Notes

1. We advise that 1 ml aliquots of 100× EmbryoMax® nucleosides (Merck Millipore: ES-008-D) should be made and stored at −80 °C. Avoid refreezing.
2. Vitamin B12 can be added directly to the PGC basal medium or alternatively to the complete PGC culture. Do not add to both as the final concentration in the complete PGC culture

must be 0.0068 μg/ml. For example, 10 μl of 5000× vitamin B12 solution can be added to 50 ml of PGC basal medium, or alternatively, it can be added to 50 ml of complete PGC culture.

3. The final concentration of BSA in 1 ml of 25 μg/ml FGF2 solution will be 0.125%, with 100 mM NaCl.
4. "FAOT" is an abbreviation derived from the growth factors used in the preparation of the medium; FGF2 (F), activin A (A), and ovotransferrin (OT).
5. "FACSOT" is an abbreviation derived from the growth factors used in the preparation of the medium; FGF2 (F), activin A (A), chicken serum (CS), and ovotransferrin (OT).
6. "FABOT" is an abbreviation derived from the growth factors used in the preparation of the medium; FGF2 (F), activin A (A), BMP4 (B), and ovotransferrin (OT).
7. We routinely use FAOT serum-free medium. Use of FACSOT or FABOT medium gives the same culture efficiency in our hands but may be more suitable for some PGC lines.
8. (a) Chicken eggs and embryos usually do not contain infectious material and so are not a biohazard. However, appropriate care must be taken and appropriate personal protection must be worn during the procedure (eye protection, gloves, protective coat).
   (b) The reusable parts of the aspirator apparatus must be rinsed (tap H_20), sprayed with 70% ethanol, and air-dried on a finely textured tissue paper before use. The same decontamination protocol must be used when finished to eliminate the possibility of bacterial and egg product contamination.
   (c) Glass microcapillaries must be safely disposed in a sharps bin immediately after use. Care must be taken not to leave used microcapillaries on bench tops.
   (d) Care must be taken not to let the mouthpiece and needle come in contact with any laboratory surfaces. Even contact with finely textured tissue paper could pose a problem. It is good practice to drape the aspirating apparatus over the microscope eyepieces so that neither end touches a laboratory surface.
   (e) Care must be taken not to touch egg contents with gloves and then onto mouthpiece. Raw egg products may contain salmonella and should be treated with care.
9. If the PGC derivation is fast growing, cells may need to be counted at 2 weeks and transferred to a 24-well plate or they will become too confluent and die by 3 weeks of culture.

10. Using less than 100,000 cells will not pellet well when cells are centrifuged after the 6-h Opti-MEM I incubation step and will also give a lower transfection efficiency. PGCs to be transfected should be in the log phase of growth and around 50–80% confluency representing approximately 120,000–200,000 cells in 500 μl of FAOT medium in a well of a 24-well plate.
11. Use of CRISPR/Cas9 and ssODN repair templates for performing small sequence changes [12]:
    (a) Design a repair template varying in length from 70–140 base-pairs.
    (b) Synthesize with Integrated DNA technologies (IDT) and purchase as desalted 4 nM ultramer™ oligonucleotides. Use other vendors if preferred.
    (c) Briefly centrifuge the tube before opening to avoid losing dried pellets and resuspend to a stock concentration of 100 μM by adding 40 μl of TE buffer in a sterile micro-safety cabinet. Briefly centrifuge the resuspended ssODN at high speed. Ultramer™ oligonucleotides may also be purchased as 100 μM suspensions in TE buffer.
    (d) If using a single ssODN repair template, dilute a small amount of the stock solution to 10 μM using TE buffer and use 1 μl per transfection. If performing two different targeting, make 5 μM aliquots of each ssODN by diluting with TE buffer and use 1 μl of each 5 μM ssODN per transfection. In our hands, 1 μM also produced a good gene editing efficiency comparable with using 10 μM. This makes it theoretically possible to design multiple ssODN repair templates to target multiple genomic locations. Do not exceed 10 μM per transfection because ssODN are toxic to cells in large amounts.
    (e) Proceed by mixing 1.0–1.5 μg of PX459 or PX458 CRISPR/Cas9 plasmid with 10 μM of ssODN if using only one repair template. If using a mixture of two repair templates, use 5 μM of each at 1:1 ratio as illustrated below:
        (i) 10 μM ssODN + 1.5 μg PX459 plasmid or
        (ii) 5 μM ssODN1 + 5 μM ssODN2 + 1.5 μg PX459 plasmid.
12. We use the PX458 and PX459 wild-type CRISPR/Cas9 vectors from Feng Zhang's lab [13]. PX458 vector expresses the eGFP protein while PX459 (V2.0) expresses the puromycin resistance protein. To increase the efficiency of ssODN-

mediated homology-directed repair, we use the HF-PX459 (V2.0) vector which expresses the high-fidelity Cas9-HF1 protein and puromycin resistance gene [12].

13. The 24-h timepoint after transfection is determined from the time the transfected cells are resuspended in FAOT culture medium and placed into the incubator set at 37 °C and 5% CO_2.
14. 2.0 μl of 0.1 mg/ml puromycin is quantitated to kill 99% of many PGC lines in 500 μl culture containing 150,000 cells after 48 h of incubation. Some PGC lines may be more or less sensitive to puromycin and may require optimization.
15. Do not isolate single-cell cultures from overgrown and highly confluent cultures as cell growth would be within the stationary and decline phases. There is a significant reduction in the number of viable cells in these phases which significantly reduces the success rate in establishing viable single-cell cultures.
16. FACSOT, FABOT, or conditioned FAOT medium may be used for PGC lines that are difficult to grow as single cells. Prepare conditioned FAOT medium as required and use immediately. To prepare conditioned medium:
    (a) Add 500,000 PGCs to 1 ml of FAOT culture medium in a well of a 12-well tissue culture plate.
    (b) Add 1.5 ml of double-distilled water to the peripheral wells of the plate and incubate for 24 h at 37 °C and 5% CO_2 in a humidified incubator.
    (c) Centrifuge the culture at 1700 rpm for 4 min to collect the culture supernatant.
    (d) Filter the supernatant through a 0.22 μm syringe filter (Merck Millipore; SLGPO33RS).
    (e) Store the filtered supernatant at 4 °C and use within 3 days.
17. Performing single-cell plating manually can be tedious, time-consuming, and inefficient. We strongly recommend the use of a cell sorting machine such as the BD FACSAria III cell sorter.

Acknowledgments

Illustrations in Fig. 1 depicting PGC derivation from embryos, CRISPR/Cas9 gene editing, PGC injection into embryos, and GE chicks were created with BioRender.com and are used under license.

References

1. Zhang G (2015) Bird sequencing project takes off. Nature 522(7554):34
2. Bravo GA, Schmitt CJ, Edwards SV (2021) What have we learned from the first 500 avian genomes? Annu Rev Ecol Evol Syst 52:611–639
3. van de Lavoir MC et al (2006) Germline transmission of genetically modified primordial germ cells. Nature 441(7094):766–769
4. Oishi I et al (2016) Targeted mutagenesis in chicken using CRISPR/Cas9 system. Sci Rep 6:23980
5. Park TS et al (2014) Targeted gene knockout in chickens mediated by TALENs. Proc Natl Acad Sci U S A 111(35):12716–12721
6. Taylor L et al (2017) Efficient TALEN-mediated gene targeting of chicken primordial germ cells. Development 144(5):928–934
7. Ballantyne M et al (2021) Direct allele introgression into pure chicken breeds using Sire Dam Surrogate (SDS) mating. Nat Commun 12(1):1–10
8. Ioannidis J et al (2021) Primary sex determination in birds depends on DMRT1 dosage, but gonadal sex does not determine adult secondary sex characteristics. Proc Natl Acad Sci U S A 118(10):e2020909118
9. Lee JH et al (2017) C-X-C chemokine receptor type 4 (CXCR4) is a key receptor for chicken primordial germ cell migration. J Reprod Dev 63(6):555–562
10. Whyte J et al (2015) FGF, insulin, and SMAD signaling cooperate for avian primordial germ cell self-renewal. Stem Cell Rep 5(6):1171–1182
11. Hamburger V, Hamilton HL (1951) A series of normal stages in the development of the chick embryo. J Morphol 88(1):49–92
12. Idoko-Akoh A et al (2018) High fidelity CRISPR/Cas9 increases precise monoallelic and biallelic editing events in primordial germ cells. Sci Rep 8(1):15126
13. Ran FA et al (2013) Genome engineering using the CRISPR-Cas9 system. Nat Protoc 8(11):2281–2308

INDEX

Thomas L. Saunders (ed.), *Transgenesis: Methods and Protocols*,
Methods in Molecular Biology, vol. 2631, https://doi.org/10.1007/978-1-0716-2990-1,

Q

R

S

T

V

Z

CPSIA information can be obtained
at www.ICGtesting.com
Printed in the USA
LVHW061922020423
743254LV00006B/1416

9 781071 629895